Risk Analysis and Control for Industrial Processes - Gas, Oil and Chemicals

Risk Analysis and Control for Industrial Processes - Gas, Oil and Chemicals

A System Perspective for Assessing and Avoiding Low-Probability, High-Consequence Events

Hans Pasman

AMSTERDAM • BOSTON • HEIDELBERG • LONDON
NEW YORK • OXFORD • PARIS • SAN DIEGO
SAN FRANCISCO • SINGAPORE • SYDNEY • TOKYO

Butterworth Heinemann is an imprint of Elsevier

Butterworth Heinemann is an imprint of Elsevier
The Boulevard, Langford Lane, Kidlington, Oxford OX5 1GB, UK
225 Wyman Street, Waltham, MA 02451, USA

Notices
Knowledge and best practice in this field are constantly changing. As new research and experience broaden our understanding, changes in research methods, professional practices, or medical treatment may become necessary.

Practitioners and researchers must always rely on their own experience and knowledge in evaluating and using any information, methods, compounds, or experiments described herein. In using such information or methods they should be mindful of their own safety and the safety of others, including parties for whom they have a professional responsibility.

To the fullest extent of the law, neither the Publisher nor the authors, contributors, or editors, assume any liability for any injury and/or damage to persons or property as a matter of products liability, negligence or otherwise, or from any use or operation of any methods, products, instructions, or ideas contained in the material herein.

Library of Congress Cataloging-in-Publication Data
A catalog record for this book is available from the Library of Congress

British Library Cataloguing-in-Publication Data
A catalogue record for this book is available from the British Library

ISBN: 978-0-12-800057-1

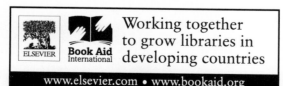

Publisher: Jonathan Simpson
Acquisition Editor: Fiona Geraghty
Editorial Project Manager: Cari Owen
Production Project Manager: Debbie Clark
Typeset by TNQ Books and Journals
www.tnq.co.in
Printed and bound in the United States of America

Contents

Foreword

At the outset, I must say that this volume by Professor Hans Pasman on "Risk Analysis and Control for Industrial Processes, A System Perspective for Assessing and Avoiding Low Probability, High Consequence Events," is a very timely and much needed treatise. It is even more important given the need for process safety and sustainable development, and the need for a rational and constructive approach to risk assessment, risk management, and control against the backdrop of globalization.

Professor Pasman has been a visionary and trailblazer in the development and application of new methods and approaches in various areas of process safety and risk assessment. Over more than two decades of our professional association and friendship, I have always been impressed by his clarity of thought and great depth of expertise. This book is another indicator of Professor Pasman's stellar contributions to the theory and practice of the diverse and complex field of risk assessment.

The hunger for energy and the need for chemical products continue to fuel the growth of the chemical and petrochemical industries. Consequently, the risks and the hazards associated with the growth continue, and the challenges posed by technology, scale, and intensity of operations grows and changes. With the increasing complexity of chemical processes, interdependent chemical infrastructure, and the need for considering diverse issues such as safety, environment, cost, and social and cultural factors, challenges to process safety and risk assessment can no longer be solved by simple approaches. Process safety is at a crossroads with systems engineering, complex systems, and engineering for sustainable development. Assessments needed to address process safety challenges most often span a complex system requiring the application of sophisticated systems analysis. A complex systems approach allows the study of parts of a system that taken together cause the whole system to behave in a certain manner and how that behavior interacts with its environment.

Process safety is very closely linked to sustainable development. Risk assessments in the twenty-first century must bring together elements of manufacturing, design, and sustainable engineering in an integrated form. Interwoven through this new paradigm is the consideration of risk in every aspect. Another important aspect of risk assessments is the ability to deal with low probability—high consequence events.

Professor Pasman has been successful in taking a refreshing and poignant look at process safety and risk management; he has applied a systems approach in a holistic manner for the analysis and control of risks inherent to the operations of the processing industry and their products. He addresses risks in the chemical industry, processing of energy carriers such as oil and gas, metal and food processing, and storage and transportation of hazardous materials. The book also provides a very comprehensive review of methods used over the years to understand and manage risks. I sincerely believe that the book has opened up a new vista and perspective on methodological improvements, necessary in the ever-increasing complexity of safe manufacturing and distribution in a competitive world.

M. Sam Mannan
Regents Professor and Director
Mary Kay O'Connor Process Safety Center
Texas A&M University
College Station, Texas, USA

Preface

Meanwhile, when the sun rises, the fog will not be harmful,

—free translation of Daniel Chodowiecki's explanation of his etching symbolizing the Enlightenment (1791)

SCOPE, MOTIVATION, OBJECTIVES, AND CONTENTS OF THE BOOK

The existence of mankind on this world in the present magnitude and growth and at a reasonable level of comfort is only physically possible thanks to the process industry providing us energy and fuels, construction and electronic materials, fertilizers and food, textiles, pharmaceuticals, coatings, drinking water, and so forth. This dependence will further grow, certainly when the standard of living in the developing countries reaches that of developed ones. However, industrialization comes with certain risks. Most dreadful are risks of *high consequence, low probability* events involving hazardous materials, which can be widely ranging in nature and effect. Many examples of major accident hazards have already been observed. High consequence means catastrophic losses, sometimes with huge loss of human life including various kinds of other harm, and losses to the means, infrastructures, and environments on which our lives depend. The human body is rather vulnerable to mechanical impact and shock, extreme temperatures, and a long list of gaseous, liquid, and solid substances that over certain concentration thresholds are lethally toxic to our system. We do not need to mention high intensities or doses of heat, nuclear, and electro-magnetic radiation. Walking into an enclosure with low oxygen concentration in the air is not healthy as well and may cause acute death. On the other hand, low probability of occurrence is very low. Indeed, the considered event may not happen in our lifetime. Measured in a probability over a certain time duration the event may be not less frequent than the impact of a comet, but this does not mean it cannot show up; it still can happen tomorrow! Yet if its precursor trail is detected, we gain the opportunity to influence and reduce its probability of occurrence.

Risk analysis as an approach and collection of methods to foster safe ways of achieving production goals has developed since the 1960s. It has roots in the nuclear power community of academia, regulators, and industry and has spread to various engineering disciplines and beyond to management, medicine, economy, and finance. Basically, nuclear power risk analysts founded the leading forum of Probabilistic Safety Assessment and Management conferences operating since the early 1990s. Even more general and perhaps less technical oriented are the aims of the Society for Risk Analysis, which is active in organizing meetings in various parts of the world. Risk management practice applying the methods of risk analysis became a must in the many aspects of business life and finance, and even further evolved to risk governance of company boards and governments.

The process industry, though, in its early development was plagued by mishaps from sometimes unknown chemical and physical mechanisms, and it was keen to establish safe approaches. Beginning in the 1960s under the auspices of chemical engineering institutions, loss prevention symposia have been organized in the United States and Europe, and some years later in Asia. These symposia have been instrumental in developing concepts and methods for hazard and risk analysis, and know-how for optimal organization and safety management, while adopting elements from elsewhere. They still give much attention to sharing knowledge on material properties, damaging mechanisms due to spills and unintentional releases, and preventive and protective measures.

This book endeavors to address all those leading or being employed in the industry and those involved in care for safety of the industry who prefer an overview and to see a timeline of developments. The book is intended especially to address students, in particular engineers, and to challenge them to advance the field further. Hopefully, the overview it offers will also be consulted by policy and decision makers, as it shows in risk prediction the strengths and weaknesses of science and engineering. For those who are less technically interested, each chapter is preceded by a summary.

The book is meant as a contribution to enhance process safety and risk and uncertainty management; it tries to apply a system approach and to cover in a holistic way the analysis and control of risks inherent to the operations of the processing industry and their products. It therefore focuses on the chemical industry, processing of energy carriers such as oil and gas, metal and foodstuff processing, and storage and transportation of hazardous materials. It briefly reviews experiences collected over the years and existing methods to understand and manage risks. The main aim is, however, to open a further future perspective of methodological improvements, necessary in the ever-increasing complexity of getting products safely manufactured and distributed in a competitive world. Knowledge of the human factor, organizational and technical aspects shall be merged and interaction in a sociotechnical system must be analyzed for improved risk control to avert mishaps.

Necessarily, the book has its limits. It does not provide a complete detailed overview of all that has been written about the subject in a technical sense as that is the objective of Lees' Loss Prevention in the Process Industries, not even in a condensed way as in Lees' Process Safety Essentials. The latter book is certainly very useful for those who want to know more on certain aspects. But, this book selects and briefly summarizes knowledge on major hazards and acute effects in a balanced way on all those aspects that are of significance for fostering process and plant safety. This approach is followed with regard to both technical and organizational aspects, including regulatory and human factor ones. All relevant aspects of which the author is aware of are touched upon with some references for those who desire further details. The book is in part established material, yet delves into new developments and methods up until the end of 2014 with a promise for improved future risk control.

With regard to examples and regulatory aspects, most of what is discussed already occurs in or is applicable to the United States and Europe. The issue of security is only mentioned in a few instances because many generic aspects of it are covered by process safety and risk reduction measures. Unfortunately, we must live with many acronyms. These have been noted and defined repeatedly; I hope this is helpful but not irritating.

The process safety "building" has many doors and rooms filled with experience, system engineering, risk assessment, management, and human factors. I have tried to connect the rooms, because so far the field consists largely of a collection of specialties.

Hans J. Pasman

Acknowledgments

Thoughts about what shall be written and how to write them need reaction from colleagues with critical minds. I was fortunate to find friends who did not necessarily agree with what I had drafted. The text with an initial Dutch flavor was first Americanized and then converted to English–English but kept in American spelling. Meanwhile, comments were made and suggestions given with regard to the content, improving its quality.

Firstly, I would like to acknowledge the invaluable support of Dr William (Bill) J. Rogers, who teaches risk management and probabilistic methodology at the Mary Kay O'Connor Process Safety Center (MKOPSC) of the Artie McFerrin Chemical Engineering Department of Texas A&M University, College Station, Texas. Bill was inspiratory, a theory provider, and supporter in systems approach to process safety and predictive risk analysis. Dealing with uncertainty is key in this problem field. I enjoyed the many discussions (and also the concerts he organized). Secondly, Dr Simon P. Waldram of Waldram Consultants Ltd, also a research fellow at MKOPSC, formerly a professor at the University of London, was my chemical-physical conscience. Apart from making critical comments, he also gave the text a true English touch. Dr Paul H. J. J. Swuste, associate professor of the Safety Science Group of the Delft University of Technology, the Netherlands, with a background in biochemistry but extensive experience in occupational safety and safety management, supported me in the long-term main lines of safety thinking and kept me straight in the human factor and organizational aspects. Further, thanks to Ms. Trish Kerin, BEng, Director IChemE Safety Center, Institution of Chemical Engineers, in Melbourne, who after her industrial experience made very useful practical comments on the manuscript.

In addition, I would like to thank Dr M. Sam Mannan, PE, CSP, DHC, Director of MKOPSC and Regents Professor at Texas A&M University, for his moral support and Dr Sonny Sachdeva and Joshua Richardson of MKOPSC who managed and obtained the copyright permissions.

Lastly, I want to thank my wife, Ina, my partner in life, who has steadfastly stuck by my side, as well as my children and grandkids who missed seeing "opa" for quite some periods.

I graduated in chemical technology and started my career at Shell, but during military service was transferred to TNO, the Dutch National Research organization. Apart from defense research, I necessarily learned much about process safety by investigating numerous disastrous accidents in the late 1960s and early 1970s by directing development of experimental methods and performance of risky experiments of various kinds to explain what happened in these accidents. I have been a member and then 10 years chairman of the European (EFCE) Working Party on Loss Prevention since its beginning in 1972, as well as helped to found the European Process Safety Centre. As well, I thank all my former colleagues at TNO, at the Delft University of Technology where I have been teaching chemical risk management, and of the Working Party for their cooperation and support. Although we have now a wealth of computational models providing the basis of system risk analysis, it is unfortunate that young generations can only obtain limited (but safe!) *experimental* experience, both in the chemical and physical senses but also in human and organizational functioning. Experimenting is expensive and models are a way of studying a problem. But, a model remains only a model that must be tested, and the mechanisms threatening process safety are complex and vary widely. The looming mishap is always hidden in the tails of the distributions.

Finally, I would like to admit that if Ms. Fiona Geraghty of Elsevier had not challenged me, I would never have started writing this book. My hope is that it will contribute to a sound understanding of how to manage industrial process risks.

Hans J. Pasman

CHAPTER

Industrial Processing Systems, Their Products and Hazards

1

Accidents will happen, but forewarned is forearmed.
Eighteenth-century English proverb

INTRODUCTORY REMARKS

Dealing with high-consequence, low-probability process industry risk sources, exposing their immediate environment to major hazards and acute effects, ought to start with discussing the dreadful tragedy caused by an accident in the Union Carbide plant in Bhopal, India, in 1984. It is certainly not the purpose here to describe this catastrophe in any detail, but this largest of all industrial disasters shall be mentioned. A toxic cloud of methyl-isocyanate escaped from a storage tank after it was heated up in a decomposition reaction with clean water (a so-called runaway reaction). The water entered the tank by accident. Due to the low wind speed and relatively cool night conditions, the growing cloud stayed low near the ground while slowly drifting from the plant to an adjacent neighborhood of migrant makeshift housing in which many slept. The cloud caused thousands of fatalities and many more suffered from toxic effects, some of which continue to this day. For complete details of this accident, see *Lees' Loss Prevention in the Process Industries*[1] or books especially about the case such as by Shrivastava[2] or Pietersen.[3] The underlying nontechnical causes of the disaster are of a nature that still may be found today, such as economic slump, a shrinking market, competition pressure, cost cuts, an almost endless row of short-staying top managers, downsizing of personnel and a shortage of competent employees, bad maintenance, miscommunication, neglected training of safety and emergency procedures, and a local government not having the strength to maintain a strict policy with respect to industrial safety and environmental protection. Fortunately, over the years our know-how to arrange preventive and protective measures has grown tremendously, and this book wants to demonstrate what has been achieved as well as what methods we can further develop to cope with the complexity of technology and organization. We need goods that industry produces, but ethics require that everything shall be done to prevent harm of employees or the public.

In this chapter, a quick "tour d'horizon" will be made qualitatively to show the effects of various hazardous phenomena that underlie the risks of processing systems

Risk Analysis and Control for Industrial Processes - Gas, Oil and Chemicals.

and their products, which we shall encounter in the rest of the book. We give brief examples of disasters caused by some known products that are indispensable to our lives, such as mineral fertilizers, car fuels, and pharmaceuticals. These substances can exhibit nasty properties when brought into certain conditions of dispersal, heat, confinement, and contamination. Of course, industry tries as much as possible to avoid using materials with distinct hazardous properties such as high toxicity and explosiveness/flammability, but lack of better substitutes, and economics, can leave not much choice on practical grounds. Despite the hazardous properties of many materials, the process/chemical industry belongs to one of the safest industrial sectors in the world. The fact that their reputation is often perceived as bad may have much to do with past disasters and that, when rare disasters do happen, they can be on a large scale. In addition, most people are not familiar with the methodology of how hazards can be identified and how well these can be controlled so as to minimize the associated risks. This book aims to contribute to a better understanding of these issues and to indicate ways to make further progress.

This chapter is written particularly for those without the substantial background in chemistry and physics that determines the hazardous potentials of substances, while for specialists it may present a quick overview of the relevant aspects. The hazard potential we will consider is one that may impact over a substantial distance from a source. It can be physical, for example, water in contact with hot, molten metal; it can be a toxic hazard to life in the environment, for example, due to a large spill of crude oil or a radioactive one by an escape of nucleides. The examples we shall describe in this chapter are mostly hazards due to energy release by a chemical (explosive) decomposition or by combustion. Details on properties and mechanisms can be found in an abundance of literature, while for further guidance, many details and references in *Lees' Loss Prevention in the Process Industries*[1] is recommended. Those who do not need all of the details and may be afraid to lose themselves in 3600 pages can opt for consulting the more modest *Lees' Process Safety Essentials*.[4] Note that throughout the book, when *process safety* is mentioned, it is meant to include plant safety. In the strict sense of the word, *process* is only the producing mechanism, while *plant* has the connotation of the area, the premises, where the processes influenced by human and organizational factors take place. In fact, our scope will be even wider because we shall also include the risks of storage and transportation.

The tour starts with the process of nitrogen fixation to form ammonia and the production of ammonium nitrate, which are the basic ingredients of fertilizers without which the world population cannot be fed. Ammonia is a gas, toxic to humans, that can explode when mixed with air; ammonium nitrate is a solid and can explode in all three types of possible, chemically energized explosions. It has produced disastrous accidents after being involved in a fire, being contaminated, or being overheated by other means. Several case histories will be briefly described that will provide insight into the complex phenomena one has to deal with and which will be further explained in Chapter 3. Then we turn to fuels, or rather energy carriers: gasoline, natural gas, liquefied petroleum gas (LPG), and hydrogen. Due to the massive quantities needed to keep the world moving and heated, one after the other of these fuels has been involved in explosions, which may display some similar effects but can also be distinctly different in the way they develop. Next, we shall turn to dust explosions and the many combustible materials that have surprised people with the vigorous way "innocent" substances such as sugar and sawdust can explode and harm victims. Further, we shall look at pharmaceutical materials and the reactor runaway events that can occur in their preparation and the damage this can cause. Finally, a few recent transportation accidents will be briefly reviewed.

For those who are interested in an historical overview of industrial disasters that have occurred, there are the lists on Wikipedia[5] and of Abdolhamidzadeh et al.[6] The latter is focused on 224 domino-effect accidents—when a mishap starts on a small scale but then escalates via surrounding chemicals, process equipment, or neighboring plants. This list provides an overview of cases from various parts of the world and ones with no fatalities but is missing nondomino cases. As part of Marsh Insurance Ltd, Marsh's Risk Consulting Practice publishes compilations of the largest 100 accidents in the hydrocarbon-chemical industry.[7] As an insurer, Marsh obtains an overview of insured property losses; examples of losses in two sectors over a period from 1971 to 2011 are shown in Chapter 9, Figure 9.1. In Chapter 3 (Section 3.6.2), a number of incident databases are mentioned.

1.1 GENERAL GLOBAL OUTLOOK

The world's population has grown numerically, and rapidly, during the last century; this is due in part to the fast-advancing capabilities of medical care. The expectation is that in due time the number of inhabitants on earth will decline; in fact, the growth rate has already fallen to a current value of 1.1% p.a. Nevertheless, between 1990 and 2010 the world population still increased by 30% or in absolute numbers from 5.3 to 6.9 billion. The projection of long-term decline is partly based on the belief that the standard of living in general will increase so that the immediate fear of starving to death in old age, when not looked after by one's progeny, disappears and the necessity to produce a wealth of descendants reduces. The trend of decreasing population growth can clearly be seen in all industrialized countries, even to such an extent that in a number of countries the population is already shrinking. However, there is only one way to raise the standard of living: industrializing and generating the power to drive it all.

Agriculture was of course a solution to mankind when the hunter-scavenger ran out of resources, but with the present size of the earth's population, agriculture without an industrial base would fall very short of being able to feed us all. In fact, our survival is entirely dependent on the process industries. The basic ingredient for artificial fertilizer is ammonia, a compound of the elements nitrogen and hydrogen (NH_3), which had been produced on an industrial scale as a side product in coal distillation, the "mother" of industrial chemistry. As the yield was very limited, other production routes were sought. The solution was direct synthesis from the rather inert gas nitrogen, which makes up about 80% of the air around us, and hydrogen to form ammonia. This reaction is called nitrogen fixation. Part of the ammonia can then be oxidized to nitric acid, with which it is then combined with the remaining ammonia to form ammonium nitrate (AN), the main constituent of mineral fertilizers. The production of pure nitrogen and hydrogen is not that simple, but it will be briefly described later in the chapter. The advent of the industrially produced mineral fertilizers together with a meticulous study of how to successfully grow plants has so far saved us from starvation. According to a United Nations (UN) Food and Agriculture Organization (FAO) publication, today already more than 40% of crops depend on the application of mineral fertilizer and this will increase to more than 80% in the future.[8] Increases in cereal crop production and the output of mineral fertilizer run in parallel. Organic fertilizers (manure) are less efficient and cannot meet the current demand.

1.2 AMMONIUM NITRATE

On April 17, 2013, a tragedy involving stored AN occurred at the West Fertilizer Company storage and distribution plant in the small town of West, Texas, about 30 km north of the city of Waco; with 15 people killed, 150 injured, and massive destructive damage to the fabric of the town's buildings around the plant. A preliminary investigation by the US Chemical Safety and Hazard Investigation Board[9] (often just indicated as Chemical Safety Board, abbreviated as CSB) concluded that about 30 tons of ammonium nitrate stored in wooden bins in a wooden warehouse building had detonated after an intense fire, which had broken out in the plant after closing hour. Also stored in the facility were large amounts of anhydrous ammonia and agricultural seeds. No sprinkler system had been installed. The incident led to a Congressional Senate hearing and to much debate about the imperative of compliance with regulatory measures and having sufficient inspection with respect to fertilizer stores throughout the United States.

This was not the first ammonium nitrate explosion and will presumably not be the last one. No informed person could say that no knowledge about the phenomenon existed. It is well known that ammonium nitrate as an oxidizing material must be stored fully separated from combustibles by fire-resistant walls, and certainly not stored in wooden bins. In case of nearby fire, adequate fire protective measures reduce consequences. A presidential US Executive Order was issued on August 1, 2013, entitled "Improving Chemical Facility Safety and Security," which called for coordination, communication, and data collection among federal, state, and local agencies to result in actions to assist communities and emergency responders for incident prevention and effective response to chemical incidents. This order will also result in an overhaul of hazardous materials regulation.

Ammonium nitrate's history is as follows. The direct-synthesis ammonia production process was developed by the German scientist Fritz Haber in Karlsruhe and scaled up by process engineer Carl Bosch in Ludwigshafen (BASF) during 1909–1913. The process uses an activated iron-based catalyst and runs at pressures up to 250 bars and at temperatures up to 500 °C. Note that when first invented, much higher synthesis pressures were needed. The development of both more-active catalysts and high-performance centrifugal gas compressors enabled the reduction in synthesis condition to its current value. This is an excellent example of being able to use technology to reduce the magnitude of the hazard (i.e., pressure) and therefore the associated risk. This enables ammonia synthesis to be inherently safer than in earlier times, a concept championed by, amongst others, Trevor Kletz.[10]

The reaction to form ammonia is not that fast, and the required reactor residence time is significant as can be imagined from the size of the reactor module shown in Figure 1.1. Part of the raw material therefore had to be recycled after separation of the ammonia as a liquid by cooling under pressure. The construction of such a reactor and its safe operation at the high pressure and temperature mentioned was

FIGURE 1.1

High-pressure reactor (steel) used for ammonia production following the Haber process. Built in 1921 by *Badische Anilin-* und *Sodafabrik AG* in Ludwigshafen am Rhein. Now reerected on the premises of the University of Karlsruhe, Germany, after Wikimedia Commons.[11]

a significant challenge for the German steel industry. Ammonia can be corrosive, is rather toxic, and moderately explosive when in a certain concentration range if mixed with air. Indeed, the central theme of this book, that of risk control of industrial operations, is already evident with one of the first large-scale synthesis processes. In fact, the German experience in gun barrel manufacture at that time (e.g., with the big Bertha howitzer) certainly had important relevance to the ability to manufacture pressure vessels and piping that could be safely used at very high pressures. The ammonia process needs pure nitrogen. Air separation by cryogenic liquefaction and distillation was developed shortly before this time by Carl von Linde. The liquefaction is achieved as a result of the Joule–Thomson effect when forcing a flow through a restriction, similar as in a home refrigerator or air conditioner. The hazard of pure nitrogen is asphyxiation when entering a space filled with it, and the hazard of pure oxygen is, among others, violent explosion when it comes into contact with combustibles.

The large-scale production of hydrogen, which is required for the process, was another challenge. Because of its small molecular diameter, hydrogen leaks from equipment develop easily and hydrogen mixed with air is very explosive. The original method of using electrolysis to split water into hydrogen and oxygen was too elaborate and uneconomical. But, bringing superheated steam in contact with glowing beds of coke (the water–gas shift reaction process) created a viable new production route. Today, the main source of hydrogen is from natural gas in which the methane (CH_4) is stripped of its hydrogen by steam reforming to produce synthesis gas and subsequently applying the water shift reaction. The end products are hydrogen and carbon dioxide, of which the latter aggravates the problem of climate change if not further processed. (It is possible to react carbon dioxide with ammonia to urea, also a fertilizer, although less efficient as fertilizer but inherently much less hazardous.) Later in this chapter, several other methods for the industrial production of hydrogen are mentioned.

Ammonia is applied in fertilizers mainly as ammonium nitrate but also as urea. The nitrate for ammonium nitrate is manufactured by first "burning" ammonia with air to form nitric oxide. This occurs in the Ostwald process by having the gas briefly contact a platinum catalyst at about 850 °C. In the subsequent cooling, it is further oxidized to nitrogen dioxide and then this is with water and air converted in trickle columns into nitric acid. The latter reacts with ammonia to form ammonium nitrate. Concentrated nitric acid had already been used in the second half of the nineteenth century to produce on an industrial scale nitrocellulose (or more properly named cellulose nitrate), known as gun cotton, and other explosive materials (among others, Alfred Nobel in Sweden). Of course, because of the direct ammonia synthesis process, ammonium nitrate availability was widespread, and as a result, the production of explosives grew tremendously during World Wars I and II. Another side effect was that the industrial production of nitrates put the people in Chile, who dug guano for nitrates, out of work.

Although the production processes described have reached industrial maturity and are now widespread, several disastrous events have occurred due to the properties of their products. Most notorious are the accidents with ammonium nitrate, a substance that can decompose with a surplus of oxygen. Some examples of accidents are described below. Ammonium nitrate (NH_4NO_3, or for short AN) is called an oxidizing substance. When stored with combustibles it will strongly enhance progression of fire. If ammonium nitrate is heated to its melting point of 170 °C, it decomposes while producing heat and gases. At those temperatures it looks like "champagne," an almost colorless bubbling liquid. Gases produced are ammonia, oxidizing nitrogen oxides (nitric acid), water vapor, and the anesthetic nitrous oxide. The latter when mixed with fuels can explode. Because the decomposition is exothermic, once this reaction has started it will sustain itself and usually will accelerate if the heat produced is not effectively removed. This process can result in a so-called thermal explosion.

The decomposition temperature can be greatly decreased if the AN is contaminated with chlorides and/or organics. Decomposition conditions cause certain compound (NPK-) fertilizer mixtures to deflagrate, which has led to incidents in which thousands of tons of stored material were involved. Decomposition and deflagration lead to generation of massive amounts of brownish clouds of toxic gases such as nitrogen oxides, hydrochloric acid, and chlorine. Deflagration is a form of self-sustaining explosive decomposition with a propagating reaction zone, which in a confined space accelerates by pressure buildup and increased heat transfer to the reaction front. A closed pipe filled with deflagrating material will in the end burst due to the high internal pressure that develops. For AN, deflagration can only be stopped by effective cooling with water, which in a large heap of fertilizer is not simple to achieve. In one incident with a deflagration in a ship's hold fully filled with fertilizer, the captain was advised to close the hatches, with the idea that a fire will extinguish when access of oxygen is closed off. Instead, the deflagration accelerated due to enhanced heat transfer to the burning front with the higher pressure. Because one of the reaction products is steam, which then condensed as water, the mass started to become slurry, which as a liquid cargo led to instability and capsizing of the ship in the middle of the Atlantic Ocean. Following various such decomposition incidents, the industry now tries to avoid formulations that are able to deflagrate.

The third type of explosion that ammonium nitrate can produce is detonation. The AN in a slightly porous form is even used as an explosive, but in that case it is usually mixed with a fuel such as diesel oil, in which case it is called ANFO. Detonation can either be initiated by a strong shock wave generated by a high explosive booster charge or by a weaker shock that is applied over a large surface area. In unintended, accidental explosions it is the latter that typically occurs. The first major AN disaster was in 1921 at the site of the production plant in Oppau near Ludwigshafen, Germany. The explosion was initiated by explosive charges deliberately set to loosen up the 4500-ton caked, hygroscopic 50/50 mixture of ammonium sulfate and AN fertilizer in a storage tower. The practice was justified after conducting detonation tests with the mixture in tubes in which detonation could not be initiated. However, due to a locally, slightly richer composition in AN, and the uncertainty margin in the test, part of the stored fertilizer detonated, with enormous consequences. Based on damage assessment, an estimated 10% of the total quantity detonated. A satisfactory understanding of the failure of the test emerged only in the 1970s, when much more knowledge about the phenomenon of detonation existed, in particular about how difficult it is to extrapolate from the behavior of a small-scale test sample to thousands of tons of ammonium nitrate. As a result of the Oppau accident, more than 500 people died and 2000 were injured, adjacent buildings were destroyed with damage occurring to the windows and roofs (i.e., the weakest elements) of structures many kilometers away (see Figure 1.2).

A second detonation disaster occurred at the harbor site of Texas City, Texas, USA, on April 16, 1947, when a fire broke out on board a ship filled with 2300 tons of ammonium nitrate fertilizer. This time, blast and fragments from the ship explosion killed 581 people and caused many domino effects to nearby ships and shore installations. The blast also initiated a fire in an adjacent moored ship, also loaded with ammonium nitrate fertilizer, but with a smaller cargo of about 1000 tons. This second ship exploded 15 h after the first one. Ten years later, American and European experts tried in vain with field experiments on Helgoland (part of the German archipelago in the North Sea) to reproduce the phenomenon of an ammonium nitrate fire-to-detonation transition. Notwithstanding this failure, the phenomenon has since reoccurred in several accidents.

FIGURE 1.2

Picture of the crater and the destruction caused by the explosion of an ammonium nitrate–ammonium sulfate mixture on September 21, 1921 at the BASF plant in Oppau near Ludwigshafen, Germany.

Taken from Wikipedia.

In the 1960s, the Organisation for Economic Co-operation and Development (OECD) set up an International Group on Unstable Substances (IGUS), which amongst other groups started to investigate the hazardous properties of ammonium nitrate. In the years following in various OECD countries, much work has been carried out. For example, in laboratory and field tests, the relationship between thickness of steel wall in a tube test of sufficient length versus the mass effect of an unconfined heap of substance was determined. This resulted, for example, in a European Union (EU)-prescribed tube test and more stringent regulations for production and, in particular, for storage and transport of AN. The industry adapted their formulations such that the probability of deflagration and detonation could be reduced to almost zero. At the same time, they tried to retain the fertilizing properties of AN essentially unchanged. This resulted in the production of nearly pure, prilled (pelletized) ammonium nitrate. If the small, hard spherical prills possess a high density, that is, with no sensitizing air bubble inclusions or cracks, this cannot be detonated. Lower density prills, however, are of detonation grade. A problem is that due to the crystal modification transition at 32 °C and the corresponding small change in density, an originally high-density prill following a few temperature cycles may degrade. The IGUS group is still active in the entire field of hazardous substances. It supports the UN ECOSOC's Committee of Experts on Transport of Dangerous Goods and the Committee on the Globally Harmonized System of Classification and Labeling of Chemicals with respect to technical aspects of tests and criteria.

A third large AN disaster took place in Toulouse, France, in a fertilizer plant on September 21, 2001; this disaster has some similarities with the Oppau event and was exactly 80 years later! Again, a large crater formed, see Figure 1.3, which caused 21 fatalities and injured thousands of people. The initiation of the event, however, was completely different. In a storage facility containing about 300 tons of fertilizer, some off-specification product was collected that somehow was

FIGURE 1.3

Crater after the ammonium nitrate detonation at the Toulouse AZF plant on September 21, 2001. The estimated explosion strength was up to 40 tons of trinitro toluene (TNT) equivalent.[12]

contaminated with sodium dichloroisocyanurate, used as a disinfectant in water purification. This substance contains the element chlorine and an organic component; hence, as mentioned before, it had exactly the right ingredients to initiate self-heating of ammonium nitrate at lower-than-normal temperatures, possibly resulting in fast decomposition and subsequently in detonation with an estimated TNT equivalent of 20–40 tons.[13] A full-scale experiment to reproduce what could have occurred was unsuccessful despite quite intensive research. We still do not fully understand the precise conditions at which ammonium nitrate becomes highly dangerous. The accident had a large impact on French regulations with respect to major hazard industries. Prior to the accident, when determining safe separation distances, the French competent authority had limited itself to consequence considerations. After the accident, it embarked on full-fledged risk analysis with consideration also of occurrence probability, a procedure earlier adopted by, for example, the United Kingdom, the Netherlands, and Scandinavian countries. In later chapters, we shall introduce a more detailed overview of some attractive aspects that were introduced in the French approach to risk assessment.

The stories about ammonium nitrate are somewhat typical for developments in general. Knowledge about the properties of chemicals has been growing over the years and can be found, for example, in Material Safety Data Sheets (MSDSs), while the founding of the Globally Harmonized System (GHS) of Classification and Labelling of Chemicals, to be discussed in Chapter 2, is a step in the direction of great progress. Although many details, and sometimes important details, remain to be revealed, safeguards against undesirable events have been incorporated into regulation, codes and standards, and best practices. But because of a variety of reasons that we shall elaborate in Chapter 4, these safeguards are not always, nor uniformly, applied. Note that the common deficiencies in knowledge include detailed insight into the mechanism and data for the complex exothermic chemical decomposition kinetics. At this time, except for combustion kinetics of hydrocarbons, important in gas explosions, one must rely on various kinds of heat and mass loss tests to determine overall kinetics characterized by reaction order, activation energy, and rate constants. This makes prediction by means of computer model simulation of the behavior of hazardous substances in various situations uncertain because most initiations, fast as in shock, by spark, friction, or impact, or slow as in self-heating, is thermal by nature. Self-heating can be the result of the material itself (the material is then designated as a self-reactive chemical) stored at too high temperature or because it comes into contact with another substance with which it is not compatible. Self-heating starts slowly but accelerates due to temperature increase. Because most reaction rates increase exponentially with temperature, the decomposition rate can achieve a high value quickly (exothermic or thermal explosion). Initially, acceleration may be due to small amounts of decomposition products accumulating, which is called auto-catalysis, or by the substance losing its crystal structure. The time to explosion is called the *induction period*, and the process is named *runaway*. The sensitivity and precision of the thermal tests needed to characterize such behavior has improved tremendously over the years. The same can be said of the equipment for analytical chemistry, yet this progress has not been sufficient to enable us to answer all the questions needed before we can make accurate predictions of thermal decomposition behavior. In Chapter 3, we shall further discuss the topic. Nevertheless, despite tens of events in the past, accidents continue to occur. On September 7, 2014, in Queensland, Australia, a truck carrying 50 tons of AN fertilizer rolled over. Fire broke out, likely due to igniting (diesel) fuel, and the load detonated, disintegrating the truck.

1.3 AMMONIA

Anhydrous ammonia, as we have seen, is a starting material for the production of ammonium nitrate; it can also be used as a fertilizer by injecting it directly into the ground, a practice common in North America. Another application is as the recirculating fluid in refrigeration systems, about which many spill incidents are reported despite guidelines for proper use. Its hazard characteristics differ markedly from those of ammonium nitrate, and it is therefore worth considering here as another example. As mentioned before, apart from being moderately explosive when mixed with air, ammonia is toxic. At normal ambient conditions, it is a gas. To store it in large quantities, it is either liquefied under pressure (about 10 bar at 25 °C) or is refrigerated (boiling point −33 °C). It can be contained in tanks of carbon steel or CrNi(Mo) steel. Stress corrosion can be avoided by a small addition of water (0.1−0.2%). Ammonia spill hazards are often underestimated, but they have resulted in a considerable number of casualties.

Despite all safety measures, loss of containment can occur as happened in the incident in Potchefstroom, South Africa, in 1973 where a horizontal cylindrical tank released 38 metric tons of anhydrous ammonia.[14] Upon catastrophic rupture of a tank and release, the ammonia will be spilled partly as vapor and partly as a boiling liquid, which will spread over the ground whilst evaporating. The vapor is visible as a white cloud by absorption of, and reaction with, the moisture in the air. Initially the cloud is cold and heavy, but gradually, while dispersing and drifting with the wind, it heats up and becomes buoyant. For the dispersion of gas clouds, computational models have been developed about which some brief observations will be made in Chapter 3. In the case of Potchefstroom, the tank cracked and a hole was formed from which mainly vapor escaped. Because of the release, 18 people died, of whom four resided in a nearby area at a distance of 150−200 m, and 65 workers needed medical treatment. The toxic properties of ammonia are expressed as a probability distribution of lethality versus concentration inhaled over time or dose. This takes the form of a probit function, which is a transformed cumulative normal distribution presented as a straight line running from 1 to 99% lethality, depending on the vulnerability condition of the exposed people. The probit function is determined by three parameters as shown in the equation:

$$Pr = a + b\,ln(C^n t) \tag{1.1}$$

where *Pr* is the probit value, *a*, *b*, and *n* are substance-specific coefficients, with *C* being the concentration in the inhaled air, and *t* the exposure time. One can imagine that it is not an easy task to determine these coefficient values for humans. Data sources are derived from animal experiments, medical considerations, and analysis of actual accidents.

The official Dutch ammonia probit figures predicted originally in 1992 (Green Book[15]) a 50% probability of lethality at 30 min exposure (LC_{50}) of roughly 5.4 g/m^3 air volume. In 2012, it was proposed[16] to modify these figures to reduce the analogous dose to about 1.8 g/m^3. Probit relations are also used to express extent of harm to people by heat and blast, as we shall see in Chapter 3. The evidence that can be extracted from an actual incident is very uncertain because there are source uncertainties, which have to be modeled for the dispersion, and there are strong cloud concentration fluctuations, which are not modeled. Furthermore, people will try to flee from the cloud, so the exposure times are very different and may tend to be short. In addition, there are local obstacles or partly sealed enclosures, which may provide some protection. Hence, even when the initial population density at various distances from the source can be estimated, counting victims does not provide data that can be used for accurate probit validation. We shall come back to this type of uncertainty when discussing the reliability of risk analysis figures in Chapter 3.

The release in Potchefstroom involved hot work on the tank but with the omission of any stress relief thereafter. Good maintenance on ammonia tanks remains a matter of continuous attention. Accidents keep on occurring. At the time of writing this text,[17] a leak at a chemical plant in eastern Ukraine released ammonia causing the deaths of at least five people and injuring 20.

1.4 PETROCHEMICALS

In the foregoing discussion, we focused on large-scale mineral fertilizer production as the world depends on this. It is just one example; the energy industry is another. And, can we live without fuels and synthetic materials derived from the petrochemical industry, such as plastics in all kinds of applications, fibers, coatings, adhesives, lubricants, and so on? No answer is needed to this rhetorical question as is clear when considering the large amounts of plastic waste produced, spilling over in the oceans' "plastic soup." Feed stock for the petrochemical industry is mostly oil and gas. These raw materials must be made more reactive to enable further processing. The two main routes are cracking the hydrocarbons at elevated temperature and collecting the unsaturated smaller molecules such as ethylene (C2-chain) and propylene (C3), and/or partially oxidizing the hydrocarbon. An example of a substance produced when in succession both cracking/distillation and oxidizing are carried out is ethylene oxide (EO, also called oxirane), which itself is used again as the starting material for a large variety of products. The boiling point of EO is 11 °C, and it is usually stored as a liquid under pressure in rust-free carbon steel tanks. Its usefulness is extensive, because it can react with the addition of many other substances. It also dimerizes and polymerizes; it can be reduced and further oxidized, all yielding different products from cosmetics and antifreeze to sterilizers. However, the material safety data sheet of ethylene oxide looks quite frightful. The gas is toxic at a concentration below where you can first smell it; when mixed with air the mixture is flammable, hence explosive, from 3% ethylene oxide in air to 100%. In other words, due to the presence of an oxygen atom in the molecule, an explosion propagates in the pure substance. This property makes it even suitable for applications such as a fuel–air explosive that makes it a dreadful blast weapon. In particular, in the 1980s and 1990s, there were serious accidents with EO, but given the properties and the amounts that are handled, its record is relatively safe, thanks to many effective preventive risk-reducing measures.

1.5 GASOLINE

If an arbitrary person would be asked whether gasoline, the car fuel, is a hazardous substance, nine answers out of 10 would probably state it is a flammable material but not particularly hazardous. Yet, in the GHS of dangerous substances (see Chapter 2, Section 2.7), its constituents are classified with the following hazard statements: (highly) flammable, toxic when swallowed, and may cause cancer. In the last few years, there have been a number of very violent explosions of vapor clouds formed in the open air after a storage tank overfilled and a massive amount of gasoline was spilled. The first of these incidents was at the Buncefield Oil Storage Depot, Hemel Hempstead, Hertfordshire, near London, UK, on the early morning of Sunday, December 11, 2005. The depot was an important hub in the fuel distribution for cars, trucks, and aircraft. The spillage had taken place during the night. It was rather cold; there was an atmospheric inversion layer and almost no wind. A huge, dense, white cloud spread over the premises as recorded on CCTV security cameras. The cloud eventually came into contact with an ignition source and exploded. The devastation was enormous both on the premises and at off-site offices and general-purpose buildings (Figure 1.4).

By pure luck and thanks to the time of the incident, nobody was seriously injured, but had it been during daytime on a normal weekday this would have been a large disaster with many victims. Due to the explosion, fire spread to 10 or more large fuel tanks, see Figure 1.5. It took days to extinguish the fires; there was a great deal of air pollution from the smoke and the soil was polluted by runoff firewater containing hydrocarbons and firefighting foam ingredients. Luckily, later the nearby aquifer did not seem to be seriously damaged. A broad investigation board was set up and an impressive investigation effort made. Much has been written about this incident and the final report[18] was published in 2008. Chapters 4, 5 and 6 will discuss nontechnical causes of accidents as also present in this case. Here, we shall describe a few physical aspects of the vapor cloud explosion as an illustration of the hazards that potentially form a threat and that should motivate plant management to implement

FIGURE 1.4

Damage at off-site buildings and to parked cars neighboring the west side of the Buncefield complex showing the blast severity.

From the Buncefield final report.[18]

FIGURE 1.5

View of the destroyed storage tanks.

From the Buncefield final report.[18]

tight risk controls. Indeed, since the Buncefield disaster, the worldwide focus on the safety integrity of overfill protection systems has deservedly enjoyed more attention. In fact, over the years at different places similar accidents had occurred.

"Unconfined" vapor cloud explosions, as they were originally named, became recognized as a phenomenon deserving research attention in the late 1960s and early 1970s when several occurred in the scaled-up plants of that time, for example, the notorious 1974 one at the Nypro plant near Flixborough, UK (see Lees[1]). From the start, there was international cooperation on the subject, but despite many balloon experiments in which explosive gas mixtures were ignited, no blast was observed unless the ignition was initiated by the detonation of a high explosive charge. This conundrum remained until a link was made with the results of experiments with deflagrating explosive gas mixtures in pipes, closed at one end and ignited at the closed end. In such experiments acceleration of the flame was observed due to turbulence by friction with the wall of the unburned gas, pushed out ahead of the developing flame. Such acceleration was associated with the buildup of pressure and, in a later stage, shock waves in the pipe. These effects possibly even generated a transition of the deflagration into a detonation given a pipe that was long enough. When the tube wall was provided with metal objects intruding into the gas space, hence amplifying drag resistance to the flowing gas (thus simulating congestion) and generating even more turbulence, the effect was much stronger. TNO in the Netherlands then conducted experiments first on a small scale[19] and later on a large scale in open space with arrays of vertical pillars with central ignition.[20] The latter tests took place with propane-air on Dow Chemical Co. premises at Mosselbanken near Terneuzen in 1982 (Figure 1.6).

During the next 20 years, with European collaboration and with EU sponsorship, one test series after another such as MERGE and EMERGE were performed and a new theory developed. The initial motivation for the sponsorship was derived from a nuclear safety program: reactors built next to rivers should be protected against vapor cloud explosions due to spills of flammables resulting from possible tanker ship collision on the river. As a result of the growing understanding, the addition "unconfined" in the original name of "unconfined vapor cloud explosion" disappeared. The theory resulted in simplified models such as TNO's Multi-Energy Method and Computational Fluid Dynamics codes such as FLACS from the Norwegian company Gexcon. Although explosion strength often became expressed

FIGURE 1.6

Field tests with propane as fuel to investigate the mechanism of blast generation in vapor cloud explosions by flame acceleration due to the turbulence created by the flow around congested obstacles in front of the expanding flame. Left: the obstacle array of concrete sewer pipes. Right: Moment of explosion.

Tests were conducted by TNO in the Netherlands in 1982.[20]

in terms of a TNT equivalent, the nature of the blast profile is very different from that of a detonating solid-phase high explosive. A gas explosion does not make a deep crater but yields relatively high overpressure and impulse effects over a larger distance. After the Piper Alpha offshore platform disaster in July 1988 (which resulted in 167 fatalities), the knowledge has been applied to the heavily congested areas typical on this type of platform. Full-scale tests on mock-ups occurred at the Spadeadam testing site of the former British Gas, now owned via Germanischer Lloyd by DNV. Quite strong blasts were observed and debate began about whether transition to detonation in the open (as opposed to confined spaces) could take place. The FLACS code results have been validated and adapted to test outcomes, and applied in many designs for safer platforms, such as the recent floating liquefied natural gas (LNG) production ships, liquefying natural gas at sea at the production location.

The Buncefield damage was overwhelming. Initially, the FLACS code could not reproduce blasts powerful enough (1 bar overpressure or more) to explain the observed damage until the most likely ignition location was identified in a pump house. Moreover, the greenery along the road leading up to the adjacent parking and neighboring buildings, in the form of trees and shrubs, was interpreted in the FLACS modeling as flame accelerating congestion.[21] This postulation was confirmed in Spadeadam large-scale experiments and field trials.[22] However, the debate whether or not such congestion induced a transition to a detonation did not conclude. This question is relevant to land use planning decisions with respect to calculation of safe separation distances. A strong improvement program on risk control of storage sites introducing safety-instrumented systems and applying standard IEC 61511 has been realized since. These systems and the standard will be briefly described in Chapter 3.

On October 23, 2009, at the Caribbean Petroleum Corporation at Bayamón near Cataño, Puerto Rico, an explosion and subsequent fire occurred, which showed similar development and characteristics to the Buncefield disaster. According to a preliminary report[23] of the US Chemical Safety Board (CSB), a gasoline tank was overfilled and a vapor cloud developed and dispersed to a diameter of about 600 m (2000 ft) before it was ignited. The blast damaged homes and businesses over a mile distant from the plant. The CSB, overloaded with work, as at the time of writing (August 2013), has not finished its final report on the accident. A few days after the event in Puerto Rico, on October 29, 2009, an explosion and fire occurred at a storage site of the Indian Oil Corporation Ltd. in Sitapura near Jaypur, Rajasthan, India. It also started with a massive loss of containment of gasoline and a spreading vapor cloud that exploded. The blast resulted in 11 fatalities and injured 45, while damage to homes was present at up to 2 km distance.[24] Further investigation revealed that despite a low degree of congestion, from very strong evidence such as from directional indicators (bent over trees, etc.), that at least part of the cloud detonated.[25] Blast overpressures of a few bars from a fast deflagration will run up in a detonation by a factor of 5–10, albeit as a pressure pulse of very short duration. (To compare, windows shatter at an overpressure of only 10–20 mbar.) Thomas et al.[26] showed with ethylene as fuel that once a deflagration to detonation transition develops in a congested area, the remainder of a cloud in a noncongested part continues to detonate, while a fast deflagration outside a congested area will slow down. More recently, Pekalski et al.[27] reported similar results with ethane, which is much less reactive than ethylene and even less reactive than some constituents of gasoline. Further research must be performed to identify those circumstances in which flammable vapor clouds can become so destructive. This is the more important issue because it concerned even clouds of saturated and thus rather low reactivity hydrocarbons. Of course, total flammable mass released and ignition delay times are important factors. Fortunately, in many instances clouds, in particular smaller ones, just drift away without being ignited or, if ignited, do not produce blast but just a flash fire.

1.6 NATURAL GAS

The main component of natural gas is methane—the lowest molecular weight, least reactive hydrocarbon. After purification from among other impurities mercury and sulfur, natural gas is a most important, relatively green fuel for heating and cooking. Also, a car can be driven on natural gas. As we have seen, it is a raw material to produce hydrogen. Companies such as Shell in its Pearl plant in Qatar convert it to diesel and kerosene via partial oxidation to carbon monoxide and hydrogen synthesis gas and the Fischer-Tropsch process. But, it can also go to methanol and gasoline. It lacks a C–C bond. However, it does not differ fundamentally from other hydrocarbons in its flammability and explosive properties. When a pipe is filled with a stoichiometric mixture with air (about 10% CH_4 in air), combustion shows flame acceleration. When the pipe is long enough, the transition into detonation occurs just the same as with other hydrocarbons, but methane requires slightly more stringent conditions. Because of the large amounts of natural gas being used and transported, knowledge of its flammable properties is critical. Explosion accidents with natural gas usually occur only in rather confined situations such as following a gas leak within a home. High-pressure pipeline explosions are initially physical explosions (i.e., ruptures due to high pressure) followed by jet flames. Natural gas is also transported and stored in liquefied form at its boiling point of 162 K, or −111 °C (−260 °F); the cold LNG is stored at atmospheric pressure.

Since the early 1970s, several series of experiments have been carried out to investigate LNG hazards in order to determine the possible risks of tankers spilling their cargo due to collision when entering harbors. In particular, when spilled on water LNG boils violently and forms a cold cloud drifting with the wind, slowly heating up and lifting off. At between 5 and 15% methane in air the cloud is flammable and can explode. Tests and models have shown that cloud length at low wind speed and high source strength may reach values of a kilometer or more. Dispersion modeling to determine the size of an exclusion zone around storage sites is required by US regulations for risk assessment studies. Over the last decade, Sandia National Laboratories in the United States performed a number of tests with LNG spills on water and on land, also to investigate the possible consequences of terrorist actions and the possible progressive damage to a laden tanker once an initial spill occurs. A brief overview of the results up to early 2013 has been presented.[28] The Mary Kay O'Connor Process Safety Center at Texas A&M University conducts studies on more fundamental LNG properties, the effectiveness of water sprays to warm and disperse the light methane, and the application of high expansion foam to boiling LNG by emergency responders so as to reduce back radiation, and hence evaporation, during a fire scenario.

1.7 LIQUEFIED PETROLEUM GAS

LPG consists for the most part of propane and additionally butane. It is stored at ambient temperature as a liquid in pressure vessels at about 9 bara. It was once a by-product of the crude refining process that produced gasoline and other fuels. Why are we including this fuel in this chapter? Is it not that different in properties from gasoline or methane? Due to its volatility, it must be kept in pressure vessels, which introduce a different hazard—the boiling liquid expanding vapor explosion, or BLEVE.[29] This type of explosion is physical, that is, it is the burst of a container because the internal pressure is too great. It most commonly occurs if the vessel is heated in a fire and as a result the vapor pressure of the contents increases. Although the name BLEVE is only 50 years old, the phenomenon is much older as it frequently occurred at the beginning of the industrial revolution when steam engine vessels exploded. The heat input increases the potential energy stored in the liquid. Once the vessel bursts, the pressure drops, the liquid becomes superheated and will boil explosively. The result is an almost instantaneous generation of a large amount of vapor and for the rest a pool of boiling liquid.

The fuel for the external fire can originate elsewhere but can also be due to a leak of the vessel itself or its connections to valves, flanges, and other fittings. Because LPG is transported by truck and railway cars, such a leak can arise when a truck is involved in a collision or overturns, or the railway cars derail. Once in a fire, occurrence of a BLEVE depends on many factors. Larger vessels must have a pressure relief valve; smaller ones have at least a bursting disc. When a valve or disc opens it can produce a jet that ignites and that may add to the chaos but does not necessarily do harm. However, due to inadequate pressure relief and depending on the degree of filling and the (corrosion, fatigue) condition of the vessel, rupture can take place in the gas space and the liquid BLEVEs. The time to BLEVE depends therefore on many factors. Hence, in quite a few cases during that induction time of 10–30 min, the firefighters arrived and started trying to extinguish the fire but were eventually surprised by the explosion. If the vessel ruptures and the fuel is released, it will be partly vapor and partly boiling liquid. The depressurizing and gas expansion causes a blast wave. However, even more destructive is the ensuing fireball. The rupture causes the metal to become hot and the fuel mixing with air will ignite. A major part of the content of the vessel will be in a burning ball, rising upwards with vortices rolling at the outer edges and mushrooming. On the ground there will also be a burning pool. Due to the turbulence the mixing with air is rather efficient, so the flame becomes very hot. A BLEVE from a large, for example, 20 ton tank, can kill people at distances of up to 100 m and seriously injure those at 200 m or more. In many instances, firefighters have been victims. Fragments of the tank (missiles) can be projected for hundreds of meters up to 1 km. On the Internet, one can find several YouTube video clips[30] of BLEVEs (see also Figure 1.7).

Past accidents have been serious, and some disastrous, the latter when the large spherical storage tanks are involved with a diameter of 25 m or more. One example is the 1966 12,000 m^3 spherical LPG tank BLEVE in Feyzin, near Lyon, France, resulting from a leak during a faulty sampling procedure. The spill caused a cloud that ignited and heated the sphere. The BLEVE caused domino effects to neighboring spheres. The disaster resulted in 15 fatalities including several firefighters and at the time made a profound impression in Europe. Another more or less similar accident but of a much larger scale with many domino BLEVEs was at a storage facility near Mexico City in 1984 that resulted in the deaths of approximately 500 people, mostly residents from nearby housing built after the plant started operating. Smaller BLEVEs of tank trucks occurred in many countries. From a March 2011 earthquake in Japan several fierce fires at refinery/oil storage sites occurred. At the Chiba refinery of Cosmo Oil, a 2000 m^3 propane sphere BLEVEd due to a pipeline leak caused by a neighboring sphere that at the time was filled with water for cleaning purposes, and because of the heavier-than-normal load, the legs failed in the quake. The first BLEVE caused a BLEVE cascade of six other spheres in the storage park of 17 tanks.

FIGURE 1.7

Fireball of a propane tank truck BLEVE at a filling station in the city of Bucheon, South Korea, in 1998; the BLEVE duration was 8 s.[31] The fire started with a leak causing a vapor cloud, which then ignited. After 15 min the propane tank exploded, followed by a butane tank a few minutes later. It is typical that tanks fail at their weakest point, usually along a weld, so the tank debris is often unfolded and flat.

If LPG (propane) just leaks from a vessel or connecting pipe, a vapor cloud can arise, hugging the ground because it may be cold and propane is heavier than air. The cloud will disperse and dilute with air over a certain distance downwind. In case of ignition, most often a flash fire will occur but at larger mass and in a congested area a cloud can explode.

1.8 HYDROGEN

It now looks like the next generation energy carrier will be hydrogen. It is not, however, certain when it will be introduced on a large scale, as much depends on economics, building infrastructure, and appropriate risk assessment. One application would be in fuel cells for automotive and home heating/cooling and electricity generation. As long as the price of natural gas remains relatively low, one would not expect the investment in a hydrogen-based economy and distribution system to be profitable. For cars, hydrogen might only be initially attractive, if long term the limited battery capacity remains and restricts electric car use to a short range. On the other hand, fuel cell technology has become mature, for which hydrogen is an attractive fuel with various "green" ways to produce it, such as from solar energy or from sewer waste being under development. It could also be produced on a large scale from natural gas and from other energy sources such as nuclear power via improved electrolysis. Blending hydrogen and natural gas and burning it in engines is also being researched, and a blend for home heating is certainly possible. The International Energy Agency (IEA), set some long-term goals; interested nations formed the International Partnership for the Hydrogen Economy (IPHE); and the International Association for Hydrogen Energy (IHAE) acts as a focus for interested scientists and developers.

A challenge with pure hydrogen is the difficulty of storing it at high density. Absorption onto specific nanomaterials would be the best option, but for the time being it will be compressed in a tank up to even 1000 bar or liquefied at −253 °C (20 K), and even hydrogenation-dehydrogenation of organic compounds is being considered. It is not certain what is the safest storage method. Cryogenic hydrogen when unintentionally released in large quantity can form an explosive cloud for a short while, although it will disperse rapidly when it warms up as it is much less dense than air. A tank containing highly pressurized hydrogen will produce a burning jet of considerable length should the tank's integrity be breached and ignition occur.

In view of the advent of the 'hydrogen economy' there have been worldwide research efforts (in the United States, Japan, and Europe, including Russia) to make hydrogen for the consumer "as safe as gasoline." Researchers cooperate in the International Association for Hydrogen Safety (HySAFE). The programs have been very broad from determining the hazardous properties of mostly compressed hydrogen and also the liquefied form. Hydrogen has a relatively wide flammable range. Its ignition energy is low, but the autoignition temperature is high. Deflagration can easily transition to detonation. Hydrogen is a very light gas, about 15 times lighter than air. Models have been developed for dispersion of jets from leaks in the open but also in confined space. Small leaks in a garage without ventilation will fill the space with a combustible mixture from the ceiling down. Various preventive and protective measures have been proposed. Much work has also been performed on protection against a jet flame from a leak. Leak-free connections are being developed. Codes, standards, and good practices have been worked on. For the automotive branch, this development also means collision trials and investigation of various scenarios involving tunnels. There is also an effort to educate young engineers and scientists in the specifics of hydrogen safety issues, for example, at the University of Ulster in Northern Ireland, UK.

1.9 DUST EXPLOSIONS

On January 20, 2014, a dust explosion destroyed an animal feed-processing plant in Omaha, Nebraska, killing two and seriously injuring 10. It was not the first and will not be the last such explosion.

It is well known that flammable gases mixed with air within a certain range of mixture ratios, above the lower or below the upper flammability limit (US) or explosion limit (Europe), deflagrate once ignited in a confined space. During flame ball expansion, pressure builds up possibly causing the confinement to rupture, shatter, and hence explode. Methods to determine explosion severity and ignition sensitivity indicators are standardized. Various safeguards such as detectors, ignition preventing measures, automatic flame extinguishing systems, and overpressure vents have been developed and applied. In Chapter 2 we shall briefly consider regulatory measures.

Less well known is that dusts of various kinds of combustible substances/materials can show the same kind of explosion properties, mostly followed by fire with consequent burns to victims who may have been present. Fine dusts derived from materials in everyday use such as sugar, flour, coffee, powdered pharmaceuticals, dusts being generated in the processing and drying of milk powder, and fine aluminum or other readily oxidizable metal particles in processing plants can all generate fierce explosions when proper safeguards are not in place. Metal dust explosions are feared for their very high temperature flames. A difference between gas and dust as a fuel is that with an organic dust preheating of the particle must occur to generate combustible gases by pyrolysis and with metals to fracture a protective oxide layer. This is a factor requiring larger ignition energy; on the other hand, the mixture ratio with air for the upper explosion limit is less sharp and critical. In addition, a notorious feature of dusts is that settled layers of dust on floors and elsewhere can be whirled up into a dust cloud with air by a small explosive blast. When it comes into contact with an ignition source, such a small dust cloud can then grow and develop into a massive explosion with subsequent fires. Many such incidents have occurred recently in the United States, for example, destroying in 2008 the Imperial Sugar plant at Port Wentworth, Georgia, killing 14 people and injuring 40. Good housekeeping and compartmentalization are the main preventive measures together with the variety of tactics mentioned above for guarding against a gas explosion. Notorious too have been the wheat silo explosions in various places in the American Mid-West. Due to frequent dust explosions in the United States, the Chemical Safety and Hazard Investigation Board (CSB) urged the Occupational Health and Safety Administration (OSHA) in 2006 to make and enforce new regulatory measures.[32] Much of the knowledge about gas and dust explosions originated from coal mine safety research. Mine explosions took many lives, and the hazard is still present.

The main laboratory test vessel for dusts and powders is a so-called 20 L sphere, which was derived from the more effort-intensive 1 m^3-vessel developed by German gas and dust explosion pioneer Dr Wolfgang Bartknecht, originally working in mine safety but later in safety of pharmaceutical product processing. The rate of pressure rise severity is interpreted through a parameter that is normalized so as to give the same results as obtained on the 1 m^3 scale.

1.10 RUNAWAY REACTIONS

The processing of fine chemicals, pharmaceuticals, and some polymerization processes such as polyvinylchloride are accomplished in batch or semibatch reactors. Many of these are multipurpose units. If a process is exothermic, there are numerous ways that the reaction can get out of control, including the wrong dosing of reactants, absence of stirring, and failure of cooling. Due to heat generated in the reaction, the temperature of the mixture rises and, as long as reactants are present, the higher temperature accelerates the reaction. So, if cooling is insufficient by an ever-increasing rate of temperature rise, a thermal runaway and explosion occurs. In a closed system, hot gaseous products will build up pressure, blow off the reactor cover or rupture the vessel, rise up, and may after mixing with air find an ignition source or self-ignite and cause a secondary vapor cloud explosion. This was, for example, the case in the halogen exchange reactor of the fluoroaromatics plant at Shell Stanlow, UK, in March 1990 (see Figure 1.8), where plant fragments were projected over 500 m, vessel parts over 200 m, and the plant was partially destroyed by the blast and the ensuing fire (six were injured and one died).[33] A pressure relief valve was present, but pressure developed so rapidly that the valve was not able to pass gas fast enough and keep the pressure at a suitable limit. In this case, somehow too much water entered the reaction mix, which initiated uncontrollable secondary reactions; most often, secondary reactions set in spontaneously if the mixture is heated to a high enough temperature. Strong exothermic secondary decomposition reactions with higher activation energy, such as oxidative conversion, take over and lead to an explosion.

The violent explosions of reactors preparing organic peroxides used amongst others as polymer initiators have been notorious, but similar events when performing nitration reactions in the production of pharmaceuticals, pigments, and explosives, have also been commonplace. A catastrophic runaway occurred at the T2

FIGURE 1.8

Runaway explosion damage within a distance of 10 m of the fluoroaromatics reactor at Shell Stanlow, UK, in 1990.[33]

Laboratories Inc. plant at Jacksonville, Florida, in December 2007, which killed four and injured 32 people. This accident has been investigated by the CSB[34] and also described by them in some crucial chemical engineering terms.[35] It concerned the manufacture of a gasoline-upgrading additive. The batch started with melting blocks of sodium and reacting it with methyl-cyclopentadiene in a solution of diethylene glycol dimethyl ether (diglyme) with the formation of hydrogen in a very heavy, closed, stirred reactor of about 9 m^3 volume. The reactor had a 10-cm-diameter emergency vent pipe closed with a rupture disc and for letting off hydrogen during the reaction in a 2.5-cm-diameter open vent pipe. Cooling was provided by letting water boil off from a jacket around the vessel. Replenishment water was supplied by a city service line. During the fatal batch, the cooling appeared not to function properly and the temperature rise could not be controlled. The damage was tremendous. The system lacked appropriate and appropriately sized safety provisions, such as an emergency cooling system. Also here, a highly exothermic but previously unidentified secondary reaction started at a few tens of degrees Celsius higher than the normal working level of almost 177 °C. Parts of the reactor having a ca. 7.5-cm-thick wall were found at various distances from the original reactor position, some even at a 1.5-km distance. The blast damaged structures at over 500 m away. Runaway reactions are not confined to reactors and have, for instance, been experienced in storage tanks, pipelines, distillation columns, reboilers, and evaporators.

Much work has been done in studying runaway reactions. Theoretical models on thermal explosion have been available for a long time, but the key problem has been establishing the kinetic parameters of the reactions as a function of concentration, temperature, and degree of conversion. Two main avenues of approach are followed: adiabatic and isothermal calorimetric analysis; in addition, by means of thermogravimetry, mass loss can be monitored. Over the years, equipment with high heat flow sensitivity has been developed for a range of sample sizes. To simulate chemical behavior on a large scale, an adiabatic device achieved by very accurate, external temperature compensation is required. In some special cases, also, the ambient pressure follows the sample pressure increase to permit the use of test cells with low thermal inertia. By way of contrast, simulation of the desired process at the correct conditions on an industrial scale can be realized by performing the reaction in a temperature-controlled reactor or reaction calorimeter on a kilogram scale, the so-called bench-scale calorimeter. In quite a few cases, matters are complicated due to reactants present in multiple liquid phases, being processed under boiling conditions with reflux present or processed under pressure.

After earlier work on pressure relief venting at, for example, ICI in the UK, since the late 1970s, there has been international cooperation amongst industry representatives, research institutes, and consultants in the Design Institute for Emergency Relief Systems, DIERS, formed under the auspices of the American Institute of Chemical Engineers, AIChE.[36] When reactor systems vent under pressure, the flow rate is often in two phases: liquid and gas/vapor, much like the way liquid can spray out of a carbonated drink can or a champagne bottle. If a solid catalyst is also present, the flow may even be of three phases. Designing a vent for two- or three-phase flow is not that simple. A distinction is made between gassy systems, vapor-dominated ones (tempered), and hybrids (combinations of the two). There are a number of factors to consider, because the resistance to the flow by the vent opening varies considerably with the constitution of the vented mass. If pure liquid would be vented, mass flow rate is maximal. Depressurizing over the opening, however, causes the liquid to swell because of growing gas or vapor bubbles, while also turbulence increases; hence, the resistance to the flow increases. Physical properties of the reaction mixture and the location of the vent determine what volume fraction (void fraction) or mass fraction (quality) of the flow at the vent will be gas. For a vapor pressure-dominated system, depressurizing the liquid will result in a partial evaporation or flash. This will cool the reactants (a tempered system). If the liquid enters the vent opening, it may be bubbly or foaming and not change in structure in the vent. But if the gas is disengaging, then the flow is churn-turbulent. This means that the gas is present mainly as large bubbles with small ones in their wake and the liquid is in a chaotic and oscillatory motion of rising and falling and not to a large extent entrained. This reduces the escaping mass flow rate further; Lees[1] provides details.

Special tests have been developed to assist in the design process, including the 10-ml Advanced Reactive System Screening Tool (ARSST)[37] (the successor to the RSST) and a larger, 120-ml test, which is either a vent sizing package (VSP), or PHI TEC apparatus.[38] For calculating the required vent diameter as a function of all parameters, the SAFIRE computer code has been developed. Obviously, the T2 management did not apply the DIERS knowledge and methodology to design their single vent to release at a sufficient rate to avert a runaway reaction. Over the last decade, with an emphasis on prevention, early warning techniques have been developed within the EU Joint Research Centre[39] to detect, with a high degree of certainty, the indication of an abnormal increase in temperature within the normal fluctuations of the temperature signal. Similar work is described by Valeria Casson and Giuseppe Maschio.[40]

Runaways are not common in continuous processes, and when they do occur, the effects are usually much less violent than in batch or semibatch reactions. This is the reason why when demand for a product increases the process is often reengineered to be continuous in nature with a much smaller reactor holdup. The first successful conversion in this respect was the production of nitroglycerine. In the batch version, many deadly accidents occurred with this powerful, but sensitive, high explosive. ICT in Karlsruhe recently converted it into a continuous (intensified) microreactor process, which makes the production even safer.

1.11 HAZARDOUS MATERIAL SPILLS IN TRANSPORTATION ACCIDENTS

In principle the type of phenomena that can occur in transportation incidents are not different from those in processing operations or in storage. Yet, an entirely different regulatory community is in charge of developing safety rules and provisions. This is mainly because in moving traffic hazards bring different legal implications, cross-border operations require international cooperation, and containing/packaging hazardous materials for moving platforms comes with its own requirements. We shall refer to this in Chapter 2. Also, with respect to risk assessments, one has to deal with different hazardous material release probabilities; we shall return to that topic in Chapter 3.

Although much of materials transport is done by truck, ship, and pipeline, there is particular concern over the derailment or collision of freight trains with tank cars filled with bulk chemicals, such as chlorine, ammonia, or acrylonitrile. The Dutch government contributed a large sum of money to have industry change operations of manufacturing chlorine at one site and using it at another, despite many years without a major accident. Sites were more than 100 kilometers apart and transport passed densely populated areas. Instead, it is now preferred to produce chlorine (and other toxic gases) where it is also used; this is another application of Trevor Kletz's inherently safer principle.[41] Due to the state of the railways, derailment occurs more often in the United States than in Europe. Notwithstanding this comment, the derailing and overturning of a train with 14 tank cars loaded with LPG near the station in Viareggio, Italy, during the night of June 29, 2009, due to a broken axle (despite recent inspection) still remains a fresh and vivid memory.[42] No BLEVEs took place, but one of the tanks was punctured by an obstacle next to the rail line. LPG escaped, formed a boiling pool and a rather dense cloud. The cloud drifted across the railway into streets, hugging the ground, until it found an ignition source about 3 min later and produced a relative long-duration flash fire. Where the gas had penetrated nearby housing through open windows and doors, some weak explosions occurred. The panic and damage in the densely populated residential area that night must have been horrendous; 33 people died.

In 2013, there were two major derailments. The first one occurred in the town of Wetteren, Belgium, in May, in which a tank car train carrying butadiene and acrylonitrile derailed, see Figure 1.9.[43,44] This occurred because the engineer-driver had not noticed points (a shunt) near a spot where repairs were being made and approached this section too fast. A tank car carrying toxic and highly flammable acrylonitrile overturned and started leaking, upon which fire broke out and spread to two other cars. Emergency responders ordered evacuation within a circle of 500-m diameter. One person was killed, and 100 were injured. It was decided to let the acrylonitrile burn out in a controlled way, but strongly contaminated fire water reached the town's sewer system where it formed cyanides, also in gaseous form (hydrogen cyanide). For this reason, later a larger area had to be evacuated. Most people could return home only after three days; those living close to the railway were in temporary accommodation for three weeks.

The second accident took place in Lac-Mégantic, Québec, Canada, in July 2013, when a train of 72 tank cars filled with crude oil and being pulled by five locomotives derailed.[45] The train was operated by a single person; it remained late at night parked

FIGURE 1.9

Freight train derailment in the town of Wetteren, Belgium, in May 2013. The overturned cars in front were loaded with acrylonitrile, which upon leaking started a fire. The firefighters let the fire burn in a controlled way so as to consume the highly toxic acrylonitrile.[44]

FIGURE 1.10

Zhu et al.[48] described the Qingdao oil leak pipeline explosion and fire accident. The damage was very extensive along a large part of the trajectory of the pipeline. The figure shows an example of damage to automobiles resulting from flying debris at one location. The authors presented details on causation, distances victims were found, and heavy damage occurring to storm drains and buildings including large-scale window fracture.

on the main line 11 km from the town, with one engine still running to keep brake air pressure up while the engineer went by taxi to a hotel. Later, fire broke out at the locomotive with the engine running; the firefighters extinguished the fire, and sometime after, during the night, the train started moving down the sloping track toward the town, gaining speed and derailing in the town center. Explosions of tank cars rupturing were heard, and a wave of burning oil flooded the town, causing 47 fatalities. The catastrophic effect further increased by the burning oil entering the (storm) sewer system. Half the town burned down. The private train operating company was bankrupted.

Buried pipelines carrying pressurized gases may explode after being heavily corroded and neglected or hit by machines working in the soil. In case the gas is a fuel, rupture will often be followed by fire, as in the case of the Ghislenghien, Belgium, 1-m diameter, 80-bar natural gas pipeline fire, in which 24 people were killed in 2004.[46] The pipe started leaking after being scratched by earthwork machinery two weeks earlier, a cloud formed, and after delayed ignition an explosion occurred. The details on underlying causes and effects are interesting to read.[47] In November 2013 in Qingdao,[48] China (see Figure 1.10), an old oil pipeline propagated an explosion over a large distance, killing 62 people. The pipeline must have been at least partly empty, permitting flame propagation along the oil surface, possibly even transiting into detonation. History shows quite a few other damaging pipeline explosions, yet, regarding the immense total length of pipelines in the world, reliability is quite high, although care in design, construction and maintenance is needed.

Accidents with hazardous materials aboard ocean ships are no exception either. The International Maritime Organization provides a regulatory framework for the shipping industry.

1.12 CONCLUSION

The description of these accidents has been a bird's eye overview meant to offer some examples of the consequences of different phenomena; we have tried to select recent examples. Three decades ago, Charles Perrow[49] gave a more general account of the hazards that have accompanied technological development and industrial activity. Trevor Kletz[50,51] published several books with examples on "what went wrong" in the UK, and Roy Sanders[52] gave an updated account on events in the American chemical industry, which served as catalysts to learn and improve. The Institution of Chemical Engineers in the UK has published the *Loss Prevention Bulletin*[53] since 1974; it is a unique collection of descriptions and analyses of process accidents occurring worldwide.

As can be noticed from the phenomena and the course of the accidents with hazardous materials that have been described, scenarios can be very diverse and therefore difficult to predict, at least with respect to details about how an event will initiate, develop, its likelihood, and what effects it will have. We shall return to these issues when discussing existing (quantitative) risk analysis methodology in Chapter 3, the longest chapter! It is long because before describing risk assessment it presents an overview of knowledge and knowledge sources of hazardous materials, hazardous phenomena, and safety measures. What has not been revealed in any detail in this first chapter is the human factor and the failing management systems underlying the majority of the causes of these accidents. These subjects will be introduced in Chapter 3 and further developed in Chapters 4 and 6. The remainder of the book will treat possible new approaches such as a consistent system approach and applying resilience principles (Chapter 5), and recently developed system-based methods and tools (Chapters 7 and 8). When applied thoroughly, these innovations contain promise to improve control and so prevent most of these accidents. Hazards and their associated risks can be reduced by well-executed risk management, saving overall costs (Chapter 9). In Chapter 10 we shall look at what regulatory innovation is considered, in Chapter 11 what role knowledge plays and what to do to enhance learning, and finally Chapter 12 will discuss risk acceptance. But first, in Chapter 2 we shall briefly review another heavily contributing factor to safe operation—namely legislation and regulation.

REFERENCES

1. Mannan S, editor. *Lees' loss prevention in the process industries. Hazard identification, assessment and control*. 4th ed., vol. 1–3. Butterworth-Heinemann; 2012, ISBN 978-0-12-397189-0.
2. Shrivastava P. *Bhopal; anatomy of a crisis*. Cambridge: Ballinger; 1987, ISBN 0-88730-084-7.
3. Pietersen CM. 25 Years later, the two largest industrial disasters with hazardous material – LPG disaster Mexico City Bhopal tragedy – the investigation and the facts – lessons for process safety – have we learnt the lessons? Apeldoorn: Gelling Publishing/SSC Rotterdam. ISBN 978-90-78440-42-0.

4. Mannan S. *Lees' process safety essentials, hazard identification, assessment and control*. 1st ed. Butterworth-Heinemann; 2014, ISBN 978-1-85617-776-4.
5. Wikipedia, the free encyclopedia. List of industrial disasters. http://en.wikipedia.org/wiki/Industrial_disasters.
6. Abdolhamidzadeh B, Abbasi T, Rashtchian D, Abbasi SA. Domino effect in process-industry accidents — an inventory of past events and identification of some patterns. *J Loss Prev Process Ind* 2011;**24**(5):575—93.
7. *Marsh, the 100 largest losses 1972—2011, large property damage losses in the hydrocarbon industry*. 22nd ed. 2012. www.marsh.com.
8. Fresco Louise O. *Assistant director general, FAO agriculture department, fertilizer and the future*. Food and Agriculture Organization of the United Nations; 2003. http://www.fao.org/ag/magazine/0306sp1.htm.
9. See the website of the U.S. Chemical Safety Board. http://www.csb.gov/west-fertilizer-explosion-and-fire-/.
10. Kletz TA, Amyotte P. *Process plants: a handbook for inherently safer design*. 2nd ed. CRC Press Taylor & Francis Group; 2010, ISBN 978-1-4398-0455-1.
11. Wikimedia Commons. http://en.wikipedia.org/wiki/File:Ammoniak_Reaktor_BASF.jpg.
12. Dobson A, Gach T, Sullivan S. AZF Ammonium nitrate explosion, web presentation, [PPT]AZF Ammonium Nitrate Explosion Final v1.3.pptx.
13. Barthelemy F, Hornus H, Roussot J, Hufschmitt J-P, Raffoux J-F. *Report of the general inspectorate for the environment: accident on the 21st of September 2001 at a factory belonging to the Grande Paroisse Company in Toulouse*. October 24, 2001.
14. Lonsdale H. Ammonia tank failure — South Africa. In: *Ammonia plant safety*, vol. 17. AIChE; 1975. p. 126—31.
15. Green Book. *Publicatiereeks gevaarlijke stoffen PGS 1, Methoden voor het bepalen van mogelijke schade*. Issued by the Dutch Government, Ministry VROM, downloadable from. 2005. http://www.publicatiereeksgevaarlijkestoffen.nl/.
16. http://www.rivm.nl/Documenten_en_publicaties/Algemeen_Actueel/Uitgaven/Milieu_Leefomgeving/Probits/Technical_support_documents/Ammoniak.
17. August 2013, see http://www.reuters.com/video/2013/08/06/ammonia-release-in-ukraine-chemical-plan?videoId=244597266.
18. Buncefield Major Incident Investigation Board. The Buncefield incident 11 December 2005, the final report of the major incident investigation board, vols. 1 and 2. Crown Copyright 2008, ISBN:978-0-7176-6270-8.
19. Van Wingerden CJM, Zeeuwen JP. Flame propagation in the presence of repeated obstacles: influence of gas reactivity and degree of confinement. *J Hazard Mater* 1983;**8**:139—56.
20. Zeeuwen JP, Van Wingerden CJM, Dauwe, RM. Experimental investigation into the blast effect produced by unconfined vapour cloud explosions. Proceedings of 4th international symposium on loss prevention and safety promotion in the process industries, Harrogate, 12—16 September, 1983, vol. 1. p. D20, 10 pp.
21. Van Wingerden M, Wilkins B. *Experimental investigation of the effect of flexible obstructions on flame propagation in vapour cloud explosions*. Bergen: CMR Gexcon Report; 2009. Ref. No.: GexCon-08-F44110-RA-1.
22. Johnson DM. The potential for vapour cloud explosions — lessons from the Buncefield accident. *J Loss Prev Process Ind* 2010;**23**:921—7.
23. http://www.csb.gov/csb-conducting-full-investigation-of-massive-tank-fire-at-caribbean-petroleum-refining-investigative-team-plans-to-thoroughly-examine-facility-safety-practices/.

24. Sharma RK, Gurjar BR, Wate SR, Ghuge SP, Agrawal R. Assessment of an accidental vapour cloud explosion: lessons from the Indian Oil Corporation Ltd. accident at Jaipur, India. *J Loss Prev Process Ind* 2013;**26**:82–90.
25. Johnson DM. Vapour cloud explosion at the IOC terminal in Jaipur. Symposium series No. 158, hazards XXIII, Copyright 2012. IChemE. pp. 556–564.
26. Thomas JK, Goodrich ML, Duran RJ. Propagation of a vapor cloud detonation from a congested area into an uncongested area: demonstration test and impact on blast load prediction. *Process Saf Prog* 2013;**32**:199–206.
27. Pekalski AA, Puttock J, Chynoweth S. DDT in a vapour cloud explosion in unconfined and congested space: large scale test. In: *Tenth international symposium on hazards, prevention, and mitigation of industrial explosions Bergen, Norway*; June 10–14, 2014. p. 847–57.
28. Hightower, M, DOE/Sandia national laboratories LNG safety research, Marine LNG Transport Cascading Damage Study Summary and Risk Management Considerations, http://www.narucmeetings.org/Presentations/Sandia%20NARUC%20LNG%20Cascading%20Damage%20Summary%20Overview%202-2013.pd.
29. The history of the name has been described by Abbasi T, Abbasi SA. The boiling liquid expanding vapour explosion (BLEVE) is fifty … and lives on! *J Loss Prev Process Ind* 2008;**21**:485–7.
30. Video clip on 1983 BLEVEs due to derailed railcars near Murdock, Illinois, U.S: one containing about 60 tons propane and another 100 tons of isobutylene. Can be found on http://www.youtube.com/watch?v=Xf3WKTwHpIU.
31. Park K, Mannan MS, Jo Y-D, Kima J-Y, Keren N, Wang Y. Incident analysis of Bucheon LPG filling station pool fire and BLEVE. *J Hazard Mater* 2006;**A137**:62–7.
32. U.S. Chemical Safety and Hazard Investigation Board. *Recommendations status change, summary*. July 25, 2013. http://www.csb.gov/assets/recommendation/Status_Change_Summary_Dust1.pdf.
33. Cates AT. Shell stanlow fluoroaromatics explosion – 20 March 1990: assessment of the explosion and of blast damage. *J Hazard Mater* 1992;**32**:1–39.
34. U.S. Chemical Safety and Hazard Investigation Board. *Investigation report T2 Laboratories, Inc*. Runaway Reaction, Report No. 2008-3-I-FL. September 2009.
35. Willey RJ, Fogler HS, Cutlip MB. The integration of process safety into a chemical reaction engineering course: kinetic modeling of the T2 incident. *Process Saf Prog* 2011;**30**:39–44.
36. Fauske HK. Revisiting DIERS' two-phase methodology for reactive systems twenty years later. *Process Saf Prog* 2006;**25**:180–8.
37. Fauske HK. The reactive system screening tool (RSST): an easy, inexpensive approach to the DIERS procedure. *Process Saf Prog* 1998;**17**:190–5.
38. Askonas ChF, Burelbach JP, Leung JC. The verastile VSP2: a tool for adiabatic thermal analysis and vent sizing applications. In: *28th Annual Conference, Orlando*. North American Thermal Analysis Society; October 4–6, 2000. p. 1–6.
39. http://ec.europa.eu/dgs/jrc/downloads/jrc_tp2528_on_line_warning_detection_system_of_runaway_initiation_using_chaos_theory_techniques.pdf .
40. Casson V, Maschio G. Reaction calorimetry and UV-Vis spectrophotometry integration aimed at runaway reaction early detection. *Chem Eng Trans* 2013;**31**:877–82. http://dx.doi.org/10.3303/CET1331147.
41. Kletz T. *Cheaper, safer plants, or wealth and safety at work: notes on inherently safer and simpler plants*. IChemE; 1984, ISBN 0-85295-167-1.

42. Landucci G, Tugnoli A, Businі V, Derudi M, Rota R, Cozzani V. The viareggio LPG accident: lessons learnt. *J Loss Prev Process Ind* 2011;**24**:466–76.
43. Facts have been reported in http://nl.wikipedia.org/wiki/Trein-_en_giframp_bij_Wetteren.
44. Additional facts and the picture have been taken from the newspaper reports of the Belgian Knack. http://www.knack.be/nieuws/belgie/explosie-trein-schellebelle-1-dode-en-17-gewonden-video/article-4000293664222.htm.
45. A detailed report is given in Wikipedia. http://en.wikipedia.org/wiki/Lac-Mégantic_derailment.
46. ARIA. *Rupture and ignition of a gas pipeline*. Belgium: Ghislenghien; July 30, 2004. No. 27681, http://www.aria.developpement-durable.gouv.fr/wp-content/files_mf/FD_27681_Ghislengheinv_2004ang.pdf.
47. HInt Dossier, gas pipeline explosion at ghislenghien, Belgium. http://www.iab-atex.nl/publicaties/database/Ghislenghien%20Dossier.pdf.
48. Zhu Y, Qian X-m, Liu Z-y, Huang P, Yuan M-q. Analysis and assessment of the Qingdao crude oil vapor explosion accident: lessons learnt. *J Loss Prev Process Ind* 2015;**33**:289–303.
49. Perrow Ch. *Normal accidents, living with high-risk technologies* (First Published by New York: Basic Books; in 1984). Princeton University Press; 1999, ISBN 0-691-00412-9.
50. Kletz T. *What went wrong*. 5th ed. Gulf Professional Publishing/Elsevier; 2009, ISBN 978-1-85617-531-9.
51. Kletz T. *Still going wrong: case histories of plant disasters and how they could have been avoided*. Butterworth-Heinemann, Reed Elsevier Group; 2003, ISBN 978-0-7506-7709-7. 13.
52. Sanders RE. Chemical process safety: learning from case histories. 3rd ed. Elsevier Butterworth-Heinemann, ISBN 0-7506-7749-X.
53. Institution of Chemical Engineers, U.K., (IChemE). Loss Prevention Bulletin. http://www.icheme.org/lpb/about-loss-prevention-bulletin.aspx.

CHAPTER

Regulation to Safeguard against High-Consequence Industrial Events

2

Lock the stable door after the horse has bolted.
Old English proverb

SUMMARY

The United States and the European Union have been the pioneers of industrial safety. Because of the accessible language in which their regulations to prevent industrial disasters and to mitigate their effects is conveyed, regulatory developments in these two regions will be briefly described. The history of regulation on major hazards in both the United States (US) and European Union (EU) is one of stepwise development and improvement, unfortunately often when governments were prompted to act in response to major incidents.

The evolution of process safety regulations in the US starting after the Bhopal, India, disaster in 1984 will be summarized first, although this was some years before industrial hazardous materials regulation had been proved necessary to combat large-scale environmental pollution. An important law issued by the US Occupational Safety and Health Administration (OSHA) in 1992 is 29 CFR 1910.119, the Process Safety Management (PSM) rule, which forms the basis for identifying hazards and taking technical and organizational measures to control the risks. This rule was instigated by a number of major accidents in the US in the late 1980s and early 1990s, amongst others the Phillips Petroleum ethylene vapor cloud explosion at Pasadena, Texas in 1989, which caused 23 fatalities and 232 injuries.[a] This PSM regulation was followed in 1996 by the law 40 CFR Part 68 Risk Management Plan issued by the Environmental Protection Agency (EPA). This latter law must be used to specify minimum safe separation distances to residential housing and vulnerable entities such as nursing homes, schools, or hospitals, as well as sensitive receptors (e.g., parks). Such calculations shall be made both for a worst-case scenario and for an alternative owner/operator chosen scenario. The calculations are

[a]One of the fatalities in this event was Mary Kay O'Connor, chemical engineer. The death of his wife prompted Michael O'Connor to make an endowment available to Texas A&M University to found a process safety center integrated within the Artie McFerrin Department of Chemical Engineering.

Risk Analysis and Control for Industrial Processes - Gas, Oil and Chemicals.

auditable, publicly available, and shall be locally discussed. Incidents must be reported to a delegated authority.

In the European Union, high-impact low-probability events are called major hazards. The Seveso incident (near Milan, Italy) dispersed dioxin into the local atmosphere and environment as a result of pressure relief venting following a reactor runaway accident. This led to drafting the Seveso Directives, which are intended both to prevent major hazards and to protect workers and citizens. The first Directive was published in 1982 (82/501/EC), with subsequent major improvements in 1996 (96/82/EC) and 2012 (2012/18/EU). The Directives require submission of a satisfactory safety report prior to the issue of a license to operate, in which the operator demonstrates that "all that is necessary has been done to prevent major accidents." Reporting of incidents is mandatory. The Directives contain paragraphs that deal with land use planning, domino effects, and inspection. The implementation of a directive in the various EU member states is mandatory but can result in significantly different laws within the EU as the same objectives can be achieved in different ways. As an example, the different approaches to safety zoning in land use planning, or in American terms stationary source siting, will be examined. We can distinguish two main categories:

- A so-called *deterministic* or *consequence-based* approach that regards only the maximum severity of consequences (so-called endpoints) as criteria for the separation distance beyond which people are safe (Germany). This resembles the US approach.
- A *probabilistic* or *risk-based* approach in which, besides the consequence severity of an event, the probability of that event is also taken into account. This approach therefore must include directional influences on the consequences, for example, as a result of wind, atmospheric stability, or topology of the local environment. Decision criteria on which a safety report is accepted/rejected are therefore also probability based (e.g., UK, France, the Netherlands).

Both approaches place substantial emphasis on applying risk-reduction measures and high safety standards and are intended to achieve the same ends. Germany, for instance, rejects the probabilistic approach because of the large uncertainties associated with outcomes.

Of course, if sufficient space is available for land use planning, the ideal *near zero-risk* to population approach could be achieved by very large separation distances. Unfortunately, this is impractical in many societies and countries. As an alternative, it would be convenient to work with *generic safety distances* depending on certain characteristics of the type of activity. The contrast between the two approaches mentioned above is less distinct than might be imagined. The deterministic approach (Germany) is now generally based *not* on a relatively unlikely, worst-case scenario of leak hole area but also permits consideration of a more likely, credible alternative. This approach introduces an element of plausibility. On the other hand, the probabilistic approach has introduced a perimeter (France) or a safety zone borderline (NL), enclosing an area within which risks are assessed, while outside of which risks are ignored. Obviously, the more that availability of space is restricted given that economic activity requires the presence of industry, the more the probabilistic approach becomes favored.

Risk communication is an important aspect of risk. In the US and France the population in the vicinity of a plant is most directly involved in the discussion about risk-reduction measures and in decision making. In France, a group that represents the workforce is required to be included in the discussion. This inclusion will undoubtedly make the process less simple, but it is anticipated that openness and participation will enhance employee motivation and better balance the risk versus the economic interest.

Due to the enormous oil spill in the Gulf of Mexico as a result of the blowout of the Macondo well in April 2010, regulation in the US and in Europe was reformed to cover process safety in offshore operations. (A brief overview of this regulation will also be given.)

The chapter will continue with a discussion of requirements for transport of hazardous materials and the Globally Harmonized System (GHS) for classification and communication of substance hazards. The GHS is relatively new and relevant worldwide as chemicals are ubiquitous and crucial for the economies of so many countries. Brief attention is given to some of its important concepts such as hazard classes and categories, signal words, and labeling. Finally, future directions are discussed in which a plea is made for improving test methods, for considering the pros and cons of "goal setting" and "prescriptive" regulation, and the necessity of improved inspection regimes and of living up to both the spirit and the letter of regulations.

2.1 SOME HISTORICAL LANDMARKS OF MAIN THEMES OF REGULATION IN THE UNITED STATES AND EUROPEAN UNION

The main themes of regulation in the United States and in Europe run in parallel, but there are many differences in the details and the timing when new laws were implemented. The goal, however, is identical: protection of the workers and citizens against hazards and risks created by operations of the processing industry and associated storage and transportation of hazardous materials. Instead of the US designation "hazardous materials," in Europe the wording "dangerous substances" is often used. Table 2.1 provides an overview of the timeline of regulatory landmarks.

The history of worker protection started much earlier than that of safeguarding the population and the environment. Concepts of occupational health and safety, backed up by legislation, first developed in the middle of the nineteenth century and became well established after World War II. This was not just protection from acute or chronic harm due to toxic chemicals but also harm resulting from environmental factors such as dust or excessive noise. There was also the need, for instance, to protect workers from moving machinery and falls from height. Occupational hygiene, also called industrial hygiene, is an even broader topic that includes

Table 2.1 Time Line of Releases of Regulation

Year	US	EU
1970	Laws on founding OSHA and EPA	
1980	Superfund or Comprehensive Environmental Response, Compensation, and Liability Act	
		National safety laws in various European countries
1982		*Seveso I Directive*
1986	Emergency Planning and Community Right to Know Act of 1986 (EPCRA)	
1990	Clean Air Amendment	
1992	*Process Safety Management Rule (OSHA)*	
1994	List of substances and threshold quantities for the accident prevention program	
		ATEX 95 Directive
1995		Clean Air Amendment
		Directive on Machinery
1996		Integrated Pollution Prevention and Control (IPPC) Directive
	Risk Management Program (RMP) Rule (EPA)	
		Seveso II Directive
1997		Pressure Equipment Directive
1998		Directive risks related to chemical agents, and founding EU OSHA
1999		ATEX 137 Directive
2003		Amendment to Seveso II Directive
2006		European Programme for Critical Infrastructure Protection (EUCIP)
		Update Directive on Machinery
2006–2008		Registration, Evaluation, Authorization, and Restriction of Chemicals (REACH), and founding ECHA
2007	Critical Infrastructure Program (Chemical Facility Anti-Terrorism Standards (CFATS))	
2010		Update IPPC
2012		*Seveso III Directive*
		Update EUCIP

consideration of stresses induced, for instance, by ergonomic and psychological factors. In this book we are mainly concerned with major hazards and acute effects.

Just as with occurrences in the present day (2014) China and other countries that are rapidly building industrial capacity, the outcry and protests from the population against the "dirty" process industry started with the large-scale pollution of air and rivers, while events involving explosion, fire, or toxic cloud dispersion provided the rationale to organize meetings and demand actions for change. Thus, regulation for an environmentally cleaner industry comes, broadly speaking, from the same concern over safety of the public at large.

In the US at the federal level, the EPA was founded in 1970 together with OSHA, the latter as an agency of the US Department of Labor. Of course, as in Europe, legal occupational safety and health endeavors started some 100 years earlier on the state level when the damaging effects and diseases that employees suffered as a consequence of industrial activities became a matter of poignant concern. The EPA's initial years were associated with the foundation of the Superfund to clean hazardous waste sites. The 1984 Bhopal disaster in India and a chemical release in 1985 at Union Carbide in West Virginia, US (also part owner of the Bhopal plant), gave momentum for measures by the US Senate to protect against low-probability, high-consequence events. This impetus resulted in passage of the federal Emergency Planning and Community Right-to-Know Act (EPCRA) to establish emergency planning and "community right-to-know" reporting on hazardous and toxic chemicals. An approach to the American Institute of Chemical Engineers also resulted in the foundation of the Center for Chemical Process Safety. In 1996, the Risk Management Plan rule was established for accidental release prevention with the Risk Management Program (RMP) rule 40 CFR Part 68, which we shall discuss in more detail in the next section. Shortly before the regulation was issued, the industry was rather pessimistic about its economic implications, but in fact, after an interim period of customization, its acceptance went quite smoothly. Next came the US Chemical Safety and Hazard Investigation Board (CSB; authorized under the Clean Air Act Amendments of 1990), which became operational in January 1998. In 2002, the US Department of Homeland Security (DHS) was created in response to the September 11, 2001, terrorist attack on the World Trade Center complex in New York. This department became responsible for emergency planning activities and in 2005 launched the Chemical Facility Anti-Terrorism Standards (CFATS), which fit into the presidential Critical Infrastructure Protection (CIP) directive of 1998.

In the Code of Federal Regulations, the 29th title is Labor, while Part 1910 of that title covers Occupational Safety and Health (29 CFR 1910). Subpart H treats hazardous materials, while the standard 1910.119 has a general nature and is called PSM of highly hazardous chemicals. The standard came into effect in 1992 after passage of the supporting Clean Air Act Amendments (1990). Standard 1910.119 had a significant impact on the industry, which will be discussed in detail in the next section.

In Europe, the development of analogous legislation ran in parallel to that in the US but was completely different in the details. To help understand the differences we need to start with a brief history of the European Union itself. A common market (European Communities with a Commission as executive) was established in 1957 with the Treaty of Rome; meanwhile with the growth in the number of member states, the first elections for a European Parliament were held in 1979 and the European Union was established in 1993 with the Treaty of Maastricht. Individual European national governments had been going through many of the same types of experiences as the United States with their process industries growing in scale and being accompanied with explosions, fires, and toxic cloud releases. Hence, national laws for environmental protection and safety had been established in the 1970s. In 1976, a runaway chemical reaction occurred at the ICMESA plant (belonging to the Roche pharmaceutical group) in Seveso, northern Italy. A pressure relief valve activated and white smoke containing approximately 3 tons of reaction products was dispersed into the local environment. It contained an estimated 2 kg of the very toxic compound 2,3,7,8-tetrachlorodibenzodioxin, which spread over an area of nearly 20 km^2 causing soil pollution that if untreated would remain for years. After several days, both wild and farm animals started to die, and then only a couple of weeks after the accident the dioxin contamination was finally identified. More than 700 people were evacuated and thousands were placed under medical surveillance. Because of the stability of the compound, tens of thousands of animals had to be slaughtered; in addition, about 450 people were treated for chloracne skin lesions. Studies over decades showed that those who had been exposed suffered from a slightly higher than expected mortality from various diseases.[1] The societal effects have been dramatic (plant director shot by activists; compensation claims and penal court trials). Finally, nine years after a clean-up scandal, the waste was burned by the Roche group. Today the area is a park in which the debris of contaminated, demolished houses was buried.

In 1982, the town of Seveso gave its name to the first Directive (law) of the EU on the Control of Major-Accident Hazards involving Dangerous Substances. Since then, this Directive has been updated twice and Seveso III became law in 2012. A directive of the EU is applicable to all EU member states and citizens; EU law has primacy over national law, and national law must be compliant with union law. The two ATEX Directives are also of direct importance for process safety. ATEX derives from the two French words *Atmosphères Explosibles* and targets the risk reduction of gas and dust explosions. The first, ATEX 95 (1994), is concerned with equipment and protective systems intended for use in potentially explosive atmospheres; and the second, ATEX 137 (1999), contains minimum requirements for improving the safety, health, and protection of workers potentially at risk from explosive atmospheres. The essential contents of these Directives will be described in the section dealing with the European Union Directives.

Several other directives relevant to environment and substance safety have been promulgated. In 1995, the Directive on Machinery was issued with the objective of reducing the number of accidents through inherently safer design; it was replaced in 2006. Following individual air and water pollution directives in the 1980s, the Integrated Pollution Prevention and Control (IPPC) Directive was first introduced in

1996 and then updated in 2010. The crux of this Directive was the introduction of the "best available technology" concept. In 1997, the directive on the approximation of the laws of the member states concerning pressure equipment or, for short, the Pressure Equipment Directive (PED) appeared (in the US such more specific rules exist often as standards or codes as, e.g., in this case the American Society of Mechanical Engineers (ASME) Boiler and Pressure Vessel Code). For protecting the health and safety of workers, the next year Directive 98/24/EC—risks related to chemical agents at work—was added to the statute book, replacing a number of more specific directives from the 1980s. At the same time, the European Agency for Safety and Health was established in Bilbao, Spain.

In 2006, a European Programme for Critical Infrastructure Protection (EPCIP) was introduced. In 2008, a Council Directive became law calling on member states to identify European critical infrastructure and to assess protective measures. In 2012/13 EPCIP was reviewed and made more practical, while risk analysis was included. So far, it focuses on pan-European infrastructures such as the electricity and the gas grid, on cyber interactions, and on increasing resilience.

A regulation that met with significant resistance from the industry was that concerning the Registration, Evaluation, Authorization, and Restriction of Chemicals (REACH; Regulations (EC) No. 1907/2006 and No. 1272/2008), December 1, 2006, establishing a European Chemicals Agency (ECHA) in Helsinki. REACH was a response to the public fear and negative attitude to "chemicals" and is intended to ban substances from the marketplace unless they have been approved. REACH was accompanied by a list of accepted test methods (Regulation (EC) No. 440/2008). ECHA's database is easily accessible and contains a wealth of substance data (more than 12,500 registered substances by June 2014).

2.2 STATIONARY SOURCE SITING (US) OR LAND USE PLANNING (EU)

As we shall see in the following sections, regulation to protect residents in areas near processing plants from so-called off-site consequences, initially thought of as pollution but later also as acute major hazards, is focused on licensing particular activities. In order to avoid the frustration of progressing projects, and even of starting construction phases, which with afterthought have to be stopped or amended, a planning application for "facility siting" and "land use" is required at the outset. In most countries, land is a scarce resource. Industrial processing activity needs the availability of transport connections and people with adequate education. Therefore, one tends to find industry most commonly near harbors or large rivers with connections to railway lines, a good road network, and the presence of a nearby population center. It is even a well-known phenomenon that people often migrate and settle where industry is established and job opportunities are good. In the absence of, or with insufficient, land use planning, the town or city quarter near the plant keeps on growing. This can have fateful consequences as with the Pemex LPG disaster in

Mexico City and the Bhopal MIC toxic cloud tragedy in India, both in 1984; more recently, the absence of good siting and land use planning greatly accentuated the fatalities and other consequences of the 2013 West Fertilizer ammonium nitrate detonation in TX, USA.

Time after time this type of encroachment into "safe separation distances" has led to conflict both because after an incident it becomes clear that safety distances were no longer adequate and also industrial expansion necessitates a move elsewhere. The demand for space for economic activity including agriculture, traffic, living, and recreation is permanently on the increase. At the same time, noise emission, air and water pollution, and risks to health, life, and property values from the presence of adjacent hazardous-material storage or processing are also all increasing. Obviously, if at an early stage of development or redevelopment a coherent plan for land use is made, then the interests of different types of stakeholders can best be served. To that end, land shall be systematically divided into sections with a certain planned end use and function (agriculture, industry, utilities, living, shopping, recreation), while a layout shall be developed for surface traffic routes (e.g., rail, road, and canal). Above-surface communication lines (e.g., for power cables, aircraft, etc.) as well as below-surface construction (for metro, pipelines, or cables) will preclude unfettered use of the ground space immediately above or below them. Because of these three-dimensional implications, land use planning is also called spatial planning. Thus, in a democracy governed by authority and with fixed restrictions for the density of buildings, their type and purpose, limits to environmental loads, perhaps even exclusion of certain types of industry, land use can be optimized.

Stationary source siting as it is called in the US or land use planning, the European equivalent, needs to occur at both the regional and local levels, the latter often being called urban planning. The regional level takes over when interests are in common among a number of neighboring local areas, such as when the premises of an industrial complex covering an area belonging to several municipalities is being considered. In the case where national interests prevail, such as in space/terrains for defense purposes, the planning is elevated to the national level.

The siting can follow various strategies. As mentioned, just like with any other regulation, the plan is subject to democratic decision making. After a proposal is submitted, there will be hearings and debate with involvement of stakeholders before a plan is authorized or rejected. Information on plans shall be public; project developers may have large financial interests that may be in conflict with other interests. After a plan is accepted and it comes to fruition, there will be licensing procedures to be initiated by the developer/owner in which on a more detailed level of activity, installations layout and construction plans must be described. Again, public discussion is possible, objections can be filed, and changes or additions proposed. Sometimes there may be time-consuming, judicial/legal court procedures in which additional information or clarifying statements must be produced or the application can irrevocably be turned down. Both misuse and frustration of democratic rights may become issues as well.

In the sections that follow, we shall consider relevant safety regulations in more detail including how in the US and particularly in Europe protection is embodied in those rules.

2.3 PROTECTION OF WORKERS AND THE PUBLIC IN THE UNITED STATES

2.3.1 OSHA—EMPLOYERS/EMPLOYEES

As we have seen above from the historical themes, the protection of employees has a much longer tradition than safeguarding the public. In the US, protection of employees is founded on the Occupational Safety and Health Act of 1970 and its General Duty Clause (29 USC § 654, 5). An important regulation for the process industries is OSHA 29 CFR 1910, which was promulgated in 1992. It contains a large variety of items of relevance to safety of industrial activity. One example is OSHA 29 CFR 1910.147, Control of Hazardous Energy rule with the lockout/tagout principle, and another 29 CFR 1910.106, Flammable and Combustible Liquids. The standard itself and a guide to the standard can be found on the Web.[2] The subpart on hazardous materials regulates work with a number of substances that are important for industrial purposes, such as hydrogen, oxygen, nitrous oxide, flammables, and explosives. However, the greatest impact, bearing in mind the risk analysis methods that we shall discuss in Chapter 3, is made by Standard number 1910.119, PSM of highly hazardous chemicals, commonly known as the PSM rule of February 1992. Because of its relevance to the theme of this book, we shall go into significant details about this part of the regulation with its 14 substantive elements, although the selection of the items presented below will be neither orderly nor complete.

OSHA Standard no. 1910.119 starts off with the area of application; this is defined in an appendix with a listing of chemicals and the threshold quantities above which the rule shall apply. The standard also applies to processes with flammable substances above certain volatility limits. Next comes definitions of terms, and in the articles that follow the demands on the employer are defined. The first demand is an employer's plan for employee participation through safety and health committees, followed by documentation of process safety information that encompasses properties and quantities of the substance inventories, processes, equipment, piping and instrumentation diagrams (P&IDs), construction materials, various safety measures, et cetera. Next, is the specification of the key part of PSM: process hazard analysis (PHA). For PHA's systematic and thorough performance, established methods of hazard identification are prescribed, such as hazard and operability study (HAZOP) and failure mode and effect analysis (FMEA). PHA also requires description of previous incidents, possible engineering and administrative controls, possible monitoring means and detection of early warning signals, and possible alarms. Further, consequences of control failure, facility siting, human factors, and possible effects on employees shall be described. It also places requirements on the way the analysis is done, for example, that it should be conducted by a team; it demands how the recommendations of the team will be put into action and that the PHA should be repeated after a period of no more than five years. This means not only an update of the tools but performing an updating implementation. A subsequent requirement is on operating procedures and the stages in a process, such as start-up and shutdown for which specific procedures are necessary. Procedures are also required for the assurance of the quality of materials used and the performance of safety systems. All procedures shall be up to date, clearly written, and easily accessible. In addition, safe work practices shall be established, such as lockout/tagout, confined space entry, and control over the entrance into the facility.

The PSM standard further prescribes training requirements, responsibilities of the employer with respect to contractors, as well as the contract employer's responsibilities. A number of basic requirements are mentioned with regard to equipment integrity and maintenance including inspection and testing. For new construction, checks shall be made on design and materials, and a prestart safety review shall be conducted. The requirements for hot work permits, control over the management of change, and incident investigation are defined. The standard then emphasizes the need for emergency planning and response, and finally dwells on requirements with respect to compliance audits. The employer can require persons involved in the PSM to sign a confidentiality agreement. A compliant program will incorporate "good engineering practice," which means that the program follows the codes and standards of ANSI, API, ASME, ASTM, and NFPA. This is known as the recognized and generally accepted good engineering practice requirement.

The PSM rule has been an excellent step forward in process safety awareness and has led to multiple implementations of safety-related measures. In Table 2.2 the elements are listed as these are given in the standard, with, for example, a still superficial elaboration of the element PHA. With all these elements, the contents of PSM effectively define a true safety management system (SMS), which we shall discuss in more detail in Chapter 3. With regard to a PSM program, the US has been in the forefront. As we shall see later also for Europe, a general complaint after an accident is that inspections are too infrequent. To mitigate this, OSHA's inspections follow a national emphasis program on highly hazardous materials.

2.3.2 EPA—PUBLIC, ENVIRONMENT

In 1994, two years after OSHA launched 1910.119, the EPA published first its List of Substances and Threshold Quantities for the accident prevention program 40 CFR 68.130. Then, in 1996, Title 40 CFR Protection of Environment, Chapter I Environmental Protection Agency (Continued), Part 68, Chemical Accident Prevention Provisions, was issued or in short 40 CFR Part 68, the RMP rule.[3] The RMP rule puts requirements on the owner, or an operator, of a stationary source (a processing facility with hazardous materials) defined in the rule. The rule was mandated by the US Congress in the 1990 Clean Air Act Amendments, section 112(r), mentioned

Table 2.2 Process Safety Management (PSM) Elements as Given in OSHA Standard No. 1910.119 and Elaboration of the Element Process Hazard Analysis (PHA)

PSM Elements	*Elaboration of e.: Process Hazard Analysis*
a,b. PSM application and definitions	**1.** Identify, evaluate, and control hazards
c. Employee participation	**2.** Methods: What-if, checklist, HAZOP, FMEA, FTA[b]
d. Process safety information	**3.** To be addressed: Hazards, previous serious incidents, engineering and administrative controls, consequences of failure, facility siting, human factors, safety and health effects of failures
e. Process hazard analysis	**4.** Performance of PHA by expert team
f. Operating procedures	**5.** Employer shall address team's findings and recommendations, document, and execute actions
g. Training	**6.** Update and revalidate every five years
h. Contractors	**7.** Retain PHAs and updates, and recommendations made for the life of the process
i. Prestart-up safety review	
j. Mechanical integrity	
k. Hot work permit	
l. Management of change	
m. Incident investigation	
n. Emergency planning and response	
o. Compliance audits	
p. Trade secrets	

earlier, while it also built on the EPCRA of 1986. The RMP rule contains eight subparts of which the following are key:

1. Hazard assessment, which includes *off-site consequence analysis (OCA)*.
2. Prevention program, which runs in parallel to the OSHA PSM requirement.
3. Emergency response program.
4. List of regulated substances and actual threshold quantities.
5. Submission of a risk management plan.

Processes covered by the rule are divided into three increasing hazard levels:

- Program 1: processes that have had neither an accident over the past five years nor an off-site death, injury, or response to an exposure. In addition, the largest calculated distance to a threshold exposure level must be smaller than the distance to the nearest public receptor, and the emergency response of the site must be coordinated with the local public emergency responders.
- Program 2: processes that fulfill neither the requirements for program 1 nor those for program 3.
- Program 3: processes that do not fulfill the requirements of program 1 but fulfill a number of additional conditions among which is being subject to the OSHA PSM rule. Hence, requirements for program 3 processes are the most stringent.

[b]See Chapter 3, in particular Section 3.6.2, for meaning of these acronyms and description of the methods.

The basis for determining potential harm to people (the population) and the environment is RMP's OCA,[3,4] which requires a specified release of hazardous material. With the release point as the center of a hazard zone, the radius of an affected circle is determined by so-called *endpoints* of the release effect. For toxic substances, endpoints are concentrations specified in a table in appendix A to Part 68 (defined in § 68.22): for explosion it is an overpressure of 1 psi (0.069 bar), and for fire a radiation intensity of 5 kW/m^2 for 40 s or a gas or vapor concentration at the lower flammable limit. Further parameters for the release such as wind speed, height of release, gas density, and others are specified in the document. Models are publicly available, they shall take account of various conditions, and they shall be recognized by industry. The hazard zone can be shown on a map with the vulnerable receptors such as homes, schools, hospitals, nursing homes, shops, offices, or parks, and sensitive receptors such as recreational areas (national or state parks, and wildlife sanctuaries.), see Figure 2.1.

OCA includes an analysis of the potential consequences of the hypothetical worst-case as well as reduced, more realistic alternative-release scenarios. It does not take account of any event probabilities, so it determines a potential for risk, the hazard level, not the actual risk. Worst-case scenarios assume the release of the greatest amount of a regulated substance held in a single vessel or pipe under specified ambient and process conditions, taking into account administrative controls that limit the maximum inventory, and accounting for the effects of passive mitigation features such as bunds and dykes if present. Detailed conditions for Program 3 processes differ considerably from those for Program 1. Alternative scenarios assume a release that is more likely to occur than the worst-case scenario, using release parameters chosen by the facility owner as appropriate for the scenario and may account for both passive and active mitigation features.

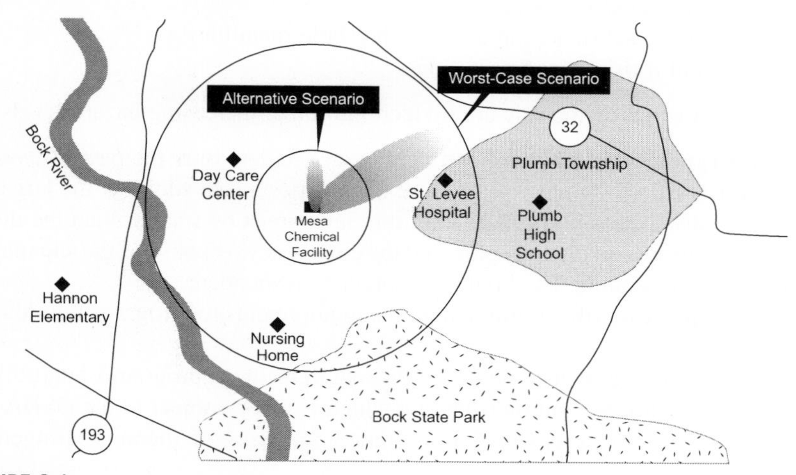

FIGURE 2.1

Hazard zone and vulnerable receptors for a worst-case and an alternative scenario according to the RMP rule (40 CFR Part 68) taken from an EPA guidance brochure.[4] The brochure even contains a list of questions that reporters may ask the facility owner or operator.

The data tell the public how far the off-site impacts of a worst-case, or alternative, scenario release from a particular facility could travel, roughly how many people could potentially be affected, and what types of public and environmental receptors could be in the path of a release. In short, the data allow members of the public to determine whether they could be harmed by a chemical release from a particular facility. Hence, via the State or Local Emergency Planning Committee (SERC or LEPC), the information should lead to an emphasis on emergency response, risk communication, and prevention efforts. The purpose of the RMP is not to generate unnecessary fear but to educate the public about hazard presence, reduction, and emergency response. In extreme cases, through hearings, members of the public can make objections that can force the owner to consider additional protection or mitigation measures.

The prevention programs follow 49 CFR 1910.119. Every three years a compliance audit report on the prevention programs shall be drawn up. The submitted risk management plan shall be the basis for a permit by an implementing authority delegated by the state. It shall contain all OCA conditions and data. The plan shall also contain a list of all unintentional releases during the last five years, while every new incident shall be investigated and documented. Documentation shall be retained for at least five years, and the OCA shall be updated every five years. When changes in the facility take place that would either decrease or increase the threshold distance by a factor of two, a revised RMP shall be submitted. The RMP can be audited by the implementing agency at any time, and a revision can be required. All information shall be available to the public.

After some initial problems, in general the implementation of the RMP rule went surprisingly well, although prior to that American industry feared that the anticipated increase of their overheads could drive them out of the market. Authorized users can explore the RMP database, which includes the reported accident.

Liquefied natural gas requires special consideration. Safety transportation at sea comes under the supervision of the US Coast Guard as for all chemical shipments. But licenses for land installations are issued by the Federal Energy Regulatory Commission of the Department of Energy. Siting is regulated by 49 CFR 193 and requires sufficient space for an exclusion zone. Calculation of exclusion distances must follow NFPA 59A (2001 edition) and are based on a thermal radiation intensity threshold from a pool or flash fire of 5 kW/m^2 and a concentration of 50% of the lower flammable limit. Details are given by Kohout.[5] Dispersion is calculated by prescribed models. In 2014 it has been proposed to add to NFPA 59A a Chapter 15 requiring a full quantitative risk analysis applying off-site risk criteria and the ALARP principle (see Chapter 12).

2.3.3 CIP—CRITICAL INFRASTRUCTURE PROTECTION

In 2006, Congress authorized the US DHS to issue a regulation for risk-based performance standards for chemical facility security. As a result, the CFATS program[6] requires since 2007 high-risk facilities to develop a vulnerability assessment and an effective security plan. While the implementation of the program progressed, in April 2013, the West, TX, ammonium nitrate detonation accident occurred (see Chapter 1). The ensuing Presidential Executive Order 13650 gave CFATS additional momentum. In May 2014, on behalf of DHS—Office of Infrastructure Protection, OSHA, and EPA, a list was published of coordinative and regulatory improvement and modernizing actions.[7] The concerted actions shall increase the safety and security of chemical facilities for workers and the public. An important measure will be reducing hazards by promoting safety technologies and alternatives. Preparedness of first responders and emergency planning committees will be strengthened.

2.4 EUROPEAN UNION DIRECTIVES AND TRANSPOSITION IN NATIONAL LAW

2.4.1 EU SEVESO DIRECTIVES

As mentioned in the Introduction section, the Seveso Directive III on the control of major-accident hazards involving dangerous substances came into authority in 2012. In Table 2.3 the three versions of the Seveso Directive, and a 2003 amendment on the second Directive, are listed and the evolution in their development is characterized with keywords. All directives start with introductory considerations before the articles are given. As shown in the table, the main change with Seveso II was the introduction of the SMS to be included in a major accident prevention policy (MAPP). The issues to be addressed in the SMS are:

- Organization and personnel: responsibilities and training.
- Identification and evaluation of major hazards.
- Operational control: adoption and implementation of procedures and instructions.
- Management of change.
- Planning for emergencies.
- Monitoring performance (note: process safety performance indicators are only mentioned in Seveso III).
- Audit and review.

The 2003 amendment was inspired by three large incidents: a mineral waste cyanide spill that polluted the Danube river in 2000; in the same year a fireworks and pyrotechnics warehouse explosion disaster in the Netherlands that killed 4 firefighters and 19 residents, injuring almost 1000 and demolishing 200 houses; and the 2001 Toulouse ammonium nitrate detonation mentioned in Chapter 1.

The safety report must demonstrate that MAPP and SMS have been put into effect, risks are identified, all necessary preventive measures taken, installations

Table 2.3 Successive Versions of the EU Seveso Directives to the Member States

No.	Year	EU Ref. No.	Number Articles	Web References	Content Keywords
I	1982	82/501/EEC	21	8	Basics: Where the directive applies; upper and lower industry tiers (1. large; 2. small material inventories); requirement to identify hazards; appointing a competent authority and submitting a notification report to that authority for tier 1 industry; emergency plans; inspections; safety measures; modifications; information to the public; accident reporting; annexes with lists of process names, substances, and thresholds.
II	1996	96/82/EC	26	9	Additional to Seveso 1: Major accident prevention policy paper (MAPP, which introduces the obligation of a safety management system); domino effect among neighboring establishments; installation modifications; land use planning; inspection requirements elaborated; cooperation for an external emergency plan; many more details, e.g., criteria when an incident has to be reported.
IIA	2003	2003/105/EC		10	Excludes offshore and other industry, but includes mineral waste facilities; additional requirements with respect to the safety report; involvement of subcontractors and public in emergency plans; demands more emphasis on long-term land use planning distances and risk approach for which it requires a database.
III	2012	2012/18/EU	34	11	Substance properties and classification on the basis of the Global Harmonized System (see Section 2.7); even more emphasis on drawing up internal emergency plans and providing data for external ones; inspection requirements further strengthened. Public consultation on major accident scenarios and participation in decision making.

safe and reliable, and emergency response prepared. Thus, it must demonstrate that off-site risk is acceptable. Several supporting and evaluating reports have been produced. Guidance has been given to drafting safety reports,[12] while also the value of safety reports has been discussed with respect to inspection.[13]

In 2008, an evaluation was made of the effectiveness of the Seveso Directive implementation in the various European member states.[14] The results showed that people were generally satisfied with the Directive and were positive about the contribution to the safety level. The Seveso Directive had not influenced European competiveness in a negative sense. It also fits well with the series of other safety directives such as the ATEX Directives (see Section 2.4.3), the PED (97/23/EC), the Machinery Directive (2006/42/EC), and the directive on the environment, previously IPPC, but since January 7, 2014, Directive 2010/75/EU on industrial emissions. A weakness noted was the diversity of implementation in national law that adds to efforts of multinational companies due to their internal safety standards, which they must adapt differently for operations in each EU member country. A list of recommendations for improvements was given including more guidance documents, some additional requirements, and the wish for a better quality of dialog among stakeholders and a greater participation of the public.

The Seveso III Directive (2012/18/EU) has been adapted for the introduction of the GHS for Classification and Labeling of Chemicals (see Section 2.7), while a long list of further improvements have also been made, e.g., with respect to detailing issues concerning inspections. As recommended, the requirements to involve the public in the decision making have become more demanding. In general, the wording has been made clearer.

Implementation in the member states does not mean that in a nation one ministry is always responsible. In some countries such as in UK there is an agency, the Health and Safety Executive (HSE), that is responsible for both worker and public safety, but in most countries worker and public safety is distributed between at least two ministries, while the emergency preparedness part usually resides within a third ministry. However, coordination is a must.

The EU Major Accident Reporting System database[15] is freely accessible and operated by the Major Accident Hazards Bureau (MAHB) as part of the Institute for the Protection and Security of the Citizen within the EU Joint Research Centre in Ispra, Italy. MAHB coordinates several activities, such as information for drafting safety reports, and runs the accident report database and the European Working Party on Land Use Planning. The latter was created by the commission in the 2003 Seveso II Amendment with the mission to reduce nonuniformity in the national implementations and therefore to obtain more transparency. The group should also establish a common database for accident scenarios, technical failure data, and other information required for risk assessments in the land use planning context. Reports have been issued, but progress has been slow. Due to court actions that claim national policies are not fulfilling the minimum requirements of directives, there is pressure also from industry to renew the effort.

2.4.2 SPATIAL PLANNING IN FOUR EUROPEAN MEMBER STATES

We shall have a look at the land use planning (LUP) practices in England (actually LUP is a delegated responsibility to Northern Ireland, Welsh, and Scottish National Assemblies), France, Germany, and Netherlands—in this "alphabetical" order. The first three countries have quite different views on the role of risk assessment in maintaining safety and have different legal systems. The UK and the US have legal systems in which besides case-based or common law made by courts on the basis of precedent, there are also constitutional, statutory (from a legislative body), and regulatory (from a government agency) laws. This legal practice differs as a system from Roman law or civil law codes, where judges do not make rules and law making is restricted to legislative governmental bodies. France has the newest major hazard legislation, and Germany is the only one of the four countries that does not take the probability of an event into account. The Netherlands is used as an example because of its relatively large process industry, high population density, and because it applied quantitative risk assessment (QRA) for land use planning on a legal basis very early on. Despite these differences, the safety performance of industry in the four countries considered is not significantly different. An interesting 2011 review of the state of affairs in Europe by the EU Working Group on Land Use Planning at the Commission's Joint Research Centre is given by Christou, Gyenes, and Struckl.[16]

2.4.2.1 United Kingdom

The HSE, which is responsible for enforcing the Health and Safety at Work Act 1974A (HSW Act) and a number of other acts and statutes, is a pivotal agency in LUP matters. HSE is also in charge of Control of Major Accident Hazards (COMAH) implementation, which is the UK implementation of the Seveso Directive. Together with the Environment Agency (EA) for England and Wales and the Scottish Environmental Protection Agency, the HSE forms the Competent Authority (CA), as specified in the Directive. The CA is responsible for the enforcement of COMAH to assure safety at major hazard sites. As a safety adviser, HSE has for years been in the forefront of conceptual and model developments, and its publications are used worldwide. Many of these can be downloaded free of charge from the HSE Website.[17]

The HSW Act refers to risks but not to hazards. A 1993 court ruling interpreted "risk" as "the idea of the possibility of danger" as cited by HSE in its policy document[18] "Reducing risks, protecting people." HSE describes *hazard* as "the potential for harm arising from an intrinsic property or disposition of something to cause detriment," and *risk* as "the chance that someone or something that is valued will be adversely affected in a stipulated way by the hazard." The policy document serves as a basis for decision making within the context of risk management. Distinction is made between broadly acceptable, tolerable, and unacceptable risk. The boundary mentioned between broadly acceptable and tolerable is an individual risk of death of one in a million per annum or 10^{-6} p.a.; for unacceptable risk to the public, this is 10^{-4} p.a.

A further fundamental principle underpinning the HSW Act is that those who create risks from work activities are responsible for protecting workers and the public from the consequences of those risks. It is also important in British law (following a ruling in 1949) that risk should be weighed against the sacrifice (in terms of cost, time, or difficulty) necessary to reduce that risk. If there is a disproportionately large effort compared to the risk reduction achieved, compliance is no longer practicable. This led to the duty to reduce risk to "as low as reasonably practicable" (ALARP), as described in the above-mentioned policy document. Supporting instruments to determine what is "reasonably practicable" are principles of good practice and cost–benefit analysis. Because elaboration can become rather complicated, HSE issued an ALARP Suite of Guidance.[19] In Chapter 12, we shall describe in some detail the criteria for achieving ALARP.

In the case of a new project, or a development, local land planning authorities have a statutory duty to consult the HSE, which bases its consultation on a risk assessment resulting in terms of "advise against" or "don't advise against" development. Planning Advice for Developments near Hazardous Installations (PADHI)[20] sets a consultation distance (CD) around major hazard sites and pipelines after assessing the risks and likely effects of major accidents. The area within the consultation zone encompasses three zones besides that with the installation containing the risk source at its center: these are the inner (IZ), middle (MZ), and outer zones (OZ). In Figure 2.2, the zoning is shown schematically for an installation (point source) and a pipeline (line source). HSE distinguishes four "sensitivity" levels of developments that are proposed for establishment in the zones: Level 1—normal working population; Level 2—general public at home and involved in normal activities; Level 3—vulnerable people (e.g., children, the elderly); Level 4—large examples

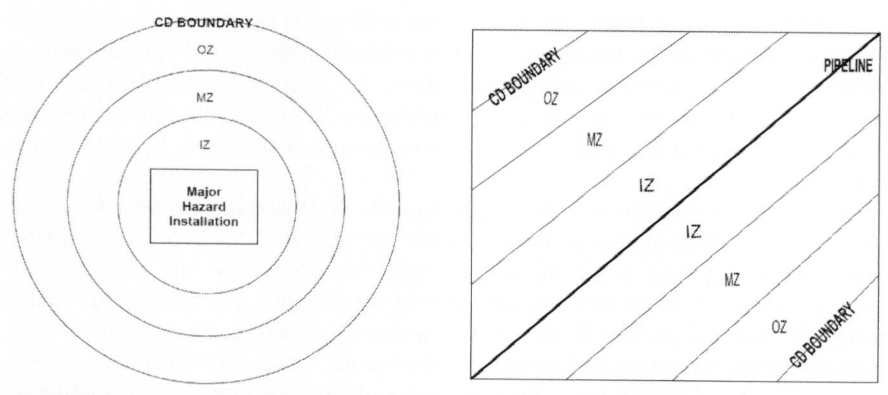

FIGURE 2.2

The consultation distance (CD) boundary comprising the installation with the risk source, *left*, and a pipeline containing a hazardous substance, *right*, the inner risk zone (IZ), middle zone (MZ), and outer zone (OZ), according to the HSE PADHI document.[20]

Table 2.4 HSE's Land Use Planning Decision Matrix in the PADHI Document,[20] Version 2011

Level of Sensitivity	Development in Inner Zone	Development in Middle Zone	Development in Outer Zone
1	DAA[a]	DAA	DAA
2	AA[b]	DAA	DAA
3	AA	AA	DAA
4	AA	AA	AA

[a] *DAA stands for Don't Advise Against.*
[b] *DA means Advise Against.*

of Level 3 and very large outdoor examples of Level 2. The four levels are further specified in tables.

A more complete overview of decision making for the various sensitivity classes is obtained from the decision matrix shown in Table 2.4.

HSE followed two approaches to determine the zone boundaries depending on the nature of the installation, its complexity, et cetera: *protection*-based and *risk*-based. Protection-based analysis is basically a consequence analysis given an event and a determination of hazard level as a function of distance from a possible event, mostly a toxic dispersion. For more complex cases, a true risk analysis is performed that calculates individual and societal risk quantitatively. We shall discuss definitions and methods of calculation of these risk metrics in Chapter 3.

In recent years and in particular after the Buncefield, UK, gasoline tank overflow and vapor cloud explosion incident in December 2005, discussed in Chapter 1, HSE's confidence in their risk analysis methodology has been growing. In 2007, HSE sent out a consultative document[21] to stakeholders (operators, developers, authorities, etc.) to probe proposals to adapt the policy and to include in the decision making that the *societal risk* of at maximum 50 fatalities including workers at the site with a probability of more than once in 5000 years should be regarded as intolerable. This inclusion was not a "hard" criterion but was guidance. It was concluded that a land use planning approach for a large oil depot must be risk based and that societal risk is an issue.

2.4.2.2 France

The Toulouse ammonium nitrate explosion in 2001 (see Chapter 1) caused a major change in the French approach to control of industrial risk and major hazards. On July 30, 2003, a new law was issued, *Loi no. 2003-699*, regulating technological and natural risks and the compensation for damage.[22] The approach of the technological part and the control of urbanization around major hazard sites were changed to be risk based. The new law requires for each top-tier Seveso II site (*installation classées à haut risque*) to have a technological risk prevention plan (TRPP; *Plans de Prevention des Risques Technologiques, PPRT*). The plan is described in an extensive guidance document[23] providing detailed information on the risk analysis method and data to be used and structure and editing requirements for the plan.

Land use planning in France is regulated in the *Code de l'Urbanisme* in which in particular Article No. 110 addresses health and safety. Before the 2003 law was in force, a worst-case scenario approach was applied with corresponding safe distances and two effect zones: *lethality* and *irreversible damage* to humans. In the application procedure by a developer of a major hazard installation, the state notifies the distances of the two zones 1 and 2 to the local authority. This is followed by a socio-economic negotiation process and an adaptation of the local land use plan restricting further rights of development given the major hazard installation. Because the new LUP strategy and procedure contain some interesting aspects of risk assessment and of risk communication, we shall dwell on it longer.

The CA (*préfet* of the *département*) receiving a request for a permit to operate a major hazard installation starts a process of land use planning, if minimum risk control requirements are met. In Article L. 515-16 the new law states that within the risk perimeter, the *PPRT* will take into account the type of risks, their severity, probability, and their dynamics. Obviously, the latter refers to the time available to warn people via an alarm and the time required for self-rescue and evacuation. Risk control is based on three principles:

1. eliminate or reduce at source the hazard giving rise to the risk,
2. limit the severity and probability of the physical effects of technological hazards (*aléas*)
3. limit the vulnerability of exposed receptors (*enjeux*).

Effect thresholds (toxic concentration—gas or vapor, radiant heat—fire, overpressure—explosion) come in four classes: significantly lethal (*seuil des effets létaux significatifs, SELS*), just lethal (*SEL* or *SPEL*), irreversible damage (*SEI*), and indirect. Also, the relative importance of the entities within the affected zone is taken into consideration, such as presence of inhabitants/residents and workers, presence of dwellings, commercial property, traffic infrastructure, public buildings, outdoor public spaces, and public utilities (electricity, gas, water, and communication installations).

From the point of view of the public authority, there are four types of action: (1) risk reduction at the source, (2) limitation of urbanization, (3) organization of emergency response, and (4) informing the public. To realize the last point, an independent committee for local information and consultation (*Comité local d'information et de concertation, CLIC*) is set up and convened by the departmental/regional authority, *préfet*, based on the *Code de l'Environnement*. The *CLIC* consists of representatives of five groups of stakeholders. The first is administration—the state—the *préfet* that has the authority for the inspection bodies for classified installations, *DRIRE* or *STIIIC*, and the Departmental Directorate for Public Works, *DDE*. The four others are local authorities, facility operators, local residents, and employees. This approximates to the "risk forum" idea of which we shall see the importance for risk acceptance later. In fact, the whole procedure for the *PPRT* starts formally at the first meeting of the *CLIC*. DRIRE and STIIIC among others provide the secretariat and acquire funding. To enable good communication, cartography is applied superimposing on the map the *aléas* (seven risk levels in different colors) and *enjeux* (receptor vulnerability). Determining the *aléas* is the task of DRIRE and STIIIC after selecting scenarios based on hazard studies by the industry. *Enjeux* are identified by the *DDE*.

Before the *PPRT* starts, a *MMR* (*Mesures de Maîtrise des Risques*) is carried out to evaluate the risks involved and to judge whether the installation in principle can be accepted in the location planned. For this evaluation, a technical risk assessment study is conducted starting with identifying accident scenarios including domino effects, while also taking into account the safety barriers installed and the effectiveness of the SMS. For the more complex cases the new ARAMIS QRA tool (2003) is used. Essential features of this tool will be described in Chapter 3. Details including calculation tool descriptions and acceptance criteria are published in a Ministerial Circular,[24] while a paper by Jérôme Taveau[25] presents an overview.

Dangerous phenomena effect intensities are determined as a function of distance. Thresholds are shown in Table 2.5. Dynamics of the phenomena are introduced to account for the possibility of evacuation. In this context effects such as, for example, explosion and fire are classed, respectively, as rapid and slow. The combination of effect intensity and number of people exposed produces an incident gravity scale, as shown in Table 2.6. In Table 2.7, as a measure of cumulative probability of a dangerous event, five classes, A to E, are distinguished. The event frequency value for class A is higher than 10^{-2}/year and for class E lower than 10^{-5}/year. If at a certain site there are four installations, each with an estimated event frequency of 10^{-4}/year (= class D), the probability is given as 4D. Probability and effect are not considered as the product of both parameters but are examined separately. Combinations of effect intensity and probability, the *aléas*, are expressed on maps in seven colors. For intensity *SELS* and probabilities >D, 5E – D, <E colors range, respectively, from red via amber, to yellow; for *SPEL* and >D, 5E – D, <E from yellow, to white and blue; for *SEIrreversible* and >D, 5E – D, <E from blue, to cyan and green, and for *SEIndirect* green.

Table 2.5 Reference Values of Thresholds for Effects on People (*Seuils d'Effets*)[23]

Effect Thresholds	*SELS*, Significantly Lethal	*SPEL*, Just Lethal	*SEI*, Irreversible	*SEI*, Indirect
Toxic concentration causing % probable lethality (LC)	LC 5%	LC 1%	Injury	–
Overpressure	200 mbar	140 mbar	50 mbar	20 mbar
Radiant heat intensity or dose	8 kW/m^2 or 1800 [(kW/m^2)$^{4/3}$].s	5 kW/m^2 or 1000 [(kW/m^2)$^{4/3}$].s	3 kW/m^2 or 600 [(kW/m^2)$^{4/3}$].s	–

Table 2.6 Gravity Scale Depending on the Effect Intensity (Threshold) and on the Number of People Exposed According to Arrêté du Septembre 29, 2005[26]

Effect Thresholds	***SELS*, Significantly Lethal**	***SPEL*, Just Lethal**	***SEI*, Irreversible**
Gravity scale	***Number of people exposed***		
Disastrous	>10	>100	>1000
Catastrophic	1 to 10	10 to 100	100 to 1000
Major	1	1 to 10	10 to 100
Serious	0	1	1 to 10
Moderate	0	0	<1

Table 2.7 Matrix for Preliminary Decision on Acceptance of an Installation Based on Probability Classes and Consequence Gravity for Land Use Planning According to the 2003 French Legislation[23]

Probability / **Gravity**	**E** **$<10^{-5}$/year**	**D** **10^{-4}**	**C** **10^{-3}**	**B** **10^{-2}**	**A** **$>10^{-2}$/year**
Disastrous	*Non-partiel/ MMR2*	*Non*	*Non*	*Non*	*Non*
Catastrophic	*MMR1*	*MMR2*	*Non*	*Non*	*Non*
Significant	*MMR1*	*MMR1*	*MMR2*	*Non*	*Non*
Serious			*MMR1*	*MMR2*	*Non*
Moderate					*MMR1*

Table 2.7 constitutes a decision matrix combining probability and gravity of the accident scenarios for a plant installation. The elements with "*Non*" (in the original document printed in a red field) refer to combinations of gravity of consequence and probability that are unacceptable for the installation. The elements coded "MMR", printed in a orange shaded field are cases that can be further evaluated in the "*PPRT*", but with additional measures of risk control—for *MMR2* more stringent than for *MMR1*. In the case where none of the investigated scenarios result in a *Non*, while fewer than five possible scenarios fall in a *MMR*, the installation will be compatible with the selected location. In such case a *PPRT* is initiated, and further risk-reducing measures at the risk source shall still be considered. New installations are treated more strictly, and *Non partiel* (a partial "*Non*") refers to those. For a site with risks by a multiple of installations, the combinations E—disastrous and D—catastrophic change from *MMR2* to *Non*. In that case, for new installations all *MMR*s turn into *Non*. The Circular[24] contains separately rules for explosives and pyrotechnics.

A major problem is that many older plants, even if originally fairly remotely sited, are now surrounded by housing. Therefore, a risk-exposure perimeter is determined around the risk source beyond which the probability of an event is lower than 10^{-5}/year. This is shown graphically on a map of the region. Within the perimeter, depending on the type of risk (explosion, fire, toxic) and the severity of it, two different zones are defined. One is called zone of control of urbanization (*maîtrise de l'urbanisation*), the other a preemption zone to avoid construction. Based on the law for the protection of the environment, so-called public utility servitude can be established on the grounds under threat. Within the zone of *control of urbanization* new constructions and extensions to existing ones can be forbidden, and the construction, use, and exploitation of existing facilities subjected to prescription. Based on the *Code de l'Urbanisme*, the land use planning law mentioned, a renunciation and an expropriation zone closest to the risk source will be drawn. In the former, relinquishment of housing (*délaissement*) will take place, meaning that on the initiative of the owner the community will buy the property at market price and thus enable relocation. In sectors with high risk to human life, expropriation of housing can be enforced. The criteria for specification of one or the other of these two types of sectors are in terms of possible severity of an event as mentioned by Cahen[27]: expropriation for *SELS*, Significantly lethal or higher, and renunciation between *SELS* and *SPEL*, Just lethal, see Table 2.5. In both cases the probability of an event shall not be extremely low, and there will be no time for seeking shelter or no shelter shall be available. If the requirements are not met, an iterative negotiation is started between the plant developer, authorities, and the public. Decision making takes place after considering cost versus benefits.

The July 30, 2003, law contained further articles concerning safety of transportation of hazardous materials in various modes (road, rail, water, and pipeline), the protection of workers and the indemnification of victims, while a second part regards natural disasters.

The whole procedure is scheduled in advance with the technical studies, the conception of a strategy for the region, the public enquiry, and the administrative completion. It is quite elaborate and refined and therefore also fairly slow, but it stimulates enhanced safety and involves all stakeholders of the industry, including the public, in the decision-making process.

2.4.2.3 Germany

The Seveso Directives have been implemented in Germany in the Major Accidents Ordinance (*Störfall-Verordnung*)[28] regulating the prevention of incidents in installations in which hazardous substances are involved and the control of their effects. The regulation does not speak of "risks" but of "dangers" (*Gefahr*), which may be best described as hazard in a particular situation. The operator is obliged to regard sources of danger of the process itself, or externally in the environment, such as those caused by weather conditions, earthquakes, etc., and those from unauthorized access. The operator shall take adequate measures to minimize the effects of an incident and is obliged to install safety systems according to the best available technology, a concept similar to measures in the IPPC Directive. No reference to cost—benefit is made. The operator shall further take measures to avoid fire and explosions, but if these do occur then measures must be taken that maintain safety integrity to avoid domino effects, while the safety integrity of the installations must not become compromised by external effects. The process installation should be equipped with reliable alarms, sufficient means of monitoring process variables, and adequate process control, while the installations shall be protected against unauthorized access. The buildings shall be well protected against collapse or other damage, and the installations shall be protected by both technical and organizational measures. The law further prescribes adequate maintenance and repairs. It is quite clear that the regulation aims to deal adequately with all imaginable dangers and attempts to remove consideration of uncertainty or probability of events. Thanks to the long industrial—chemical tradition in Germany, the industrial safety knowledge, and the safety culture starting with pressure vessel safety at the end of the nineteenth century, relatively few major events have occurred that initiated discussion of the appropriate approach to safety. But, as an accumulation of safety and environmental protection measures may become counterproductive, and because risk assessment offers an economic cost-benefit option, while the probabilistic approach becomes more reliable, in due course, the situation may change.

Germany is a federal state. Besides the federation level there is the hierarchy of states (*Länder*), regions (*Bezirke*), and districts (*Kreise*); therefore, many tasks are decentralized. States and regions make spatial plans (*Raumordnungspläne*). Regulation is by a number of statutes at the federal and state level. According to Adams,[29] the precautionary principle established in the German constitution has had a strong influence in the development of the European environmental policy. The principle originated in the 1970s in Germany as *Vorsorgeprinzip* and was originally formulated as "the principle of taking care before acting." The polluter-pays principle (*Verursacherprinzip*), the proportionality principle between costs and gains (*Wirtschaftliche Prinzip*), and the common-burden principle (*Gemeinlast Prinzip*) were also at the basis of the German constitution before being discussed and eventually adopted at the European level.

On the federal level there is the Spatial Planning Act (*Raumordnungsgesetz, ROG*),[30] which provides the principles, definitions, boundary conditions, interests of public and private parties, and the effects of other acts, such as protection of the environment. It also gives guidelines with respect to the planning process. Further guidance on the federal level is given by the Federal Building Code and by the Federal Land Use Ordinance (*Baunutzungsverordnung, BauNVO*).[31] The latter shows a systematic distinction among zones with different functions. Areas for dwelling, for a village, mixed areas, areas for economic city center activities, for light and heavy industry, et cetera, are all distinguished. All further land use planning regulation is decentralized to the state level and is applied by municipalities.

Based on the *Störfall-Verordnung*, mentioned above, documents SFK/TAA-GS-1k-EN (abridged, English) and SFK/TAA-GS-1 (extended, German)[32] provide land use planning guidance in accordance with Article 12 of the Seveso II Directive.

The documents present a list of relevant hazardous materials and calculation methods for minimum recommended separation distances between a certain quantity of a chemical and the population. It is written from the perspective of establishing a new site. If details about the installation to be built are not fully known, a table with nominal safety distances is given. If details become known and conditions specified, models and data are provided to determine final safe distances.

The calculation of dispersion of a toxic or flammable volatile material is prescribed in a VDI guideline. Based on historical data collected in the ZEMA data base,[33] a maximum leak cross-sectional area of 490 mm^2 is assumed for a release, but this may be reduced to a minimum of 80 mm^2 depending on the actual situation. A dispersion calculation is performed according to VDI guideline 3783, sheet 1 neutral and light gas, sheet 2 heavy gas, at 3 m/s wind speed, and with immediate ignition. Also, the threshold values applicable to exposed people are specified: for toxicity the ERPG-2 value (Emergency Response Planning Guideline 2: 1 h exposure without irreversible effects) is specified. For heat radiation intensity 1.6 kW/m^2 (causing low percentage first-degree burns) and for blast overpressure 0.1 bar (window fracture 100% at about half that pressure) are used. (Explosives and ammonium nitrate are regulated under separate ordinances.) Distances range from 50 m, for example, for methanol (vapor) to 130 m for methanol (fire) to just over 1400 m for acrolein (2-propenal) or 1340 m for chlorine.

As would be expected, regulation on worker protection comes under the federal ministry for Work and Social Affairs in Berlin, while the states have their own ministries. A special German organization involved in occupation health and safety is the German Social Accident Insurance Institution (*Berufsgenossenschaft*) for statutory accident insurance and prevention, which has various branches. Of particular relevance here is the branch on raw materials and chemical industry (BG RCI) in Heidelberg providing, among other resources, training courses and technical advice.

2.4.2.4 The Netherlands

The Netherlands is open to the sea and has modern harbor facilities attracting a large oil refining and chemical process industry, but it also has a high population density. Due to various mishaps in the late 1960s, the population started to object to further expansion in the 1970s. Eventually, this concern led to the development of QRA.[34,35]

With the first National Environmental Policy Plan in the middle of the 1980s, the Dutch government informed parliament and the people about plans to adopt a risk approach for licensing of major hazard activities. Acceptance criteria were proposed for individual risk of death of new installations of $<10^{-6}$/year and of existing installations of $<10^{-5}$/year. (Death means immediate decease, hence instantaneous in the event or shortly after. Death is selected as criterion because its definition is clear, although injury may be very relevant as well.) Societal risk probability for an event with more than 10 fatalities (among residents, not workers or passers by) shall be less than once in 100,000 years, if more than 100 fatalities then less than once in 10^7 years, and for more than 1000 fatalities then less than once in 10^9 years. Societal risk applies only to inhabitant residents near the plant and not to workers or to people passing by.

The Seveso Directive is concerned with safety inside and outside plants and with emergency response. In the Netherlands the responsibilities for safety are divided between three ministries: "internal" safety, or rather occupational safety and health, resides with the ministry of Social Affairs and Employment (SZW), "external" safety to the public with the ministry of Infrastructure and Environment (IenM), and emergency response with the ministry of Security and Justice (VenJ).

The Arbo law 1998[36] regulates employee occupational health and safety, but in addition an employer is obliged to perform (qualitative) risk identification in combination with making an inventory of all risk-reducing measures and an evaluation. It resembles to a certain extent the OSHA PSM rule. Regulations for land use planning are the Spatial Planning Act (*Wet Ruimtelijke Ordening, Wro*[37]) and the Environmental Protection Act (*Wet Milieubeheer, Wm*[38]). The former decentralizes the planning in principle to the municipalities and requires a renewal of the land "destination of use" plan (*bestemmingsplan*) every 10 years, but the higher-level body of the province can overrule it. In matters of national interest, the national government can also intervene. A comprehending Environment Act (*Omgevingswet* 2015) is in the making.

Seveso II was implemented in 1999 with the Ordinance on Major Hazards (*Besluit risico's zware ongevallen, Brzo* 1999[39]). The lessons learned from the fireworks explosion in Enschede in 2000, which caused 22 fatalities, injured many, and resulted in massive material damage, brought this kind of installation (with potentially similar hazards to those in the process industry) under the same Seveso II regulation with the Ordinance on External Safety of Installations (*Besluit externe veiligheid inrichtingen, Bevi* 2004[40]), which is also relevant for spatial planning under the *Wro*. This ordinance widened the sphere of application of *Brzo* extensively, which means that more installations must apply a standardized model for QRA. More details of the model will be given in Chapter 3.

In the same *Bevi* ordinance, individual risk was renamed as "location-based" or "location-bound" risk (LR) to avoid confusion, since it is the risk for continual exposure at a certain location with respect to the risk source. Many interpreted it wrongly as the risk of an individual person as compared to that of a group. The existing limiting value of *LR* risk of 10^{-6}/year has been confirmed, although for temporary, tolerable existing situations (for 3 years) or new situations for limited vulnerable receptors (dispersed dwellings and buildings or terrains where people gather during part of the day for work or otherwise, and buildings/infrastructure of high public interest/value), the limiting value may be 10^{-5}/year. But the ideal guidance value is always 10^{-6}/year. For societal or group risk (GR), the criterion mentioned above became a guidance value and the CA is accountable if it wants to allow a higher value. The idea behind the latter is that the governing body can weigh in its decision making whether in an emergency response it can cope with the societal distress associated with the incident. Emergency responder organizations must give advice in this matter. The problem, however, is that although many have become familiar with the criterion, they do not have a good feeling for comparing low event probability figures. To facilitate their planning, municipalities can designate a safety zone borderline or safety contour in which the LR does not exceed the threshold. For this purpose also, fixed safety distances to certain types of risk sources such as LPG refueling stations are applied. Accumulation of risks from different sources at one location is not considered, but domino effects by an incident are assessed. A risk map[41] is available on the Web for those who want to know which risk sources are in their neighborhood.

The Act on the Transportation of Dangerous Materials of 1995 (*Wet vervoer gevaarlijke stoffen, Wvgs*[42]) in principle excludes transport through build-up areas and tunnels or other vulnerable infrastructures and can specify permitted routes. The fourth National Environmental Policy Plan of 2001 was introduced to define main transportation routes. In principle it specifies the same system of risk thresholds and guidance values as for stationary installations, although the values are expressed per unit length and per year. A similar regulation holds for pipelines.

Lately, one hears the wish to make decisions less fixed-criterion dependent and to assess for consequences and probability separately. Also, there is a plea for having greater public involvement, considering costs-benefits, and introducing a stronger challenge for industry to reduce risk. In Chapters 9, 10, and 12 we shall return to interpretation of risk figures and to problems of risk-based decision making given the assumptions and data are uncertain.

2.4.3 EU ATEX (*ATMOSPHÈRES EXPLOSIBLES*) DIRECTIVES

Although rather specific and not pertaining just to major hazards as defined by the Seveso Directives, the ATEX 95 and 137 directives have been a blessing in reducing the number of gas and dust cloud ignitions, explosions, and ensuing fires. Therefore, we take a brief look at the essential elements of both directives. The ATEX numbers 95 and 137 refer to the articles of the 2002 consolidated version of the Treaty Establishing the European Community (Rome Treaty, March 25, 1957) on which they are based.

The oldest, ATEX 95,[43] issued by the European Commission Directorate Enterprise and Industry, has come with a separate guidance document[44] as it involves a large part of the European industry, not only the process industry. The underlying objective is to allow free trade by harmonizing existing standards and requirements. These health and safety requirements for integrated explosion safety are listed in Annex II of the Directive and are specific with respect to equipment that could act as an ignition source in (potentially) explosive atmospheres and to systems and devices meant to prevent gas and dust explosions or reduce their effects. The definition of an explosive atmosphere is a mixture of a flammable gas, vapor, dust, or mist with air under atmospheric conditions in which after ignition, combustion spreads to the entire mixture, whereas with dusts not all material must be consumed.

Equipment is classified into groups and categories: Group I equipment is for use in underground mines and Group II for all other applications. Group I comes in only two categories for respectively very high and high levels of protection. By way of contrast, Group II has a third category for normal levels of protection. This applies to situations where an explosive atmosphere is unlikely to be present and, if so, it will only be for a short duration. The guide describes the types of equipment and devices in detail and presents examples. Machinery may have an explosive mixture inside, but as long as an internal ignition of this does not propagate to any external explosive atmosphere, it may meet the requirements. Details become rather complicated. Ignition hazards are distinguished from different sources and in different situations. This includes flameproof enclosures, where an internal explosion should not ignite an external cloud. The procedures to be used to assess conformity to the requirements of the directive are defined. When the results are satisfactory, then the product should display the CE marking. Further, an ⟨Ex⟩-plate is given with information of the notified body that tested the equipment and details of the group, category, gas, or dust with which it may be safely used. One of the other pieces of information to be on the ⟨Ex⟩-plate is the temperature rating. Because fuel—air mixtures can autoignite at hot surfaces, a maximum surface temperature is prescribed for a class. Dusts can self-heat at even lower temperatures and ignite. See Figure 2.3 for some examples.

Before ATEX was introduced, there were several existing standards. Beside, for example, the older British BS and German VDI standards, international ones designated as EN (by CEN and CENELEC), and IEC standards had been established. Products bearing their markings are widely distributed. The introductory effort in European companies for compliance with ATEX 95 from approximately 2000 onwards was significant. Non-European manufacturers also had to complete the same testing, certification, and labeling if their products were to be sold within the EU. It took much hard work, at significant cost, to update the inventory and reclassify and renew equipment, but efforts to follow the ATEX 95 Directive are now providing the long-term payback of reduced explosion frequency.

ATEX 137[47] is more oriented on the gas and dust explosion phenomena themselves as they can appear in workplaces rather than on equipment. It addresses

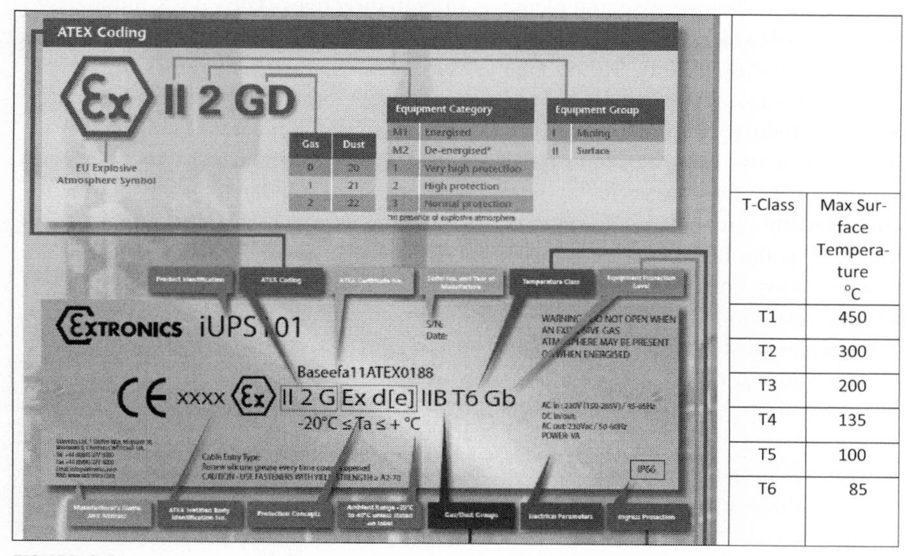

T-Class	Max Surface Temperature °C
T1	450
T2	300
T3	200
T4	135
T5	100
T6	85

FIGURE 2.3

Top and bottom left: As an illustration, the main ATEX coding is shown as part of the 2014 Extronics Wallchart with below an example of marking for equipment that can be used in the presence of defined combustible gas and dust hazards. Table right: temperature classification as a function of maximum surface temperature. For reading the small print please consult the complete wallchart at the Extronics website.[45]

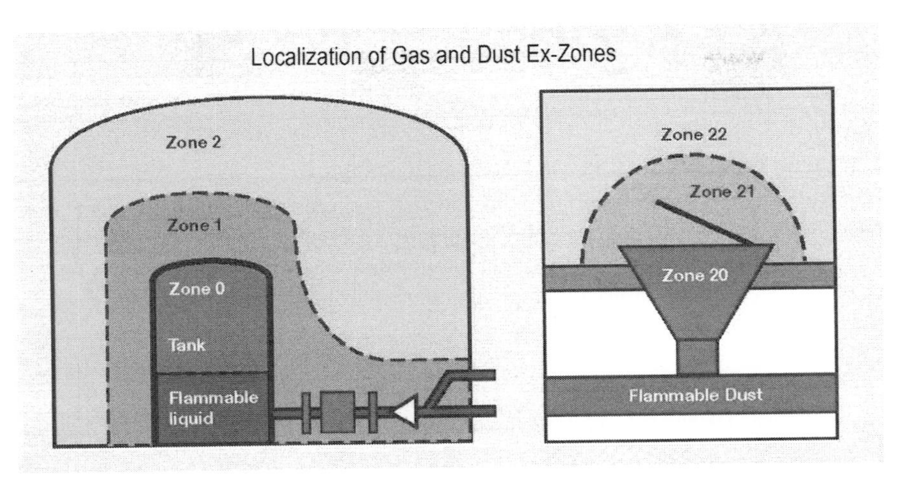

FIGURE 2.4

Illustration of zones around a tank with a volatile fuel (left) and of a hopper around which combustible dust can disperse (right).[46]

explosion prevention and protection for the consequences. Hence, it is concerned with avoiding explosive atmospheres or reducing their likelihood, and the same for ignition sources and with mitigation of their effects. Endangering health and safety of workers is the criterion for action, so ATEX 137 requires a risk assessment. The classification of locations with explosive atmospheres into three hazard zones is fundamental: these zones are designated 0, 1, and 2 for gases and vapors and 20, 21, and 22 for dusts; see Figure 2.4. In zones 0 or 20, an explosive atmosphere in air is present permanently, for long periods, or frequently, and only category 1 equipment is allowed within these zones. In zone 1 or 21, the explosive atmosphere is likely in normal operation to be present occasionally, and equipment categories 1 or 2 are permitted; and in zone 2 or 22, the explosive atmosphere is not likely but if it is present, it will be of only short duration. For the latter zones, equipment of categories 1, 2, or 3 are all appropriate. Where necessary, appropriate signs must be erected at the boundaries of zoned areas and workers must be trained appropriately.

2.5 OFFSHORE AND GAS SAFETY

The countries bordering the North Sea who have offshore activity—UK, Norway, and the Netherlands—developed *safety case* regulation following the Piper Alpha disaster in 1988 and the publication of the Lord Cullen report. Before a license to operate is granted, the UK HSE version of the assessment principles[48] of this regulation requires the operator/owner of an activity to demonstrate that:

1. the management system is adequate to ensure compliance with statutory health and safety requirements; and for management of arrangements with contractors and sub-contractors,

2. adequate arrangements have been made for audit and for audit reporting,
3. all hazards with the potential to cause a major accident have been identified, their risks evaluated, and measures have been, or will be, taken to control those risks to ensure that the relevant statutory provisions will be complied with.

The UK Safety Case guidance documents and manuals can easily be found on the HSE Website.[49] A Norwegian guidance on risk and emergency preparedness analysis is also available.[50] Following the Deepwater Horizon platform catastrophe, governments worldwide updated and refined their policies concerning offshore activities and their effects on the safety of people and the environment. We shall briefly review the developments in the US and EU.

2.5.1 DEVELOPMENT IN THE UNITED STATES

In Chapter 4, the Deepwater Horizon accident, or Macondo catastrophe, of 2010 will be analyzed and the relevant reports referenced as an example of how present-day accidents can develop. Here a brief description will be given of the rather fundamental changes that have been made in organization of oversight and in regulation. This is based on, amongst others, the report to the president.[51] Obviously, besides drilling the regulation also concerns production.

First, we need to mention the creation in 2011 of two separate entities under the US Department of the Interior, which before then had been under the single roof of the Minerals Management Service (MMS). In June 2010, this was renamed the Bureau of Ocean Energy Management, Regulation and Enforcement. The new offices are the Bureau of Ocean Energy Management and the Bureau of Safety and Environmental Enforcement (BSEE). BSEE is responsible for the process safety aspects. BSEE[52] launched regulatory reforms to achieve both enhanced drilling safety and workplace safety under federal regulation 30 CFR 250. The Drilling Safety Rule came with two so-called National Notices to Lessees and Operators and Pipeline Right-of-way Holders (NTLs) of which one is on an emissions inventory (already expired) and the other is on the use of blowout preventers pursuant to 30 CFR 250. The Drilling Safety Rule in the federal regulation implements the improvement measures that were recommended to the department by a committee of independent experts following the Macondo catastrophe.

The Workplace Safety Rule is mainly on safety and environmental management systems (SEMS). Quoting from the law, it contains the following elements: 1. Job Safety Analysis (JSA) provides additional requirements for conducting a JSA. 2. Auditing requires that all SEMS audits must be conducted by audit service providers, who have been certified by a BSEE-approved accreditation body (AB). 3. Stop Work Authority (SWA) creates procedures that establish SWA and makes responsible any and all personnel who witness an activity that is creating imminent risk or danger to stop work. 4. Ultimate Work Authority (UWA) clearly defines requirements establishing who has the UWA on the facility for operational safety and decision making at any given time. 5. Employee Participation Plan (EPP) provides an environment that promotes participation by employees and management in order to eliminate or mitigate hazards on the outer continental shelf (OCS). 6. Reporting Unsafe Working Conditions empowers all personnel to report to BSEE possible violations of safety or environmental regulations and requirements and threats of danger. Chang and Barrufet[53] explained Shell's PSM measures to maintain safe operation of its Gulf subsea production systems. They compared SEMS with PSM and concluded, for example, that advent of SEMS is an "opportunity to adopt industry wide process safety and integrity management best practices such as risk based assessment and fitness for service validation when extending field life on aging subsea infrastructure." Consideration is being given in the US to adopting a safety case regulation as is common in the North Sea countries and now in the EU as a whole, as mentioned below, and in Australia.

2.5.2 EU OFFSHORE DIRECTIVES

In 2004, the EU issued a directive on offshore oil and gas, but this directive focused on protection of the marine environment. In 2013, a directive[54] on offshore safety was launched with the objective of preventing major accidents in offshore oil and gas operations; it obviously based much of its content on the national regulation of the North Sea countries mentioned above. The content of the directive is quite similar to the American regulation. It asks member states to require operators to perform risk management and to assess their capabilities, including financial ones. Before new drilling is allowed, the public shall be heard and a safety report (or safety case) on major hazards shall be drafted, submitted, and, if adequate, will be accepted by the licensing authority. A CA shall be established that can enforce compliance to the regulation. A European Maritime Safety Agency shall provide member states technical and scientific assistance. Documents to be delivered by the operator or owner are a MAPP report, a SEMS description, a design notification and an independent verification of the design, and a report on major hazards. In addition, in the case of change and/or dismantling, an emergency response plan shall be delivered, a notification of well production or of relocation of operation, and any other document that a member state will require.

2.6 TRANSPORT OF HAZARDOUS MATERIALS

Transport is in principle a worldwide, border-passing activity. Obstacles to the principle of free trade need to be removed, and the United Nations Economic and Social Council (ECOSOC) Transport of Dangerous Goods Sub-Committee issued the first recommendations on the transport of dangerous goods in 1956. Later the document was split in two: Model Regulations[55] and Manual of Tests and Criteria[56] (the Orange Books). Note from the title of the subcommittee that it is about dangerous goods and not only substances by themselves. Thus, what is being considered is the hazard presented by the system of substance and package, together forming the article or good. For example, even very sensitive, explosive substances can be transported safely if the quantity and manner of packaging are properly matched. Bulk transport is regulated separately.

The classification of the substances is very important. Permitted combinations of materials for storage or transport are defined according to the type and severity of risk in terms of classes, hazard divisions, and packing groups I, II, and III (ranging from high to low danger), or compatibility groups. Classes and hazard division keywords are:

1. Explosives: divisions 1.1 mass explosion capable, 1.2 projection hazard only, 1.3 fire hazard, 1.4 not hazardous, 1.5 very insensitive, mass explosion capable (e.g., ammonium nitrate–based fertilizers), and 1.6 extremely insensitive articles, not mass explosion capable (e.g., so-called insensitive munitions);
2. Gases: flammable, nonflammable and nontoxic, and toxic;
3. Flammable liquids;
4. Flammable solids, self-reactive substances and solid-desensitized explosives, solids liable to spontaneous combustion, and solids that on contact with water emit flammable gases;
5. Oxidizing substances and organic peroxides;
6. Toxic and infectious substances;
7. Radioactive materials;
8. Corrosive substances; and
9. Miscellaneous.

Tests and criteria serve to classify a "dangerous good" into one of the categories. A substance is tested for propensity of a type of explosiveness (detonation, deflagration, or thermal explosion), toxicity, et cetera. In addition, if it is capable of a certain adverse effect, also sensitivity to initiation and the severity of effect is determined, albeit depending on the test in a rather crude sense. Prescribed test methods are chosen and specified such that in principle equipment can be procured and tests performed in less-developed countries. Based on the test results, flow chart decision-making schemes are provided from which assignment to a class follows. A substance includes mixtures and solutions and cannot belong to more than one class, although a substance may be, e.g., both corrosive and toxic. In such a case, the prescribed precedence of hazards shall be held, for which a procedure and a table are provided. Governments and international organizations are responsible for the classification, which should be made in the country of product origin. The recommendations also contain requirements for the construction of packaging and containers.

The technical backup of the subcommittee, in particular, with respect to tests and criteria with respect to physical hazards by explosion and heat, is provided by a group of experts who call themselves the International Group of Experts on the Explosion Risks of Unstable Substances.[57] The group was formally established by the Organization for Economic Cooperation and Development (OECD) in 1962 to stimulate safe, free cross-border traffic of goods. Later it was made independent of the OECD organization but maintained its way of working. Members are governmental experts with ties to the national representatives in the subcommittee, but the meetings are free to experts from industry and universities so that the continuous flow of problems with new substances or test improvements is discussed by experienced people. There are quite a few well-known tests developed long ago, which with present instrumentation could be improved, but with which so much data is collected that experts are reluctant to change or abandon the test definition. Moreover, in different countries, there may be other test methods for the same property with no one country being keen to replace its own favored method. So, amongst "other methods" the group's activities are on standardizing methods and harmonizing them so that on the basis of defined criteria the same decisions will be taken. The results trickle down into the policy discussions in the subcommittee via the personal contacts.

Transport of hazardous materials is governed by international rules, which are based on principles of the UN recommendations. The rules differ as the mode of transport changes (acronyms are explained in the references): road ADR,[58] ADN,[59] Inland Waterways, river Rhine ADNR,[60] sea and waterways IDMG,[61] railways RID[62] under the Intergovernmental Organization for International Carriage by Rail (OTIF), and air ICAO.[63]

2.7 GHS, GLOBALLY HARMONIZED SYSTEM OF CLASSIFICATION AND LABELING OF CHEMICALS

Several EU directives have been concerned with "hazardous materials" (or in EU terminology: "dangerous substances") classification, packaging, and labeling. The oldest is Council Directive 67/548/EEC, which has been amended several times. This Directive also introduced in Annexes III and IV a series of numbered risk and safety phrases (shortened: R-and S-Phrases), which, respectively, are brief characterizations of the hazardous properties of the substance considered and protective safety recommendations when handling them. Annex V described test methods. As of June 1, 2015, this is all superseded by the GHS, which is summarized in the following section.

On the initiative of the United Nations Environment Programme (UNEP) in the early 1990s, the United Nations Conference on Environment and Development (UNCED), which was held in Rio de Janeiro in 1992, started sustainability actions following the Brundlandt report. UNEP, like the International Union of Applied and Pure Chemistry, was quite active at that time because of the toxic properties of chemicals spread throughout the world by industry, importers/exporters, and sales organizations. Following the conference, the Inter-Organization for the Sound Management of Chemicals (IOMC) was established in 1995 by UNEP, International Labour Organization (ILO), Food and Agriculture Organization of the UN (FAO), World Health Organization of the UN (WHO), United Nations Industrial Development Organization (UNIDO), and the OECD, the participating organizations, to strengthen cooperation and increase international coordination in the field of chemical safety. The United Nations Institute for Training and Research joined the IOMC in 1997 to become the seventh participant (see Figure 2.5).

FIGURE 2.5

Logos of the organizations participating in the IOMC program to set up the Globally Harmonized System of Classification and Labeling of Chemicals.

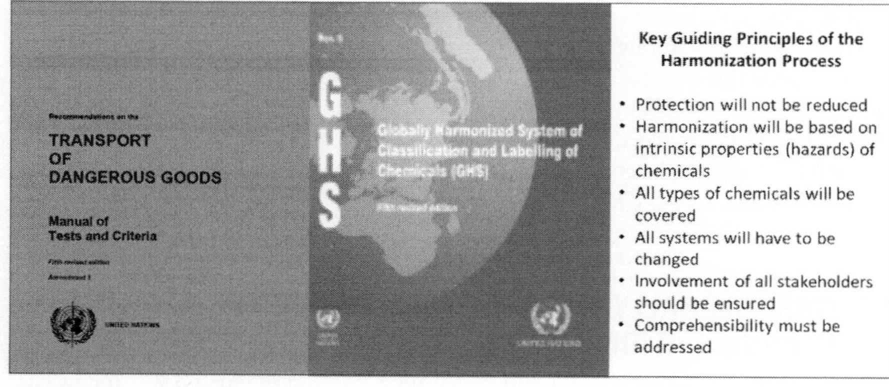

FIGURE 2.6

From left to right: the cover of the "Orange Book" or UN Manual of Tests and Criteria,[56] of the "Purple Book" of the Globally Harmonized System,[69] and at the far right is a table with the key guiding principles of the harmonization process agreed to in the Inter-Organization Programme for the Sound Management of Chemicals.

Finally, worldwide contact was made and on a consensus-basis agreement reached on, for example, harmonizing principles as seen in the table of Figure 2.6. The agreement led to establishment of the ECOSOC's Committee on the GHS of Classification and Labeling of Chemicals (GHS in short) and a distribution of tasks. All chemicals are identified by their Chemical Abstracts Service Registry number, while the labeling is from a set of characteristic symbols. The GHS (Purple Book) is now accepted worldwide. US OSHA[64] and EU[65] have incorporated GHS into their systems; the OSHA guide provides interesting background information. Figure 2.7, and the further explanation is taken from this guide. More about the history and the content of the GHS when it was still in the development stage can be found in an article by Winder et al.[66] Only when in use for a sufficiently long period can the real gain of the system be evaluated.

GHS contains harmonized elements of criteria and requirements, and is set up as building blocks of classification and communication requirements to facilitate a variety of regulatory schemes applicable to the transport, workplace, consumer, and agriculture (pesticides) sectors. Classification is in classes, and categories and communication is by labels and signal words.

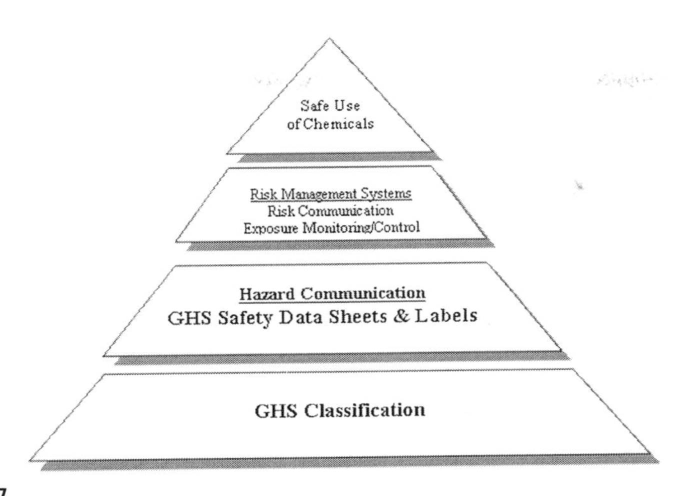

FIGURE 2.7

GHS classification is intended to lead to reduction of risk in the use of chemicals in all stages of their life cycle, from R&D via production, transport, use, and disposal.[64]

It all has to start with hazard classification of the chemicals (under the GHS no articles/goods are classified, just substances and mixtures according to their intrinsic hazardous properties). In GHS, hazards are expressed as "endpoints," terminology derived from toxicology. The GHS endpoints cover physical, health, and environmental hazards.

Physical hazards are based on the existing criteria used by the UN Model Regulation on the Transport of Dangerous Goods. Test methods have been delegated to the Transport Dangerous Goods subcommittee. The physical hazard classes are listed in Table 2.8.

The explosives class distinguishes divisions 1.1–1.6, as specified previously in Section 6. Oxidizing substances provide oxygen, which can cause or contribute to combustion severity of other materials. Flammable liquids are in four categories depending on their flash point: $<23\ ^\circ C$ and boiling point $\leq 35\ ^\circ C$; f.p. $<23\ ^\circ C$ and b.p. $>35\ ^\circ C$; f.p. $\geq 23\ ^\circ C$ and b.p $\leq 60\ ^\circ C$; f.p. $>60\ ^\circ C$ and b.p. $\leq 93\ ^\circ C$. Self-reactive substances come in seven types (A–G), which all show explosive reaction to some extent: Type A can deflagrate and detonate; B cannot do that in a package,

Table 2.8 GHS Physical Hazards Associated Substance Categories

• Explosives • Flammable gases • Flammable aerosols • Oxidizing gases • Gases under pressure (compressed, liquefied, refrigerated, dissolved)	• Flammable liquids • Flammable solids • Self-reactive substances • Pyrophoric liquids • Pyrophoric solids • Self-heating substances	• Substances that, in contact with water, emit flammable gases • Oxidizing liquids • Oxidizing solids • Organic peroxides • Corrosive to metals

but can exhibit thermal explosion in a package; C does not do either of these; D either detonates partially, or deflagrates slowly, or shows some effect when heated under confinement; E, F, and G show low reactive responses in various degrees. Pyrophorics spontaneously ignite when exposed to air. Self-heating substances such as coal or milk powder react exothermally with oxygen at higher temperature, self-heat, and may ignite. Organic peroxides are distinguished in the same types as self-reactive substances.

Health and Environmental hazards are summarized in Table 2.9.

For acute toxicity, substances with a health hazard are assigned to one of five toxicity categories on the basis of LD_{50} (oral or dermal lethal dose at 50% probability) or LC_{50} (inhalation lethal concentration at 50% probability). Skin corrosion means irreversible damage to the skin, which comes in three categories, while skin irritation refers to only reversible damage. Respiratory sensitizer means inducing hypersensitivity of the airways following inhalation. Skin sensitizer means inducing an allergic response following skin contact. Germ cell mutagenicity is either known or presumed (category 1) or suspect (category 2). The same holds for carcinogenicity and reproductive toxicology. For both target organ systemic toxicities and aspiration toxicity there are two categories: known and presumed.

In the case of environmental hazards, the acute aquatic toxicity is divided into three categories in stepwise increasing quantity ranges, and chronic toxicity is divided into four categories. Experimentally derived test data are preferred. Where no experimental data are available, validated quantitative structure activity relationships for aquatic toxicity and log K_{OW} (the octanol–water solubility partition coefficient of a substance) may be used in the classification process, but a measured value of the bioconcentration factor will take precedence, if available. Test methods are not prescribed and can be freely selected as long as they are scientifically sound and validated according to international procedures and criteria already referred to in existing systems. In the case of mixtures, experimental values are preferred, but GHS also defined a number of so-called bridging principles to make an estimate on the basis of the properties of the pure components of the mixture.

The GHS label elements are product name or identifier (identify hazardous ingredients, where appropriate); pictogram (see Figure 2.8, top part); signal word;

Table 2.9 GHS Health and Environmental Hazards

Health Hazards	Health Hazards	Environmental Hazards
• Acute toxicity • Skin corrosion/irritation • Serious eye damage/eye irritation • Respiratory or skin sensitization • Germ cell mutagenicity	• Carcinogenicity • Reproductive toxicology • Target organ systemic toxicity—single exposure • Target organ systemic toxicity—repeated exposure • Aspiration toxicity	• Hazardous to the aquatic environment • Acute aquatic toxicity • Chronic aquatic toxicity • Bioaccumulation potential • Rapid degradability

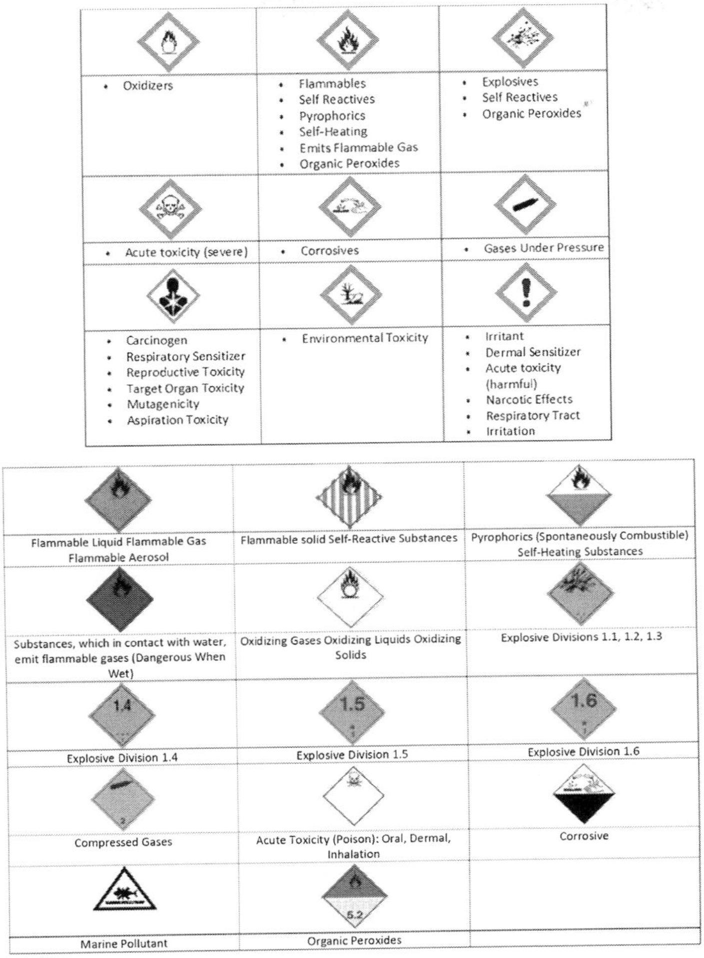

• Oxidizers	• Flammables • Self Reactives • Pyrophorics • Self-Heating • Emits Flammable Gas • Organic Peroxides	• Explosives • Self Reactives • Organic Peroxides
• Acute toxicity (severe)	• Corrosives	• Gases Under Pressure
• Carcinogen • Respiratory Sensitizer • Reproductive Toxicity • Target Organ Toxicity • Mutagenicity • Aspiration Toxicity	• Environmental Toxicity	• Irritant • Dermal Sensitizer • Acute toxicity (harmful) • Narcotic Effects • Respiratory Tract • Irritation

Flammable Liquid Flammable Gas Flammable Aerosol	Flammable solid Self-Reactive Substances	Pyrophorics (Spontaneously Combustible) Self-Heating Substances
Substances, which in contact with water, emit flammable gases (Dangerous When Wet)	Oxidizing Gases Oxidizing Liquids Oxidizing Solids	Explosive Divisions 1.1, 1.2, 1.3
Explosive Division 1.4	Explosive Division 1.5	Explosive Division 1.6
Compressed Gases	Acute Toxicity (Poison): Oral, Dermal, Inhalation	Corrosive
Marine Pollutant	Organic Peroxides	

ACUTE ORAL TOXICITY - Annex 1					
	Category 1	Category 2	Category 3	Category 4	Category 5
LD_{50}	£ 5 mg/kg	> 5 < 50 mg/kg	³ 50 < 300 mg/kg	³ 300 < 2000 mg/kg	³ 2000 < 5000 mg/kg
Pictogram					No symbol
Signal word	Danger	Danger	Danger	Warning	Warning
Hazard statement	Fatal if swallowed	Fatal if swallowed	Toxic if swallowed	Harmful if swallowed	May be harmful if swallowed

FIGURE 2.8

Reproduced for illustration purposes[62]: top, GHS pictograms and physical hazard classes; middle, the same for transport; and bottom, pictograms for acute health hazards.

physical, health, or environmental hazard statements; supplemental information; precautionary measures and pictograms; and first aid statements. The signal words are *danger* for more severe hazards and *warning* for less severe ones.

2.8 FUTURE DIRECTIONS

2.8.1 IMPROVEMENT OF TEST METHODS

Accurate and deep knowledge of the hazards of materials is fundamental to optimum regulation. Characterizing and determining properties of materials depend considerably on test methods producing a complete picture that can be interpreted without ambiguity. This in turn depends on how accurately we can bring the material under the right conditions to react and how accurately we can make observations of those reactions. Reaction can be by the substance in pure form, in contact or mixed with other materials, and on receptors. Indeed, one needs to be able to reproduce in a test exactly, or as closely as is feasible, the conditions that lead to phenomena causing the adverse consequence in an accident situation. This holds for explosive properties and ones that are initiating or supporting fire and toxicity. Hence, applications of improved instrumentation and simulation methods are critically important. Although it is an essential way to make progress, carrying out research on new or improved test methods usually does not attract funding nor is it regarded as "academically respectable." And once an improved test method does become available, preparedness to adopt it is usually rather lukewarm. The argument to maintain the status quo concerns loss of previous experience and the accumulated data. Only enduring, close international cooperation of independent expert parties who trust each other can produce the necessary improvements.

2.8.2 PRESCRIPTIVE VERSUS GOAL-SETTING REGULATION

For years there has been a debate about the advantages of goal setting or performance-based regulation versus the older, more direct prescriptive way of regulation that defines what shall, or shall not, be done. This holds not only for the process industry but also for a large variety of industrial activities such as engineering and construction of airplanes, ships, trains, buildings, bridges, or dams, which in their use and loss of function may lead to dramatically high consequences, albeit as rare events. In the competitive world, right beside knowledge and the right technical know-how, many other human characteristics such as stupidity, envy, deceit, fraud, greed, corruption, mistrust, negligence, indifference, or sabotage can lead to disaster. Safety regulation is only one means to counter failure but has proven to be able to make significant contributions to safety.

On the other hand, economics are a powerful driver. With higher complexity of technology, installations, and work processes, prescriptive regulation will lead to more detailed rules, which on a lower level may become very complex, confusing, or even conflicting. At the same time, for a certain requirement there may be a variety of solutions and depending on conditions, each of which may be attractive in specific circumstances. The aim of performance-based regulation is to enable a free choice of available solutions in order to obtain the same desired goal of reliability and safety. The problem is then for a governing authority being held accountable for safety and security to maintain oversight, to inspect, and to control. Performance-based regulation requires, though, a higher level of insight from the inspectors having to make the assessments and judge the trust among parties. For setting goals and simply checking whether the goals are right and the means to achieve them adequate, the authorities must trust the competence and the intentions of the leaders of industry. Much of this boils down to culture, the defining metrics of which are still underdeveloped. So, goal-setting regulation shall be favored, but it depends a great deal on local conditions, levels of education, attitudes, and the liability regime as to what extent this will be able to deliver levels of safety that are acceptable to all parties. Practice will see a mix of both approaches such that countries with a higher extent of mutual trust, more ingrained experience, and more temperate mood tend to have more goal-setting regulation while others will stick to more prescriptive regimes. Although there is much literature on this topic, Brian Meacham published a paper on the subject from an engineering point of view with some interesting examples from the world of construction.[67] We shall further dwell on inspection in the next section and on "prescriptive versus goal setting" in Chapter 10.

2.8.3 INSPECTIONS ON COMPLIANCE

Inspections are considered the best approach for enforcing safety regulations. For this, a regulator must be strong, competent, and well resourced to do the necessary inspections. However, after disastrous mishaps time and time again, authorities in various countries have to admit that inspection has failed. On the other hand, when budget cuts are discussed, inspection will not be spared. To cut costs over the last decades, part of the basic inspection function has been outsourced to certifying organizations. Not without reason in the evaluation of the Seveso II Directive, the wish to reinforce inspection efforts had high priority; and in the US, as the CSB reports reveal, OSHA planned and performed actions following serious accidents, such as the BP Texas City vapor cloud explosion in 2005 and the Imperial Sugar dust explosion in 2008. Note that in the case of the West Fertilizer ammonium nitrate detonation in Texas in 2013, the last inspection had taken place more than 25 years before the accident. Chief inspectors complain that young knowledgeable inspectors are hard to find, and when inexperienced candidates become available and sent on courses for training, industry has been known to entice them away from the regulator by offering better salaries and more attractive working conditions. Performance-based regulation was seen as a means to reduce the regulatory burden, however, as mentioned, inspection must then be conducted at a higher level of competence. Moreover, in many countries safety regulation with respect to workers and residents is divided between two different ministries, while emergency response resides in yet another ministry. Thus, inspections were performed by separate ministerial sections, perhaps independent of regulators per se but not coordinated over the various ministries. Under pressure to increase effectiveness and efficiency of the inspection effort, harmonization of inspection is taking place—an effort that is also supported by the industry. Where not easily allowed earlier, after an accident surprise inspections have also become a possibility. This kind of inspection has the obvious advantage of an unadorned observation but the disadvantage of not being able to meet the right people. Third-party inspection by certified bodies, such as DNV, Lloyd's, Bureau Veritas, and others, paid for by the companies they inspect, can help. Yet, for that purpose a method needs to be found to guarantee true independence of assessment.

The need to improve the inspection effort in the EU led in 2008 to an initiative of organizing workshops or seminars, plant visits, and issuing a series of reports on the topic of inspection.[68] In succession, five volumes have been brought out on the following topics: petroleum storage sites, petroleum refineries, enforcement of Seveso II, the value of safety reports for inspection, and risk management in industrial parks. The last two volumes were published in 2012.

- The first volume, on storage, was clearly inspired by the Buncefield disaster and focuses on the SMS and technical measures. It also demonstrates the use of the Belgian PLANOP tool for layer of protection analysis, which we shall briefly discuss in Chapter 3.
- The second volume, deals with refineries and integrity management, safety performance metrics, inspection strategies, and human factors. Especially the latter requires much more attention than it has so far obtained. Inspection suffers from inconsistency and differing levels of knowledge of inspectors in certain areas; recommendations are provided.
- The third volume is on compliance drivers and barriers to enforcement. Reputation appears to be a driver for many companies. Companies with competent health, safety, and environment staff are able to comply better than smaller companies without such support. The knowledge level of inspectors is very important. Regulators should pay more attention to enforceability when they create regulation. Taking advice from inspectors and industry representatives in the process can help. Enforcement is hampered by resource restrictions, inconsistencies in regulation, and unclear, "soft" rules.
- The fourth volume is on the value of safety reports for inspection. Inspection needs to determine if the safety report demonstrates a coherent and convincing case and assesses the reality of risk management in a company relative to what the safety report states. Time devoted to drafting the safety report varies greatly among countries, and the most difficult conundrum is the number of scenarios to be covered and how these should be presented. (This issue will be revisited in the next chapter on risk assessment.) Also, the reporting on risk management by inspectors appears not to be easy, and it was recommended to take a few example scenarios and go through the entire process of risk assessment, measures installed, and looking for evidence concerning whether emergency planning had included it.

- The fifth volume covers industrial parks and domino effect sites, where there can be the risk of an event in one plant triggering a domino effect in a neighboring one, of which owners/operators are not aware. There appears to be about 400 domino effect sites in the EU, of which 80% are in industrial parks. A number of requirements emerged from the discussions: Inspections shall target the safety of the park as a whole, and not only individual establishments. Common utilities shall be identified. A legal entity shall take responsibility for common utilities and emergency planning. Establishments shall be compelled to exchange information. Inspection checklists are needed. In large parks, competent authorities shall communicate and perform joint inspections. Domino effects can even mutually occur between Seveso and non-Seveso facilities. Legal instruments for industrial parks are lacking.

2.9 CONCLUSION

As this chapter indicates, over the years regulation has grown rapidly and covers many topics. Industrial safety reports with periodical updates, or safety cases with respect to offshore installations, have led to major safety improvements. (There is a trend, e.g., in Australia, to use the designation "safety case" also to indicate the submission of a "safety report."). In the US, following the Macondo catastrophe and multiple refinery accidents, there have been calls to also introduce a safety case regime.

One could ask, if every employer and employee would follow the rules, are accidents then still possible? We all know that due to competition and other causes, people do not always follow the rules and that violation is sometimes flagrant. But even, if all intentions are good and no violations occur, accidents will happen—as we shall see in the following chapters. Nonetheless, the safety professional will always maintain that all industrial accidents are avoidable.

Plant operators and owners have primary responsibility for safety, but evidence clearly shows that effective regulation is a must and that effective inspection on compliance appears to be a necessity. Interaction of regulators and industry is a subtle matter in which the question can always arise of "who owns the risk?" This is particularly so in the case of new technology: "Is it the best available (*and* safe) technology?"

REFERENCES

1. The information is taken from various sources: Lees, Loss Prevention in the Process Industries, see Chapter 1; Wikipedia. http://en.wikipedia.org/wiki/Seveso_disaster; Corliss M. Dioxin: Seveso disaster testament to effects of dioxin; 1999. http://www.getipm.com/articles/seveso-italy.htm.
2. OSHA Standard 29 CFR 1910. https://www.osha.gov/pls/oshaweb/owadisp.show_document?p_table=STANDARDS&p_id=9760. Guide to the standard: OSHA 3132, Process safety management. https://www.osha.gov/Publications/osha3132.pdf.

3. APPENDIX A: 40 CFR PART 68. http://www.epa.gov/osweroe1/docs/chem/Appendix-A-final.pdf. An amendment to the submission schedule and data requirements was published in 2004 as 69 FR 18819 (to be obtained from the US Government Printing Office).
4. EPA 550-B-99–015. *EPA guides to chemical risk management, evaluating chemical hazards in the community: using an RMP's Offsite consequence analysis*. May 1999. www.epa.gov/osweroe1/docs/chem/oca.pdf
5. Kohout AJ. U.S. regulatory framework and guidance for siting liquefied natural gas facilities—a lifecycle approach. In: 15th Annual international symposium, Mary Kay O'Connor process safety center, October 23–25, 2012, College Station, Texas, pp. 760–92.
6. Shea DA. *Implementation of Chemical Facility Anti-terrorism Standards (CFATS): issues for congress*. Congressional Research Service; April 2014. 7–5700. www.crs.gov. R43346
7. Executive Order 13650. *Actions to improve chemical facility safety and security—a shared commitment, report for the president*. May 2014. https://www.osha.gov/chemicalexecutiveorder/final_chemical_eo_status_report.pdf
8. Commission of the European Communities, Council Directive 82/501/EEC on the major-accident hazards of certain industrial activities. bookshop.europa.eu/...directive-82-501-eec.../CDNA12705ENC_001.pdf.
9. Council Directive 96/82/EC of 9 December 1996 on the control of major-accident hazards involving dangerous substances. http://eur-lex.europa.eu/LexUriServ/LexUriServ.do?uri=CELEX:31996L0082:EN:HTML.
10. Directive 2003/105/EC of the European Parliament and of the Council of 16 December 2003. http://eur-lex.europa.eu/LexUriServ/LexUriServ.do?uri=OJ:L:2003:345:0097:0105:EN:PDF.
11. Directive 2012/18/EU of the European Parliament and of the Council of 4 July 2012 on the control of major-accident hazards involving dangerous substances, amending and subsequently repealing Council Directive 96/82/EC. http://eur-lex.europa.eu/LexUriServ/LexUriServ.do?uri=OJ:L:2012:197:0001:0037:EN:PDF.
12. Major Accidents Hazard Bureau. *Guidance on the preparation of a safety report to meet the requirements of Directive 96/82/EC as amended by Directive 2003/105/EC (Seveso II)*. EC DG Joint Research Centre; 2005. EUR 22113 EN, http://ipsc.jrc.ec.europa.eu/fileadmin/repository/sta/mahb/docs/GuidanceDocuments/EUR22113EN_1__NewSafetyReportsGuidance.pdf.
13. Gilbert Y, Aho J, Ahonen L, Wood M, Lähde A-M. *The role of safety reports in preventing accidents, key points and conclusions,* Seveso inspection series volume 4. A Joint Publication of the European Commission's Joint Research Centre and the Finnish Safety and Chemicals Agency (TUKES); 2012. JRC73519, EUR 25490 EN, ISBN 978-92-79-26264-7, ISSN 1018-5593, http://dx.doi.org/10.2788/45867, 2012. 138 p. http://ipsc.jrc.ec.europa.eu/?id=817. http://ipsc.jrc.ec.europa.eu/fileadmin/repository/sta/mahb/docs/SevesoInspectionSeries/SevesoInspectionSeriesVolume-4.pdf
14. F-Seveso. *Study of the effectiveness of the Seveso II Directive, EU-VRi.* contract no.70307/2007/476000/MAR/A3. 08 29, 2008. http://ec.europa.eu/environment/seveso/pdf/seveso_report.pdf.
15. Web address for eMARS is https://emars.jrc.ec.europa.eu/.
16. Christou M, Gyenes Z, Struckl M. Risk assessment in support to land-use planning in Europe: towards more consistent decisions? *J Loss Prev Process Ind* 2011;**24**:219–26.
17. http://www.hse.gov.uk/.

18. Health & Safety Executive UK. *Reducing risk, protecting people: HSE's decision-makings process*. HSE Books; 2001, ISBN 0-7176-2151-0. http://www.hse.gov.uk/risk/theory/alarpglance.htm.
19. ALARP Suite of Guidance. http://www.hse.gov.uk/risk/theory/alarp.htm.
20. Health & Safety Executive. PADHI, HSE's land use planning methodology. Published by HSE, Version March 2008, © Crown copyright, last updated May 2011. http://www.hse.gov.uk/landuseplanning/padhi.pdf.
21. Health & Safety Executive U.K. *Proposals for revised policies to address societal risk around onshore non-nuclear major hazard installations, CD 212*. HSE Books; © Crown copyright 2007. http://www.hse.gov.uk/consult/condocs/cd212.htm.
22. Loi no. 2003-699 du 30 juillet 2003 relative à la prévention des risques technologiques et naturels et à la réparation des dommages, JO 31/07/03. http://www.legifrance.fr.
23. Ministère de l'Ecologie. du Développement et de l'Aménagement durables. Le plan de prévention des risques technologiques (PPRT), Guide méthodologique, downloadable in three badges from: http://www.ecologie.gouv.fr/Les-Plans-de-Prevention-des.html.
24. AIDA. Circulaire du 10/05/10 récapitulant les règles méthodologiques applicables aux études de dangers, à l'appréciation de la démarche de réduction du risque à la source et aux plans de prévention des risques technologiques (PPRT) dans les installations classées en application de la loi du 30 juillet 2003. INERIS. http://www.ineris.fr/aida/consultation_document/7029#7030.
25. Taveau J. Risk assessment and land-use planning regulations in France following the AZF disaster. *J Loss Prev Process Ind* 2010;**23**:813−23.
26. Arrêté du 29 septembre 2005 relatif à l'évaluation et à la prise en compte de la probabilité d'occurrence, de la cinétique, de l'intensité des effets et de la gravité des consequences des accidents potentiels dans les études de dangers des installations classées soumises à autorisation NOR: DEVP0540371A Version consolidée au 08 octobre 2005. http://www.legifrance.gouv.fr/affichTexte.do?cidTexte=JORFTEXT000000245167&dateTexte=.
27. Cahen B. Implementation of new legislative measures on industrial risks prevention and control in urban areas. *J Hazard Mater* 2006;**130**:293−9.
28. Zwölfte Verordnung zur Durchführung des Bundes-Immissionsschutzgesetzes (Störfall-Verordnung − 12. BImSchV), Störfall-Verordnung in der Fassung der Bekanntmachung vom 8. Juni 2005 (BGBl. I S.1598). http://www.gesetze-im-internet.de/bundesrecht/bimschv_12_2000/gesamt.pdf.
29. Adams M. The precautionary principle and the rhetoric behind it. *J Risk Res* 2002;**5**:301−16.
30. Raumordnungsgesetz vom 22.12.2008 (BGBl. I S. 2986), zuletzt geändert durch Art. 9 G v. 31.7.2009 I 2585. http://www.gesetze-im-internet.de/bundesrecht/rog_2008/gesamt.pdf.
31. Baunutzungsverordnung, Verordnung über die bauliche Nutzung der Grundstücke. In: der Fassung der Bekanntmachung vom 23. Januar 1990 (BGBl.1 S. 132), zuletzt geändert am 22. April 1993 (BGBl.1 S. 466). http://www.gesetze-im-internet.de/bundesrecht/baunvo/gesamt.pdf.
32. SFK/TAA Major Accident Commission. Technical committee for plant safety at the federal ministry for environment, nature conservation and reactor safety, short version of the guidance SFK/TAA-GS-1, recommendations for separation distances between establishments under the major accidents ordinance and areas requiring protection within the framework of land-use planning − implementation of §50 federal pollution protection law (BlmSchG), SFK/TAA Working Group Land-Use Planning, SFK/TAA-GS-1k-EN. http://www.hrdp-idrm.in/live/hrdpmp/hrdpmaster/idrm/content/e5783/e6127/e11287/

infoboxContent11289/sfk-taa-gs-1k-en.pdf; SFK/TAA-GS-1, http://www.kas-bmu.de/publikationen/sfk/sfk_taa_gs_1.pdf.

33. ZEMA Data Base. Zentrale Störfallmelde- und Auswertestelle im Umweltbundesamt. www.umweltbundesamt.de/zema.
34. Bottelberghs PH. Risk analysis and safety policy developments in the Netherlands. *J Hazard Mater* 2000;**71**:59–84.
35. Pasman HJ, Reniers G. Past, present and future of quantitative risk assessment (QRA), and the incentive it obtained from land-use planning (LUP). *J Loss Prev Process Ind* 2014;**28**:2–9.
36. Arbeidsomstandighedenwet, 1998. http://wetten.overheid.nl/BWBR0010346/geldigheidsdatum_28-08-2013.
37. Wet ruimtelijke ordening, 2006. http://wetten.overheid.nl/BWBR0020449/geldigheidsdatum_28-08-2013.
38. Wet milieubeheer, 1979. http://wetten.overheid.nl/BWBR0003245/geldigheidsdatum_28-08-2013.
39. Besluit risico's zware ongevallen, 1999. http://wetten.overheid.nl/BWBR0010475.
40. Besluit externe veiligheid inrichtingen. http://wetten.overheid.nl/BWBR0016767.
41. The Dutch risk map with English explanatory text can be found in the website. www.risicokaart.nl/en/.
42. Wet vervoer gevaarlijke stoffen, 1995. http://wetten.overheid.nl/BWBR0007606/geldigheidsdatum_28-08-2013.
43. Directive 94/9/EC of the European Parliament and the Council of 23 March 1994 on the approximation of the laws of the member states concerning equipment and protective systems intended for use in potentially explosive atmospheres. http://ec.europa.eu/enterprise/sectors/mechanical/documents/legislation/atex/.
44. DG Enterprise and Industry, European Commission. *ATEX guidelines, guidelines on the application of Directive 94/9/EC of the European Parliament and Council of 23 March 1994 on the approximation of the laws of the member states concerning equipment and protective systems intended for use in potentially explosive atmospheres*. 4th ed. September 2012. http://ec.europa.eu/enterprise/sectors/mechanical/files/atex/guide/atex-guidelines_en.pdf.
45. Adopted from Extronics Wallchart. http://www.extronics.com/media/208315/atex_wallchart_15_12_2014.pdf.
46. Adopted from ATEX wall chart. http://www.poweroilandgas.com/2011/07/atex-iec-reference-for-explosive.html.
47. Directive 1999/92/EC of the European Parliament and of the Council of 16 December 1999 on minimum requirements for improving the safety and health protection of workers potentially at risk from explosive atmospheres (15th individual Directive within the meaning of Article 16(1) of Directive 89/391/EEC). http://eur-lex.europa.eu/LexUriServ/LexUriServ.do?uri=OJ:L:2000:023:0057:0064:en:PDF.
48. U.K. Health and Safety Executive. *Assessment Principles for Offshore Safety Cases (APOSC)*. Issued March 2006.
49. U.K. Health and Safety Executive Safety cases. http://www.hse.gov.uk/offshore/safetycases.htm.
50. Norsok Standard Z-013. *Risk and emergency preparedness analysis*. Rev. 2, 3rd ed. October 2010., www.standard.no/PageFiles/18398/z013u3.pdf.
51. *Deep water, the gulf oil disaster and the future of offshore drilling, report to the president, national commission on the BP deepwater horizon oil spill and offshore drilling*. January 2011. www.gpo.gov/fdsys/pkg/GPO-OILCOMMISSION/pdf/GPO-OILCOMMISSION.pdf.

52. Relevant regulation can be downloaded from the BSEE website: http://www.bsee.gov/About-BSEE/BSEE-History/Reforms/Reforms.aspx.
53. Chang E, Barrufet M. Process safety management for subsea oil and gas production system in the gulf of Mexico. In: 16th Annual int'l symposium Mary Kay O'Connor process safety center, October 22–24, 2013, CS TX. 15 p.
54. Directive 2013/30/EU of the European Parliament and of the Council of 12 June 2013 on safety of offshore oil and gas operations and amending Directive 2004/35/EC. http://eur-lex.europa.eu/LexUriServ/LexUriServ.do?uri=OJ:L:2013:178:0066:0106:EN:PDF.
55. United Nations, Recommendations on the transport of dangerous goods, model regulations. 12th ed. http://www.unece.org/fileadmin/DAM/trans/danger/publi/unrec/English/Recommend.pdf.
56. United Nations, Recommendations on the transport of dangerous goods, manual of tests and criteria. 5th revised ed. New York and Geneva; 2009, ST/SG/AC.10/11/Rev.5. http://www.unece.org/fileadmin/DAM/trans/danger/publi/manual/Rev5/English/ST-SG-AC10-11-Rev5-EN.pdf.
57. For more details on history, membership, working groups, meetings of IGUS see http://www.igus-experts.org/.
58. ADR. Accord Européen relatif au transport international des merchandises dangereux par route. Genève, 30 September 1957 (1959, 171). Implementation of Directive 94/55/EG.
59. ADN. European agreement concerning the international carriage of dangerous goods by inland waterways. http://www.unece.org/trans/danger/publi/adn/adn_e.html.
60. ADNR. Accord Européen relatif au Transport International des Marchandises Dangereuses par voie de Navigation du Rhin.
61. International Maritime Dangerous Goods (IMDG) Code by IMO (International Maritime Organization). http:/www.imo.org/safety/dangerous goods.
62. RID. Règlement concernant le transport international ferroviaire des marchandises dangereuses. Bern, 9 mei 1980. Implementation of Directive 96/49/EG.
63. ICAO (International Civil Aviation Organization). Technical instructions for the safe transport of dangerous goods by air (Doc 9284). http://www.icao.int/safety/DangerousGoods/Pages/technical-instructions.aspx.
64. United States Department of Labor, Occupational Health & Safety Administration. *A guide to the globally harmonized system of classification and labelling of chemicals (GHS)*. 2009. http://www.osha.gov/dsg/hazcom/ghs.htm.
65. Regulation (EC) No 1272/2008 of the European Parliament and of the Council of 16 December 2008, on classification, labelling and packaging of substances and mixtures, amending and repealing Directives 67/548/EEC and 1999/45/EC, and amending regulation (EC) No 1907/2006.
66. Winder Ch, Azzi R, Wagner D. The development of the globally harmonized system (GHS) of classification and labelling of hazardous chemicals. *J Hazard Mater* 2005; **A125**:29–44.
67. Meacham BJ. Accommodating innovation in building regulation: lessons and challenges. *Build Res Information* 2010;**38**(6):686–98.
68. Seveso inspections series – vol. 1, necessary measures for preventing major accidents at petroleum storage depots. Key Points and conclusions, EUR 22804 EN, ISBN 978-92-79-06197-4, ISSN 1018-5593, © European Communities, 2008, 96 p.; vol. 2, Improving major hazard control at petroleum oil refineries. Key points and conclusions, EUR 23265 EN, ISBN 9789-279-084263, http://dx.doi.org/10.2788/69413, LB-NA-23265-EN-C, © EC,

2008; 44 p.; vol. 3, Enforcement of Seveso II: an analysis of compliance drivers and barriers in five industrial sectors: Key points and conclusions, EUR 23249 EN – 2008, 126 p.; vol. 4, see 10; vol. 5, Larsen RG, Olsen AL, Wood M, Gyenes Z. Chemical hazards risk management in industrial parks and domino effect establishments. Key points and conclusions for Seveso Directive enforcement and implementation, JRC77563, LB-NA-25-664-EN-C, ISBN:978-92-79-27993-5 (print), ISBN: 978-92-79-27992-8 (pdf), ISSN 1018-5593 (print), ISSN: 1018-5593 (online), http://dx.doi.org/10.2788/7438, 141 p. http://ipsc.jrc.ec.europa.eu/?id=817.

69. Globally Harmonized System of Classification and Labelling of Chemicals (GHS), Fifth Revised Edition, United Nations, New York and Geneva, 2013, ST/SG/AC.10/30/Rev.5.

CHAPTER

Loss Prevention History and Developed Methods and Tools

3

A poor tradesman blames his tools
After a 14th Century French proverb

SUMMARY

Since the 1960s, major contributing factors such as improvements in technology and in management, and recognition of the roles of leadership and culture, have increased the level of process safety. First, a brief historical survey will show the evolution of technical know-how, the start of interest in human factors, awareness of the role of management, the introduction of safety management systems (SMS), and the development of a safety culture. Then, after a section on terminology and on analysis of how accidents occur, we come to discussion of the present body of knowledge. Progress made relies on insight and knowledge in which one can distinguish four areas:

1. *Properties of hazardous materials*, *test methods*, and *computational tools* to describe and predict hazardous phenomena, such as toxic cloud dispersion, fire, and explosions;
2. The *system safety* shared with other engineering disciplines with its principles of inherent safety, fail-safe constructions, and analytical tools to investigate reliability, availability, and maintainability of equipment;
3. Design of *safer process technology* and of an operations SMS with its feedback loops and taking account of human and organizational factors (HOFs); and
4. *Probabilistic quantitative risk assessment* to support rational decision making and risk management. Its tools are continuously being further developed, and it is making use of the growing body of knowledge generated in areas (1)–(3), while also giving some direction to these areas.

In operating a process installation, in principle everything can go wrong. "Murphy's law" expresses this in an alternative way: "If anything can go wrong, it will."[1] In a plant we have to be aware of chains of events as cascading or escalating domino occurrences and associated effects. The same is true of course in transporting hazardous materials. Much has been learned and gained in loss prevention in the last 50 years. Hence, after mentioning one of the most renowned accident models and accident investigation tools, it will be shown how the role of leadership and organization, the advent of safety management and the SMS, and the importance of giving full attention to human factors and safety culture have made contributions to improved safety levels. These concepts are first treated in a condensed form but will be revisited in this and later chapters in more detail highlighting new directions. The remainder of this chapter will give in a panoramic fashion a characterization of the physical and chemical phenomena associated with hazards of processes and process materials, methods of investigation, concepts of system safety, and traditional methods of systematic hazard and risk analysis. Risk analysis is mostly meant in a predictive sense, but it is also used to be descriptive of what went wrong and how it went wrong in the past. We will show the weak sides of predictive risk analysis and then new promising approaches to it will be introduced later in this book.

To begin, after explaining basic concepts, such as hazard and risk and elementary accident models, the knowledge base will be discussed in a very condensed way. Emphasis will be placed on explaining the phenomena of leaks, spreading of gas and vapor clouds, toxicity, and qualitative and crude quantitative measures relating to various types of explosions and fires. For more detailed quantitative information, references will be provided. Effects on people and other receptors are given in a convenient form of probability distributions, namely probits. Although their blast effects may be similar, explosion mechanisms differ vastly. Understanding helps prevention and because explosions are most damaging, this subject is further discussed, including the stages of initiation and ignition. The chapter will show what we know and can do as well as where uncertainties are and knowledge and data are lacking.

Principles of system safety such as "inherent safety" and renowned tools such as failure mode and effect analysis, and analysis by means of fault and event trees, and their combination into bow-tie diagrams, are briefly described. This part of the chapter concludes with an overview of the trends in process design and their effects on maintenance.

The next section explains risk assessment in terms of six steps and describes the associated tools. Identification of risk and definition of possible scenarios is still the most difficult step. Step 5 is on risk reduction, which is discussed in considerable detail because of the importance of the subject. The two most successful analysis tools—hazard and operability (HAZOP) study and layer of protection analysis (LOPA)—are given special attention. Because risk analysis is crucial in developing insight into how well we are prepared, we shall end with a brief evaluation of the methodology of today. Although risk assessment of process plants, offshore platforms, and transportation of hazardous materials has been performed for many years, the methodology is far from perfect. Improvements will be proposed.

As mentioned, all of this information serves as preparation for subsequent chapters. After describing in Chapter 5 some evolutionary changes in society and their influence on safety, the sociotechnical model will be described, which more or less demands a systemic safety approach and resilience engineering. Consequently, in Chapter 6 we need to dive deeper into human factors, human error, organization, safety culture, and management. Then finally, in Chapter 7 we come across newer tools that can be used in analysis and control, so that in Chapter 9, after brief comments on cost aspects, we are prepared for improved, risk-informed decision making with respect to process safety and risk control.

3.1 BRIEF HISTORY/EVOLUTION OF LOSS PREVENTION AND PROCESS SAFETY

Recently, Kidam et al.[2] and Kidam and Hurme[3] analyzed more than 364 chemical process installation accidents using the Japanese Failure Knowledge Database, finding that in most cases the primary cause was technical. Accidents occurred due to failure of auxiliary systems, design error (materials, dimensions, etc.), chemical reactivity and incompatibility, operating beyond the equipment design limits, and underestimating thermal expansion phenomena. From an equipment point of view, piping is the largest contributor—no surprise, but in the end seeking for root causes it turns out to be the lack of knowledge, not learning from previous incidents, or just human error. Broadribb[4] with quite some experience in the process industry concludes that most accidents are due to human factors, often as we shall see later in detail as key cause and with failing management as the most important underlying cause. One can repeatedly observe these facts emerge in the history of loss prevention knowledge development.

Although industrial safety as a topic has a much longer history, loss prevention in the process industries as a "derived discipline" began in the late 1960s when engineers organized meetings to exchange views. In the United States (US), from the 1960s onwards there has been an annual Loss Prevention symposium under the aegis of the American Institute of Chemical Engineers (AIChE), while in Europe there is a triennial event organized by the Working Party on Loss Prevention (and Safety Promotion in the Process Industries) of the European Federation of Chemical Engineering (EFCE), which organized its first symposium in 1974. A recent overview of the European activities was published by De Rademaeker et al.[5] In the early 1970s, as we have seen from examples in previous chapters, the accident rate with explosions, fires, and toxic dispersion was higher than today and deemed by all parties to be unacceptable. This was particularly true during the period when industry was scaling up in size, in the 1970s in Europe but which occurred earlier in the US. Although for many the efforts to gain a higher safety level had reasonable success over the years, in hindsight the efforts were rather slow. In Figure 3.1 there is an approximate indication of the time when new contributing factors in process safety developed. Progress still continues for all factors.

Improved materials, new insights, innovations, new codes and standards, improved design methods, and quality assurance, in part driven by competition, have resulted in more reliable and safer installation components. Process instrumentation and

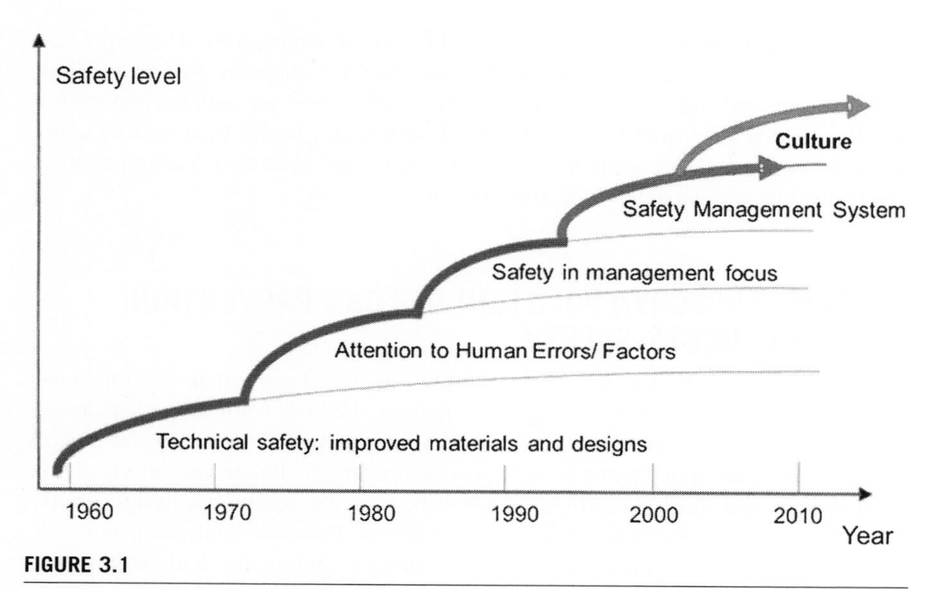

FIGURE 3.1

Contributing factors in improving loss prevention performance in the process industry. Of course, work on industrial safety in general started much earlier, but it focused on personal worker safety. In Chapter 6, historical aspects of occupational safety and health aspects shall be mentioned.

equipment exhibitions, such as the annual June ACHEMA event in Frankfurt for the more chemical-oriented industry or the oil- and gas-related conference and exhibition events in the Houston, TX, area, are greatly assisting the technical safety development. Progressive thinkers about accident causation and how to prevent mishaps developed concepts years before these ideas impacted the higher-up contributing factors shown in Figure 3.1. In Chapter 5 we shall mention names such as Charles Perrow and Jens Rasmussen whose sharp analyses have been most revealing and stimulated improvements in safety management.

It was only in the early to middle 1970s that loss prevention engineers began to target nontechnical issues, such as the human factor in process plants. This interest started with behavioral science and had more to do with reducing job injury and doing work more efficiently than with real loss prevention. Within the nuclear safety community in the 1980s, the approach appeared to consider human error as a component failure to which a probability value could be assigned. We shall discuss this approach further in Chapter 6, but for now it is sufficient to just note the development and the feeling of dissatisfaction that the approach evoked. Since then several variants and methods that take account of context and influences have been proposed. However, in the late 1980s and early 1990s, the work of psychologist, James Reason of Manchester University, made a great impact in the field. His ideas materialized in a layered model of various categories of causal factors

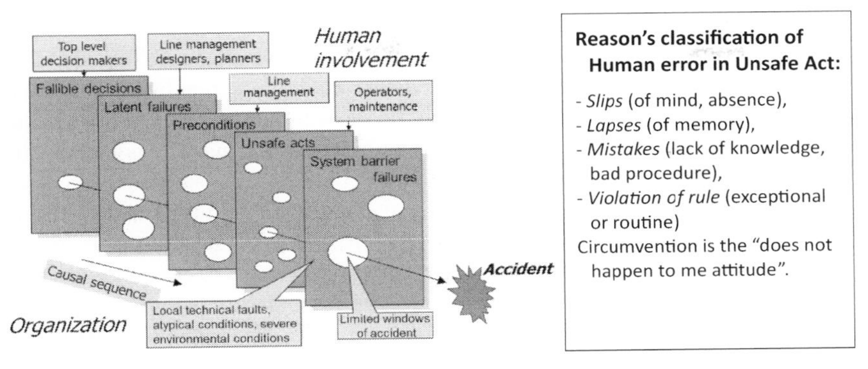

FIGURE 3.2

Swiss cheese (epidemiological) model of accident occurrence after Reason[6] in 1990.

of active and latent failures in defenses that upon coincidence allow the accident to happen, as shown in Figure 3.2. Later, the layers were depicted as slices of a Swiss (Emmentaler) cheese with the holes representing failures that when lined up trigger the accident. Some holes are permanent, as a dormant failure; others are temporary. (We shall see later how continuously monitoring a system's equipment and organization can prevent holes.) The model is typified as epidemiological (analyzing cause, distribution, and control of "disease"); later we shall discuss other types of accident models.

Active failures may be physical, for example, loss of system integrity by a burst of a pressurized component or a failure of a protective device, or they may be a human unsafe act, for example, an operator failed to keep the process within its safe window of operation or maintenance did not perform a repair well. Latent failures result from preconditions, for example, of too low a workforce competence because lack of training, unfit ergonomics, or creating time pressures to achieve a high production rate and granting a bonus. These conditions take many forms and can increase risk in many ways. There are also the latent failures of designers, planners, and construction personnel, and fallible decisions of top and line management, for example, on precursor, audit, and other indicator information. Years before a mishap takes place these conditions or decisions can leave certain constraints or cause system weaknesses that go unnoticed until it is too late. Each possible aligned series of holes forms a failure cause-and-effect chain. There is a dynamic aspect to this model. If at a certain moment as a result of changing conditions or human actions the holes line up, a chain of consequential events develops, resulting in an accident with possibly catastrophic effects. More often, only some of the holes line up and a precursor event appears (which is always important to investigate!).

This model has been used by many. After the disastrous 1988 offshore accident on the Piper Alpha platform in the North Sea that killed 167 workers, University of Manchester, UK (James Reason) and University of Leiden, NL (Willem Wagenaar, Patrick Hudson, and later Jop Groeneweg) cooperated on further developments sponsored by Shell E&P (Koos Visser). This led in the 1990s to the Tripod tools[7] for accident investigation (Tripod Beta) and safety culture measurement (Tripod Delta), which are still commonly used. Tripod Beta determined for the first time the effect of latent failures on accident occurrence. However, as we shall see later in this chapter and in Chapters 5, 6, and 7, methods of accident investigation and of determining the effect of HOFs further evolved.

Meanwhile, with rising standards of living, the public rightfully demanded an ever-increasing level of safety and security, while at the same time the economic consequences of major accidents increased. In addition, due to the large amounts of hazardous materials present-day society needs in association with its energy, materials, and food requirements, the potential of disaster has not diminished. This growth has been further exacerbated by increasing population densities around centers of economic interest where industry has tended to settle, for example, such as harbors and other good transportation links. The drive for further improvements in process safety was at the time therefore strong.

3.2 ORGANIZATION, LEADERSHIP, MANAGEMENT, SAFETY MANAGEMENT SYSTEM, CULTURE

Initially, better designs and higher quality materials contributed mostly to the increased safety of process plants, as mentioned above. A decade later, the focus shifted to human behavior and recognition of the importance of the influence of management. The critically important role of wise leadership and good management in loss prevention had become clear, not least as a result of the experience of failures. If those at the top of a company are not sincerely convinced that safety is of primary importance to the health of the company and do not actively encourage staff and workers to work safely, safety actions on the work floor will remain ineffective. Effective leadership at each organizational level is crucial for the overall atmosphere in a company and in the end determines the safety culture and climate. Safety has to be managed and is a line management responsibility.

Management is just a means to get people together to achieve desired goals using available resources as effectively and efficiently as possible. Usually, a good leader is also managing an activity well. Management can use an organizational framework of procedures and work processes including required resources, called a *management system*, to achieve a desired level of quality and safety. The second half of the 1980s was also the period in which quality assurance systems were introduced.

As we saw in Chapter 2, the OSHA 29 CFR 1910.119 Process Safety Management rule was launched in the early 1990s; it required companies to establish a SMS. In Europe the need to introduce such a SMS was provided by Lord Cullen's public inquiry report into the Piper Alpha disaster. Safety management systems focus on safety but have analogous elements to the quality management system under standard ISO 9000 or an environmental management system under standard ISO 14001. The elements may have different wordings and formats, but the basic

Table 3.1 1989 Center for Chemical Process Safety (CCPS) Guideline[8] SMS Elements

No.	Safety Management System Element
1	*Accountability*, i.e., clarity in objectives (who is responsible for what, which lines of communication)
2	*Process knowledge and documentation*, records of design criteria and management decisions
3	*Critical project review and design procedures* for new or existing plants, expansion, and acquisition
4	*Process risk management*, including encouragement of clients and suppliers to conform
5	*Management of change* of technology, facility, or organization, both temporary and permanent
6	*Process and equipment integrity* (reliability, materials, installation, inspection, maintenance, alarms)
7	*Human factors* (error assessment, task design, man–machine interface, ergonomics)
8	*Training and performance* (development of programs, design of procedures, manuals)
9	*Incident investigation* (near-miss reporting, accidents, follow-up)
10	*Standards, codes, and laws* (internal and external of company)
11	*Audits and corrective actions* (examination of technology, procedures, practices by internal or external experts)
12	*Enhancement of process safety knowledge* by research and improvement of predictive techniques

contents do not differ. The Center for Chemical Process Safety (CCPS) of the AIChE in New York was the first to define and describe in detail 12 essential elements of a SMS in one of their guideline series of books[8] in 1989. These are compiled in Table 3.1 with some added explanatory text.

Later, the element "communication in all directions" was added to the 12 elements because it appeared that after accidents communication at handover between shifts was often inadequate. Note that in these 1989 CCPS guidelines there is no explicit mention of hazard identification and characterization, failure scenario development, or risk assessment, which are prerequisites for process risk management. In 2007, CCPS published greatly extended guidelines[9] that are now risk based. The latter means that all SMS elements that contribute to safety or performance deterioration are to be viewed from a risk perspective, hence the importance of considering both consequences and likelihood in the case of a mishap. This is readily apparent from the 21 technical chapters bundled into four main thematic parts, called "pillars of accident prevention," as shown in Figure 3.3.

1. ***COMMIT to PROCESS SAFETY***
   Process Safety Culture
   Compliance with Standards
   Process Safety Competency
   Workforce Involvement
   Stakeholder Outreach

2. ***UNDERSTAND HAZARDS AND RISK***
   Process Knowledge Management
   Hazard Identification, Risk Analysis

3. ***MANAGE RISKS***
   Operating Procedures
   Safe Work Practices
   Asset Integrity and Reliability
   Contractor Management
   Training and Performance Insurance
   Management of Change
   Operational Readiness
   Conduct of Operations
   Emergency Management

4. ***LEARN FROM EXPERIENCE***
   Incident Investigation
   Measurement and Metrics
   Auditing
   Management Review and
   Continuous Improvement

FIGURE 3.3

2007 CCPS guidelines for risk-based process safety,[9] with four "pillars of accident prevention" and 21 technical chapters, all provided with performance indicator suggestions.

Continuous improvement is a crucial element in all management systems. Improvement can be realized in cycles with different time constants, as shown in Figure 3.4. *Auditing* by in-house or external parties with enforcement is a must. Guidelines for good auditing are available; Cameron and Raman[10] give a brief account. An audit can be performed at different levels and shall be structured in a protocol. There are different types of audits depending on the life cycle phase of the plant. It will be technical at design/construction; precommissioning at construction; behavioral, walk-through, and operational during operations. Also, at a higher level there will be specialist audits starting at plant commissioning and management audits on SMS functioning. In addition, in the design and operations phase regulatory compliance audits are needed and a decommissioning audit will be the final one. An audit report will likely recommend corrective actions. For assurance, management shall monitor actions being realized.

The introduction of a SMS is one issue, but measuring its effectiveness is another. A safety performance indicator in use since the early days was the number of personal safety incidents per unit of time. Worker lost time was recorded as a matter of routine, so just adding whether it was because of illness or an accident at work provided useful information; this became known as lost time injury frequency or rate (LTIF or LTIR). These, and the broader total recorded injury rate (TRIR), were usually expressed per million hours worked but can sometimes be expressed on a different basis. These indicators have declined steadily since the 1960s as indicated in Figure 3.5. The same is true for the fatal accident rate (FAR) metric, defined as the number of fatalities per 100 million working hours. Despite many improvements, major accidents in the process industry are not decreasing with an analogous convincing trend.

FIGURE 3.4

On the basis of a policy document specifying objectives (among other things), the SMS elements are combined here in the rectangular blocks underneath. In the EU Seveso Directive this would be included in the Major Accident Prevention Policy (MAPP) document. The blocks to the right depict the correction processes with their time constants mentioned.

This figure is taken from a project by the European Process Safety Centre, EPSC, published by the Institution of Chemical Engineers, IChemE, in 1994.[11]

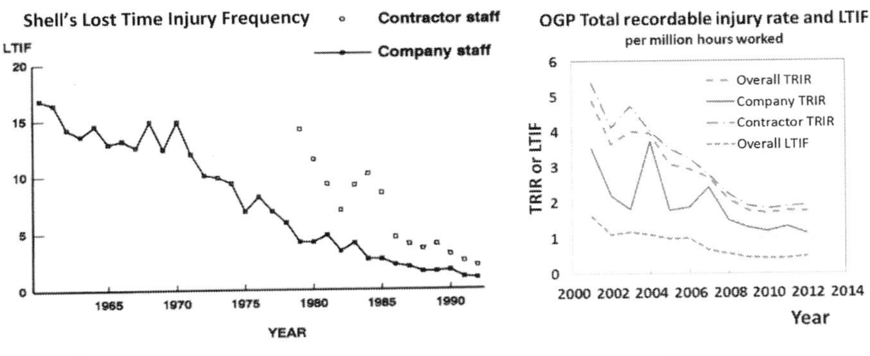

FIGURE 3.5

Left: Trend of lost time injury frequency (LTIF) metric of the Shell company from 1960 to early 1990s. The metric is the number of individual cases of an injury causing absence of a worker of one day or more per million hours worked by the total workforce. The trend of more than 30 years was published at the 8th European Loss Prevention symposium in 1995.[12] In 2012, Shell's LTIF had further decreased to 0.34. Notice a delayed effect for contractors having to catch up with the standard; Right: Total recordable injury rate (TRIR) and LTIF after data published by the International Oil & Gas Producers in 2013.[13] TRIR encompasses fatalities, lost workday cases, restricted workday cases, and medical treatment cases.

At this point we have to distinguish between *personal* safety and *process* safety. The former embraces, for example, the avoidance of worker slips, trips, and falls that may be daily occurrences, while the latter concerns the much bigger, knowledge intensive, prevention of rare, major hazard incidents for processes. Such events may even threaten the continued existence of a company such as happened to Union Carbide, which lost value after the Bhopal disaster and was taken over a number of years later. Regrettably, the downward trend in personal injury rates is not replicated with the major accident rates as illustrated in Figure 3.6.

The steady improvement of the personal safety LTIF metric gave a false impression of the situation with respect to process safety. The contrast became clear after the Esso Longford gas plant explosion accident in Australia in 1998 and the BP Refinery isomerization unit vapor cloud explosion in Texas City, TX, in 2005. In particular, the latter accident attracted much attention. Management had been cutting costs, for example, by limiting maintenance and renewal of the plant, which seemed justified by the ever-decreasing LTIF figures. The BP incident, which resulted in 15 deaths and 170 injured people, was first extensively investigated by the US Chemical Safety and Hazard Investigation Board. Triggered by its findings, a panel was established under chairmanship of the former US Secretary of State James A. Baker. This panel investigated management systems, how they worked in practice, and the safety culture within all BP refineries in the US.[15]

In Chapters 4, 5, and 6, we shall elaborate further on the panel's findings, but it suffices here to state that BP top management had an unrealistically positive perception of the performance level due to the decreasing injury rates. So, the panel recommended the introduction of lagging and leading indicators of process safety performance. Lagging, also called outcome indicator metrics, could for instance be

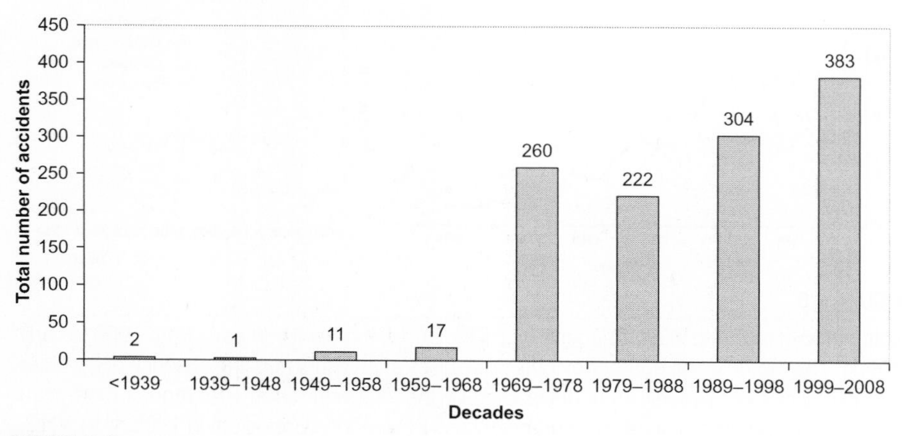

FIGURE 3.6

Global total number of accidents in the downstream oil industry per decade recorded in the TNO FACTS database.[14] Unfortunately, no overall capacity figures are available to normalize the data.

derived from the number of incidental releases and other undesired upset events over a certain period causing injury or damage. By way of contrast, leading or activity indicator metrics are derived from the degree to which those elements of the SMS with the objective of preventing failures are implemented or comply with an action that has been agreed to. Not falling clearly into either a leading or lagging category are accident precursors or near misses, which sometimes do cause damage but often do not. It is vital to investigate, learn from, and take corrective action following near-miss events. The characteristics of lagging and leading indicators have been extensively discussed in a series of articles provoked by the introductory paper of Andrew Hopkins, emeritus professor in sociology, Australian National University, Canberra.[16]

Monitoring performance by means of indicators is a logical follow-on from establishing a management system as it will enable the "check" stage in the PCDA management cycle of Dr Deming; see Figure 3.7.

The nuclear power industry in the US and the United Kingdom (UK) preceded the process industry in introducing process safety performance indicators. Since the beginning of the 2000s, various organizations, such as Organisation for Economic Co-operation and Development (OECD), Health & Safety Executive (HSE) UK, CCPS, API, and others, have issued guidelines for defining safety performance metrics for the process industry. These guidelines have helped introduce standardized indicators globally. On behalf of industry associations in the International Council of Chemical Associations, the European Process Safety Centre (EPSC) and the European Chemical Industry Association (Cefic) organized a conference in Brussels, in early 2012, to discuss possibilities and problems with introducing

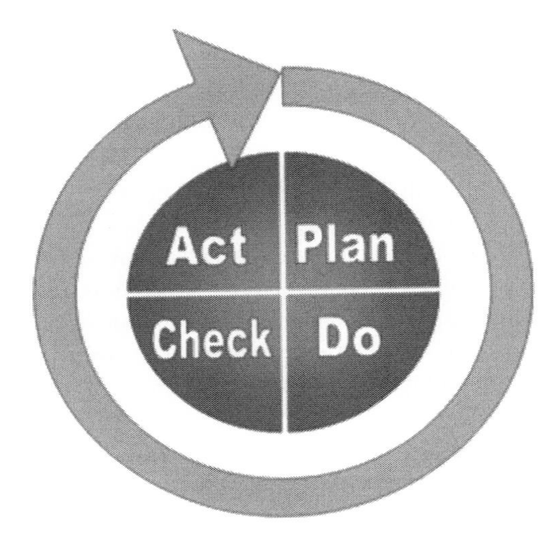

FIGURE 3.7

One of the many representations of Deming's management PDCA cycle. Obviously, monitoring performance shall occur in the "check" stage following which corrective action can take place. "Plan" shall also comprise "prediction."

indicators and promoting their use. The aim is to establish an international organization for standardization (ISO) standard for lagging indicators, while the choice of leading indicators is up to individual companies. This is partly because which leading indicators are most appropriate can depend on local organizational structure, responsibilities, and other conditions.

Criteria for lagging indicators will be proposed. Indicator values shall be derived from the number of unintentional Losses of Primary Containment of a hazardous substance, or energy, that cause:

1. Lost time injury (≥1 day) and/or fatality, or
2. Fires or explosions resulting in ≥€ 25,000 of direct cost, or
3. Substance release ≥defined release threshold quantity, which will be defined by the Globally Harmonized System (GHS) hazard class.

Three main groups of leading indicators can be distinguished:

1. Mechanical integrity (MI) indicators (inspections, controls, and test schedules of system components)
2. Indicators monitoring action item follow-ups of process hazard analysis (PHA) or of audits, monitoring the number of times procedures of management of change, or issuing of work permits are breached and functioning of other SMS element fails, of challenges to safety critical technical systems, and of recognized precursor and near misses, including unusual system and component behavior, which reveal new or previously unidentified scenarios that can lead to failure.
3. Training/competence indicators (quality testing: % people trained, number of completed roles in process safety (PS)).

CCPS's publication on process safety performance indicator metrics provides a wealth of possibilities.[17] In fact, as we shall see later in Chapter 6, so many indicators can be created that indicator aggregation is needed to maintain an overview for system management.

The last concept appearing in the evolution of loss prevention (Figure 3.1) is *safety culture*. It appeared suddenly on safety agendas, almost as a panacea. Other related concepts are safety climate and safety attitude, which are observed manifestations of the safety culture. Much has been written about safety culture, but it remains a concept difficult to define. Frank Guldenmund[18] presents an extensive overview, parts of which are published in the journal *Safety Science*. The cover figure of his dissertation of "blind monks examining a white elephant" (Figure 3.8) clearly expresses the difficulty of exactly describing the concept. It shows, however, how the elephant can be measured and characterized through information aggregation.

There is very little written about the observations of personal actions within an organization that are necessary in order to decide intuitively whether there is a sincere desire to take safety seriously or not. Safety culture is not an isolated construct; it stems from, and is embedded in, the organizational culture. Perhaps the oldest publications on the subject are by Dov Zohar[19] of Technion, Haifa, Israel (in 1980), who is still active in the field.[20] His focus is on safety climate, and he

FIGURE 3.8

Painting by known Japanese artist Hanabusa Itcho, seventeenth century, of blind monks examining an elephant (Wikipedia). This source also mentions several versions of the tale about it, the oldest an Indian one. The men touch and grope the elephant, but on the question of what it is that they feel, they all give a completely different answer, followed by a heated discussion.

developed questionnaire methods to obtain an objective measure of it. Safety climate is considered to be observable and measurable; culture is the underlying fabric of how humans think and react. He began a lecture for a Houston audience a few years ago with the warning, this is "difficult learning material for chemical engineers."

Again, it was the nuclear power community that first launched this concept following the Chernobyl disaster in 1986. Safety culture then picked up, and the number of papers published has been growing ever since. Further accidents such as with the Columbia space shuttle and the BP isomerization unit in Texas City made sure that it remained center stage in the spotlight. It is quite obvious that leadership attitudes to a large extent define the culture within an organization. Zohar also stresses that there need not be any relationship between good safety culture in a company and the presence of an abundance of safety warnings, slogans, and artifacts/paraphernalia. The level of safety culture is the result of the deeper drivers and motivation in the human mind. When he measures safety climate, Zohar probes through all levels of the organization supervisor—subordinate conversations. Alternatively, Patrick Hudson et al.[21] sponsored by Shell E&P developed an approach, and a set of training tools, to determine and improve the state of safety culture. The approach is called Hearts and Minds. It is freely available through the UK Energy Institute.[22] Figure 3.9 illustrates the levels of safety culture and the degree of maturity of management in a very convincing way. Paramount is the employees' perception of the value they are personally receiving by participation in the safety program. We shall return to safety culture in Chapter 6.

The conclusion so far is that the human element is the overruling factor in process safety. However, in order to communicate about risks and control them, the remainder of this chapter shall be dedicated to phenomenology of hazards, abstraction, and methods of quantification.

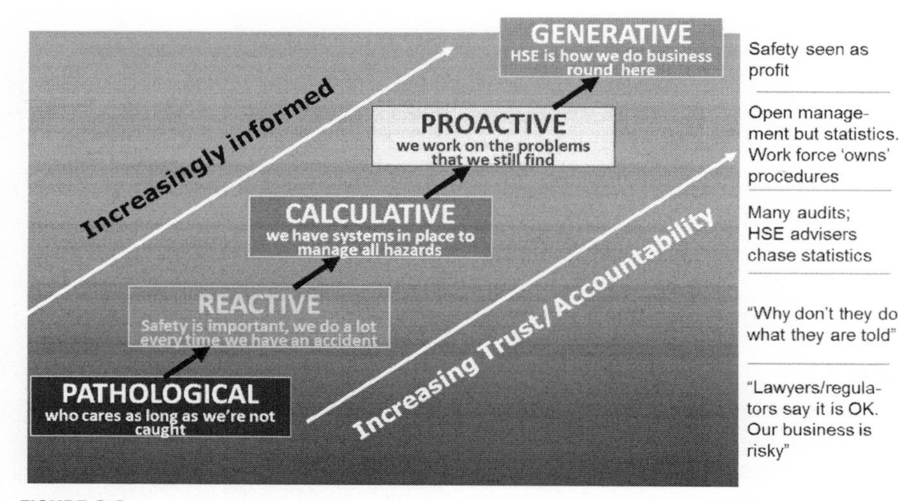

FIGURE 3.9

Culture ladder representative of the management maturity progress of the Hearts and Minds program and tool kit for safety culture improvement, originally developed for Shell International E&P by Patrick Hudson[21] and accessible via the Energy Institute, UK.[22] The text at each step level, including that at the right-hand side, is either typical as qualifying management statement or qualifies the state of the organization.

3.3 HAZARDS, DANGER, SAFETY, AND RISK

The concepts mentioned above all have different but related meanings, and in many languages no word equivalent to *hazard* exists. Similarly, for many people there may be no differentiation between the words *safety* and *security*. Therefore, some brief, perhaps oversimplified definitions are given below:

- *Hazard* is an inherent capability (energy, property, activity) for harm or damage to something that is valued; *danger* is a hazardous situation prone to result in harm or damage.
- *Safety* is a state of being protected against harm and other consequences of failure, while *security* is concerned with being protected, for instance, against terrorist or other criminal acts. Safety can be inherent in the absence of hazards; it can be engineered by applying measures and it can be procedural by developing and implementing rules. Much the same can be said of security.
- According to ISO 31000, Risk Management Principles and Guidelines,[23] risk is defined as the effect of uncertainty on objectives. In the present context, *risk* is interpreted as the combination of the severity of the consequence or impact resulting from the hazard causing it and the likelihood of that event. Both are associated with uncertainty, but are indicators of a lack of safety. *Consequence* means loss, which in principle can be expressed in monetary units, and

likelihood is the chance of occurrence of that consequence event; probability is a more exact term, and when an event occurs with a certain probability repeatedly over a certain time period, for example, a year, one speaks of a frequency or an event rate. Hence, in principle, risk can be quantified by providing a measure for a lack of safety. Safety itself is not quantifiable, although with sufficient preventive and protective measures taken, one can judge and declare a situation safe; based on certain material properties and amount, a (release) hazard is quantifiable in a relative sense (hazard classification).

In other words, a risk is the result of a potential (explosion, fire, toxic or corrosive spread) that in a specific situation by some event, by a chain of events, or by a multiple of parallel events (the trigger) can become uncontrolled and do harm or damage to exposed people or vulnerable assets. The entire course of cause—consequence events from an initiation, through intermediate events, and ending in an outcome event is called a *scenario*. For any risk, a scenario can be described; distinct ones are a *worst-case* scenario and a *maximum-credible* accident. Part of the intermediate events in a scenario can consist of failure of preventive barriers intended to forestall the potential of becoming uncontrolled, or of protective barriers designed to deflect the potential to a safe outcome or location, or at least reduce its power in order to attenuate effects on exposed targets, which is called mitigation. In fact, a main preventive barrier is formed by a feedback loop of an operator supervising the process (in which many automated controllers may also act) and keeping it within an envelope of safe working conditions. At the base of the scenario can be many possible root cause failures, which may not be known at all to the operator but involve HOFs. The safety control of a process plant is therefore analogous to other complex technical installations, such as in a car or an aircraft.

Systems in which humans and machines interact in a complex way are called *sociotechnical systems*. We shall consider safety as a system property in Chapter 5. For a more comprehensive and elaborate set of definitions, we refer to the ISO 31000 standard and the corresponding ISO Guide 73 vocabulary.[24] Comments on this relatively new standard are given by Terje Aven.[25] We shall return to this much discussed area in Chapter 12, as it still generates many questions in the context of risk communication and decision making.

3.4 ACCIDENT INVESTIGATION TOOLS

It is rather obvious that it is important to investigate the causes of an accident. Without the "what, how and why" no correction action or future prevention is possible. As we have seen, a SMS goes a step further and requires that an incident must be investigated, and this includes a near-miss event. The purpose is to learn and improve. Even better is to identify precursors such as defects, disruptions, and deviations that under perhaps slightly different conditions can initiate a cause—consequence chain to serious effect. Precursors are closely related to a failure process, and certainly when they prove to be possibly affecting the function of safety critical components and to be occurring in larger numbers, they should be corrected. The American safety pioneer Herbert W. Heinrich,[26] creator of the falling domino stone accident causation model (1941), also devised the accident pyramid (1929). The pyramid metaphor tells us that for one real accident there may be many near misses and even more precursors. So, while fortunately accidents do not occur that often, for prevention it is critically important to pay attention to and learn from near misses and precursors. In Chapter 7, we shall see how their frequency of occurrence can be used to predict a rare serious event probability.

As mentioned, the Swiss cheese model presented in Figure 3.2 has been applied as a guide to accident investigation, while later in the 1990s a formal method (Tripod Beta) was derived from it. An older (1975), very detailed but most useful and complete American accident investigation tool is the Management Oversight and Risk Tree, MORT, proposed by Johnson,[27] Systems Safety Development Center of the US Department of Energy. It branches out as a fault tree down from a top event of various kinds of damage via a large variety of types of technical and organizational management oversights that may be less than adequate (LTA). The ones that are LTA allow the accident to occur by a combination of a potentially harmful energy flow or environmental condition, inadequate barriers and controls, and exposure of vulnerable people and objects. In Figure 3.10 the top part with the part below the accident are shown. The tree is, however, much too large to show! Long before the process safety community looked at management decisions as causes of accidents, Johnson's MORT proved to be an advanced tool for finding failing controls and mechanisms leading to accidents. Application of the approach to improve safety spread only slowly. Figure 3.11 illustrates in a simple way the basic elements of an accident scenario.

A 2001 HSE report[28] provides a useful overview of principles and cause—consequence charting, while discussing some 20-odd methods that have been developed for accident investigation. Many so-called *root cause analysis* (RCA) methods have characteristics that are reminiscent of the MORT approach and are computer supported. They can also be applied in a proactive a priori, rather than a posteriori, sense. A simple qualitative fault tree is always helpful when one has to make a quick diagnosis with an investigation team. Chris Johnson[29] wrote a thousand-page handbook of incident and accident reporting. Joseph Saleh et al.[30] of the Georgia Institute of Technology produced a review of accident-causation models with emphasis on a multidisciplinary approach—psychologically, socially, and from a systems point of view, and the cross-fertilization with PRA. In a later paper,[31] differences between near misses, precursors, "accident pathogens," and warning signals are considered, and the concept of "accident pathways" and a dynamic time- and discrete event-driven modeling are proposed. Noteworthy is the RCA approach, IPICA_Lite, by Milos Ferjencik,[32] University of Pardubice, Czechoslovakia, following a similar line of thought with respect to a cause chain, while it ties underlying causes to failure of subelements of the 21 main SMS elements defined by CCPS as shown in Figure 3.3.

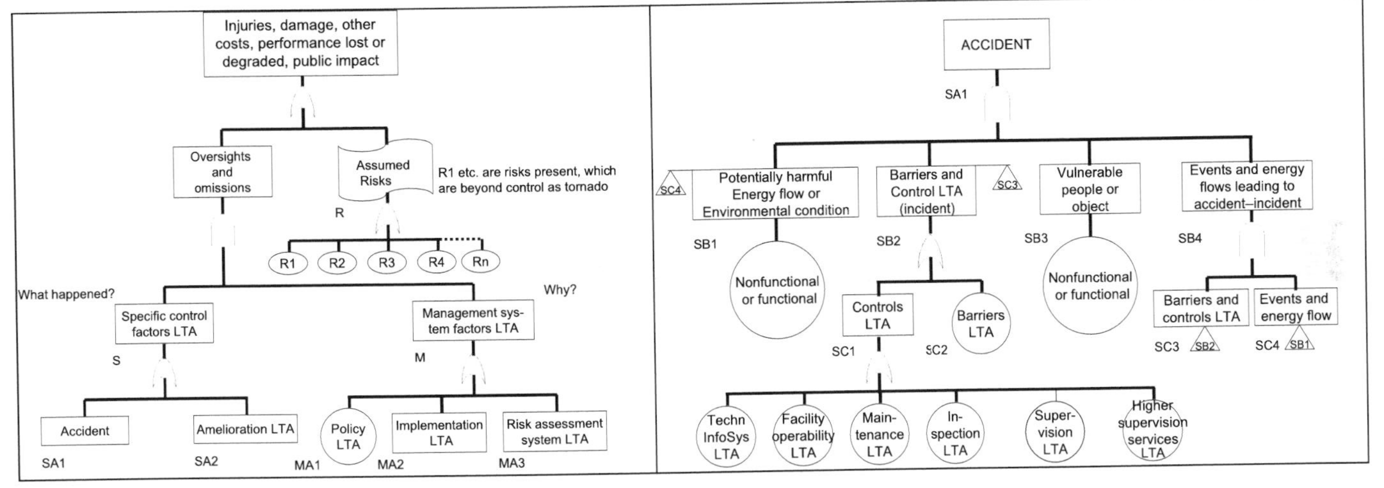

FIGURE 3.10

Left: The top part of William G. Johnson's Management Oversight and Risk Tree[27] showing what management could have done to minimize damage. LTA means less than adequate. At the bottom left is the actual accident. In the middle right are the events beyond control ("acts of God") but that could be predicted. Right: The accident part of the tree further expanded showing the hazard potential (energy/release), barriers, exposed receptors, and a trigger. A full tree requires at least one square meter of paper.

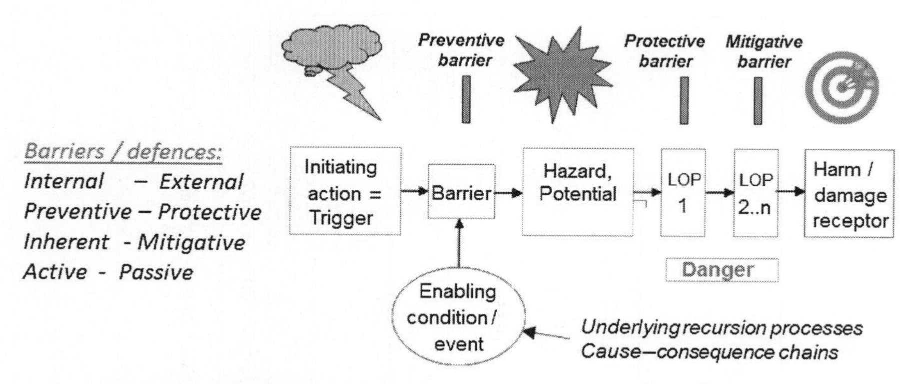

FIGURE 3.11

Illustration of elements of a simplified accident scenario including hazard, a barrier preventing a trigger from setting the hazard off, layers of protection (LOP), and a vulnerable receptor. The conditions and events undermining the preventive barrier are sometimes rather intricate cause—consequence and dependency chains. Types of barriers are listed, as well as technical and organizational measures, which form the basis for the barrier function.

In Chapter 5, we shall discuss the more comprehensive systemic approach of Jens Rasmussen's Accimap and Nancy Leveson's STAMP method (with the associated references).

3.5 KNOWLEDGE AND TOOLS: HAZARDOUS SUBSTANCE PROPERTIES, SYSTEM SAFETY, PROCESS TECHNOLOGY

Safety depends very much on knowledge of what can go wrong and how to prevent this from happening. Overall, the field is broad, so a safety "specialist" may be T-shaped that is broad and aware of the many technical, human, organizational, and management aspects, but with a deep-down specialization in one or more aspects of the field. The three main knowledge areas in process safety (viz. properties of substances, system safety, and process technology and operation) are shown in Figure 3.12. This is against a background of knowledge sources required to accomplish conceptualization, design, construction, and operation of large technical installations. In the center, a diffusive circle is sketched representing knowledge of risk analysis methodology and data. Risk analysis of a plant, and an assessment of the risks revealed, is a key foundation of decision making and risk management and comprises a host of methods and data. Risk analysts make use of the underlying data from the three main areas and also raise challenges for further research in these areas. All four knowledge hubs will now be discussed.

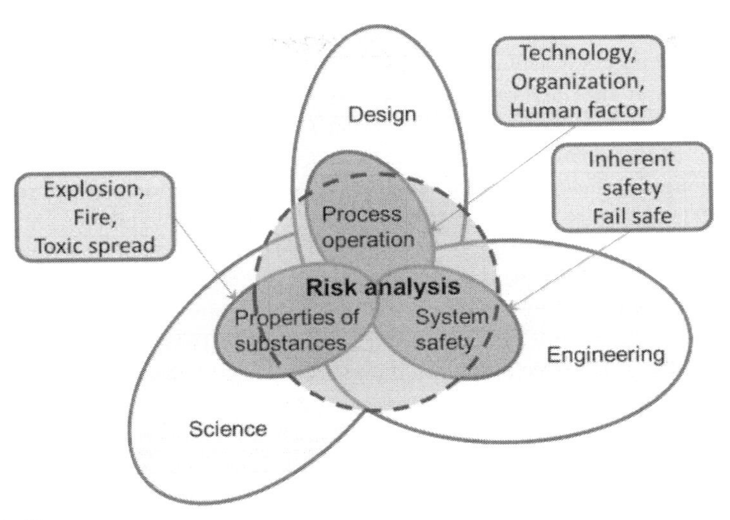

FIGURE 3.12

Knowledge fields in process safety against the background of general sources for technology creation; risk analysis overarches the three basic process safety fields.[33]

3.5.1 PROPERTIES OF SUBSTANCES, UNINTENTIONAL RELEASE PHENOMENOLOGY, AND DISPERSION

The field *properties of substances* pertain to a broad spectrum of substances and investigation methods. The latter serve to determine chemical and physical properties of solids, liquids, gases, and dust/aerosols, with respect to leak and dispersion, decomposition mechanisms and reaction kinetics at various conditions, combustion and explosive behaviors, and sensitivities to various stimuli. Further, possible harm and damage by fire and explosion, toxicity or corrosiveness of substances, and effect of radioactive radiation have to be known. Here, because of short-term effects of interest to risk analysis of release events, we shall focus on acute harm from hazards. Chronic or long-term effects are of interest to occupational safety and health and of environmental protection, and are more particularly the work field of industrial hygiene.

The instruments and tools to determine hazardous properties are tests, theoretical/analytical models, and computational methods. The latter can be on various scales, for example, atomic scale with quantum mechanics models, molecular scale with molecular dynamics, microscale with direct numerical simulation by computational fluid dynamics (DNS CFD), and macroscale with large eddy simulation (LES) CFD. While testing can be expensive and mixtures can have near limitless varying compositions, the approach of quantitative structure–property relationship (QSPR) has evolved. In this, the contribution of molecular structural groups to a certain property as, for example, the flash point of a flammable liquid, is determined by regression applying statistical methods or even better by (artificial) neural networks (ANNs) (nonlinear) to measured values of known pure substances. ANNs cope with nonlinearity and can produce higher correlation coefficients. QSPR is performed with the objective of predicting properties of substances or mixtures for which no experimental test results are available.

Many countries have developed standards on how to determine certain properties such as flammability or explosion limits of gases and dusts, required vent opening areas to mitigate explosion effects, maximum explosion safe gap, and many more. In Europe, standards bear the designation EN, short for European Norms, while in the US, depending on application area, the National Fire Protection Association (NFPA), American Society for Testing and Materials (ASTM) and the American Petroleum Institute (API) which provide among others hazardous property testing standards and norms. A major step forward has been the development of the GHS of classification and labeling of hazardous substances in conjunction with the United Nations (UN) test methods, as discussed in Chapter 2. Today, much data can be found on the Internet in material safety data sheets and as a result of the European REACH project on the European Chemical Agency data bank with much background information.

For many years the well-known, symbolic sign of hazardous properties classification in the US has been the NFPA 704 diamond,[34] seen on trucks and rail tankers transporting material or on bottles or other packaging containing a material. An example is shown in Figure 3.13, top left. There are a few notes to be made with respect to the diamond classification. Being formulated by the NFPA means that the signs are of special interest to firefighters. It may be accompanied by any recommendation for personal protection equipment that should be used. The newer GHS has different pictograms, as illustrated at the top right of Figure 3.13. GHS is global, is law, and supersedes other hazard classification standards by June 1, 2015. Principles, concepts, signal words, and pictograms of GHS can be found in Chapter 2.

Toxicity and also flammability are usually a greater hazard if the substance is capable of higher mobility. Therefore, gases are to be found in the highest blue and red NFPA hazard classes 4 rather than solids. Discharges of hazardous chemicals to the atmosphere involve a host of physical processes depending on physical state—gas, liquid, two-phase flow, cryogenic, solid, and the environment of the release—soil, water, elevated release, weather, air humidity, presence of bunds/dykes, and confinement by buildings. Much work has been done to describe and model the processes occurring during and just after the release, such as pool formation, boiling, evaporation, cloud formation, and drifting and dispersion of the cloud, and also to validate such models with laboratory and field experiments.

The models are needed for describing the *source terms* of scenarios in risk assessment, which determine the physical effects of a release. In particular, dispersion of gas and vapor clouds has been studied extensively, and still is, to determine concentrations as a function of time at selected locations, the effectiveness of water or steam curtains to enhance dispersion, and for optimally locating gas detectors, for

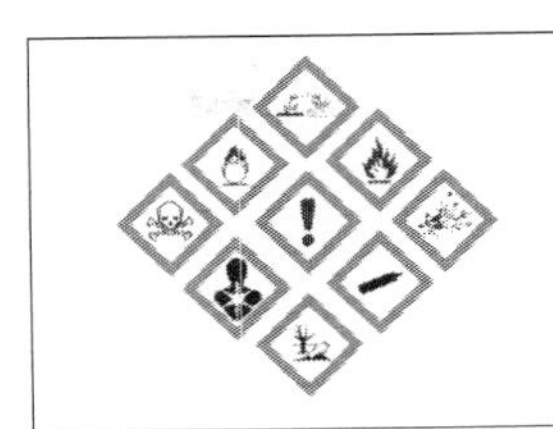

From top to bottom and left to right

Corrosive
Flame over circle: Oxidizers
Flammable, pyrophoric, self-heating, etc.
Skull & crossbones: Acute toxicity
Exclamation mark: Attention hazard
Explosive; Self-reactive; Organic peroxide
Health hazard
Gas under pressure
Environment: Aquatic toxicity

Health risk: toxicity (blue)

Symbol Number	Meaning	Example
Blue - 0	Does not pose a health hazard. No precautions are necessary.	water
Blue - 1	Exposure may cause irritation and minor residual injury.	acetone
Blue - 2	Intense or continued non-chronic exposure may result in incapacitation or residual injury.	ethyl ether
Blue - 3	Brief exposure may cause serious temporary or moderate residual injury.	chlorine gas
Blue - 4	Very brief exposure may cause death or major residual injury.	carbon monoxide

Flammability (red):

Symbol Number	Meaning	Example
Red - 0	Will not burn.	carbon dioxide
Red - 1	Must be heated in order to ignite. Flash point exceeds 90°C or 200°F	mineral oil
Red - 2	Moderate heat or relatively high ambient temperature is required for ignition. Flash point between 38°C or 100°F and 93°C or 200°F	diesel fuel
Red - 3	Liquids or solids that readily ignite at most ambient temperature conditions. Liquids have a flash point below 23°C (73°F) and boiling point at or above 38°C (100°F) or flash point between 23°C (73°F) and 38°C (100°F)	gasoline
Red - 4	Rapidly or completely vaporizes at normal temperature and pressure or readily disperses in air and readily burns. Flash point below 23°C (73°F)	hydrogen, propane

Reactivity (yellow):

Symbol Number	Meaning	Example
Yellow - 0	Normally stable even when exposed to fire; not reactive with water.	helium
Yellow - 1	Normally stable, but may become unstable at elevated temperature and pressure.	propene
Yellow - 2	Changes violently at elevated temperature and pressure or reacts violently with water or forms explosive mixtures with water.	sodium, phosphorus
Yellow - 3	May detonate or undergo explosive decomposition under the action of a strong initiator or reacts explosively with water or detonates under severe shock.	ammonium nitrate
Yellow - 4	Readily undergoes explosive decomposition or detonates at normal temperature and pressure.	TNT, nitroglycerine

Special hazards (white):

Symbol Number	Meaning	Example
White -OX	oxidizer	hydrogen peroxide, ammonium nitrate
White - ~~W~~	Reacts with water in a dangerous or unusual way.	sulfuric acid, sodium
White - SA	simple asphyxiant gas	Only: nitrogen, helium, neon, argon, krypton, xenon

FIGURE 3.13

Left: NFPA 704 substance hazard diamond with its meanings in the table; example shown is that for lithium hydride. Right: Pictograms of the new GHS labels (see Chapter 2).

example, on an offshore platform. *Gas dispersion* could be discussed in Section 3.6.3 on consequence analysis, but in order to better understand the following sections on toxics and vapor cloud explosion, it will be treated here. A special problem in predicting dispersion is that many chemicals disperse as heavy gases. Heavy toxic and flammable gases are particularly hazardous because their properties of staying long near the ground and in low-wind speed condition diluting slowly.[35] Even gases that at ambient temperature are lighter than air, such as ammonia and methane (natural gas) that are often stored as refrigerated liquid (LNH3, −33.35 °C) or cryogenic (LNG, −161.5 °C), on release boil off immediately producing cold heavy gases, which after some time warm up, mix with air, and rise. Because of expansion in the leak hole and consequent cooling due to the Joule−Thomson effect, sometimes two-phase flow or condensation occurs, and when in free air rainout takes place. In some cases the substance, such as hydrogen fluoride, reacts with moisture in the air, which necessitates more complex descriptive models. Another special case is carbon dioxide, which when released in supercritical state under high initial pressure, sublimes upon discharge.

In a number of countries such as the US and Germany (see Chapter 2), regulation to determine safe distances is based on concentration thresholds. Hence, there is significant interest to be able to predict this distance with accuracy and confidence. Initially, only a neutrally buoyant gas dispersion model was available, the Gaussian plume model. It was empirically validated and was called Gaussian because the concentration profiles are shaped like the bell form of a normal probability density function or distribution. Later, in the 1980s, as a second generation, integral heavy gas models appeared, but the spread in their results was large, at least an order of magnitude. Wind tunnel and field experiments were necessary to study the effects of atmospheric turbulence and the way wind entrained the volatile chemical and diluted the cloud. These models were suitable for flat terrain only.

In the 1990s, more sophisticated CFD software became widely available. This can simulate these subprocesses in some detail and take into account the three-dimensional effects of buildings and terrain topology. But the spread in predictions was still large because of the complexities in the physics, in particular the turbulence effects, and the approximations in the models. CFD subdivides a domain or space of interest into many cells. Then, within each cell it solves the Navier−Stokes flow describing conservation equations of mass, momentum, and energy in partial differential form, together with the equation of state for the fluid and equations describing turbulence and energy generation by chemical conversion and dissipation. Solving the set of equations is accomplished by various finite difference techniques for various applications ranging from atmospheric dispersion, engineering unit operations (drying, extracting, mixing) to reactive flows (propagating flames, chemical reactors) and high-velocity projectile penetration.

A CFD code leaves sufficient parameters undefined, which have to be user specified, sometimes empirically. In the middle of the 1990s, an EU Model Evaluation Group was established, which started project SMEDIS, Scientific Model Evaluation of Dense Gas Dispersion Models, resulting in a protocol.[36] This protocol verifies the equations and validates the computational results of both integral and CFD codes against selected, representative field experiments. This protocol helped to increase the quality of the outcomes, but they are still far from being perfect, certainly in built-up environments, and deviations by at least a factor of two in concentrations as a function of distance are rather common. This is also true for the more recent CFD codes as shown by Plasmans et al.[37] Very low wind velocity conditions form another limitation, in particular with gases that are heavy relative to air. Improvements are continuing.

All the above subjects are extensively treated in *Lees' Loss Prevention in the Process Industries*.[38] Much less detailed but still covering the field comprehensively is Crowl and Louvar's textbook.[39] Commercially available software tools for gas dispersion calculations for given substances and conditions are available, such as DNV's PHAST UDM model; TNO's EFFECTS adapted the SLAB model, while the vapor cloud explosion CFD code FLACS of Gexcon AS is also validated for dispersion. Concentration predictions for validated codes shall for the most part be accurate within a factor of two. The simple, free downloadable computer code ALOHA (see Section 3.6.6.6) developed by the National Oceanic and Atmospheric Administration is distributed by Environmental Protection Agency in the US for the primary purpose of obtaining quick, albeit rough, estimates for emergency response warning and alarms. All these codes are easily traceable via the Web.

Jürgen Schmidt,[40] University of Karlsruhe, Germany, edited a compilation of a variety of process and plant safety applications of CFD, which besides dispersion cover subjects such as explosion, flow (also two-phase) from high-pressure relief valves, water-hammer, LNG spills and dispersion, large-scale hydrocarbon and peroxide pool fires, fire scenarios and smoke migration within structures, and dynamics of plant upsets and disturbances in distillation.

3.5.1.1 Toxicity

Acute toxicity data are required for risk assessment and emergency response. Various systems have been developed by different organizations. For the purpose of risk assessment, acute toxicity data have been collected from various sources, such as wartime actions, accidents, in vitro tests, tissue tests, and animal trials supported by physiologically based pharmacokinetic (PBPK) modeling. For a substance to have a toxic effect, it has to have mobility. Asbestos fixed in a substrate such as cement is not hazardous as long as it is not loosened from its setting as in a fire and then inhaled as dust (although in apparently "fixed" condition some migration may also take place). PBPK is concerned with transport of the agent within the body, absorption at the organ, tissue or fluids where it is damaging, and the local metabolism kinetics. These are all important factors in determining the potential impact of the hazard.

Because response depends on dose and concentration of the agent, results are presented as cumulative normal probability distributions of lethal response fractions (R) versus exposure intensity. The latter is a combination of concentration of the substance and exposure time. As shown in the graph of Figure 3.14, the cumulative distribution response function is linearized by transformation into probits (probability unit or Pr) by means of the equation:

$$R = \frac{1}{\sqrt{2\pi}} \int_{-\infty}^{Pr-5} e^{-u^2/2} \mathrm{d}u \tag{3.1}$$

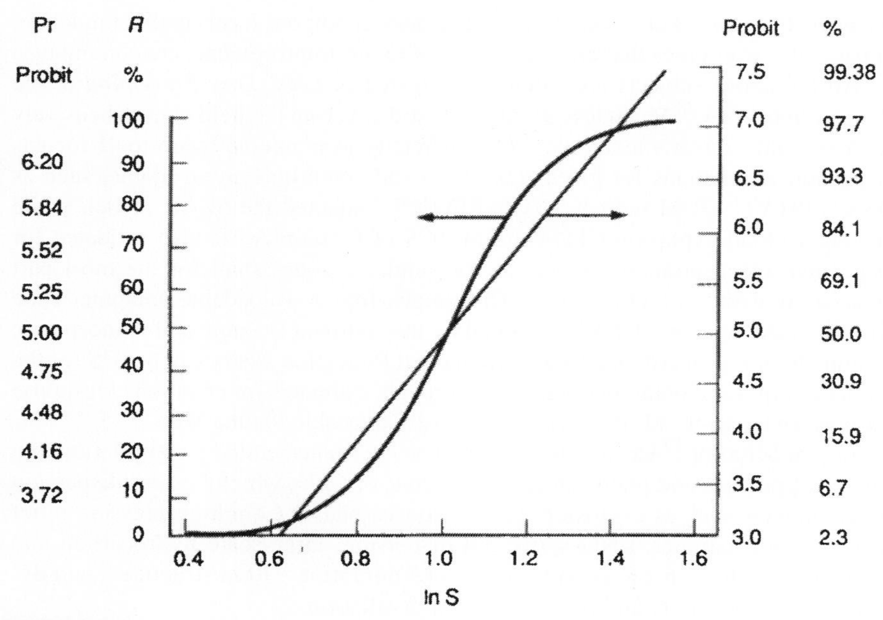

FIGURE 3.14

The graph shows response fraction, *R* and corresponding probit values, *Pr* versus logarithm of dose, *S*, or rather $C^n t$. In risk assessments usually only the 50% point is used instead of the entire distribution. (This is unfortunate because information is lost that otherwise could have been used to predict, e.g., conservatively.)

This equation illustrates the relation between probits and probability values R in percent. Toxicity data mentioned are then fitted to the linear probit equation in the form we saw already in Chapter 1:

$$Pr = a + b \ln(C^n t) \tag{3.2}$$

Constants a, b, and n in this equation are substance specific, whereas C is the toxic substance concentration and t the exposure time. A probit of value 5 is lethal for 50% of the population exposed. The concentration at this 50% fraction is known as LC_{50} and the dose, $C \times t$, as LD_{50}. Probit 2.67 corresponds to a lethal dose for 1% of the exposed population and probit 7.33 for 99%. Probit tables are readily available. All this is for an average person. The range of human susceptibility to an agent is large. The elderly and children are usually more vulnerable. Because of this and because of uncertainties in the average value determination, overall uncertainty in outcomes is substantial. Through international cooperation, probit coefficients are still being updated. For illustration purposes, values used in risk assessments for a few substances are shown in Table 3.2; these are taken from a 2009 reference manual of the Dutch Institute for Public Health and the Environment, RIVM.[41] By means of

Table 3.2 A Sample of Probit Equation Coefficients for Use in Risk Assessments

Substance (*C* in mg/m^3; *t* minutes) (mg/m^3· 24.5/mol. weight in g)· 10^{-6} = ppm	*a*	*b*	*n*	ERPG1 ppm	ERPG3 ppm	AEGL3 ppm 1 hour exposure	TLV TWA ppm
Acrylonitrile	−8.6	1	1.3	10	75	28	2
Ammonia	−15.6	1	2	25	750	1100	25
Chlorine	−6.35	0.5	1	20	20	20	
Phosgene	−10.6	2	1	N.A.[a]	1.5	0.75	0.1
Carbon monoxide	−7.4	1	1	200	500	330	25
Nitrogen dioxide	−18.6	1	3.7	1	30	20	0.2
Hydrogen cyanide	−9.8	1	2.4	N.A.	25	15	–
Hydrogen fluoride	−8.4	1	1.5	2	50	44	0.5
Hydrogen sulfide	−11.5	1	1.9	0.1	100	50	1
Sulfur dioxide	−19.2	1	2.4	0.3	25	30	–

[a] *N.A. = Not appropriate.*

the coefficient values, LC_{50} concentrations at any selected exposure time can be calculated at $Pr = 5$. RIVM is continuously updating information and published a status report[42] in 2013.

For emergency response purposes with respect to volatile releases, threshold values for alarm warnings and evacuation have been developed. Two main systems exist: Emergency Response Planning Guidelines (ERPGs) and Acute Exposure Guideline Levels (AEGLs). Both systems are intended for the protection of the population at large, and both distinguish three levels of concentration. ERPG[43] defined these levels as follows:

- ERPG-1 is the maximum airborne concentration below which it is believed that nearly all individuals could be exposed for up to 1 h without experiencing other than mild transient adverse health effects or perceiving a clearly defined, objectionable odor.
- ERPG-2 is the maximum airborne concentration below which it is believed that nearly all individuals could be exposed for up to 1 h without experiencing or developing irreversible or other serious health effects or symptoms that could impair an individual's ability to take protective action.
- ERPG-3 is the maximum airborne concentration below which it is believed that nearly all individuals could be exposed for up to 1 h without experiencing or developing life-threatening health effects.

ERPG levels are established by the American Industrial Hygiene Association. AEGLs are issued by EPA. For AEGL[44] there are slight differences in definition as the concentrations limits have been reversed and are now given as "above which," et cetera; another difference is the time period for which a value is given—AEGLs are for 10 min, 30 min, 1 h, 4 h, and 8 h. Thus:

- AEGL-1 is the airborne concentration, expressed as parts per million or milligrams per cubic meter (ppm or mg/m^3) of a substance above which it is predicted that the general population, including susceptible individuals, could experience notable discomfort, irritation, or certain asymptomatic nonsensory effects. However, the effects are not disabling and are transient and reversible upon cessation of exposure.
- AEGL-2 is the airborne concentration (expressed as ppm or mg/m^3) of a substance above which it is predicted that the general population, including susceptible individuals, could experience irreversible or other serious, long-lasting adverse health effects or an impaired ability to escape.
- AEGL-3 is the airborne concentration (expressed as ppm or mg/m^3) of a substance above which it is predicted that the general population, including susceptible individuals, could experience life-threatening health effects or death.

Data can be found on the websites given in the references above. The number of substances for which ERPGs or AEGLs have been established is not large, less than 150. Additionally, the US Department of Energy issues temporary emergency exposure limits,[45] which are given according to a standard procedure, which is less elaborate than the one for ERPG or AEGL but with the advantage of encompassing 3000 chemicals. However, different organizations introduced other measures, such as short-term exposure limit (STEL) in the Occupational Health and Safety (OSH) domain, and immediately dangerous to life or health defined by the US National Institute for Occupational Safety and Health for exposure to airborne contaminants. See Table 3.2 for some 2013 ERPG and AEGL values.

Nonacute but chronic toxic limits for work environment are known in the US as threshold limit values (TLV), which are established by the American Conference of Governmental Industrial Hygienists (ACGIH).[46] (Companies must use the latest data.) A distinction is made between 8 h, 40 years' time weighted average (TWA) concentrations exposure (see Table 3.2), short-term exposure limits (STEL, 15 min) and ceilings (TLV-C). In Europe, with the same objective the Scientific Committee on Occupational Exposure Limits (SCOEL)[47] publishes recommendations. For example, AGCIH (2012) and SCOEL quote for ammonia TWA 25 and 20 ppm, and STEL 35 and 50 ppm, respectively. Another source is the OECD EHS Newsletter. For a number of toxic substances a no observed effect level (NoOEL) can be found. Quite a few others show a hormesis response, which means that very low concentrations of toxic substance can have a beneficial effect on a receptor rather than be detrimental. It is suspected that hormesis holds also for radioactive radiation. The International Commission on Radiological Protection[48] recommends 1 mSv/year (milli-Sievert per year) nuclear radiation level as safe to the public, and to workers 20 mSv/year averaged over 5 years with no 1 year higher than 50 mSv.

Nonfatal injury data, of importance in risk analysis for emergency response planning (triage classification) and estimation of impaired self-rescue abilities (mobility hampering), are lacking.

3.5.1.2 Explosion types and mechanisms

Gas and vapor explosions often followed by prolonged fires are notorious for the damage they cause in refineries and other process and storage plants. Also, fires in warehouses can do much damage. Let us first consider various types of explosions, because the mechanisms can be rather confusing. An overview is presented by Abbasi et al.[49] and Figure 3.15 is adopted from their work. A main distinction shall be made on the basis of the energy supply for an explosion. It can be physical when the material has a high internal energy potential, such as a fluid under pressure or a chemical when involved in an exothermic reaction of the substance or mixture.

The type of physical explosion with the most impact is the boiling liquid expanding vapor explosion (BLEVE). In a BLEVE energy is supplied as heat to a vessel filled with a liquid; it may just be water, which after reaching its boiling point (BP) will rise in pressure. If this continues, then at a certain point in time the vessel wall ruptures, and a BLEVE occurs producing a physical blast. Steam boiler explosions have been frequent. The strength of the blast depends on many factors such as vessel strength, filling degree, and quite a few others. Because the heating may be due to a fire resulting from a leak in a fuel line, many BLEVEs occur with a flammable material, for example, with propane or butane-filled cylinders. We have seen

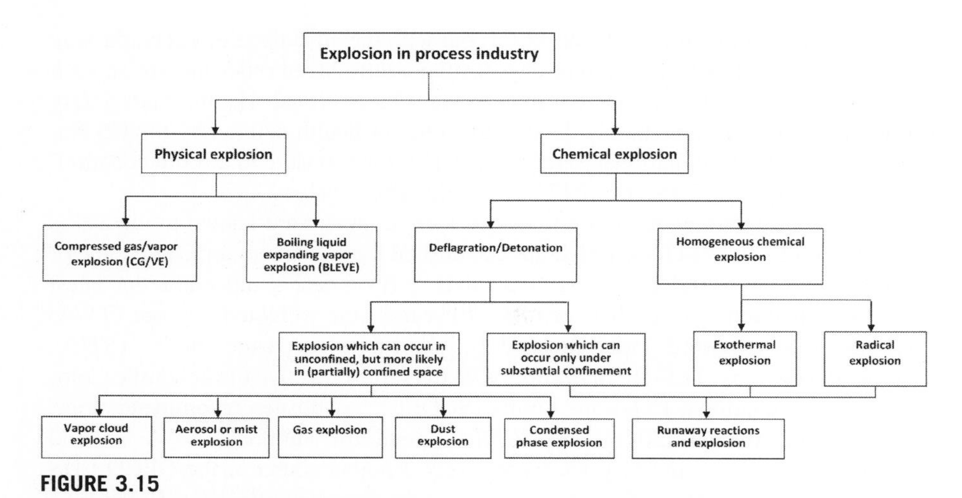

FIGURE 3.15

Various types of explosions that can occur in industrial plant, near leaking pipe lines, during tanker transport on road or rail, at collision accidents in tunnels, and in ship holds.

the BLEVE phenomenon of liquefied petroleum gas (LPG) in Chapter 1. As mentioned there, a common secondary effect with a flammable liquid is that at rupture beside a boiling pool a cloud of hot gas ignites, and a rising flame ball develops, which can be even more hazardous than the blast. Numerous firefighters have succumbed due to a BLEVE, which often develops 10–30 min after a fire causes it. For risk assessment, a distinction is made between a cold and a hot BLEVE, the former due to puncturing of a tank in a collision or by a fragment and the latter by a fire resulting in a higher pressure at failure.

The rapid phase transition is a rare and relatively weak phenomenon, which occurs, for example, when LNG containing some ethane and propane spreads over water and explosively vaporizes.

Considering ammonium nitrate accidents in Chapter 1, we have seen how a chemical explosion can follow three possible mechanisms. It depends on whether the reaction is taking place throughout the substance or in a propagating reaction zone. The first type, the one having a reaction throughout the substance, is called a homogeneous explosion in which everywhere the reaction rate is essentially equal. When, however, as in most cases, rates are highest at the hottest location, often the center, it is called an (*exo-*) *thermal explosion*. In solids and liquids the phenomenon is called a *runaway* reaction. All so-called self-reactive substances are capable of this type of explosion albeit in different degrees of intensity, but also many mixtures usually consisting of an oxidizing agent and a reducing (fuel) component. Typically, quite a few self-reactive substances show acceleration of decomposition (that is increased instability) over time, or rather the products of some initial decomposition accumulate and then accelerate further decomposition. This effect is called *autocatalysis*; for example, the thermal decomposition of sodium hydrosulfite follows this type of mechanism. It can occur via the gas phase, for example, by acidifying gaseous decomposition products. Melting, hence loss of crystal structure, also accelerates decomposition.

Radical explosions occur in some gas mixtures when propagation by radical reactions is accelerating, because the reactions produce a net increase in radical concentration. A known example is the rather violent explosion of hydrogen and chlorine, or of hydrogen and oxygen, where initiating radicals can be generated by UV radiation in sunlight. Other less violent examples are hydrocarbon-oxygen cool flames due to peroxide formation and decomposition. In many instances, the distinction between deflagration and exothermal explosion is not very sharp. Depending on conditions, a developing thermal explosion may lead to deflagration of the remaining material. Confinement and external heating from a fire influence the combustion mechanism.

The other two types of explosion, deflagration and detonation, have very different zone propagation speeds due to the underlying mechanism of energy transfer to the substance just ahead of the reaction front, which has to heat up to start reacting. In both cases, propagation will cease due to lateral energy loss if the charge diameter becomes too small. In gases and dusts the fuel–oxidizer mixture ratio where no flame propagation occurs anymore is called the lower and upper flammability (US) or explosion (Europe) limits. *Deflagration* is the slower type; it may progress at a velocity as low as 10 cm per hour in some compound fertilizer formulations and stays subsonic as energy transfer is by heat flow. In gases and dust–air mixtures, one has to distinguish flame laminar *burning velocity* relative to the unburnt mixture (which for many mixtures is of the order of 0.02–0.05 m/s but in reactive cases can reach 3 m/s) and *flame speed* to which the turbulent, expanding, hot, reacted gas is contributing and that can go up to, for example, 600 m/s. At stoichiometric composition, all fuel and oxygen can react to H_2O (water vapor) and CO_2 (carbon dioxide). Optimum explosion condition is just over to the fuel-rich side. Hydrocarbon–air mixtures expand at maximum roughly nine times in volume at standard conditions, while for the hotter combustion of metal powders the analogous factor may be as high as 12. If vessels are connected by a pipe, a phenomenon can occur known as pressure piling. A deflagration in one of the vessels pushes unburnt mixture through the pipe to the second one, so that a deflagration there starts at a higher pressure and hence produces a higher pressure transient, which may be enough to rupture a vessel wall.

Detonation is fast, even supersonic, as a result of energy transfer by compression in a reactive shock wave. Detonation velocities range between 1500 and 10,000 m/s depending on density, speed of sound, and energetics of the material. Pressures can be up to hundreds of kilobars for solids and liquids, but for gases or dust–air mixtures initially at atmospheric pressure the maximum pressure will not exceed some 20 bar (with pressure wave reflection this maximum may at least double). Energy released in or before the Chapman–Jouguet (C–J) plane behind the shock front contributes to the detonation velocity. Together with the equation of state, the C–J condition describes in gross terms the shock and detonation physics. Ammonium nitrate, mentioned in Chapter 1, is relatively slow reacting, so in the C–J plane only part of its energy is released, and its detonation velocity and pressure are therefore much lower than for high explosives, such as TNT. Further details can be found elsewhere, for example, in Lees.[38]

In almost all solid and liquid cases, *deflagration* burning rates increase with pressure. This is because pressure compresses the reaction zone, which steepens the temperature gradients and enhances heat transfer to the fresh material ahead of the zone. As reaction products are almost always at least partly gaseous, confinement always tends to increase the explosion intensity. In gases, as a result of kinetic effects the effect is opposite: the burning rate reduces with pressure.

All types of explosions have their own specific testing methods. Simple test methods have been listed by the *UN Manual of Tests and Criteria*[50] (see Chapter 2 and the summary given in Table 3.3 top part). The tests must be considered in connection with the test schemes and criteria for classification. The lower part of Table 3.3 mentions "Screening tests" and "Fundamentally based tests" without specifying brand names or acronyms under which some of the methods are known. The latter are methods designed to measure a single property by creating a well-defined explosion type such that tests are highly sensitive, accurate, and precise. More detail on thermal stability testing of solids and liquids, in particular for the benefit of small and medium enterprises (SME) with respect to the runaway hazard, is available from the EU HarsNet[51] (Thematic network on hazard assessment of highly reactive systems) project. HarsNet is integrated into the European process safety S2S net,[52] which was set up particularly for SMEs.

Gas and dust explosion testing is according to ASTM and EN standards. For designs and safety measures of installations in which gas or dust explosions can occur, it is therefore important to determine explosion indices such as explosion pressure and rate of pressure rise and their maximum values depending on fuel−oxidizer ratio, flammable limits, minimum ignition energy, minimum oxygen concentration, and maximum safe steel gap that a flame cannot pass. In principle, all except the last one can be found by tests performed in a 1 m^3 vessel. At ignition, in the center a flame ball (outside reacting flame sheet/zone−inside hot combustion gas) expands while pressure rises. A 20-L spherical vessel is, however, a better manageable alternative with respect to cleaning and laboratory space. Ignition conditions are then adapted to produce an equivalent result to the 1 m^3 vessel when applying the cube root volume ratio normalization correction. Depending on the case the method may lead to biased results. The DECHEMA Chemsafe database[53] is an extensive source of explosion and other data. In addition, project Safekinex[54] produced explosion indices at elevated pressure and temperature. An overview of dust explosion characteristics and results of recent research can be found in Lemkowitz and Pasman.[55]

Another almost forgotten, and not explicitly mentioned, type of explosion is typically of a reactive liquid such as a nitrated lower hydrocarbon, for example, nitropropane, when confined in a steel pipe and ignited. The flame front, acceleration-generated compression waves move faster in the wall of the pipe than in the liquid so that energy radiates into the liquid ahead of the reaction zone and contributes to an increased propagation velocity and violence. This type of condensed phase explosion, able to cause significant damage, is called *low-velocity detonation*.

Table 3.3 Overview of Reactivity and Explosion Test Methods

Phys. State Substance	Explosiveness	UN Classification Test Methods	
Solid–liquid	• General reactivity: • Sensitivity to stimuli: • Deflagration: • Detonability: EIDS and ANE[a] tests:	• Heated tube (Koenen); time to pressure; thermal stability; bonfire; Dutch and US pressure vessel; ballistic mortar; BAM Trauzl • Impact (various drop hammers); friction; packaged drop; cap sensitivity • Dewar test • Tube tests; UN gap test; internal ignition (DDT); French and US DDT Specially designed tests: heat, shock, projectile impact, confinement	
Phys. State Substance	**Explosion Type**	**Screening Tests**	**Fundamentally Based Tests**
S–L	Runaway, thermal explosion (induction period, activation energy)	Differential thermal gravimetry, differential scanning calorimeter	Isothermal heat flow calorimeters, adiabatic calorimeters (AC), pressure compensated AC, bench scale calorimeter
Gas–dust/air	Auto- and minimum ignition temperature or energy	*Gas*: ASTM E 659-78 test *Dust*: Godbert–Greenwald; 5-mm layer glowing temperature; combustion spreading	Instrumented ASTM derivative test; minimum ignition energy test
Solids, liquids	Deflagration (propensity, zone velocity, severity)	Cook-off bomb; open tube tests; open trough tests	Instrumented bomb tests; constant pressure, burning velocity measurement
Gas–dust/air	Deflagration (explosion limits and severity)	20-L vessel tests; 1 m^3 vessel test (ASTM and EN standards)	Instrumented vessel tests measuring flame expansion
Solids, liquids	Detonation (detonation velocity and pressure)	Tube tests with various diameter and confinement strength; Gap test	Instrumented tube tests; field tests; deflagration-detonation transition
Gas and dust/air	DDT, Deflagration–detonation transition	Run-up (congested) pipe tests	Smoke foils, laser optics instrumented tubes

[a] *EIDS is extremely insensitive detonating substance; ANE is ammonium nitrate emulsion.*

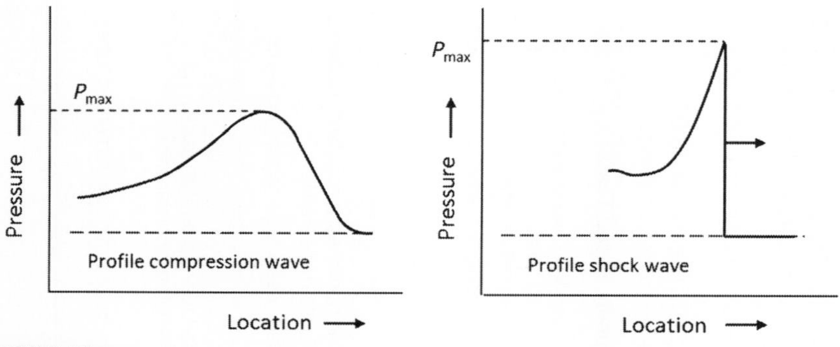

FIGURE 3.16

Schematic profiles of a (isentropic) compression wave and a shock, here both followed by a rarefaction in which pressure decreases.

A transition from an accelerating deflagration into a detonation is possible within substances in all three possible physical states. In gas, acceleration is achieved by turbulence resulting from interaction with a wall or obstacles to the flow of unburned gas pushed ahead by the flame, further reinforced by flame instabilities. In liquids it is turbulence at the surface and gas bubbles that play a role, and in solids it is a pressure increase or hot product gases penetrating a granular or porous unreacted mass. In all three cases, gas dynamic feedback causing acceleration to high propagation speeds is associated with compression waves, which steepen up at elevated pressure to form an intense shock wave.

In Figure 3.16 profiles of both types of waves are schematically shown, while in Figure 3.17 the transition process is depicted for a gas. In liquids and solids a similar process can occur. Lack of recognition of the importance of this phenomenon and the increase of probability of the transition through increase of item dimension and/or scale-up of total quantity led, for example, to the conditions of the May 2000 fireworks explosion disaster (23 killed) in the city of Enschede, the Netherlands. Storage of smaller diameters and quantities of so-called consumer pyrotechnics (fireworks) in the middle of a city is far less risky than the larger-sized professional ones.

As mentioned at the outset of this section, *thermal explosions* can occur with all unstable substances decomposing exothermally after some source of activation energy has been supplied. Unstable or instable substances with large endothermic heats of formation can explode violently after a certain induction period, whether these are gaseous such as acetylene or ozone, liquid such as several organic peroxides (see Figure 3.18), or solid such as metal azides, et cetera. The effects of accumulating contaminations are notorious, for example, with copper azide or locally formed peroxides, which suddenly by some action can explode and initiate a larger, more hazardous situation in a plant. Initially, slow self-heating processes (of the order of a degree Celsius per day), are also treacherous when after many days they suddenly accelerate.

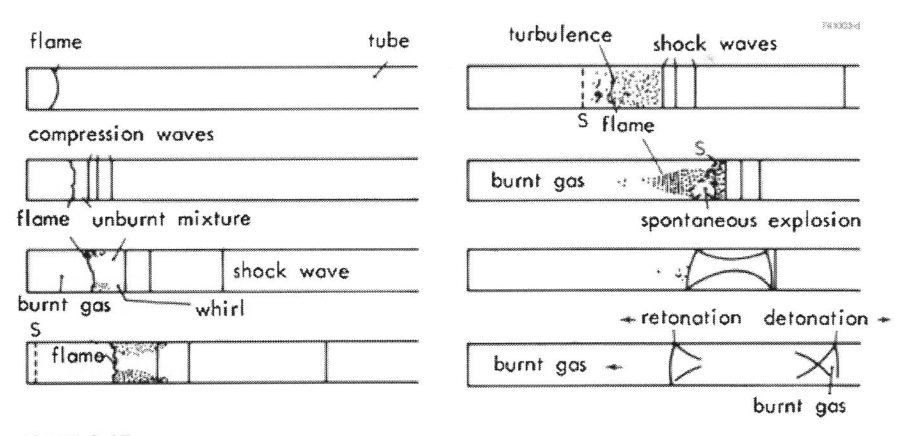

FIGURE 3.17

Generalized presentation of the gas dynamic process of transition from deflagration into detonation in a gas mixture in a pipe, based on the early experimental work of Oppenheim and coworkers.[56] Acceleration of the flame causes compression waves to run ahead and to steepen up into shock waves. At passage of one of the waves, the pressure becomes high enough for the fuel—air mixture to self-ignite and sustain the shock with the energy released.

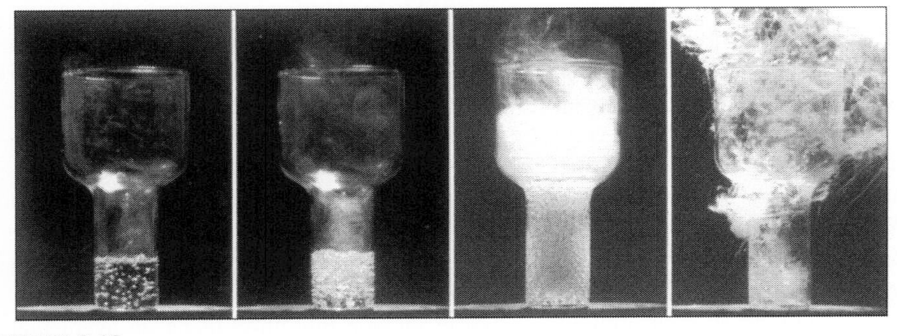

FIGURE 3.18

Example of a simple test with a liquid organic peroxide showing four successive stages, at increasingly shorter time intervals, after bringing it initially to a temperature where it starts to react (early 1970s at TNO, the Netherlands).

Self-heating is tempered by heat loss. In liquids and gases temperature equalizes as a result of stirring and mixing. Heat loss is proportional to the temperature difference over the vessel wall, but the heat generation by the reaction can increase exponentially with the fluid temperature following the Arrhenius law. According to the Semenov model (for details, see Lees[38]) given a sufficiently high cooling rate a maximum of two stable operating temperatures can be achieved, an upper and lower temperature. However, at the unstable upper one a perturbation can cause runaway, because the cooling will at higher temperature not be able to exceed the heating. For cooling, nonstirred fluids depend on natural convection, which is both relatively slow to increase and difficult to model. In gaseous hydrocarbon rich—air mixtures, this can give rise to radical induced cool flame. Solids lose heat by conduction. For solids, Frank-Kamenetzky and Merzhanov (see Lees[38]) have provided a variety of analytical models. Today, the basic heat balance equations can be solved through finite element or finite difference models.

An example of the possible pressure build-up effect of *reactor runaway* is given in Chapter 1. It occurs mainly in batch and semibatch reactors due to a large variety of causes such as incorrect dosing sequence, cooling and/or stirring failure, and reaction initiation delays, for example, because of stratification. In semibatch dosing, a required reactant is fed over a defined time period, and this can be used to avoid the problem of an accumulation of unreacted materials with significant potential for a thermal runaway. Although the primary reaction producing the intended product may be exothermic, the more serious hazard is usually the potential for a secondary, often oxidative and very exothermic, decomposition reaction with a higher activation energy. Methods are available to estimate reactor stability by determining the ratio of Damköhler (dimensionless measure of reaction rate) and Stanton (dimensionless cooling capacity relative to reaction time) numbers of the mixture (see Lees[38]). Even better is to determine the safety margin or "distance" between the maximum temperature of the synthesis reaction (*MTSR*) and the onset temperature of the secondary reaction. The latter is indicated by the Stoessel criticality diagram that defines two safe (green), two unsafe (red) and one possibly unsafe (orange) types of processes.[57] For this diagram, besides the *MTSR*, the boiling temperature of the reaction solvent (T_b) and the onset temperature for an adiabatic process that results in decomposition within 24 h (ADT_{24}) must be known. The latter requires experiment. If T_b is higher than *MTSR* and much below ADT_{24}, the process is inherently safe; if T_b is close to ADT_{24} it is still safe. But if $MTSR > T_b$ risk is higher, certainly when *MTSR* comes close to ADT_{24}. The latent heat of evaporation shall then be higher than the heat of reaction. The fifth case with $T_b > MTSR$ and both close to ADT_{24} presents a problem and requires additional thermal management safety measures.

Measures to stop an imminent runaway include inhibition using an agent that halts or slows down the reaction, quenching the reaction in situ with a large volume of diluent that will act to reduce both the concentration and temperature, and bottom venting the reactor contents into a large external vessel containing quench fluid. Much research has been done to detect a possible runaway very early on, as reported by Valeria Casson.[58] In Chapter 1, Section 1.10, the role of the AIChE design institute of emergency relief systems (DIERS) in designing reactor pressure relief has been described.

Unstable substances are by definition self-reactive; there are, however, reactive substances that are stable alone but that react violently with other materials, such as sodium with water or hydrogen with oxygen. There are many mixtures with associated reactions that can run away to high pressures, see Bretherick's handbook.[59] When intimately mixed at a temperature at which they still do not react, reaction can be initiated by a source of heat or a catalyst. In runaway reactions as in other reactor accidents, this is usually a source of a thermal nature such as a flame, heater, or leaking steam valve. In other instances, a hot spot of mechanical friction, or impact, or an electric or electrostatic spark will be sufficient. Gas and dust explosions are examples where such initiation is common, but there are others such as pyrotechnics, the thermite reaction between aluminum and iron oxide, iron in chlorine fire, et cetera. In those cases, there is most often intermolecular contact between an oxidizer and a fuel, while for most high explosives and propellants the contact is intramolecular. The oxygen balance is a characterizing measure that can be used as an indicator of the propensity to explode. High explosives and quite a few propellants can be initiated by a sufficiently strong shock wave.

Thermodynamic codes to calculate an explosion potential such as CHETAH[60] help to predict reactive chemical hazards, while CHEETAH[61] does the same but is primarily meant for predicting high explosive performance. Computational fluid dynamics (CFD) codes, for example, as described by Oran and Boris,[62] to simulate reactive flow and flame propagation including flame-turbulent gas flow obstacle interaction helped greatly to increase understanding of the complex processes occurring. These latter situations are meanwhile most often handled by LES.[63] Expertise is dispersed within various communities; in the past, even vapor cloud and dust explosion experts were rather isolated and in separate communication networks. However, the free distribution of expert information was improved by the conference platform of the International Symposium on Hazards, Prevention, and Mitigation of Industrial Explosions.

3.5.1.3 Ignition mechanisms and static electricity

The burning or explosion of gas—air mixtures is easily initiated by various types of ignition. The most obvious is ignition by an electric spark or flame. Ignition energies for near stoichiometric mixtures fall to roughly 0.5 mJ, for example, for propane-air. This is a very small amount of energy requiring a special apparatus to generate a controlled ignition, as shown in various work led by Rolf Eckhoff.[64,65] Not only the amount of energy, the dose, but also the rate of energy supply, hence power, is important. A spark generates a plasma with a short lifetime and a blast wave. However, the circuitry of which the ignition electrodes are part has a capacitance and induction that determine the time the discharge lasts. Also, the distance between electrodes is important. The minimum will be reached at a time of discharge and a sufficient discharge volume in which chemical conversion can build up sufficiently to propagate spherically outwards. Sparks can also be of a nonelectrical nature such as being mechanically induced by friction or impact. Ignition can also be by hot surfaces or fast compression of a gas (adiabatic compression) depending on the autoignition temperature of the mixture involved. Hydrogen and methane need a relatively high temperature to be ignited; longer chain hydrocarbons ignite more readily. "Hot work" (welding, cutting, grinding) on vessels and pipes in which residuals present of hydrocarbons have caused many fatalities. In the US, the Chemical Safety Board (CSB) has warned many times to maintain hot work permit discipline.

Dust—air mixtures require an energy dose over a much longer time to initiate explosion, because sufficient heat has to penetrate into the particles for volatiles to evaporate and start a reaction in the case of organic materials or break an oxide layer and initiate combustion when it concerns a metal powder. In addition, dust dispersions are usually created in turbulent air, which also affects the initiation process. Hence, in dust explosion tests one uses pyrotechnic igniters. On the other hand, when conditions are right (a dry dust with small diameter particles), the ignition energy may also come down to the milli-Joule range. Settled organic dusts may ignite due to self-heating at sufficiently high ambient temperatures as can occur in dryers. It should be stressed again that all mixtures that can burn in air can also explode, when dispersion conditions are right, so tests are essential.

Flammable liquids have a *flash point*, which is the temperature where the vapor pressure is sufficiently high to produce a concentration in air to enable successful ignition in a recognized (ASTM) standard test device. Combustion reaction stops once the ignition source is removed. Only above the fire point will there be sustained flame. Initiation of a deflagration of a reactive solid or liquid material able to decompose exothermally usually needs more heating than a spark can deliver, unless it concerns highly sensitive substances or mixtures. Due to their high reactivity, many peroxides can be ignited at very low temperatures. Pyrophoric substances such as sufficiently fine metal powders or white phosphorus spontaneously ignite when exposed to air. Hypergolic substances ignite when just mixed with each other. Materials capable of sustaining a detonation can be initiated by a shock wave from a high explosive "booster," which itself is initiated with a detonator containing a sensitive primary explosive easily transiting deflagration into detonation and in turn initiated by a glowing wire or another thermal device.

A hazard for all flammables may arise from static electricity. When two different materials are in contact, they exchange charge. If one of the materials is isolating, a static charge remains after it separates by moving (streaming current). Notorious were the ignitions of pumped kerosene or gasoline before conducting additives were developed. Also, the rubbing of particles along a surface as in pneumatic transport of dusts results in an accumulation of charge. By the same token, charge accumulation can occur when a person takes off a pullover or walks over a carpeted floor. Martin Glor[66,67] and coworkers performed much work on static electricity and charge-preventing safeguards in the context of process safety. Discharges can be of different types: an incendive spark is the most common; a brush discharge between a metal and a plastic or other nonconducting material is more dispersed in space and hence less intensive but still able to cause ignition, at least of the more sensitive gas mixtures. Coronas at a metal tip are even more homogenously spread discharges and are nonhazardous. Propagating brush discharges are more powerful and typical for pneumatic transport of dust along a duct with a plastic inner wall that accumulates charge but with a conducting and earthed outer wall producing a compensating opposite charge. Discharge may occur through a hole in the plastic wall or at a duct exit. Lightning is the last type of electrostatic discharge and is well known from thunderstorms, but this type can also appear in dust clouds as, for example, in silos. Volume resistivity, surface resistivity, overall resistance to earth, capacitance of conductors, surface charge density, and breakdown voltage are all measurable parameters for which thresholds have been defined in standards. Phenomena can now be modeled in 3-D, which will be helpful to adapt designs in order to prevent ignition hazards. Proper grounding is imperative. Choi et al.[68] discharged plastic granules by ionizing air.

3.5.1.4 Fire and fire protection

Fire and combustion expertise reside in a completely different community, although the physics and chemistry are very similar to those of explosion. Combustion for the purpose of hazard analysis is, for example, testing the rate of flame spread over the surface of a material with the cone calorimeter. Fire development in enclosures such as rooms, corridors, and tunnels has been studied intensively over many years to enable predictions to be made. Supported and coordinated by the US National Institute of Standards and Technology (NIST) in Gaithersburg, MD, a large international group of researchers worked together in an effort resulting in the Fire Dynamic Simulator (FDS). An overview of the methodology is given by McGrattan et al.[69] The computational fluid dynamics code used for this purpose simulates various kinds of fire: fires in enclosures, but also pool fires, and combustion of materials with different behaviors in a fire. FDS was used to simulate the fire in the World Trade Center buildings after the terrorist aircraft collision attack on September 11, 2001.

Determination of the effect of protective measures such as fire-rated walls, fire proofing of crucial equipment such as pipe rack supports or LPG spherical tank legs, zoning, detectors and deluge or sprinklers, and smoke abatement is important for risk assessments. Most victims in a fire are the result of hot smoke inhalation. Smoke is oxygen deficient and contains toxic products such as carbon monoxide and hydrogen cyanide, and these disorient people before they cause unconsciousness. Suitable construction material selection less easy to ignite and producing less toxic smoke can be decisive. Many studies have been done for the purpose of efficient evacuation of people from buildings, for example, those by Galea[70] on model EXODUS. Therefore, accurate smoke development simulation in time and in composition is important as an input to evacuation models that in turn serve to design escape routes. Another aspect is the threat to firefighters formed by a cloud of partly combusted, hot products, which accumulate, for example, adjacent to a ceiling and which as a result of a sudden supply of fresh air can self-ignite and produce a so-called "flashover." Soot formation in smoke is a rather complex process occurring via acetylene and aromatics, but there is an interest to model it. Less smoke means a higher radiation intensity and therefore a higher probability of burns and ignition of secondary fires across a certain distance. Research results on fire spreading, firefighting, and fire protection are well debated at international conferences, for example, a triennial symposium organized by the International Association for Fire Safety Science. Various countries have built large-scale test facilities to investigate the effect of fires on furnishings, building materials, and construction elements.

3.5.1.5 Explosion blast, projected fragments

Understanding the mechanisms explained above helps to model the effects that cause damage to vulnerable receptors of different natures. One of the first problems that arose in the 1970s–1980s was the strength of blast waves from vapor cloud explosions. Overpressure blast tables for the high explosive TNT had been published, for example, by Wilfred E. Baker,[71] so initially, when the first industrial vapor cloud and other explosions took place, it was common to express in TNT equivalents the strength estimated from the observed damage. The equivalent unit was based on a dimensionless combustion energy–scaled distance, $\overline{R}$ (Sachs scaling):

$$\overline{R} = R \Big/ \sqrt[3]{E/p_0} \tag{3.3}$$

This is the distance from the center of the source R divided by the cube root of the quotient of combustion energy, E and atmospheric pressure, p_0. However, in case of vapor clouds, these equivalent figures grossly overestimate the effect in the center of the cloud if the equivalent was based on the far-field effects, such as windows broken. For prediction, another uncertainty appears, namely the fraction of the fuel contributing to the explosion combustion energy, often taken as 10% of the cloud's total fuel content.

The first to consider blast by rapid vapor cloud deflagration were Roger A. Strehlow and Wilfred E. Baker[72] in the late 1970s. In 1985, Van den Berg[73] published TNO's multi-energy method, based on the combustion energy, E of a cloud part producing blast due to congestion, an assumed initial blast strength at the edge of that cloud part (grades 1–10), and solving the gas dynamic conservation equations for the surrounding air by a 2-D finite difference code. Plotting the resulting blast parameter against $\overline{R}$ appeared to be a good step forward; see Figure 3.19. This may be done for a multiple of blast-producing cloud parts. Later, Quentin A. Baker and M. J. Tang made improvements to the original Baker–Strehlow model. Sari[74] made a comparison. These types of simplified models produce a very general and uniform blast strength prediction not taking local conditions, buildings, greenery, and terrain topology into account; only computational fluid dynamic codes can simulate such specifics of an environment. The best validated CFD code for simulation of vapor cloud explosion blast is FLACS from Gexcon[75] in Norway. FLACS has been further developed for dispersion prediction, while a version named DESC is adapted to simulate dust explosions including the effect of whirling up settled dust ahead of the flame. Useful information on gas explosions and loading of structures, including plant equipment, can be found in a handbook by Bjerketvedt et al.[76]

Lethality data as a function of blast overpressure, like toxic fatalities, also vary widely. A human body can sustain quite high overpressures, but there are secondary effects, such as being blown over with resultant injuries. A probit response function that produces a value of 50% lethality at 0.9 barg, mentioned in a recent HSE document,[77] is: $Pr = 5.13 + 1.37 \ln P$, with P in barg. Inside structures, blast pressures of only one-third of this strength produce a similar effect. This may be due to wave

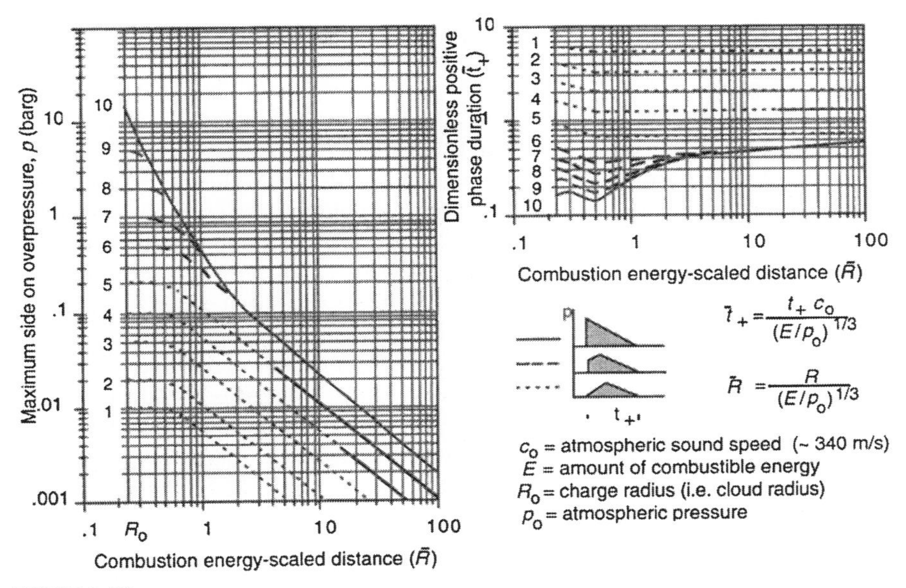

FIGURE 3.19

Multi-energy method blast curves presented as overpressure (left) and pressure duration (right) as a function of combustion energy—scaled distance to the cloud center. Different explosion intensities are indicated by the grades 1–10. The solid line at grade 10 represents a detonation and is in the range shown not appreciably different from a high explosive blast. The fine dashed lines grades 1–5 represent deflagrations at lower flame speeds and 6–9 deflagrations at higher flame speeds.

reflections, impact from flying debris from doors or ceilings, impact from flying glass fragments, and the body hit by blast falling against walls and floor.

For a certain level of structural damage response, above a threshold overpressure, P and impulse, $I = \int P(t)dt$, with blast duration, t, the combination of both is determining. Considering a structure as a mass-spring system (a single degree of freedom), energy transfer from an oscillating load to the structure is optimum if the time constants of oscillation and responding match (ringing in the eigenfrequency). Although a shock wave is certainly not a perfect oscillator, depending on the time duration of loading, T, it can be shown that deflection becomes maximum at twice the value under a static load when T is sufficiently long (the dynamic load factor = 2), while the deflection decreases with decreasing T. This analysis explains the use of the $P - I$ response diagrams of structures: a given damage can be achieved by a minimal pressure provided the time duration is long enough or by a minimal impulse in which decreasing T is compensated by increasing overpressure. Alonso et al.,[78] characterizing building damage as minor, major structural, and collapse, and using observed damage of the 1974 Flixborough VCE relative to distance to cloud explosion center, derived VCE $P - I$ diagrams and corresponding probit equations for the three damage categories. Relationships for overpressure damage to process equipment can be found in the *Gas Explosion Handbook*[76] and in an article by Zhang and Jiang.[79]

Fragment projection occurs, for instance, when a pressurized tank ruptures due to an internal gas explosion, a runaway reaction, or BLEVE. The projection direction is estimated from the location of the weakest point of the vessel and is usually not very accurate. An idealized optimum angle for projection is the ballistic 45° angle. The projection velocity depends on the energy transfer from the propelling gas and is uncertain. The air drag depends on shape and surface area of the fragment. Yet, bulky fragments from exploding vessels have been found to be projected over distances as much as 1 km. Further data on this and previous aspects are found in *Lees' Loss Prevention in the Process Industries*[38] and in a more concise but instructive form also in the Yellow Book.[80] Considering all available data, Tugnoli et al.[81] developed probability distributions for initial projection directions.

A threat to people is not only building collapse but also glass shards flying around after a blast impact. Many parameters are relevant, and therefore window breakage, starting from about 1 kPa (10 mbar) blast overpressure, covers a very broad range of overpressure values with most windows breaking at about 10 kPa. In addition, the range of values causing lethality mostly by skull fracture is large, while available data contain considerable uncertainty. A rough impression can be obtained from the Green Book[82] probit equation, $Pr = -15.7 + 2.2 \ln P_s$ with overpressure on the window P_s in Pa; the person is presumed to be 1.75 m behind the window and therefore has 5% probability of being hit when present somewhere in the room. Additional injury will result from fragment penetration. Details can be found in Lees.[38] Window safety films and other measures may reduce these effects. No data are available about the extent of injury as a function of blast strength or fragment mass and velocity.

3.5.1.6 Flame radiant heat

Radiant heat from pool fires, BLEVE fireballs, torch or jet fires, and flash fires also need modeling. If one goes into details, the analysis quickly becomes very complicated. In the case of a pool fire, there is firstly the heat release rate, which depends on the feedback of the heat of the flame to the liquid surface thus producing the evaporation that feeds the burning rate. In fireballs due to turbulence, the processes are even more complex. Torches and jet flames depend greatly on the mixing of entrained air with the fuel jet, but these can be studied in fairly stationary conditions. Burning jet impingement on equipment is a source of domino effects. In flash fires of a flammable cloud, flame is traveling throughout the cloud in which by flow due to expansion and lifting of hot gas, vortices arise that enhance mixing of unburned fuel and air, which greatly influences the path of the flame. The modeling complexity increases further when the combustion kinetics and the production of soot particles are included.

Engineering solutions to this combustion behavior were derived in the 1980s and 1990s by defining an overall parameter, the surface emissive power (SEP). This is the radiant heat flux emanating from a flame surface, and this parameter can be measured remotely. To determine the radiant heat intensity from a flame on the surface of a receptor, the view factor and the distance between the flame and receptor must be known. The view factor, F_{view}, is determined by the geometric condition of flame (width, altitude, aspect angle relative to the receiving surface) versus receptor and the distance and orientation of the receptor. The distance and the moisture content in the air also determine the radiation transmissivity, τ. The radiation flux on a receptor given a situation, q'', is therefore:

$$q'' = SEP \times F_{\text{view}} \times \tau \quad [\text{W/m}^2] \tag{3.4}$$

The *SEP* value is highly dependent on the clarity of the flame, which is influenced by the reactivity of the fuel, the access of oxygen, and the amount of soot produced. Fires of hydrocarbons in pools with low flame height have a relatively high *SEP* (about 100 kW/m^2). Due to soot formation, the *SEP* value decreases with taller flames and larger diameter pools to one-half. Methane contains four hydrogen atoms and burns as a blue flame with a relatively high *SEP* value. LNG pools therefore have high *SEP* values (150–200 kW/m^2) unless the pool becomes very wide. BLEVE fireballs have the highest *SEP* (300–350 kW/m^2).

For radiant heat lethality, HSE[83] made an evaluation of probit correlations (many have the same original source), which calculate an average lethal thermal dose ($t \times q''^{4/3}$) using the equation $Pr = -12.8 + 2.56 \ln(t \times q''^{4/3})$ with t the exposure time in seconds and q'' radiation intensity in kW/m^2. At 50% lethality, the probit thermal dose is 1000 $(kW/m^2)^{4/3} \cdot s$, while for the 1% vulnerable part of the population the dose is 500, and for (offshore) workers 2000 $(kW/m^2)^{4/3} \cdot s$. Casal[84] cites the same equation but expressed in W/m^2. He also addresses the issue of protection of clothing or deterioration by its ignition: $Pr = -9.43 + 2.56 \ln(t \times q''^{4/3})$ when converted to kW/m^2, and the effect on the exposure time of fleeing (5 s to start running; speed, 4–6 m/s). He further reproduces the probit for first-degree burns:

$Pr = -12.03 + 3.1086 \ln(t \times q''^{4/3})$, while $Pr = -15.3 + 3.1086 \ln(t \times q''^{4/3})$ is expected to hold for second-degree burns, all with q in kW/m^2. In risk assessments, it is assumed that people located in a cloud at the time of a flash fire or an explosion will be killed. Phani Raj[85] proved (in person) that an intensity level of 5 kW/m^2, the threshold for the LNG exclusion zone in the US, is bearable.

The older *Green Book*[82] also quotes the above equations (but in W/m^2), and in addition ignition of combustible materials (wood, plastics) expected to occur at a critical temperature of about 400 °C and 15 kW/m^2, glass at 120 °C and 4 kW/m^2, and failure of steel at 500 °C and 100 kW/m^2.

3.5.2 SYSTEM SAFETY

System safety can be understood as a systems approach to safety and risk control in operation. In Chapter 5, we shall consider integral analysis of system safety. Here, we shall first discuss principles of how to design and engineer a safe system. In Section 3.5.3 when dealing with asset integrity, briefly, analysis methods will be mentioned of how long a system will function without *failure* of a component. The latter requires the concepts, methods, and distributions of reliability, availability, and maintainability. These are the domain of the discipline of reliability engineering. Finally, in Section 3.6 on risk analysis, we shall discuss how to identify critical failures, what consequences may occur, and how often to expect failures. For identification, systematically tracing causes of possible deviations of observable process variables (*faults*) is crucial.

Knowledge of safety principles is shared among many engineering disciplines in which safety is important, such as nuclear, aerospace, and civil engineering. Möller and Hansson[86] discussed principles of engineering safety, and we shall follow their line of thinking. They group the principles into four main categories:

1. Inherently safe design
2. Safety reserves
3. Safe-fail (going much further than "fail-safe")
4. Procedural safeguards

To category 1. Inherent safety principles with regard to processes were formulated in the late 1970s by Kletz[88] with his motto: "What you don't have, can't leak." The first four principles are process *Intensification* (hence, smaller quantities), *Substitution* (e.g., with less toxic or less flammable solvents), *Attenuation* (e.g., by less aggressive or more moderate conditions), and *Limitation of Effects* (e.g., by using or storing smaller quantities and having fewer or simpler protective entities). Figure 3.20 illustrates the point. Hence, it starts with the selection of materials and the process and plant designs for which the hazard potential, often associated with reduced inventory or holdup, is reduced. Inherent safety in its absolute sense with elimination of all hazards is not realistic, but some individual hazards can be totally eliminated, for example, the stairs in a home for the elderly or in a children's nursery, the lead oxide (toxic) pigment from white paint, or changing a formulation to eliminate a flammable hydrocarbon solvent and at the same time protecting the environment as in the case of water-based emulsion paints. As seen in Chapter 1, many common substances, essential to our existence and not able to be replaced by others, have properties that require much care when producing, storing, and handling them. As we cannot discard working with these substances, an absolutely inherently safe situation can seldom be achieved. But indeed, we can make designs inherently *safer* by drastically reducing risk. Hence, within the risk margin one can then still speak of safe design. Process intensification limits quantities but compensates with greater efficiency by increasing "driving" forces. Technology enabled innovative concepts in chemical engineering by combining in one step what in the past were seen as separate unit operations, for example, such as reactive distillation or extraction. Alternative intensifying reaction "drivers" other than temperature and catalysts, such as microwaves, supersonic waves, and ultraviolet irradiation have also been applied. In addition, microreactors (with a very small inventory) or spinning disc

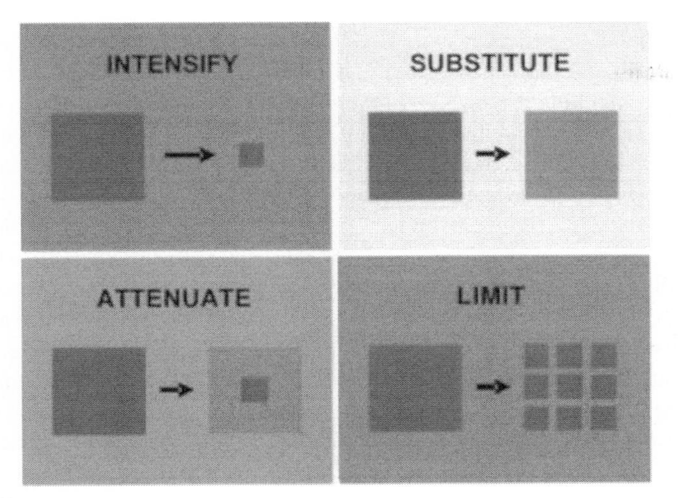

FIGURE 3.20

Graphical representation of some of the late Professor Trevor Kletz's inherent safety fundamentals which he developed based on an Appendix in Kletz and Amyotte[87] entitled "An Atlas of Safety Thinking". In colors the dark square is in red and the light one in green, see also Pasman and Rogers[89].

(Trevor Kletz was previously with former Imperial Chemical Industries (ICI), leading its very successful advanced process safety activities, and then taught at Loughborough University, UK. He died in 2013.)

reactors are becoming more practical. There have been several attempts to develop a quantitative measure of the degree of inherent safety with Khan and Amyotte[90] proposing an integrated inherent safety index.

What can also be included under the headings of "attenuate" and "limit" are measures as prescribed by the ATEX Directive of establishing zones around installations or equipment containing flammables. To prevent gas and dust explosions in such locations only certified, explosion-safe equipment may be installed and operated. Also, zoning in plant layout and many design measures in equipment are taken to prevent and to protect against spreading of fire.

To category 2. Safety factors have traditionally been applied in engineering to allow for the presence of unforeseen duties or loads and still to have safe performance. However, competition forces cost cutting, which also means reducing or taking away margins of safety. Improved knowledge also makes it possible. On the other hand, risk assessments have their limitations, and uncertainty remains about what can happen during a life cycle, particularly a long one. So, by removing or reducing safety factors, resilience is reduced. In Chapter 5 we shall discuss the concept of resilience more extensively.

To category 3. According to the authors, "safe-fail" is different from "fail-safe." Safe-fail means the design should be such that the item cannot fail. Reliability is the crux. Whereas fail-safe implies that if a component fails it shall fail in a safe state or leave the system in a safe state. Fault and error tolerant are somewhere between these two. There are good examples of the safe-fail principle, but just as with inherent safety, safe-fail can certainly not be applied universally. Reliability is achieved by a combination of possible measures. The first measure is to make use of historical failure data when designing the item and trying to do better. Then, testing will update knowledge, and applying redundancy will increase continuing functionality. However, when redundancy is applied, Möller and Hansson stress that the redundant units should be sufficiently separated (segregation) in location so as not to be subjected to the same environmental disturbances to which they are susceptible and thus to avoid simultaneous failure from a common cause. In addition, the redundant units should have sufficient diversity, for example, of different design, and not be subject to common cause failure, for example, by being dependent on the same services (power, air, or cooling). The last two points immediately bring to mind the diesel engines that should have produced the power to shut down the Fukushima nuclear reactors following the 2011 earthquake and tsunami (see Chapter 4). These diesel power generators were redundant but failed because of lack of segregation and diversity.

To category 4. Procedural safeguards are indispensable, but require rather stringent discipline when high long-term reliability is required. We shall return later to this topic when dealing with layers of protection, both in this and following chapters.

Engineers share a wide variety of tools to systematically analyze the safety of an installation on a detailed level. The most important of these tools include failure mode and effect analysis (FMEA) of components and logic tree analysis: fault tree (FT), event tree (ET), and master logic diagrams (MLD). FMEA and FT have their roots in reliability engineering. An MLD has similar characteristics to an FT. MLD is suitable to be applied on a higher system level than FT for analyzing complex systems of several functionally independent but interacting subsystems. Details can be found, for example, in Modarres et al.[91] FTs and ETs were later combined to form a bow-tie diagram, all of which will be explained shortly. Furthermore, a very important tool is hazard and operability study, a team effort investigating possible hazards resulting from operational deviations from design intent. We shall consider HAZOP further in the section on risk assessment.

3.5.2.1 Failure mode and effect analysis

FMEA, or FMECA when criticality is included, is a qualitative method used to systematically analyze an equipment unit or part of it for possible failures and to identify the effect the failures will have on its function and on the system it is part of. As a method, it began in the 1950s in the military and spread via the aerospace and aeronautics industry to other engineering sectors. It is by preference a team activity performed by, for example, an operator, a maintenance specialist, a process engineer, and a product expert, and is under the control of an experienced team leader. The failure modes can be found by checking potential errors in the "6 M's": man, material, machine, measurement, method, and environment ("milieu"). FMEA is organized suspicion. Personnel other than team members shall be asked to comment on the first result of an FMEA or FMECA. The results are described in a format as in Table 3.4.

Table 3.4 Sample Table for Failure Mode Effect and Criticality Analysis (FMECA)

Component	Failure		Effects		Criticality			
No. I.D.; function	Error mode	Causes	Local	System	Severity	Frequency	Detection Method	Remarks

3.5.2.2 *Fault tree analysis*

Fault tree analysis (FTA) is an investigation of how a selected "top" fault (abnormal condition) or failure event, for example, an unintended/undesired release of a hazardous material, can be resolved into its causes. It is a deductive analysis progressing downward to prior events that could have either caused the top event/system failure or allowed it to occur via failing components in subsystems that work together or replace each other in function. Fault trees are modeled with Boolean AND- and OR-gates or operators in a logic tree (component or subsystem function is on = true, or off = false Boolean logic). The subsystems are either core processing functions or controls. The deductive defect tracking analysis proceeds downward to ever more subsystems and components that together form the system. Finally, failing base components will be reached that can be regarded as nondecomposable units for which failure rates are known or initially can be estimated from generic data for similar units. The latter data enable bottom-up quantification of the tree and finally calculation of the probability per unit of time, for example, a year, or frequency of the top event.

Because the fault tree is usually shown with the top event at the apex, it looks like a tree (with low branches); see Figure 3.21. An AND-gate means that all composing components must fail or other undesired events must occur for the AND-gate to become true; while for an OR-gate, one or more failures or other events to occur will be sufficient for the OR-gate to become true. In an exclusive XOR, it is either/or, but not both, events to occur. This technique stems from 1961 (Bell Laboratories); graphic symbols are standard. Cut sets are the unique combinations of component failures that cause system failure. The cut set is minimal if no basic failure can be left out. Limitations include that only two states (Boolean) for each component or event are considered—accounting for temporal effects, for partial functioning, and application of spares and repairs is not possible. Some of these limitations can be overcome but require a more complex setup. There are many literature references that can be consulted for further information, again see Lees,[38] and there are several supporting software distributors.

3.5.2.3 *Event tree analysis*

Event tree analysis (ETA) is an inductive logic tree with branching consequences out from an initiating critical event, for example, a release, to various possible intermediate consequences and then on to end points of final major hazard phenomena, such as explosion or fire. The event occurrence is quantifiable in terms of probability of the initiating event and of the probability pairs at each branch

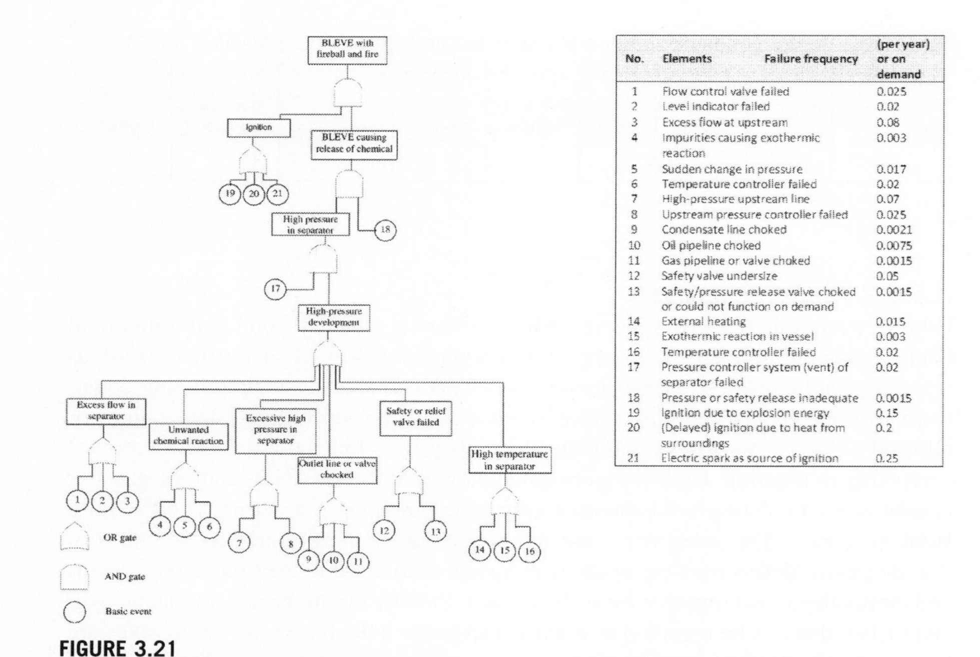

No.	Elements	Failure frequency (per year) or on demand
1	Flow control valve failed	0.025
2	Level indicator failed	0.02
3	Excess flow at upstream	0.08
4	Impurities causing exothermic reaction	0.003
5	Sudden change in pressure	0.017
6	Temperature controller failed	0.02
7	High-pressure upstream line	0.07
8	Upstream pressure controller failed	0.025
9	Condensate line choked	0.0021
10	Oil pipeline choked	0.0075
11	Gas pipeline or valve choked	0.0015
12	Safety valve undersize	0.05
13	Safety/pressure release valve choked or could not function on demand	0.0015
14	External heating	0.015
15	Exothermic reaction in vessel	0.003
16	Temperature controller failed	0.02
17	Pressure controller system (vent) of separator failed	0.02
18	Pressure or safety release inadequate	0.0015
19	Ignition due to explosion energy	0.15
20	(Delayed) ignition due to heat from surroundings	0.2
21	Electric spark as source of ignition	0.25

FIGURE 3.21

Example fault tree for an offshore platform separator rupture event resulting in a BLEVE, according to Khan.[92] Given the failure rates of components 1–21, the expected frequency of the top event can be calculated.

node. An example is shown in Figure 3.22. If the events concern actions or active failures occurring in a certain sequence producing a range of results, it is called an event sequence diagram.

3.5.2.4 Bow-tie

A bow-tie is a fault tree rotated 90° clockwise, with the fault tree top event merging with the critical or initiating event of a follow-on event tree. Its name derives from a gentleman's bow-tie; see the caption to Figure 3.23. In that way, a combination was obtained that for an overview appeared to be very useful because all preventative barriers (at the left for the critical event) and protective ones (at the right for the critical event) could be shown. The cause–consequence chain from each fault tree base level event to each event tree branch end point forms a risk *scenario*. The bow-tie[93] was used in the 1990s by analysts in the Shell Company but existed earlier as it was already mentioned at an ICI training course in 1979. It found its way into the EU ARAMIS project where bow-tie was applied extensively during 2002–2003. We shall discuss ARAMIS in further detail in chapter subhead 3.6; then we shall also explain the types of release events and dangerous phenomena mentioned in the legend of Figure 3.23. Subsequently, the use of bow-tie analysis spread quickly.

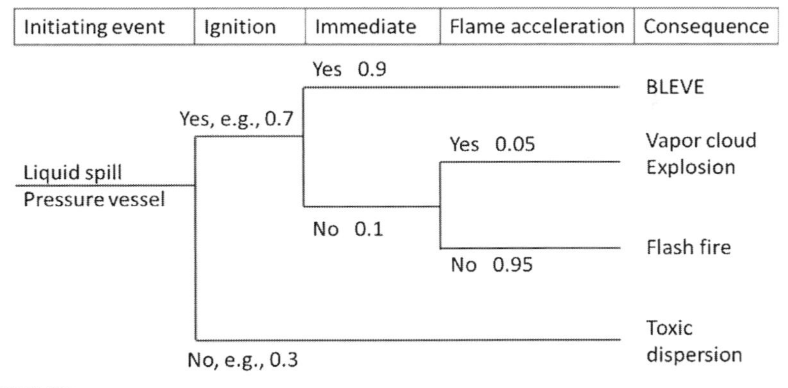

FIGURE 3.22

Example of an event tree for a pressure vessel rupture causing a flammable toxic liquid, boiling spill that may result in a BLEVE fireball, a vapor cloud explosion, a cloud flash fire, or a toxic cloud. At each node a probability point value can be assigned, as shown in the example.

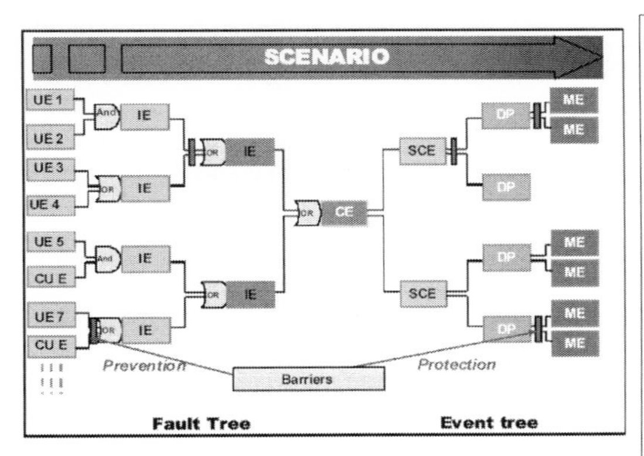

LEGEND (see also Section 3.6.2.3)

UE = Undesired Event, e.g., human act
CU E = Current Event condition, direct cause
IE = Intermediate Event, e.g., pump fails
CE = Critical (Initiating) Release Event, 12 types: leak, start of fire, etc.
SCE = Secondary CE, escalation
DP = Dangerous Phenomena, 13 types: Vapor, cloud explosion, pool fire, jet fire, etc.
ME = Major Event, 4 types: overpressure, heat radiation, toxic load, and missiles.
Barriers are Preventative (or Pro-active) left of CE, Protective or Mitigative (also called Reactive) to the right of CE.

FIGURE 3.23

Example of a schematic bow-tie (▶◀) showing preventive and protective barriers.[94]

3.5.3 PROCESS OPERATION

Process operation is the third and last basic safety knowledge field to be mentioned; it encompasses knowledge of application of unit operations, of the technology of plant installations, their design, engineering, construction, operation, and the organizations that will enable all this. Human factors and the organization have already been briefly described in the subchapter on organization, leadership, management, SMS, and culture, but will be more extensively discussed in Chapters 5 and 6. Process control is an essential part of process operation, but this will be treated in Chapter 8.

Process technology is a wide field. Standards, codes, and best practices are essential for safety but not sufficient. Adequate procedures are another point. Safety considerations concern not only design and operational requirements of an installation but also the layout of a site. Layout can make a great difference when explosion or fire occurs in terms of spreading, access of emergency responders, and (self-)rescue of employees. Control rooms, office buildings, and workshop locations should be selected with due consideration for the safety aspects involved. Risk analysis can be applied to optimize a layout situation, as we shall see in Section 3.6.

The entire installation life cycle, from development to construction, commissioning, operations including maintenance, decommissioning, demolition, and site remediation, requires safety reviews. Corrections at a late stage are expensive, and it is much better to adapt early on. Also, in the life cycle, many environmental aspects will be encountered. The ISO 14000 family of standards is very helpful in this respect. ISO 14001:2004 deals with environmental management systems, and ISO 14040 and others of the 40-series deal with life cycle assessment.

In Figure 3.24 an example of the initial life cycle stages and the timing of the reviews are presented. Design and construction of a plant occurs in stages. After the R&D stage, a conceptual process design (CPD) is made, followed by basic plant design, and then by an engineering design stage and a construction period. Although

Procedure for the design and operation of safe chemical plants
Safety analysis team performs technical safety audits for new plant or for major modifications

Design history	Team review stages
• Laboratory development	
• Pilot plant	
• Conceptual process design	
Initial project presentation	**Concept certificate review**
• Basic plant design	• Technical safety concept
• License planning	• System safety analysis
	• Safety report (regulatory)
Permit application	**Design certificate**
• Cost estimate	• Detailed P&ID safety review, HazOp, FTA , Consequence analysis
Appropriation request	
• Detailed engineering design	**Safety certificate**
• Procurement/ construction	• Acceptance tests
	Acceptance certificate
• Start-up	• Continuous updating of safety documentation
• Production	

FIGURE 3.24

An example of the various stages and the timing of the reviews in a process/plant development project. The terminology is borrowed from an early Bayer concept. (Basic plant design is now elsewhere known as Front-End Engineering Design or FEED; conceptual design is also indicated as pre-FEED; Detailed Engineering, Procurement, and Construction Management as EPCM; acceptance certificate as Pre-Start Safety Review or PSSR, start-up as commissioning, while decommissioning and demolition require own safety studies.)

a number of starting conditions are already settled during the R&D stage, in the CPD phase quite a few decisions remain to be made, such as on chemicals, type of operations, operational conditions, and main equipment types and sizes. Trade-offs must then be made to optimize product yield against the lowest energy requirements, minimal environmental impact, maximum safety, lowest investment capital requirements, and lowest operational expenditures. Because time to market is often a major issue, automation and computational support to speed up the design process and decrease man-hour costs is imperative. Therefore, a future design method should include rapid automated hazard identification to aid and guide the designer toward safer and more cost-effective solutions, despite possibly limited personal experience in safety matters. The interaction of the designer with such a knowledge-engineering unit, together with additional analysis, tests, and supporting studies, could form a growing body of process-specific safety knowledge. Such a knowledge-engineering approach would greatly shorten the overall design process because the conventional but cumbersome trial-and-error cycles of making a design with subsequent hazard analysis by applying index methods, that is, HAZOP-ing, and making modifications where necessary, should be avoided. These methods are slow, labor intensive, and not sufficiently effective. Safety considerations should thus be integrated into the core activities of the process developer and the plant designer.

Innovations that have as a main objective higher efficiency by moving from a batch to a continuous process, applying process intensification, or seeking energy savings, can each make a contribution to a higher safety level. Energy saving can be realized, for example, by applying membrane technology to avoid high-temperature operations. Other possible contributions are reduction in maintenance by use of improved materials and techniques to abate corrosion and thus assure better asset integrity, and using leak-free connections to protect the environment. Simulation models and techniques, such as computer-aided design, facilitate more efficient design and engineering practices and can also help to identify safety problems before they emerge in practice. The same is true for process control equipment and its operational implications. The full digitization and new control module technology forecast an improved overview of process operations. We shall return to these subjects at a later stage in Chapters 7 and 8.

Maintenance and maintainability are important, and boundary conditions for them are already laid down in the design, such as:

- How easily can components be reached for inspection, diagnosis, and servicing?
- How rapidly can components be tested, repaired, and reinstalled?
- How much protection is provided against various types of corrosion, et cetera?

Asset integrity in an aging plant plays an important role with respect to safety. Several explosions and fire incidents in US refineries over the last decade have been caused by eroded and corroded pipe ruptures. Quite effective nondestructive test methods based on a variety of physical principles such as X-ray, ultrasound, and eddy current are available to detect imminent failure. Indeed, condition-based monitoring and risk-based inspection (RBI) help to avoid nasty surprises. In particular, RBI is an efficient tool to guard the integrity of pressure equipment; see also Section 3.6.6.4. Leading indicators shall be applied to identify and follow trends.

Reliability, Availability, Maintainability, and Sustainability of equipment have become branches of engineering. Abundant literature on this topic is available; see, for example, Ebeling.[95] For industrial applications, more attention should be given to availability, which represents the total probability of failure on demand (*PFD*), or total unavailability due to latent failure during time between tests, T_0, and time to diagnose, repair, or replace the unit with estimated time T_R. So the expected downtime, M, or mean unavailability due to downtime, $M/(T_0 + T_R)$, should be estimated for critical components as part of a system approach. Then the system approach probability of failure on demand, or unavailability to include both the random failure contribution and the downtime contribution, assuming exponential failure probability increase in time of the component behavior, within $\lambda T_0 << 1$, is $PFD = (1/2)\lambda T_0 + M/(T_0 + T_R)$ with λ as failure rate. For units with a two-parameter Weibull failure function (bathtub curve), the expression is $PFD = (\lambda T_0)\beta)/(1 + \beta) + M/(T_0 + T_R)$, which yields the exponential behavior expression for shape parameter $\beta = 1$.

Despite all, operational care mishaps do occur. It is therefore important to install risk-reduction technology. In particular for those devices is *PFD* and its degradation over time, which is important. This will be discussed in Section 3.6.5, after we have seen how to analyze risks.

3.6 RISK ANALYSIS TOOLS, RISK ASSESSMENT

3.6.1 WHAT IS RISK ANALYSIS AND WHAT PURPOSES DO THE RESULTS SERVE

In general, following Stanley Kaplan and John Garrick's 1981 triplet definition,[96] technical risk analysis consists of a systematic search for what can go wrong, what likelihood it will have, and how severe consequences will be. Further assessment will encompass an appraisal of the risk, its associated uncertainty as represented by its distribution, whether it is acceptable, or if not what can be done to reduce it. It therefore focuses on the confrontation with undesired, unlikely, but damaging events. In a process plant, offshore operation, storage facility, or transport phase, an undesired event involves spill of hazardous material with a threat of injuring or killing people, damaging structures and the environment, and, in the end, causing financial loss. Hence, it starts with a breach or hole in a containment system, and the cross-sectional area of this will determine the release or spill rate. Such an event can itself be the result of a chain of cause-and-effect events designated a scenario; most likely there will be a number of management, organizational, employee-induced, technical, and environmental causes and conditions contributing, and happening to coincide and interact. In the Swiss cheese concept, shown in Figure 3.2, these are called fallible decisions, latent failures, preconditions (management/organizational/procedural, but also workplace environment, weather, earthquake, flooding), unsafe acts, and barrier failures. All of these when combined form *scenarios*. In the following, we shall first consider "spontaneous" spill scenarios, which means those scenarios that are due to equipment breakdown as a result of wear, degradation, or hidden defects. Spills may also be due to operator error. External causes of spills can result from impact by blast, fragments, or fire caused by an earlier release or by the presence of combustibles elsewhere in the plant igniting. These follow-on, domino effects are highly varied. Further causes are natural phenomena such as earthquake, flooding, storms, and lightning or electromagnetic interference, and finally there is intentional sabotage or terrorist acts. These will be considered when relevant and will be treated explicitly in Section 3.6.6.5 in the context of protective measures.

In principle, risks can be quantified. This is accomplished by quantitative risk assessment (QRA), also called probabilistic risk assessment (PRA). Safety cannot. Descriptive QRA is applied in hindsight on accident data, but assessment is mostly performed in a predictive sense to enable, or at least support, *rational decision making about how safe is "safe enough."* UK's "as low as reasonably practicable" (ALARP) principle roughly states that if a risk level is above tolerable or even below it, there shall be efforts by adding such measures to reduce it until the residual risk becomes acceptable or benefits grossly outweigh the costs (see Chapter 12 for further explanation). Risk assessment is applied to make a design and an operational plant safer, to optimize facility siting, to perform emergency planning, and to assess risks that sites present to residential areas and in particular to vulnerable entities therein such as schools, hospitals, and care centers. In the end, cost–benefit consideration will be included. Balancing of risks, costs, and benefits is known as *risk management*. Much has been written about it. Ian Cameron and Raghu Raman[10] (2005) published a comprehensive treatment of the subject applied to process systems.

A risk assessment of an on- or offshore installation can be performed to different levels of depth. Qualitatively, it starts with finding out what hazards can be identified and where they are located. Because it appears that there are so many hazards, while resources to consider these hazards and to devise measures to control them within bounds are limited, prioritization is needed. Prioritization requires a certain degree of quantification, and for that, two factors play a role: *severity* and *likelihood*. One then arrives at a semiquantitative method, which classifies the two parameters in terms of, for example, a five-point scale. The risk points, thus found, can be plotted in a risk matrix, as shown in Figure 3.25, with the highest risks in the top-right corner.

The completeness and level of detail of information available depends on the stage of a project, as described in the previous section: conceptual process development, basic plant design or front-end engineering design, and detailed engineering design. Risk assessment usually starts with screening the risks qualitatively, then in a next stage semiquantitatively, and if a full QRA must be carried out, it will be done after an engineering design has become available. Once a plant is running, an operational risk assessment can be performed taking increased account of human

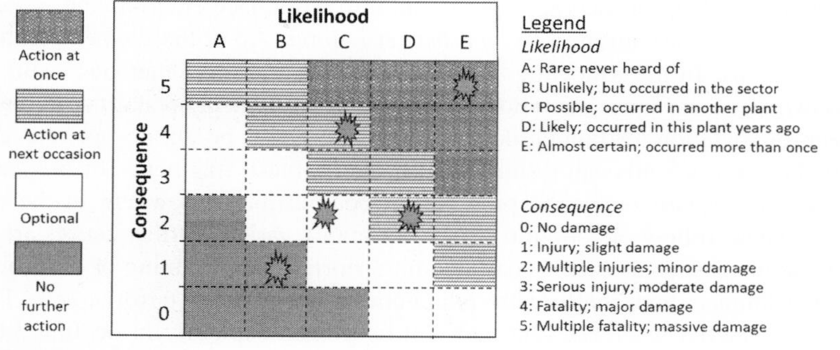

FIGURE 3.25

Qualitative risk matrix with five risk dots in various action priority fields.

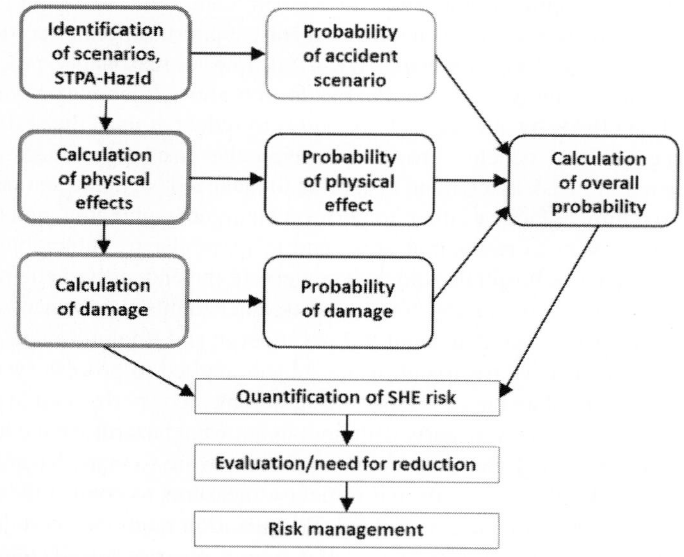

FIGURE 3.26

Basic flow scheme to perform a risk assessment. In the colored print the edges can be seen and refer to the cover colors of the Dutch QRA guidance books[97]: Identification = Purple Book; probability of scenario = Red Book; physical effects = Yellow Book[80]; Damage = Green Book[82]. "SHE risk" stands for Safety, Health, and Environment risk.

failure influences (operations, maintenance). In dynamic operational risk assessment emphasis is on degradation effects.

Although in many instances a superficial qualitative look or a semiquantitative analysis suffices, to obtain a risk assessment for realistic cases with significant losses at stake, full quantification is required. Figure 3.26 provides a basic scheme. Note that the probabilistic nature is not limited to the occurrence of the undesired event.

Occurrence of a particular scenario, the consequence effects embedded in that scenario, the damage as a result of the consequences, and the exposure and the vulnerability of receptors are all of probabilistic nature. This is because there are many possibilities present, each with differences in likelihood depending on conditions and because of the variability in their susceptibility properties. In principle, distribution functions can describe the variability, but one always must be alert for outliers!

International standard IEC/ISO 31010 Risk management—Risk assessment techniques[98] provide an excellent overview with examples of existing risk analysis tools in general for various engineering disciplines. The ones most commonly used in the context of process safety, onshore and offshore alike, will be briefly discussed, together with some information on the history of their development.

A QRA study can be visualized as being performed in six steps. The steps and applied tools are presented in the fish-bone diagram of Figure 3.27. First, hazards must be recognized and scenarios of events identified that lead to potential upsets. Identification of scenarios leading to undesirable events appears to be the easiest step for students new to the field to perform. However, it turns out to be the most difficult step, and a source of tremendous variability in QRA outcomes when the study is performed by different analysts. Indeed, the many factors of influence on a scenario make that rather obvious. Because under the Seveso Directives quite a few countries in Europe apply risk analysis for land-use planning and licensing purposes, this variability has worried the European Commission. In the early 1990s and later around 2000, two multinational EU projects were conducted, both on a virtual ammonia storage plant. The first one focused on the variance in dispersion model outcomes that led to the corrective action of EU SMEDIS and code validation as mentioned in Section

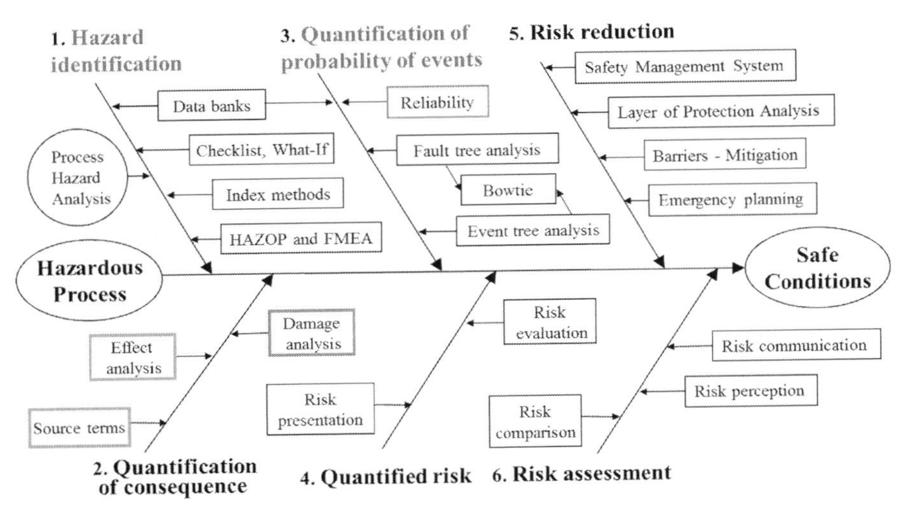

FIGURE 3.27

Ishikawa or fish-bone scheme of the six steps of a quantitative risk assessment (QRA), showing various tools to perform the process. Not shown is the preparatory work of establishing the actual objective, the stakeholders, the evaluation criteria and other context.

3.5.1.1. The second project, called ASSURANCE[99] (ASSessment of Uncertainties in Risk Analysis of Chemical Establishments), addressed the core of the variability and found that scenario definition is the largest QRA uncertainty factor. We shall consider results of this project in more detail later when looking back on the value of the methodology. However, following ASSURANCE, during the years 2002–2004 under directorship of Olivier Salvi and Bruno Debray,[100] a 3-year EU project ARAMIS was run with the objective of improving existing risk analysis codes. In the following, we shall include the main features of ARAMIS and where it helped to create progress. More about the history of QRA in relation to land-use planning has been described by Hans Pasman and Genserik Reniers.[101]

3.6.2 STEP 1. HAZARD IDENTIFICATION AND CHARACTERIZATION

The tools to identify process hazards are shown in the first "bone" of Figure 3.27. Hazard identification (HAZID), or in OSHA's PSM rule (Chapter 2) also called PHA or in Europe sometimes HAZAN, is performed by a team. It starts with making an inventory of hazardous materials present, properties, quantities, and collecting process data sources, such as process descriptions including heat and mass balances, a process flow diagram, a piping and instrumentation diagram (P&ID), a computer-aided design (CAD) model of the installation, plot plan or layout plan of the site, and environmental information. The nonscenario-based HAZID study tools comprise also checklists and searches in an incident database. The scenario-based follow-on tools are "What-if" question sessions (also called Structured What-If Technique -SWIFT), application of index methods such as the Dow Fire & Explosion index, and the more labor-intensive hazard and operability study (HAZOP) study, failure mode and effect analysis (FMEA), qualitative fault tree analysis (FTA) and event tree analysis (ETA) analyses, and operator task analysis.

Over the years several incident databases have been created, but because of costs some ceased to operate. In the 1970s, one of the first was the Failure and Accidents Technical Information System (TNO FACTS), which contains over 24,500 accident reports; it is now operated by the Unified Industrial & Harbour Fire Department in Rotterdam-Rozenburg. In France, the ARIA database is operative with over 40,000 accidents on record. EU's MAHB at JRC, Ispra, Italy, maintains the electronic Major Accident Reporting System database in which, besides accident reports of OECD and UNECE countries, (compulsory) reports of EU countries on Seveso Directive breaches are collected. Entries are given by substance, process, country, year, et cetera. Confidentiality can be assured. CCPS in New York operates the Process Safety Incident Database. There are still several other databases; an overview of them is given by the Norwegian University of Science and Technology (NTNU) in Trondheim.[102] However, no single database provides all of the following highly desirable features: being free at the point of use for anybody, anywhere, including root causes and lessons learned and being able to be searched on the basis of these; making available as pdfs or hyperlinks relevant background information (i.e., investigations, reports, papers, details of relevant publications). It is perhaps partly because of this lack of an easy access to incident information that repeats of major accidents continue to occur around the world, each time being the subject of surprise, anger, and possible devastation.

FMEA was introduced in the previous section. Index methods and HAZOP will be discussed in more detail in the next sections.

3.6.2.1 Dow's index methods

The Dow Fire & Explosion Index (F&EI)[103] and Chemical Exposure Index (CEI)[103] are well known and widely used. The former is the oldest and in fact consists of a self-contained risk study of a plant installation. The Dow Chemical Company condensed a large amount of knowledge and experience into these indexes and decided to make them available throughout the process industries with no charge, as a contribution intended to improve safety standards. Distribution of the user manuals for the Dow indexes is by the AIChE. The starting point is the NFPA material factor, which is a measure of the hazard of the substance involved and which falls between 0 and 40 in a matrix of instability/reactivity and flammability. This material factor is then multiplied by general and special process hazard penalties and credit points depending on the type of processes and their operating conditions of temperature and pressure—each of these may be the sum of several contributions. Then, if the resulting index value is above 100, the degree of hazard is judged in many companies to be unacceptably high, and risk-reducing measures are required. The positive contributions of these measures are expressed through additional credit factors. The process unit risk analysis summary is concluded by calculating the potential area of damage, the value of equipment within this area and the maximum probable property damage, and the business interruption. The Dow Index is a valuable tool at the early design stage of a project, as it makes clear which process units should be considered as (most) hazardous and where alternative or protective measures must be contemplated. However, it fails to consider details. Besides the Dow F&EI, the Dow CEI is a measure of the relative acute toxicity risks due to liquid spill, evaporation, and dispersion of a vapor cloud.

3.6.2.2 HazOp[a], hazard and operability study

One of the most successful hazard identification methods is HazOp, which evolved as a method within ICI in the UK in the early 1970s. It started as an operability study on a new process, but while doing the study hazards became recognized and analyzed.[105] It was developed over the years to a well-proven team-based method structured to identify facility hazards during process design, completion, or planned modifications. There is a great deal of literature on HazOp; we shall follow a guide by Crawley, Preston, and Tyler.[106] HazOp is essentially a qualitative procedure in

[a]Nowadays, hazard & operability is usually abbreviated to HAZOP, all in capitals; it is not the acronym H.A.Z.O.P. but a contraction of Haz & Op, so in this section in recognition of the roots we shall keep it as HazOp, in older articles also written as hazop and Hazop. In the rest of the book we shall conform to the current notation.

Table 3.5 Standard HazOp Guide Words and Their Generic Meanings

Guide-word	Meaning
No (not, none)	None of the design intent is achieved
More (more of, higher)	Quantitative increase in a parameter
Less (less of, lower)	Quantitative decrease in a parameter
As well as (more than)	An additional activity occurs
Part of	Only some of the design intention is achieved
Reverse	Logical opposite of the design intention occurs
Other than (other)	Complete substitution—another activity takes place
Other useful guide words include:	
Where else	Applicable for flows, transfers, sources, and destinations
Before/after	The step (or some part of it) is affected out of sequence
Early/late sequence	The timing is different from the intention
Faster/slower	The step is done/not done with the right timing

which a small team consisting, for example, of a process engineer, a control instrument expert, a risk analyst, and an operator, under an experienced chairman, examines a proposed design by generating questions about it in a systematic manner. For this purpose a number of "guide words" indicating potential deviations of a process parameter from the design intention that question the (often hidden) possible causes and the effect on process behavior, hence causes and effect of a possible fault. Commonly used guide words are given in Table 3.5. The HazOp study is function driven. Thus potential safety and operability problems are identified and appropriate actions can be taken. A poorly operable process is usually also unsafe.

Process parameters can be chosen from a wide variety, but the first five of the set mentioned hereafter are the most frequently used: flow, pressure, temperature, mixing, stirring, control, pH, level, sequence, viscosity, signal, reaction, start/stop, composition addition, separation, services, operate, maintain, time, communication. Meaningful combinations shall be selected, as suggested in Table 3.6.

Table 3.6 Examples of Meaningful Combinations of Parameters and Guide Words

Parameter	Guide words That Provide a Meaningful Combination
Flow	None; more of; less of; reverse; elsewhere; as well as
Temperature	Higher; lower
Pressure	Higher; lower; reverse
Level	Higher; lower; none
Mixing	Less; more; none
Reaction	Higher (rate of); lower (rate of); none; reverse; as well as/other than; part of
Phase	Other; reverse; as well as
Composition	Part of; as well as
Communication	None; part of; more of; less of; other; as well as

The questions arise creatively through interaction (brainstorming) among the team members. Sometimes a guide word first approach is chosen; other times preference is given to calling the parameter first. The team secretary or scribe keeps track of the findings in a format shown below:

Number	Parameter	Deviation	Cause	Consequence	Safeguard	Action	On Person

For the exercise, the P&ID is sectioned in analysis nodes to enable paying attention to required details. The optimum order of discussing the sections can be an issue. A consequence in one node may by its impact be a cause for a more severe consequence in a next node. To assist in the identification of hazardous deviations, the team will usually find it helpful during the exercise to compare the proposed design with relevant engineering standards. In using the HazOp method, the need for action is decided semiquantitatively based on the team's experience and judgment of the seriousness of the consequences, together with the expected probability (frequency) of the occurrence. In a situation in which uncertainty remains, analysis using quantitative techniques may be helpful, such as prioritization of actions to reduce risk. Although identification is carried out vigorously and to a simple extent fault tree paths are probed, very detailed fault analysis is not normally a systematic part of the HazOp procedure. Its main purpose is to identify the main hazards and operability problems. In-depth analysis of high hazard/low probability events is the domain of the next steps in risk analysis. Commercial software packages are being used to record the actions and conclusions of the HazOp team; this is normally done as the HazOp proceeds with the computer screen being projected so that all team members can see it and agree on the documentation being produced by the secretary. In small HazOp studies, the roles of chairman and secretary may sometimes be combined.

In addition to its open-ended approach, a fundamental strength of HazOp is the brainstorming and thinking together among members of the study team. Its success depends on the degree of cooperation among individuals, their experience, competence, and the commitment of the team as a whole. A guarantee cannot be given that during a study all possible potential problems have been unearthed. To maintain the team's focus (it is often not a highly inspiring activity!) and not to interfere too much with the members' other work, time devoted is often limited to half-day. Despite various trials to automate at least part of the procedure, IT support is mainly constrained to administrative aspects. The main complaint is therefore the time involved performing a HazOp and therefore the costs (e.g., 5 × 20 h for an average P&ID, hence for a plant 1–8 weeks). Since a HazOp is repeated once every 5 years, the search for more efficient methods is continuing. In Chapter 7, we shall discuss HazOp developments with respect to automation, embedding it in a system approach in the blended Hazid method, which promises great advance, while at the end of Chapter 8 its application to noncontinuous operations, such as start-ups, shall be explained.

3.6.2.3 Identified hazards/events to be analyzed in depth

The methods reviewed in Section 3.6.2, in particular HAZOP and FMEA, produce the constituents for building numerous, rather detailed scenarios of combined or propagating failures and malfunctions leading to a loss of containment (LOC) event. Making quantitative calculations for each of these would require an enormous effort, and this is often not necessary because many, perhaps even the majority of, potential scenarios can be eliminated or at least reduced in likelihood. This can mostly be achieved in a qualitative way by risk-reduction measures such as installing another barrier/layer of protection. For those that remain, a selection has to be made on the basis of priority. Priority may follow, based on contribution to the overall risk, from expected consequence weighed by likelihood and made visible in a risk matrix, as explained above. If a safety instrumented system (SIS) must be selected and a full quantitative layer of protection analysis is required, a detailed risk analysis may be necessary for a particularly critical scenario to estimate an initiating event frequency and expected consequences, as will be discussed in Section 3.6.6.2.

In contrast to a "classical" approach to QRA by considering a detailed scenario, for land use planning purposes and assessment of general risk level to people both inside and outside a plant, a simpler top-down approach can be followed as, for example, applied in the Netherlands. For the latter, an inventory of all forms of containment (tanks, reactors, columns, pipe works) is made and of pumps, compressors, and safety valves so that possible released quantities can be estimated. Next, upon failure, one can assume three release rates: full rupture of the containment and instantaneous release; a release from a large leak over a period of, for example, 10 min; and a semi-continuous release from a small leak. Although the release rates decrease with these forms of leak and with that, also the severity, the leak frequency will increase.

As mentioned at the end of Section 3.6.1, scenario identification obtained in the EU ARAMIS project received special attention. Delvosalle et al.[107] reported their approach. The selection of scenarios consists of two steps, MIMAH and MIRAS. MIMAH is "Methodology for the Identification of Major Accident Hazards" and concerns the worst possible accidents when no safeguards are present. The second step, MIRAS is "Methodology for the Identification of Reference Accident Scenarios" or RAS, and represents realistic cases. Delvosalle et al.'s article gives an example of ethylene-oxide storage.

MIMAH is based on the bow-tie representation of scenarios within the center of a critical event (see CE in Figure 3.23), which for a liquid or gas leads to an LOC and for a solid to a loss of physical integrity (LPI) by a chemical or physical change of state. After collecting information, one identifies potentially hazardous equipment in the plant for storage, transport, processing, and pipe networks, distinguishing these in among 16 categories. So, a listing can be made of hazardous substance, its quantity and physical state, and the equipment category that it occupies. On the basis of properties and threshold quantities, a selection of cases is made. Next, from a set of 12 critical event types for each case selected, relevant events are picked, which leads to a CE. The 12 CE types are: decomposition; explosion; materials set in motion (*a.* through entrainment by air, *b.* by liquid); start of fire (LPI); breach on the shell (*a.* in vapor phase, *b.* in liquid phase); leak (*a.* from a liquid-filled pipe, *b.* from a gas-filled pipe); catastrophic rupture; vessel collapse; and collapse of roof.

Also, deductively for each CE a generic fault tree (FT, in the bow-tie from the CE to the left) is built over four levels down, via necessary and sufficient conditions for a CE, to direct causes (e.g., corrosion), detailed direct causes (e.g., why corrosion), to the lowest level consisting of generic, undesirable events of human behavior and organizational deficiency. From 14 generic FTs, so developed, through adaptation a specific tree is derived for each case defined above. Subsequently, an event tree for each CE is constructed. The CE causes secondary CEs, for example, pool formation or vapor cloud, and tertiary ones, for example, pool fire and flash fire, to dangerous phenomena (DPs) (see Figure 3.23) of which 13 are defined: pool fire, tank fire, jet fire, vapor cloud explosion (VCE), flash fire, toxic cloud, fire, ejection of missiles, overpressure generation, fireball, environmental damage, dust explosion, and boil over with resulting pool fire. Finally, major events are defined as the damaging effects on exposed receptors: people, structures, and the environment, through radiant heat, blasts, projectiles, and toxic clouds.

MIRAS then selects safety systems and procedures from the MIMAH bow-ties, after further information has been collected on failure rates. On both the fault tree and the event tree, safety functions and barriers must be identified, and their performance assessed, after which the frequency of the critical event and of the dangerous phenomena can be calculated from bottom-up computation of the FT. Subsequently, a rough categorization is made of consequence severity into four classes in which, for example, a tank fire is class 1, a pool fire class 2, a flash fire class 3, while a vapor cloud or toxic cloud are both class 3 or 4 depending on quantity or for toxic cloud, toxicity of species present. Frequency and consequence class results are plotted in a risk matrix and bow-ties located in the red and yellow zone selected for further analysis. The lower bound of the yellow zone stretches from 10^{-3}/year for class 1 severity to 10^{-7} for class 4. From the foregoing, it is clear that the ARAMIS method falls somewhere between a detailed scenario approach and a rough top-down one. Because of the effort involved, its application may be limited to high-risk installations.

3.6.3 STEP 2. QUANTIFICATION OF CONSEQUENCE

When in Section 3.5.1 we presented "Properties of substances and unintentional release phenomenology," an overview of source terms, spread mechanisms, and effects were discussed. From Figure 3.28, the diversity of release paths and effects can be seen.

As mentioned in Section 3.5.1, for model calculations of each phenomenon, commercial software can be used (e.g., DNV's PHAST and TNO's EFFECTS). A more approximate result can be obtained applying the ALOHA software (see Section 3.6.6.6), while suitable computational fluid dynamic codes can yield higher resolution three-dimensional results. Insight into the physical and mathematical basis of the models can be gained with the aid of, for example, Lees,[38] Crowl and

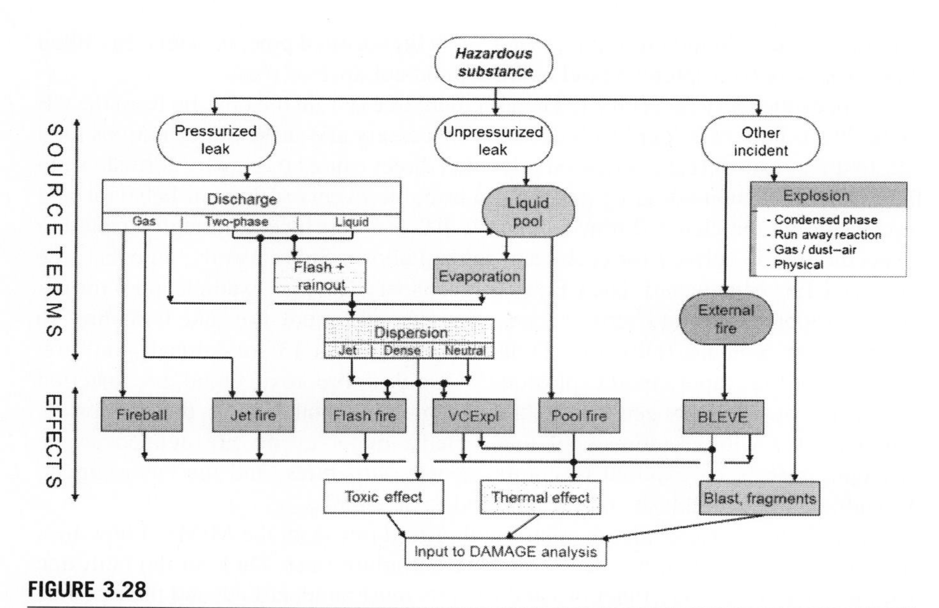

FIGURE 3.28

Logic diagram for consequence analysis of releases of flammable substances (*adapted from Pitblado and Turney.*[108]). Jet fire is similar to torch fire. Tank fire is more locally restricted than pool fire. Dust explosion is in effect a combination of flash fire and blast.

Louvar,[39] the Dutch Yellow Book[80] and the compilation by Joaquim Casal,[84] Universitat Politècnica de Catalunya. Given the intensities of the effect as a function of distance from the risk source, and assuming a geographical and spatial relationship between the risk source and susceptible threat receptors, then by means of the literature referenced in Sections 3.5.1.1−3.5.1.6, probability of injury, fatality, structural damage, and environmental damage can be determined. In principle, the potential receptors closest to a release point will be the plant's operating personnel, and the operating site and the equipment that it contains. In the context of domino effects, a topic that we shall address in Section 3.6.6.5, Cozzani et al.[109] published threshold data on plant equipment vulnerability to escalation vectors resulting from a primary release such as blast or fragment impact, fire engulfment, or more distant radiant heat effects. Zhang and Liang[67] improved the probit correlations of the effect of blast on equipment.

For determining major effect severity, the ARAMIS project refers also to sources like those that are mentioned above. However, where other models are subsequently used to calculate the damage to receptors directly, ARAMIS collects results first in an aggregated severity index mesh and separately determines the vulnerability of receptors in the area of interest as a function of threat intensity level. In that way, for a process activity one is able to more easily try out various locations before plans become fixed. For determining an index value of a dangerous phenomenon, the intensity range of interest for a major event is spread over index values from 0 to 100.

The range of interest is determined by a low and high threshold of the probit response function. Just as with the intensity, the index value is also a function of distance to the risk source. For each DP in a reference accident scenario (RAS) event tree, the severity indices are multiplied by the expected frequency of the DP, and the resulting values are summed to a RAS index. Next, the RAS indices are multiplied with the frequency of a RAS and again summed to an aggregated risk index that can be mapped on a mesh and overlaid on a receptor area map, for instance, provided on a geographical information system infrastructure.

3.6.4 STEP 3. QUANTIFICATION OF PROBABILITY OF EVENTS, FAILURE RATES

If one asks about the likelihood of an undesired event, the first thought will be to estimate the probability of a deviation of a planned path of operation assuming that this planned path will be safe given the specified conditions. (In Chapter 5, we shall find that this assumption does not always hold.) A deviation can be initiated by an error event initiated by people or by failure or malfunction of equipment. If one digs deeper, it will appear that every deviation is caused by an incorrect human intervention, as we shall see later, but here for the purpose of "traditional" risk analysis we shall consider equipment failure as the primary cause of mishap and human error as a contributing factor or underlying cause.

Indeed, in process plants MI is an important issue. A large fraction of piping and other containment is under pressure and/or elevated temperature and fluids can contain aggressive ingredients. Leaks can occur directly or indirectly due to a large variety of failing components: gaskets, pipes, pumps, valves, controls, vessels, tanks, et cetera. Apart from stress resulting from faulty operations exceeding ultimate strength, failure can occur by propagation of fracture due to material defects, wrong design, fatigue (e.g., by vibration), sagging, erosion, and corrosion, which all can reinforce each other. Corrosion alone comes in quite diversity: stress, pitting, at welds, under insulation, and microbiological or bacterial, known as microbial corrosion to name a few. Also, hydrogen embrittlement of steel, or attack at elevated temperature by hydrogen sulfide and sulfur, or active nitrogen cause failures. Although there is much preventative knowledge available, degrading conditions may persist for a long time unnoticed (see Section 3.6.6.4). All of the above can appear in steady-state operation, but in addition there may be transient pressure and temperature excursions causing explosion or shock (e.g., water hammer). It can therefore be concluded that determining failure probability with any accuracy is rather illusory.

However, for the time being and for the risk analysis performed here, we shall restrict ourselves to the objective, physical context of probability as is the case in reliability engineering. A failure rate, also called failure frequency or failure probability over a year, can be determined by conducting a trial. When a number of items are running of the same make, age, and load, records of the times at which they fail, the conditions, and the modes of failure must be kept. Then, depending on the number and the time duration of the trial, assuming a distribution function and applying routine statistics, a mean fail rate value with confidence limits can be calculated, see, for example, Modarres et al.[91] Usually an exponential distribution and hence a constant failure rate as the single parameter is assumed, but reality can be different as is found from descriptive statistics of the collected data (e.g., a fail rate following a bathtub curve against time, which can be fitted with a two-parameter Weibull distribution). This is apart from maintenance and repair actions, which can complicate the picture. Alternatively, a failure rate of a larger component or subsystem can in principle be calculated from the reliability of its parts by applying a fault tree. As small items can be tested more easily in larger numbers, this can be done easily for electronics or for important parts in a nuclear power reactor, for example, a pressure relief valve (PRV), although with the expense of significant effort. However, for a common installation in a process plant, this is not an option that promises success. In the first place, there is a lack of component reliability data. Even if generic data are provided by a supplier, the values may start deviating systematically because of the specific environment in which they are operating. The success rate will increase with the size of the company's databank. Hence, in a multinational operating company with well-standardized installations and reliable maintenance practices, it will certainly be possible to collect fairly good data, although even then, designer, construction worker, and operator influences cannot be excluded. Considerable uncertainty remains.

Therefore, risk analysts use historical data and sometimes even expert estimates of failure rates of integral higher-level components as pressure vessels, tanks, piping, valves, pumps, heat exchangers, et cetera. One of the oldest sources is the Purple Book[110] of the Dutch QRA Guidance series. This collection is the result of working groups with industry-government representation meeting under the auspices of the Committee for the Prevention of Disasters, itself founded in 1964. The published data originate from the 1980 and 1990s The history of these data has been described by Beerens et al.[111] in the ARAMIS project issue of the *Journal of Hazardous Materials* and later commented upon by Pasman.[112] For example, for pressure vessels, to a significant extent nonrelevance of original data, exclusion of human factor influence, and a bias in view of risk acceptance criteria affected the data. Later, after safety cases on offshore installation in the North Sea became obligatory, the OREDA database[113] became a source for common components. Robert Taylor's reports contain many plant-specific data.[114] Taylor[115] also reported on failures resulting from design errors and sometimes only manifesting themselves after years of operation. Establishment of the European Working Group on Land Use Planning in 2003 (Chapter 2) was aimed by the European Commission at collecting reliable data for risk assessments but did not generate satisfactory results. Although efforts continue to improve the situation further, for the time being, the HSE collection for risk assessments[116] is rather clear and useful. Although not fully finished, it contains or will contain not only failure rates but also event data such as ignition probabilities and probabilities of lightning strikes, earthquakes, and flooding. In particular, ignition probability is important because many flammable vapor spills may either ignite immediately or not at all. But, when after a few minutes of cloud dispersion delayed ignition occurs and the total quantity of flammable material between explosion limits is over one ton, according to Vilchez et al.[117] a destructive vapor cloud explosion is possible (Concerning ignition, see further the fourth suggestion at the end of Section 3.7.).

As regards the influence of human factors, the HSE collection gives some advice, but this still remains a difficult area. Auditing management quality[118] and how to allow for the effect of its results on failure rates[119] was tackled in the ARAMIS project and later discussed by Pitblado et al.[120] In Chapters 6 and 7, we shall return to this particular subject with some new insights on human reliability, the influence of management on it, and the effect of dynamic interactions of humans with plant installations.

3.6.5 STEP 4. QUANTIFIED RISK

Various commercial software packages perform the calculations described in previous sections. Harm to people resulting in injury or death is of primary concern, followed by damage to structures and the environment. In many accidents, there are only injuries and no fatalities. However, injury probit equations are sorely missing, at least for toxic and blast consequences. This is because injury is not easy to define with respect to severity level, chance of recovery, and residual capabilities. Hence, so far injury risk has not been presented. Lethality is of course much more serious as a consequence, but it does have the benefit of being well defined, so that risk of harm to people is usually expressed as the probability of being killed. In accidents, the ratio between the number of injuries and fatalities varies widely, so this also does not give a clue.

Most direct is the value of *individual risk* at a certain location with respect to the risk source (IR, or rather LIRA, localized individual risk per annum, IRPA). This expresses the probability of death at that location due to the presence of the risk source over a certain period of time, usually 1 year, when continually exposed. It is typically used to express risk of the public. If consequence severity values are calculated for a meshed area around a risk source, points of equal risk can be connected to produce individual risk contours at various levels, for example, 10^{-4} per annum or per year (abbreviated as /year), 10^{-5}, 10^{-6}, or 10^{-7}/year. If directional effects are absent, the contours are circular; if a dominating wind direction exists, or directional blast or fragment projection occurs, contours become ellipsoidal or even irregular. In the case of assessment of fire risk in a building or risk of a worker in a plant, IRPA is used. This measure is defined as the average value of the risk measured over a year of a person being killed by fire when residing in that building or of a worker being exposed during that time in his plant. With respect to a worker, IRPA will be related to the FAR, although that metric is often used descriptively (in hindsight) and for a site or sector; see also, below under societal risk.

Societal or group risk (GR) is derived from the frequency of an event and the expected total number of people instantaneously harmed to a certain level by that accidental event. For the reasons set out above, that harm level is usually death. Because

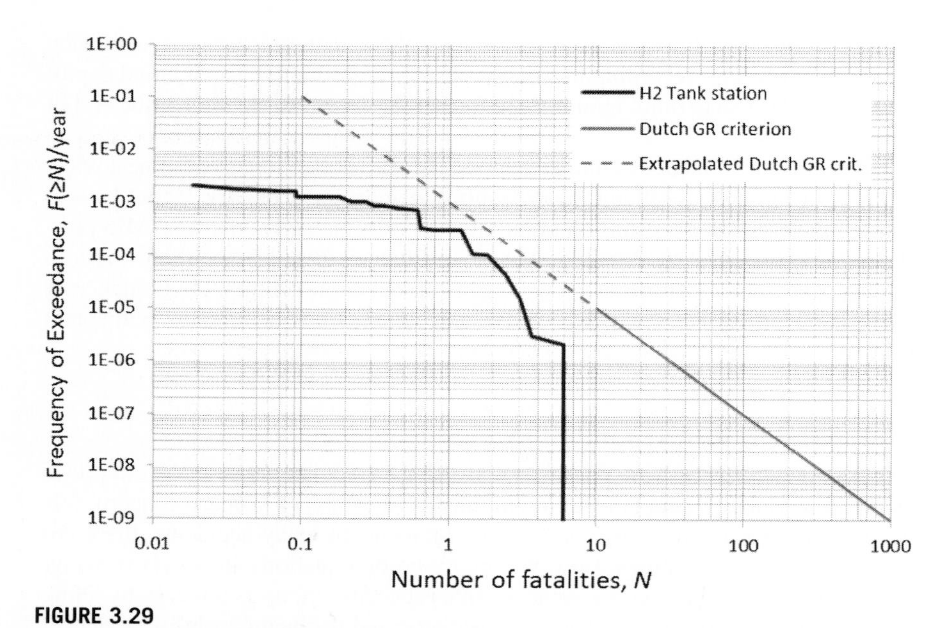

FIGURE 3.29

Group-risk plot for a possible future hydrogen tank station and the official Dutch straight-line guide criterion or advisory limit, which if surpassed from below (not the case here) requires the competent authority and the local mayor to consider whether a license is justified in view of possible emergency consequences and response measures.

the physical and emotional impact on society as a result of an accident increases sharply with the number of victims, GR is a measure of societal disruption. As seriously damaging accidents occur less frequently than less severe ones, the societal risk resulting from a plant site can be presented as the plot of the cumulative frequency of the exceedance of a number of fatalities ($F \geq N$) versus the number of fatalities (N). For an example, see Figure 3.29, which was calculated for a hydrogen distribution tank station making use of Yellow Book[80] models (Section 3.6.3) and the free GeNIe Bayesian network software instead of commercial codes. Bayesian causal networks will be treated in Chapter 7. The station is supplied by compressed gas and not liquefied hydrogen.[121] Group-risk plotting accumulation starts with the scenario with the highest number of fatalities, and the curve breakpoints are the result of the successively more frequent but less severe scenarios. For this hydrogen station case, the individual risk circular contour of 10^{-6}/year had a radius of 15 m. For calculating GR, data shall be available of the population density within the area of interest. Given the effect intensities for a certain case, this area can be predetermined by calculating the 1% lethality borderline. The distribution of people within the area is assumed to be homogeneous.

In the Netherlands, TNO and RIVM in a cooperative project investigated the feasibility of displaying GR on a map. An example is shown in Figure 3.30. Because

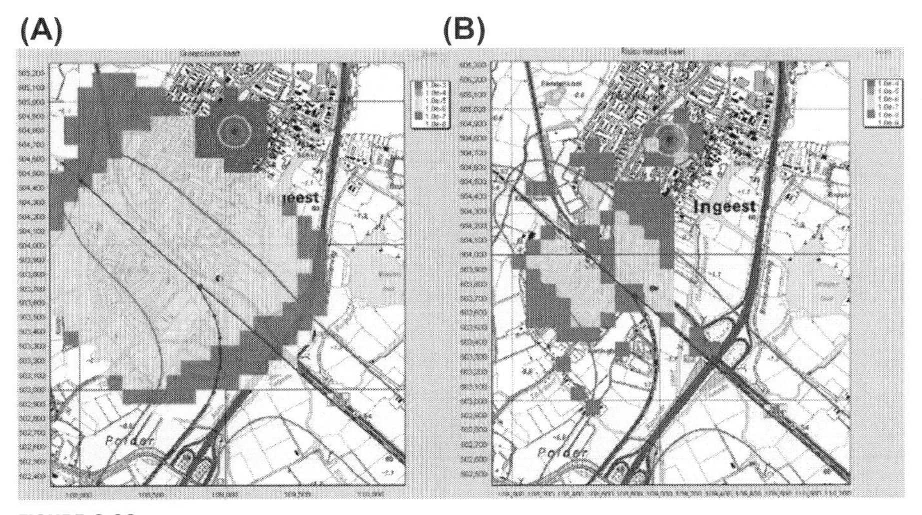

FIGURE 3.30

TNO/RIVM[122] representation of societal risk as a location-specific value. The example concerns a railway shift yard and a stationary risk source. The small circles represent individual risk contours around a static risk source. The map mesh size is $50 \times 50\ m^2$. Population density data are embedded in each cell. Left: A group risk $F-N$ curve is calculated for each cell taking into account only risk sources that affect the cell but counting fatalities due to this set of sources in the whole area. Right: the same, but only counting fatalities in the 50-m^2 square cell. Subsequently, the position distance of the calculated $F-N$ curve versus the norm is translated into color: Left: red (or in grayscale print dark area middle top with circle and dot) is above the norm guide value; orange (not present in the example) a factor 10 below; yellow (outlined diagonally dashed area center left) a factor 100 below, light green (periphery lightly vertically dashed) safe, and green (periphery dark) very safe. Right figure serves to find more easily the hazard "hot spots." The coloring proved helpful to decision makers.

some prefer a figure instead of a plot, Hirst and Carter[123] looked at alternative possibilities. A rather convincing measure is the expectation value, $EV = \sum_{N=1}^{N_{max}} f(N)N$, also called potential lives lost, *PLL*, in which *f* is frequency of the scenarios considered for which a corresponding number of *N* fatalities is calculated. For the case mentioned above, *PLL* is approximately 0.0005/year, or 500 cpm (chances per million per year). A further measure is the risk integral, *RI*, in which the number of fatalities is given a higher weight as perceived emotionally by the population at large. The formula then becomes $RI = \sum_{N=1}^{N_{max}} f(N)N^a$, where for exponent a, risk aversion parameter, a value of 1.4 is proposed; this change would mean that *RI* is now nearly 600 cpm compared to 500 cpm with $a = 1$ for a risk neutral calculation. Given that conditions are homogeneous, from the *PLL* over a whole plant area or a site, an estimate of FAR (per 100 million working hours) can be obtained. For this calculation, the population density is that of the workers in the plant, while the number of fatalities multiplied with 100 million hours and then divided by the exposed working hours per annum for all exposed staff produces a *FAR* value for the scenarios considered.

3.6.6 STEP 5. RISK REDUCTION

3.6.6.1 Means of risk reduction

Risk reduction can be realized by having a highly reliable organization and a SMS to guard and preserve that state, as we have discussed in Section 3.2. In addition, safeguards or barriers are required. For this aspect, the present section will go into some more detail than previous ones because of its importance with regard to process safety. Barriers come in many forms. An overview of possible barrier types and examples as presented in the ARAMIS project is reproduced in Table 3.7. Barrier Nos. 1, 3, 4, 8, 9, and 10 are preventive (or proactive) types, which will be found to the left of the critical event (CE) in the bow-tie of Figure 3.23 in Section 3.5.2.4. The remaining barriers, Nos. 2, 5, 6, 7, and 11 are protective (or reactive) ones to the right of the CE. An additional significant distinction is between passive and active barriers. The former, for example, bunds, are continuously present in the same state and reduce the effects of a release; the latter are barriers that are triggered by a process variable when an upset is imminent and which will actively bring a process back into a safe state. Examples are a PRV, a flow-limiting device, or a block valve. It is also possible to avoid an unsafe state, for example, by prohibiting backflow with a check valve or an unsafe sequence of actions by means of various types of interlocks. Mitigative barriers such as water sprays just reduce effects and therefore provide only partial protection. After a barrier functions, the process may continue (with a preventive barrier) or shut down (with a protective one). The measures to prevent ignition of a flammable gas or dust cloud as specified in the EU ATEX Directives described in Chapter 2 form a special group.

As we have seen, bow-ties present an effective system overview of where in cause—consequence chains barriers are present. For this to visually be true, they should not contain too much detail data. Barriers to protect against severe consequences or to mitigate effects may be qualified as *safety critical*. As reliability of any measure is limited, these shall be arranged in independent layers of protection to achieve a sufficiently high safety level. The analysis of overall performance of such a system, the layer of protection analysis (LOPA), is described in the next section.

Finally, emergency response is also part of risk reduction. As we shall see, it appears in the layers of protection as it can be regarded as an independent layer. However, effective emergency response requires training, preparation, and testing. This is one of the purposes a risk analysis will serve. Therefore, we shall finish the description of this step with a short discussion of emergency response planning.

By introducing risk-reducing measures, we must check that risk is not transferred to another mechanism or to other locations, for example, extinguishing a fire of a leaking toxic flammable will become a toxic cloud risk with effects at much greater distances. A warning for another potential drawback of relying on measures is that

Table 3.7 Barrier Types as Formulated in the ARAMIS Project, According to Guldenmund et al.[118] (with a few additions)

Barrier	Examples	Detect	Diagnose/ Activate	Act
1. Permanent–passive, MORT[a] control	Pipe/hose wall, anticorrosion paint, tank support, floating tank lid, viewing port in vessel	None	None	Hardware
2. Permanent–passive, MORT barrier	Bund, dyke, drainage sump, railing, fence, blast wall, lightning conductor, bursting disk	None	None	Hardware
3. Temporary–passive, put in place (and removed) by person	Barriers round repair work, blind flange over open pipe, helmet/gloves/safety shoes/goggles, inhibitor in mixture	None	None (human must put them in place)	Hardware
4. Permanent–active	Active corrosion protection, heating/cooling system, ventilation, explosion venting, inerting, vacuum/pressure purging, sweep-through and siphon purging system	None	None (may need activation by operator for certain process phases)	Hardware
5. Activated–hardware on demand, MORT barrier or control	Pressure relief valve, interlock with "hard" logic, sprinkler installation, p/t/level control,	Hardware	Hardware	Hardware
6. Activated–automated	Programmable automated device, control system or shutdown system	Hardware	Software	Hardware
7. Activated–manual, human action triggered by active hardware detection(s)	Manual shutdown or adjustment in response to instrument reading or alarm, evacuation donning breathing apparatus or calling fire brigade on alarm, action triggered by remote camera, drain valve, close/open (correct) valve	Hardware	Human (s/r/k)[b]	Human/ remote control

Continued

Table 3.7 Barrier Types as Formulated in the ARAMIS Project, According to Guldenmund et al.[118] (with a few additions)—cont'd

Barrier	Examples	Detect	Diagnose/ Activate	Act
8. Activated–warned, human action based on passive warning	Donning personal protection equipment in danger area, refraining from smoking, keeping within white lines, opening labeled pipe, keeping out of prohibited areas	Hardware	Human (r)	Human
9. Activated–assisted, software presents diagnosis to the operator	Using an expert system	Hardware	Software–human (r/k)	Human/ remote control
10. Activated–procedural, observation of local conditions not using instruments	(Correctly) follow start-up/shutdown/ batch process procedure, adjust setting of hardware, warn others to act or evacuate, (un)couple tanker from storage, empty and purge line before opening, drive tanker, lay down water curtain	Human	Human (s/r)	Human/ remote control
11. Activated–emergency, ad hoc observation of deviation + improvisation of response	Response to unexpected emergency, improvised jury-rig during maintenance, fight fire	Human	Human (k)	Human/ remote control

[a] *MORT means Management Oversight and Risk Tree, see Section 3.4.*
[b] *Cognitive effort to carry out these tasks—skill (s), rule (r), or knowledge (k) based, see Section 6.1.*

operators may not be aware of measure degradation until it is too late—"the fallacy of defense in depth." Hence, sufficient testing is a must.

With respect to transportation, an overview of possible risk-reduction strategies and measures is given in a 2008 CCPS Guideline.[124] This guideline includes security considerations.

3.6.6.2 Barrier effectiveness analysis, layer of protection analysis

Because individual barriers may have only limited effectiveness, it is important to analyze the overall protection. Although a QRA can serve this purpose, it is usually considered too costly. Instead, as a team effort, a mini-QRA or rather order-of-magnitude QRA developed, called LOPA, which is performed focusing on one type of *critical release event*, hence one single scenario of a specific initiating event and consequence. It means that a LOPA must be preceded by a PHA or risk identification to single out the scenario that deserves further analysis. The method of choice to do this is a HAZOP. So, indeed, a LOPA is a risk analysis with the limited scope of a single scenario. Protection layers are special barriers designed such that if attacked they function in sequence as layers of defense, and events will terminate only if the attack is fully stopped. The layers must therefore remain independent of each other in their functioning. Various types of layers are distinguished as shown in Figure 3.31.

As a process safety and risk control tool, LOPA has become as successful as HAZOP and is applied worldwide. The first LOPA papers appeared in 1997 by Dowell III[126] and Huff and Montgomery.[127] Although the latter did not mention the acronym LOPA, they described the quantitative order of magnitude evaluation well. In 2001, CCPS[125] published a guideline on how to perform a LOPA. Although in the beginning the LOPA terminology was not fully established, over the years it became more

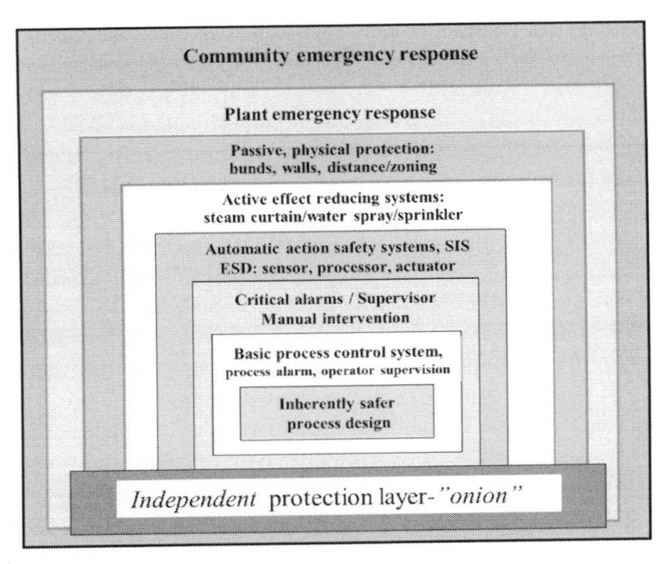

FIGURE 3.31

Various "layers" in plant operation. Shown are normal process control (alarms may be activated and corrections made), abnormal course of process developing (critical alarm activated) requiring intervention or shutdown (trip); if situation develops faster and more serious the safety layer of emergency shutdown can be activated (manual or automated), while in addition active safety systems shall protect for larger damage (safety instrumented systems – SIS – and actively effect reducing devices); in case the latter are breached, permanently present passive layers shall mitigate effects, and ultimately emergency response shall act.

After Wikipedia

developed and ingrained. Initiation begins with a system to which the inherently safer design approach has been applied: minimized extreme toxic inventories, et cetera. Next will be a reliable basic process control system (BPCS), which in a modern plant has the form of a distributed control system (DCS, see Chapter 8). A BPCS failure is one of the causes of an initiating event, for example, onset of a reactor runaway or overflow of a tank during filling. A protective system is a dedicated add-on designed to compensate for the shortcomings of a BPCS. Barriers can be temporary or permanent, but protection layers that we are addressing here are always meant to be permanent. A safety instrumented system consists of a sensor, logic and an actuator. The critical situation is often too high pressure; the actuator is in that case a pressure relief valve. Actively release effect reducing devices are for example water deluges or steam curtains. Passive are bunds, dykes, walls and zones. Plant emergency response operations backed by the community emergency response are recognized in the CCPS guideline as independent protection layers. Of course, emergency responders will rush to the scene only after being alerted, but they are permanently on stand-by. The system of layers around the risk source has been pictured as an "onion".

LOPA as a concept was derived from a CCPS Guideline on Automation issued in 1993. The process industry was then in the middle of introducing automated safety devices. Technology was enabling it and it reduced the chance of human error. Standards ANSI/ISA-S84.00.01 (American National Standards Institute/The Instrumentation, Systems and Automation Society) and IEC 61508 (International Electrotechnical Commission) defining safety integrity levels (SILs), were developed. Basically, potentially hazardous plants are safeguarded by alarms, trips, and other devices, such as control valves. These safeguards form the BPCS, supervised by operators. These controls serve to ensure stability, optimize performance, and suppress the influence of fluctuating external disturbances. In addition, in the 1990s for a variety of reasons a class of highly reliable SIS was developed to bring a process to a safe state in the event of a process upset. SISs are also called Emergency Shutdown System (ESD), Safety Interlock System, or Safety Related System (SRS) and if extremely reliable also a High Integrity Protective System (HIPS, or sometimes HIPPS for high integrity pressure or pipeline protection system). The reasons for introduction were fundamentally economic ones. Plant design and operation had been optimized and also become more complicated, and plants needed to be operated closer to the safety limits. Other contributory reasons were that more hazardous processes were introduced, operations became more remote (e.g., offshore), and the workforce was downsized. In fact, these automated SISs control the plant as a last resort backup independent of the BPCS. The requirements for their reliability are extremely demanding.

As shown in Table 3.7 the protective elements can consist of a person acting while following a procedure, or of hardware components. In addition, the (micro-) processor or logic solver can be in the form of software, in which case it is known as a programmable logic controller (PLC). As mentioned an SIS is a combination of a sensor, for example, a pressure transducer, a PLC, and an actuator, for example, an interlock or a PRV, forming together a protection layer SIS. Because it is programmable, as opposed to the older relays or hardwired trip amplifier, the whole is also called a programmable electronic system (PES). An SIS performs a safety-instrumented function (SIF). For example, a SIF prevents a vessel from bursting if an initiating event occurs such as loss of control of a gas-producing reaction, for example, due to incorrect dosing of feed or catalyst.

In the second half of the 1990s, at about the same time as the introduction of LOPA method, it became possible to determine the reliability of electronic circuits, programmable devices, and other components with a high degree of confidence, which offered needed dependability for the SISs. This development prompted further drafting of international standards on requirements for safety-instrumented protection systems for use in the process industry with respect to their safety integrity level (SIL) and the safety life cycle of the plant. The standard for the process industry is IEC 61511 (3 SILs) published in 2003;[128] this is derived from the more general standard, IEC 61508 (4 SILs) of 2001, and with an American equivalent, ANSI/ISA 84.00.01 (2004)[129] (first published in 1996). These standards were the first to be based on a probabilistic basis, as they require a (semi-) QRA to be carried out applying reliability figures. Previously, one simply applied a barrier to protect against a hazard. The SIL levels defined in Table 3.8 specify both the *probability of dangerous failure on demand* (*PFD*) of an SIS designed to function in low demand mode, and of *probability of dangerous failure per hour* ($PFH = \lambda_{AVG}$), in high demand mode. The former are powered only in an emergency, while the latter are (almost) continuously powered and operative and may have a latent probability of dangerous failure. Such failure goes unnoticed when not tested and hence in the case of demand will lead to mishap. (In contrast to dangerous failures, safe failures transition a system into a safe state.)

Initially, various companies developed their own approach while applying LOPA. The working group of industrial practitioners organized by CCPS collected and sifted through the material obtained from experience, which resulted in the CCPS guideline[123] of 2001. Briefly, after a HAZOP team has identified a particular serious release scenario that could occur, a separate team will carry out a LOPA to be performed by first categorizing the possible consequence of the release due to the initiating event. The consequence severity will depend on the properties of the released material, the size of the release, and the possible occurrence of exposed personnel casualties (injury or fatality). The latter adds to the uncertainty, because

Table 3.8 Safety Integrity Levels 1–3 (SILs) of Safety Instrumented Systems (SISs) as Specified in Standard IEC 61511 (λ_{AVG} is the independent, constant failure rate or rate of occurrence of failure, ROCOF, to be distinguished from the conditional failure rate, given past survival time, *t*)

SIL	Probability of Dangerous Failure on Demand, *PFD*	Probability of Dangerous Failure λ_{AVG} (h^{-1}), *PFH*
	Low Demand SIS	High Demand SIS
3	$\geq 10^{-4}$ to $< 10^{-3}$	$\geq 10^{-8}$ to $< 10^{-7}$
2	$\geq 10^{-3}$ to $< 10^{-2}$	$\geq 10^{-7}$ to $< 10^{-6}$
1	$\geq 10^{-2}$ to $< 10^{-1}$	$\geq 10^{-6}$ to $< 10^{-5}$

the likelihood of personnel being present must be estimated in addition to estimation of the likelihood of the release. Therefore, as described in the CCPS LOPA guideline of 2001, the category classes are first arranged in a matrix of orders of magnitude of quantity of material possibly released. In addition, the material property can be classified as combustible, flammable below atmospheric BP, highly toxic or flammable above the BP, extremely toxic below the BP or highly toxic above the BP, or extremely toxic above the BP. For further steps, the type of damage to the plant can be estimated as can the time for which the plant will be not operational while repairing the damage. Also, qualitative estimates of human harm among workers, for the public outside the plant, and damage to the environment can be made, not to mention the damage to the reputation of the company. This can all be converted into estimates of orders of magnitude of financial loss, as we shall see when we discuss full QRA.

Then comes an estimate of the likelihood of the initiating event. The event can be primarily due to mechanical failure because of wear, corrosion, vibration, internal explosion, hydraulic hammer, defect or design error of some component, operational human error, maintenance fault, or an external event such as earthquake, lightning, storm, or a violent event in an adjacent installation, all of which the BPCS cannot contend with. From the variety of possibilities above covering only the most frequent ones, it is already clear that uncertainty associated with an estimate is high. As in all risk assessments, an analysis can be developed only to a limited extent from first principles but must rely heavily on historical data, conditions, and experience. The CCPS guideline provides order of magnitude initiating event frequency estimates.

The initiating event forms the base of an event tree. An example event tree for a polymerization reactor prone to runaway is shown in Figure 3.32. When a critical alarm sounds, a supervisor must manually stop the reaction. In case of the failure that operation, pressure shall be relieved to a safe place through the operation of a SIL1 SIS relief valve, or in the case that the valve sticks and a reactor flange bursts, gas sensors will trigger a water curtain/deluge system to disperse the flammables that escape. But fire can still break out, and in this case an emergency alarm will be sounded and the firefighters called. When this action is too late, an explosion may follow. If the functioning of later LOPA layers is independent of the failure of the previous ones (independent protection layers), and common cause failure probability is assumed to be minimal (and therefore can be ignored), then evaluation of the event tree probabilities is easy. The initiating event frequency can be multiplied with the failure probabilities of the layers to estimate the failure rate of the ensemble. (Given uncertainties, this will not be a "conservative" estimate.)

The question of how many layers must be added shall be answered by the probability of failure on demand (*PFD*) of the layers, often also indicated by its reciprocal, the risk-reduction factor (*RRF*). Suppose the initiating event frequency is 10^{-1}/year and the consequence category without protection layers is 5. Then if one wants to reduce the frequency of risk to 10^{-6}/year, one could add four layers of protection: three with $PFD = 10^{-1}$ ($RRF = 10$) and one with $PFD = 10^{-2}$ ($RRF = 100$). Together these provide five "credits to close the protection gap," if

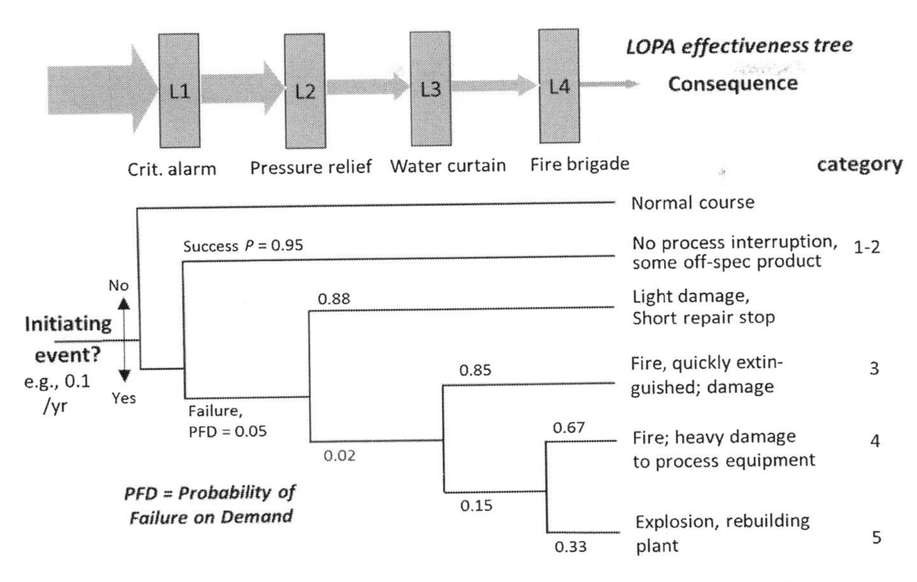

FIGURE 3.32

LOPA event tree example of a polymerization reactor with consequences specified (although each failure and release may have various possible consequences, hence an own sub-event tree). Overall failure probability of the system given the initiating event frequency and fully independent protection layers will be: $0.1 \times 0.05 \times 0.02 \times 0.15 \times 0.33 = 5.0.10^{-6}$/year.

one can assume independence among the layers. This type of assessment may be somewhat crude but does not require sophisticated computing and calculations are easy. The CCPS 2001[125] guideline provides approximate data and examples.

Protection layer analysis – optimizing prevention (PLANOP) is internationally relatively unknown but an interesting and practical method to analyze and curb hazards of chemical process installations by considering the effect of all safety barriers and measures in a plant, not only SISs. The method has been developed and perfected over more than 10 years by Peter Vansina and Koen Biermans[130] of the Belgian department of monitoring chemical risks of the Federal Public Service for Employment, Labour and Social Dialogue in Brussels. As LOPA, it is a follow-on to HAZOP. It is not replacing LOPA or any other method, but its strength is structuring, integrating, and managing the many details of all measures and barriers of an entire plant. Hence, it is a systematic approach, while the Web-based model run by Python guides users and supports them with suggestions and checklists. It provides an overview of structure and assumptions made by listing all substances, quantities, and possible reactions, including those with oxygen and water, which can always be present. It then identifies scenarios by first making a site breakdown in sections and subsections, and filling in what can go wrong (cause trees) by selecting from suggestion lists and adding as desired. Subsequently, with other suggestion lists, safety measures and barriers are introduced. In case of LOPA, layers are formed and overall effectiveness calculated. The model enables users of different disciplines of a company to share a common model. So far, only Dutch and French language versions are available.

Wang and Rausand[131] proposed practical approximate formulae for analyzing the overall hourly reliability (PFH) of an SIF/SIS in a mixed setting, accounting with emphasis on high-demand components for common cause failures, degradation, downtime and mean-time-to-repair effects for a voted group of identical channels within a proof test-time interval.

3.6.6.3 Further issues with LOPA and SIF/SISs

Superficially, a protection layer looks like a simple, straightforward system whose performance is easy to evaluate. Some theoretical complications are clearly explained by Jin et al.[132] For making the calculations, Markov state transition models are applied or more conveniently Bayesian Networks (BNs), as will be explained in Chapter 7. But in practice there appears many more pitfalls and complications of which the lack of independence of layers is one. In the following, a brief survey will show the ways to assess the overall effectiveness of a protective system, the various complications, and what can be done to improve the situation, also in view of a cost/benefit analysis.

The standard IEC 61511 considers the safety life cycle, meaning that in all phases from concept, via design, implementation, operation, maintenance, and decommissioning of the facility the SIS safety functions shall be regarded in the context of the safety of the total system. Periodic testing and maintenance are critical, which shall be covered by a SMS. Hence, all steps shall be verified, documented, auditable, and audited. A drawback is that layers can trigger from time to time, causing spurious trips, which lower the usefulness. A malfunctioning fire alarm in a building is a simple example—if it regularly gives false alarms, then people start to ignore them and no longer respond as required.

As explained in the previous section, the SIL level necessary for the operation depends on the risk presented by the installation (which in the standard is called equipment under control, EUC) and the risk reduction that must be realized to achieve a risk level below the set point of tolerable risk. The standard contains a type of risk matrix, called a risk graph (Figure 3.33), which may give guidance to decision making depending on the type, severity, and expected frequency of occurrence of damage that can be expected if no layers of protection are installed.

There are various qualities of protection layer equipment (SIL levels, brands), which each has its associated price. One way of increasing overall reliability of a layer is by making use of redundancy and a preconfigured voting or polling system, for example, action can be initiated on the basis of a one-out-of-two error detection, 1oo2, or a 2 out of 3 detection, 2oo3, et cetera; another option is simply to buy a system with a higher SIL rating. To make a choice, various cost factors have to be weighed beside the desired overall reliability. Hence, after having performed a semiquantitative risk assessment, a cost–benefit analysis is required. A natural next step is to search for more precise data on failure probabilities of various components and costs. The latter can be divided into costs of ownership of the protection layers and the probabilistically estimated costs of consequences. The former consists of costs of investment (capital cost) and operational expenses for testing and maintenance. Also, once a layer functions, there will always be some cost/damage. Even if no person is hurt, the plant will be shut down with costs of cleanup, lost production, replacing parts, repairing damage, business interruption, and credibility loss of the plant stakeholders. In general, consequence costs will increase with each

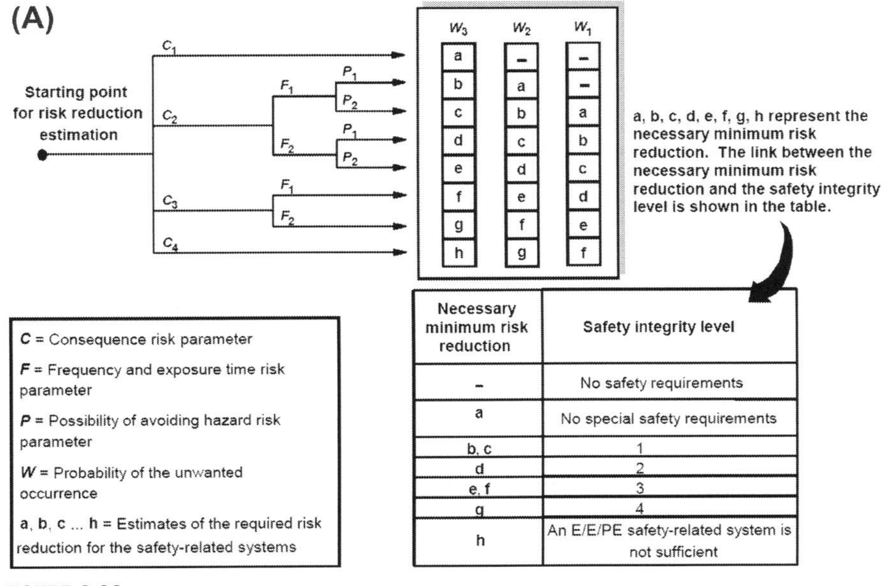

W_3	W_2	W_1
a	–	–
b	a	–
c	b	a
d	c	b
e	d	c
f	e	d
g	f	e
h	g	f

Necessary minimum risk reduction	Safety integrity level
–	No safety requirements
a	No special safety requirements
b, c	1
d	2
e, f	3
g	4
h	An E/E/PE safety-related system is not sufficient

FIGURE 3.33

(A) Example of the risk graph of IEC 61508, functional safety of E/E/PES safety-related systems (2010 version, IEC 61508-5 ed.2.0) illustrating the general principles only. The graph serves as a qualitative aid to decision making for what SIL level in a certain case will be necessary. The corresponding table with data relating to the risk graph is shown in Figure 3.33(B). Copyright © 2010 IEC Geneva, Switzerland. www.iec.ch.*

*The author thanks the International Electrotechnical Commission (IEC) for permission to reproduce information from its International Standard IEC 61508-5 ed.2.0 (2010). All such extracts are copyright of IEC, Geneva, Switzerland. All rights reserved. Further information on the IEC is available from www.iec.ch. IEC has no responsibility for the placement and context in which the extracts and contents are reproduced by the author, nor is IEC in any way responsible for the other content or accuracy therein.

Figure 3.33(B) Table with data relating to the IEC 61508 risk graph in Figure 3.33(A). EUC is Equipment under control. Copyright © 2010 IEC Geneva, Switzerland. www.iec.ch.

(B)

Risk parameter		Classification	Comments
Consequence (C)	C_1	Minor injury	1 The classification system has been developed to deal with injury and death to people. Other classification schemes would need to be developed for environmental or material damage
	C_2	Serious permanent injury to one or more persons; death to one person	
	C_3	Death to several people	2 For the interpretation of C_1, C_2, C_3 and C_4, the consequences of the accident and normal healing shall be taken into account
	C_4	Very many people killed	
Frequency of, and exposure time in, the hazardous zone (F)	F_1	Rare to more often exposure in the hazardous zone	3 See comment 1 above
	F_2	Frequent to permanent exposure in the hazardous zone	
Possibility of avoiding the hazardous event (P)	P_1	Possible under certain conditions	4 This parameter takes into account - operation of a process(supervised (i.e. operated by skilled or unskilled persons) or unsupervised); - rate of development of the hazardous event (for example suddenly, quickly or slowly); - ease of recognition of danger (for example seen immediately, detected by technical measures or detected without technical measures); - avoidance of hazardous event (for example escape routes possible, not possible, or possible under certain conditions); - actual safety experience (such experience may exist with an identical EUC, or a similar EUC or may not
	P_2	Almost impossible	
Probability of the unwanted occurrence (W)	W_1	A very slight probability that the unwanted occurrences will come to pass and only a few unwanted occurrences are likely	5 The purpose of the W factor is to estimate the frequency of the unwanted occurrence taking place without the addition of any safety-related systems (E/E/PE or other technology) but including any other risk reduction measures
	W_2	A slight probability that the unwanted occurrences will come to pass and few unwanted occurrences are likely	6 If little or no experience exists of the EUC, or the EUC control system, or of a similar EUC and EUC control system, the estimation of the W factor may be made by calculation. In such an event a worst case prediction shall be made
	W_3	A relatively high probability that the unwanted occurrences will come to pass and frequent unwanted	

FIGURE 3.33

(Continued)

successive layer failing, since one can expect that event severity is maximum when unimpeded.

The simplicity of the scheme shown in Figure 3.32 is deceptive. As for all risk assessments, LOPA results are rather sensitive to the assumptions and the approach adopted by the analysis team. After the Buncefield gasoline tank overflow accident in 2005, which was followed by a violent vapor cloud explosion destroying the major part of the tank park and its adjacent office buildings, described in Section 1.5, the UK HSE[133] had LOPA studies performed on tank overflow protections for 15 different tanks. The results varied widely. Many assumptions were badly founded, initiating event frequencies were derived in a variety of ways, and the independence of layers not adequately investigated. Human factors, for example, in the timing and type of operator response, introduced large variability. All of this is apart from the effectiveness of testing, maintenance, and lack of availability. We shall need a method that provides more flexibility and will allow greater insight into causation, also with respect to confounding causes (common cause failure), greater detail where necessary, and inclusion of variability of data in predictive models. When occurring at the time of the same demand, common cause failure of components reduces the reliability of a system and can in its simplest form be accounted for by adding to the current failure rate the product of a beta factor and the failure rate. The standards give guidance. Also, here BNs (Chapter 7) can help.

A great deal can be written on input values for specifying a SIL design through to obtaining verification and satisfactory functioning after the system has been installed. Vendor data do not include systematic failures and are usually overoptimistic. Systematic failures are unpredictable and predominantly caused by human errors in specifying, designing, configuring, installing, calibrating, and maintaining the system. As regards the offshore industry, much PDF and SIS failure rate data has been collected by SINTEF in Norway in a cooperative effort with industry and is condensed in the PDS methods, handbook, and tool.[134] Due to the uncertainty, it makes sense to include a safety factor and to design for a risk-reduction factor of 50 when one only needs a factor of 10 to reach a SIL1 level.[b]

Apart from designing and selecting an optimum configuration of protection layers, there are some other points of concern. First, are some aspects in the design not covered so far, such as the set point of the variable that triggers action of the layer. Besides miscalibration, erratic signal fluctuations and drift of the sensor, environmental effects of corrosion, fouling and weather on the sensor and actuator, the trip setting also has to take account of SIS response times and process lag times. This

[b]To estimate the contribution of the component to the overall risk in a system approach, a systematic method must be applied to calculate an appropriate *RRF* or safety factor with confidence limits. Further, to identify the required risk reduction needed for this contribution the particular application conditions for which the risk assessment has been performed, shall be taken into account. Employment of a generic safety factor, however, will possibly result in either a greater and more costly *RRF* than required or in an *RRF* that is too small and therefore much more costly due to more likely upsets and possible losses.

is, for example, for safeguarding against overpressure relative to the design limit of the containment. A SIF shall preferably be verified by a separate team to reduce bias. In this connection, and as the final element, it should also be emphasized that a weak link in the SIS chain is scarcity of good reliability data for the PRV. PRVs can fail by sticking due to fouling or corrosion, or being undersized for a worst-case duty. Bukowski and Goble[135] report proof test data during the useful life of PRVs, defined as the period in which in the bathtub curve the failure rate parameter of the exponential model can be considered constant as an approximation that must be periodically verified. Results of three data sets show a useful life of 4–5 years and a failure rate between 10^{-7} and 10^{-8}/h. Another aspect is to investigate what effect the action of the protection layer will have on the rest of the equipment. There may be feed streams that must be shut off to avoid secondary effects. And downstream obstacles or restrictions should be avoided in the relief line.

A second aspect is the life cycle management, which starts with inspection, testing, and preventive maintenance. Manufacturer's data on failure probability may be acquired under cleaner conditions than existing in a plant. Based on extensive experience, Lucchini[136] described the following practical facts about SIF reliability. So-called SIL certificates can be of variable quality. However, online testing and health diagnosis can be difficult to perform, and the performance of valves usually determines the final result. Proof testing must be done periodically with a suitable frequency based on the reliability data and contribution to the overall risk; traditionally the interval was up to 12–18 months, but the trend with new designs is to extend that period. Too long a test interval T_p can result in a long period of dangerous failure (rate λ_d). The appropriate T_p for a component can be easily calculated based on the minimum reliability required for the component over the interval T_p. As long as the product $\lambda_d \cdot t << 1$ where t is time, and with assuming λ_d is constant (exponential behavior), the average PFD value over the interval T_p between two tests is[c]:

$$PFD_{avg} = \left(\frac{1}{T_p}\right) \int_0^{T_p} \lambda_d t \mathrm{d}t = \frac{1}{2} \lambda_d T_p.$$

[c]The inequality is the basis for the estimation of the exponential cumulative failure probability distribution, cdf, $F(t) = 1 - \exp(-\lambda t) \sim \lambda t$, which defines the range for an appropriate approximation of PFD for exponential behavior with constant λ given $\lambda t << 1$. As mentioned in Section 3.5.3, a more realistic and flexible expression for PFD, also for $\lambda t << 1$, is $(\lambda t)^{\beta}/(1 + \beta)$, where β is the Weibull-shape parameter that reduces to exponential behavior for $\beta = 1$. Without a risk-informed testing schedule, PFD is likely to be used improperly and dangerously for any magnitude of λt in a similar way to the use of other untested approximations or heuristic "rules of thumb" used in practice. Therefore, the adoption of this Weibull version of PFD is to be favored; this makes it less likely that it will be misused and more likely that in actual practice a cost-effective test strategy will be developed. Such test strategy shall be based on the contribution of a critical component failure to the overall risk, the minimum reliability required for the component between tests, and the minimum acceptable availability (considering maintainability) for the component to respond to system demand.

Layer failure rate λ_d must be determined from the actual, true failure rates of the layer components. Further, it is the parameter of an exponential distribution with mean time to failure, $MTTF = 1/\lambda_d$, and standard deviation, $1/\lambda_d$, implying that the smaller the failure rate the larger the uncertainty. In addition, it is only an approximation because measured over longer times, λ_d will not truly be constant. As mentioned before, the uncertainty leads us to propose a *RRF* of 50 where strictly speaking a factor of 10 would be sufficient. On top of that is that too frequent testing can degrade component availability to respond to system demand and will also affect production negatively. The preventive maintenance interval to optimize availability is based on the mean number of failures during a component design life for a required reliability level and the mean system downtime for maintenance. The probability of initial failure (*PIF*) in which the valve sticks shut is important for a PRV. Bukowski and Goble[135] proposed the following relation: $PFD_{avg} = PIF + (1 - PIF) \cdot \lambda_d \cdot T_p/2$. Lucchini[136] stressed the importance of distinguishing systematic SIS failures (which are mostly design related), from random PRV failures. This distinction requires different testing tactics and metrics, such as valve partial stroke testing, which has been shown to be helpful in some cases of random PRV failure. This type of testing lengthens the required proof-testing coverage intervals and alleviates the drawback of production shutdown. Additional analysis on partial testing can be obtained from Jin and Rausand.[137] The latter author published much more about SIS reliability analysis, also by making use of Markov state transition approach and Petri Net, to be explained in Chapter 7.

With failures and repairs, component availability becomes an issue, and bypassing is unavoidable. For that purpose, a SIL2 layer can be considered as only giving SIL1 protection; hence downgrading credits results in bypass time available for repairs.

Another point is that LOPA analysis results shall be stored and be retrievable, and all events and incidents occurring in a protection system shall be recorded. This sounds simple but it is not. Over the years in a plant with many protection systems, installed problems lead easily to proposed changes. The safety management rules for "management of change" shall be carefully followed; previous documentation is precious and shall be consulted. Assumed values in the original analysis may be replaced by later, more accurate historical data obtained in the same environment. It will also help with the auditing process. Any activation of the SIF shall be investigated, learned from, and recorded as a lagging indicator of poor performance.

Bridges and Clark,[138] coauthors of the guideline CCPS 2001, also mention a series of other problems that originate in the areas of organization and human factors. Some organizations use LOPA without implementing a SMS and following the rules of testing, maintaining, and record keeping. Others apply LOPA where it is not necessary and mix PHA/HAZOP team work and LOPA, which should be performed after PHA. Consequences are often not defined sufficiently, so risk may be over- or underestimated, which affects the choice of a protection layer. In a later paper, Bridges[139] and even more recently, Schmidt[140] elaborates on the human aspect in

LOPA. Human error occurs at any stage and in any aspect of the life cycle of a protection layer system, but quantification is cumbersome. Independence of layers is an issue in particular when human elements are involved. A basic rule is not to consider members of a particular crew or team as independent. Local culture can have a strong influence on layer failure probability. If checklists or procedures are involved, these shall be clear and document every step the operator shall perform. Also, operators shall be adequately trained and retrained periodically to increase their proficiency. Mostly errors do not occur when a component of the system is taken out of service (lockout/tagout procedure) but are more likely to occur when the component is returned to service. A bypass of an isolation valve that is accidentally left open after testing the valve falls more or less into the same category. Based on experience, Schmidt suggests the following generic probability values, *P*, on human error: When the task is to be executed under high-stress conditions and is nonroutine, $P = 1$; for a routine task or low-stress, nonroutine task, $P = 0.1$; when properly executing a written procedure, $P = 0.01$; the same with independent review of the various checklist steps, $P = 0.001$. He further distinguishes administrative controls from operator responses; the former is a procedural measure routinely performed to prevent a hazard with PFD_{avg} at best 0.1; the latter is a (procedural) response to a (rare) initiating event. In such cases the probability of human failure to respond will depend on the difference, Δt, between the time available to respond and the nominal average time required for the response action. With Δt smaller than 5 min, PFD_{avg} will be almost 1; $PFD_{avg} = 0.1$ with a response time of at least 10 min, but some take this figure as up to 20 min; $PFD_{avg} = 0.01$ for Δt between 45 and 90 min and $PFD_{avg} = 0.001$ if Δt is between 5 and 10 h. This translates into a maximal $PFD_{avg} = 2.37t^{-1.3873}$ (this is apart from the *PFD* value of the alarm). Independence is essential for both types of activities. The system shall be audited in all its aspects. We shall see this again in Chapter 6 on human error as Rasmussen's SRK rule and time–reliability correlation. A further source on this topic is the work of Kumamoto and Henley[141] offering a more elaborate, but still practical relation—the human cognitive reliability model providing a nonresponse probability.

Pasman et al.[142] mentioned two other points. The first is that because of the growing variety of systems/components offered for procurement, optimum selection becomes more difficult. The second point is that in a large organization, a lack of communication, and hence cooperation among various specialists in different departments, can degrade optimum SIS functioning. An example would be automation and instrumentation on the one hand, and testing and maintenance on the other, both dealing with different aspects of the same layer but failing to communicate adequately. Rahimi and Rausand[143] provide guidance on how to prevent an unacceptable increase of the common cause beta-factor during the operational phase by HOFs and by environmental conditions. HOFs are usually introduced by improper management of change in manufacture, specification, installation, operation, and maintenance. To control the effect, the authors propose a management system.

After a successful LOPA and having installed the required equipment, a satisfactory feeling of being safe might be engendered, but the analysis usually only covers the steady-state process and not abnormal situations, such as start-up and shut-down. A significant fraction of accidents occurs in these latter situations.

3.6.6.4 Risk-based inspection and reliability centered maintenance

Risk-based inspection was developed in the late 1990s and has become a common good (see API RP 580 and RP 581 where it may also appear with the name "Asset management"). RBI adapts risk analysis to the process of decision making upon maintenance inspection of equipment. It concentrates on MI of pressure-containing parts, not on instruments, electronics, or controls. Pressure equipment has in the past been responsible for a majority of LOC incidents. Due to material degradation, equipment deteriorates over time and hence the risk of failure increases.

Time, effort, and resources are saved by initially calculating a critical condition for a certain asset (vessel, tank, pipeline), developing an inspection plan, and following it up with inspections, which become more frequent when the parameter becomes closer to its critical condition. The inspection plan will depend also on the potential consequence of a failure, that is, low, medium, or high risk. Inspection instrumentation applying ultrasound and X-rays has become quite sophisticated. A Weibull two-parameter reliability model is assumed. As mentioned earlier, corrosion can be in the form of general thinning, local corrosion, pitting, stress corrosion cracking, hydrogen attack, increased temperature hydrogen damage, and others such as brittle fracture. These phenomena are influenced by environmental factors, in particular the nature of chemicals (contaminants) in contact with the metal. It may take a long time before effects becoming noticeable. Examples are sulfidic and microbial corrosion and high-temperature hydrogen attack (for serious accidents due to corrosion, see, for example, US CSB reports on Richmond[144] and Tesoro[145] refineries). Of course, one has to consider type and rate of deterioration, effectiveness of detection in the inspection procedure, and tolerance built into the equipment. Criteria are rather conservative in view of uncertainties, so in general RBI results in a larger number of inspections. The inspections are performed by specialized sections or companies.

Reliability centered maintenance (RCM), sometimes called availability centered maintenance,[d] was developed in the late 1960s, when the Boeing 747 aircraft was introduced and maintenance became a significant cost factor. RCM starts by describing the functions of each part of equipment. This can be a primary function (e.g., pump increases fluid pressure) and one or more secondary functions (e.g., fluid containment, also satisfactory to safety, health, and environment, or SHE). It then states the functional failures, the associated components, failure modes of interest,

[d]Reliability, availability, and maintainability are complementary concepts. Reliability-based methods represent random failures, but availability-based methods include both random, generally latent failures and failures due to testing, maintenance, and replacement, so they should be treated separately.

and the potential effects and consequences by using the results of an FMEA(CA). Four types of consequences are distinguished:

1. Hidden failure (unrevealed failure) consequence in protective devices, only being revealed when there is a demand;
2. Safety, health, and environmental consequences;
3. Operational consequences, production losses;
4. Others.

However, in the MI program, an inspection, test, and preventive maintenance plan (ITPM) is added. The ITPM tasks shall be prioritized. Further, prior to a detailed equipment analysis, a PHA shall be carried out. This shall identify the hazards resulting from safety-related system failures, causes of the scenarios leading to incidents with those hazards, and safeguards against these scenarios. The risk part of the analysis can be completed with a logic tree analysis and a summing of the ITPM responsibilities and frequencies for each failure mode. Hence, RCM is more general and broader than RBI, and by the addition of the ITPM has also become risk based. RCM is described in technical SAE standards.[146] Remote condition-based maintenance by way of the Internet is a further alternative. In a fully implemented system approach, contributions of all of these methods combine instead of focusing on only one or the other as a panacea that is complete in itself.

3.6.6.5 Domino effects and risks arising from natural and intentional causes

Above minimum threshold distances, space between equipment items offers effective protection against domino or knock-on, and co-occurrence effects of a failing unit initiating a LOC of an adjacent unit. This is certainly true when the items are large, such as storage tanks. There is some confusion whether the definition of "domino" is limited to an effect from site-to-site, plant-to-plant, or is any knock-on increasing overall effects. We apply the latter. Protection against domino effects is enhanced by passive barriers, such as bunds and dikes (impoundment) of good quality to prevent pool spreading and even overtopping in the unlikely event of a tank rupture. With respect to fire, NFPA 30-2008,[147] but also FM Global, Property Loss Prevention Data Sheets,[148] provide many protection recommendations against flame propagation and explosion by flame arresting, venting, fire proofing, safe separation distances, and other measures. Recommendations for tank construction are provided by the American Petroleum Institute in API standards. Di Padova et al.[149] developed a risk-based approach.

Also with respect to limiting and suppressing propagation of gas and dust explosions in installations, many practical measures are in use. Much of it exists in compartmentalization or isolation of sections of an installation by fast-acting or rotary valves and flame suppression by extinguishing agents triggered by sensors. An overview is given by Pekalski et al.[150]

For a long time, risk assessment of process plant including domino effects has been ignored, but this has changed. Bernechea et al.[151] provided a useful literature overview and proposed a model to determine the contribution to overall risk from a site by domino effects on storage tanks. An interesting, new simulation approach to escalation via domino effects is proposed by Rad et al.[152] Given all possible release scenarios for all plant equipment items containing a certain threshold quantity of hazardous material, domino effects will result in an increase of the frequency of failure of an item compared with the situation of that item failing due to its own mechanism. The requirement in the definition of "domino" of an intensity escalation of the effect resulting from a secondary release introduces an escalation probability value problem of the primary item on the secondary one. A secondary item may in any case increase the calamity not only because of a larger inventory but also because for an item a multiple of possible release scenarios may exist with a range of intensity levels depending on where failure occurs (e.g., in the vapor or liquid space of a vessel). Hence, in a risk analysis, first all possible scenarios must be defined with their potential intensities. Then, applying the data on probability of failure of equipment by blast, fire, and fragments as a function of distance, secondary frequency increases must be derived for those scenarios that create a higher intensity than the primary one, et cetera. In principle, this will result in a series of ever-smaller higher intensity level frequency increments, so it shall be truncated. The third-level frequency increase is already small. The authors developed software for the calculation. This approach would also facilitate a plant layout analysis to reduce the risk of domino effects. Also, cost consideration stimulates modular construction of plant. As modules are designed without much regard of their future "neighbors", potential domino effects shall require later enhanced attention.

In principle, this would apply also to possible domino effects of plants within a cluster of neighboring companies. Reniers et al.[153] made a plea based on applying game theory to consider the overall risk-reduction gain that can be obtained when the managements of the different companies would collaborate and tune their own effort of risk reduction to that of the others to obtain a satisfactory overall result with the best cost-benefit ratio.

Apart from protective measures against effects of threats emanating from other parts of an installation, protection may be needed against natural threats or "natural-technological" (NaTech) events such as earthquake, flooding, and lightning. Valerio Cozzani and coworkers (Universities of Bologna, Pisa, and Napoli, Italy) investigated both consequences and event probabilities. A general overview by Krausmann et al.[154] concludes that 5% of industrial accidents involving large amounts of hazardous materials are due to NaTech. These events require specific structural measures for risk reduction, while early warning is also important. Because of the considerable challenges, emergency responders must prepare for them. Effects of risk from seismic,[155] flooding,[156] and lightning[157] events have been investigated separately.

As mentioned in Chapter 2, based on the critical infrastructure protection directives in the US and the EU, the process industry is subject to security measures. These include making an inventory of hazardous materials, assessing risk, and controlling more stringently access to plant sites. Although public safety counts for both, because of strategic material availability interests, measures in the US are more stringent than in Europe. Internal social control (which will be well when safety culture is high) is a noncostly, effective means to reduce security risk.

3.6.6.6 Emergency response planning

In the case of a developing emergency or disaster, responders have to be activated. They are recruited from three types of organizations: firefighters, medical staff, and police. Their main task is to bring the event under control as soon as possible, to save as much life as possible, and to limit damage. So, although a crisis staff may be established under the local mayor, the field command is normally in the hands of a firefighter. Police will facilitate access to the site, restore order, keep spectators at a distance, and open an investigation into the cause as soon as possible in view of the possible commission of criminal act. The medical staff takes care of the casualties, assesses priority of treatment of injured, and organizes transportation to nearby hospitals. All of these assignments and measures must function flawlessly as if the crews were doing them on a daily routine, therefore training is paramount.

Emergency preparation starts in the industrial organization itself. After having made risk analyses based on experience, the size and outfit of the plant emergency response team must be determined and a number of possible scenarios defined. An aspect that also deserves significant attention is evacuation and self-rescue. In the first place, this will be important for plant personnel but may also be critical for nearby residents. Evacuation requires emergency communication, possibly temporary refuge (muster area), for example, in case of collective evacuation by lifeboat, and (fire/blast wall protected) escape routes. Protection shall be in particular be from smoke. Next will be cooperation between the plant's emergency response personnel and that of the local community, or, in the case of major hazard, regional forces. Normally, cooperation will already exist, because for preventive fire safety for licensing, measures will have been prescribed and recommended to the plant's management. In a later stage, these measures will have been inspected by the local or regional firefighting organization.

In a next step of preparation, scenario analyses are performed as a team effort. The scenario developed in an analysis will be inspired by QRA results with respect to an initiating main event, but a QRA study does not go as far as modeling the course of follow-on events over time till consequences are under control and, e.g., fires are extinguished and leaks sealed. Because success of disaster control depends heavily on the deployment of sufficient resources at the right locations and times, accurate time sequencing in a disaster scenario is vital. Moreover, in a plant many scenarios are possible, so a representative one has to be selected for the exercise. On the other hand, it would be an illusion to predict exactly how a disaster scenario will develop, in particular with regard to the late effects. But the value of the team working on a scenario analysis is the same as for a LOPA team. It is the brainstorming and the development of common thinking that is key in preparation. In this case, the responders' actions are chosen by realizing what problems to expect and how to solve them. However, the eventuality is always likely to develop differently from that which has been imagined when developing the scenario.

Of course, it helps greatly to support exercises with simulation models for fire dynamics (FDS), and modeling or practicing muster and evacuation on offshore platforms and from buildings with, for example, EXODUS (Section 3.5.1.4). Also, in case of accident in a chemical plant or fire in a building complex and the hazard arising of toxic vapors or carcinogenic smoke (e.g., asbestos particles in smoke) reaching residential areas, the commander shall have contingency plans ready for evacuation of population. He shall need gas detection equipment, simulation software, and meteorological information to determine, based on AEGL limits or other criteria (Section 3.5.1.1), the area borders of the 4 D's: Detection, discomfort (AEGL-1), disability (AEGL-2), and death (AEGL-3). Known freely available software is US EPA's CAMEO[158] (Computer-Aided Management of Emergency Operations, including data on chemicals) with dispersion module ALOHA (Areal Locations of Hazardous Atmospheres) and MARPLOT (Mapping Applications for Response, Planning, and Local Operational Tasks). Depending on dispersion speed and concentrations, the commander must make the decision of alarming and then staying inside the home with windows and doors closed and ventilation stopped, or evacuating. The latter causes a great deal of stress, which itself can make victims. At the same time, the commander must keep an eye on social media such as Twitter to keep track of reactions.

Regarding injured victims, they must be collected and classified according to the seriousness of their injury and the urgency of treatment. For selection, triage classes can be used: Class T1 are patients that have acute, life-threatening wounds. They must be stabilized immediately and then treated further in a hospital within 1 h. Class T2 patients need hospital treatment but within 6 h to avoid complications. Class T3 injured do not need to be hospitalized. Hence, optimal lifesaving measures have limited time available. On the other hand, self-rescue may be delayed by injuries hindering or even preventing walking, or by having to take shelter because of smoke or toxic gas, therefore, self-rescue also has its limitations. Measured in time, it may be a matter of minutes to reach a safe haven or not. Hence, management of emergency response works under the stress of limited time to perform life-saving operations.

Ben Woodcock and Zachary Au[159] have emphasized the dependence of the success of emergency response on human factors. This holds for onshore but even more for offshore installations. Because emergency response is a safety critical operation, an error can have dramatic effects, and because the operation exists mainly of human activity, there are many opportunities to make errors. In addition, the activities must be executed under severe time pressure and stress under often harsh conditions with many hazards around. The authors describe the present offshore emergency response organization and procedures, as well as what went wrong in a number of cases. They make a plea for design of an improved incident command center close to, but separate from, the Control Room, and provided with highly reliable communications with the control room operator (CRO) and other operators. The CRO shall have optimal situation awareness. He/she shall trigger the muster alarm and shall notify the incident commander. Design shall also provide well-demarcated and relatively protected escape routes for egress and routes/ladders for evacuation/abandonment from muster area(s) to lifeboats and rafts. For optimum procedures, task analysis and performing a HAZOP on task description to identify issues will help. It is obvious that staff training is important and that with rehearsals sufficient routine develops.

3.6.7 STEP 6. RISK ASSESSMENT

The ultimate purpose of a risk assessment is to enable an optimum decision to be made whether the task is to select provisions for an installation or a site to result in a sufficiently high safety level. There will always be a residual risk, but the question is, will this risk be tolerable? Tolerable in this context means a willingness to accept and live with a risk in view of the expected benefits of the activity, and it is therefore different from acceptable, as explained by Le Guen.[160] In design and also during operations, questions can arise: Is our safety level sufficient? In an iteration process of steps 5 and 6, mentioned in Section 3.6.6.3, the effect of risk reduction measures can be assessed against residual risk. But, for the intention to answer the question "How safe is safe enough?" there will be a need for decision criteria. Criteria are not simple to give, but a risk matrix can help. In the IEC standard mentioned above, the risk graph in Figure 3.33 is used for the limited scope of a decision about a required SIL level with respect to a certain consequence. Also, the FAR for workers involved in the activity can be used for decision making. In some countries, the government has set criteria for the risk level outside the site's premises, for example, an individual risk (IR) of 10^{-5} or 10^{-6}/year or, as we have seen, for a defined societal/GR.

It will be prudent for management to stay below prescribed levels. A way to go is to consider an order of magnitude risk matrix with, for example, consequence classes plotted on the ordinate and frequency on the abscissa, as shown in Figure 3.34. The consequence classes are elucidated in the table at the right-hand side. The figures attached to the classes run parallel with the rough cost estimates in units of thousands of US dollars. Besides repair, costs include business interruption and

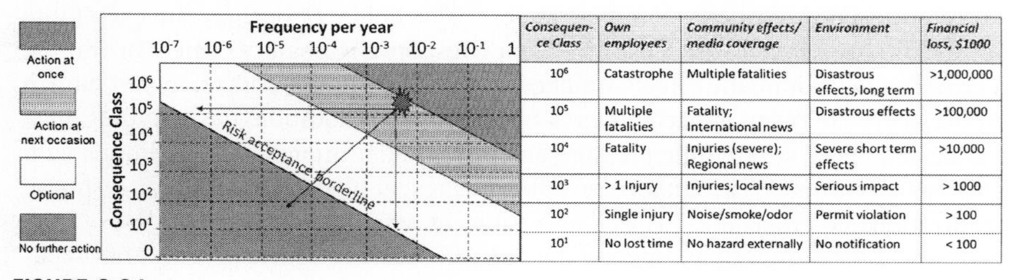

Consequence Class	Own employees	Community effects/ media coverage	Environment	Financial loss, $1000
10^6	Catastrophe	Multiple fatalities	Disastrous effects, long term	>1,000,000
10^5	Multiple fatalities	Fatality; International news	Disastrous effects	>100,000
10^4	Fatality	Injuries (severe); Regional news	Severe short term effects	>10,000
10^3	> 1 Injury	Injuries; local news	Serious impact	> 1000
10^2	Single injury	Noise/smoke/odor	Permit violation	> 100
10^1	No lost time	No hazard externally	No notification	< 100

FIGURE 3.34

Risk matrix example showing order of magnitude classes of consequence and likelihood (expressed as an annual frequency), and the resulting risk with acceptability indications. Risk acceptance and other borderlines are arbitrarily chosen depending on what a company decides to establish.

liabilities. The more severe levels of 10^5–10^6 will be accompanied by adverse reputation damage for the company concerned and will cause shareholder value losses. Indeed, with the Deepwater Horizon demise, costs exceeded for the first time $10 \cdot 10^9$ or 10 billion USD.

The slope of the borderlines between areas in the diagram is the risk aversion parameter ($-\beta$). In Figure 3.34 it is chosen to be -1 (the utility relation for a decision maker with neutral attitude toward risk, see Chapter 9). In the case of fatalities, societal disruption grows disproportionally larger with a higher number of victims. People's attitude in that case is not neutral but risk averse, and a slope of -1 will be judged as insufficiently stringent. Fifty fatalities from a single accident arouse a much deeper emotion than one, the latter occurring on a daily basis in the traffic in the streets. A simple utility expression is $f \cdot c^{\beta} = k$, where f is the frequency, c the consequence, and k is the risk acceptance threshold. Exponent β is to describe risk aversion, for example, equaling 1 for being risk neutral, or 1.2 for monetary risk, and 1.4 (HSE UK) or 2 (NL) for fatality risk. An example is the criterion for fatality risk from collapse of structures in ISO 2394, 1998[161] in which cumulative probability for societal risk is $P(N_d > n)n^{\beta} < A$, where A is the risk acceptability threshold, β the risk aversion parameter, n the number of fatalities, and N_d is the number of fatalities within 1 year from a single accident.

The risk acceptance borderline of the "safe" green area is relatively severe with respect to level of frequency with the lowest point of the order of 1 in 10 million years at the highest loss indicated on the consequence axis. Benefits calculated as risk-reduction gains by adding barriers or making process and storage less hazardous can be balanced against costs of investment, operation including tests and maintenance, and potential losses. A more detailed consideration of the costs of accidents versus the costs of safety measures will be given in Chapter 9.

Of course, reality is more complex than presented by the matrix. As we shall discuss later in more detail, due to uncertainties any risk dot as in Figure 3.34 has rather large confidence limits. Also, risk tolerability is subjective and depends on circumstances, so that decision making against a fixed criterion is simplistic. Weighing and judgment will be the decision maker's task. Further, one has to consider other possible large mishaps that can occur at the same site, which will potentially increase the frequency. Whether risks from various scenarios shall be accumulated depends on what the analysis is used for. If it is to inform the public, accumulation is desirable. It is also necessary to account for domino effects. These can even come from neighboring sites in the case of plant clusters.

In the UK the HSE has not yet decided whether it will introduce a GR criterion. Anyway, the British industry has to comply with the requirement of ALARP, or in clearer terms to make the risk "as low as reasonably practicable." Risk has to be reduced with regard to industry best practices until costs disproportionally outweigh benefits, but future developments in knowledge and technology still needs to be considered. The first years after introduction the interpretation of this criterion left room for debate, but later HSE gave guidance how to comply. In Chapter 12, we shall expand on the content of ALARP because the requirement supports much the idea of continuous improvement of process safety and it is gaining in importance. Several countries adopted the principle. ALARP has been included in the Australian Model Work Health Safety bill. The Norwegian offshore industry follows ALARP too, as well as the new EU Directive on Offshore Safety 2013/30/EU. In the US, after a few serious accidents the principle started to be promoted by, among others, the Chemical Safety and Hazard Investigation Board, see Chapter 10, last part of Section 10.1.

The Dutch GR-guide criterion is as shown in Figure 3.29. It should be noted that in the Dutch practice, GR comprises only immediate fatalities (i.e., instantaneous and with some delay) among residents, not industrial workers nor fatalities arising from road users passing and hit by the disaster effects. A "how to use it" procedure is available for decision makers, but the implication is that in practice it is insufficiently followed. Generally, the concept of the $F-N$ curve in terms of the cumulative frequency of accidents, F, with fatalities greater than or equal to a threshold value of $N=10$ and of how rapidly F drops as N is increased, is not understood sufficiently well.

For transportation routes of hazardous materials (ship, road, railway, or pipeline) and comparison of which one is the safest, the same risk measures as for static risk sources can be used, that is, individual and GR. The latter can be calculated per kilometer distance traveled with the highest value over the trajectory considered. Boot[162] proposed to improve the approach by plotting GR location-specific values the way it is presented in Figure 3.29 and taking account of risks in loading and off-loading activities as well as risks accumulating from other local static risk sources.

Value-at-risk is a concept that has been used extensively in the financial world (at least before the 2008 crisis) and that was investigated by Prem et al.[163] for application in the type of risks that are considered here. It is defined as the expected portfolio loss at some confidence level over a certain time horizon and is interpreted for process safety as the cumulative probability of losses over a certain threshold value that can be insured or accounted for. However, an assumption has to be made about the shape of the distribution and the threshold cut-off value. The statistics law for large numbers, or rather the possible occurrence of a rare catastrophic event (Nassim Taleb's[164] Black Swan), spoils the application of the concept for process risks. Most organizations have been fortunate enough not to sustain an extremely large loss, but on the other hand, who could have predicted the order of magnitude of the losses suffered by the demise of the Deepwater Horizon offshore platform in the Gulf of Mexico in 2010?

In public perception and therefore their acceptance of risk, magnitude of consequence plays a far more important role than probability. In the late 1970s and 1980s, experts in risk communication such as Paul Slovic and Baruch Fischhoff warned the technical risk analyst community and decision makers (in their extensive oeuvre) for being too optimistic about "playing around" with probabilities of once in a million years and therefore declaring them as risk acceptable. People want certainty, not something vague when a new facility or pipeline is built close to their homes, although what we may be dealing with here may be not so much the fear of being harmed by an incident but rather the fear of value depreciation of their asset. Discussions about objective and subjective risk, on voluntary risk (mountain hiking) or involuntary risk (being exposed to a hazardous material storage next door, which in acceptance level can be 10,000 times lower than voluntary risk) or every day comparisons with being killed by lightning or driving a motorcycle do not help much. Psychometric study results on risk perception help to reveal the facts, but give only weak guidance about how to improve perception of risk. Slovic and Weber[165] also observed social amplification of risk turning the public against an industry sector in which calamity was caused by only one member of that sector, that is, risk aversion can be indiscriminate. They recommended further analysis of the interplay between emotion and reason. Fischhoff et al.[166] analyzed the various meanings that risk can have within the context of decision making. He urges first to obtain an agreement among parties on the definition of risk before a decision in the debate about acceptance of technologies can be made. Risk can be offset against utility, but a risk definition requires a variety of value judgments. For a definition of risk, the consequences of the risk shall be clear. Hence, risk is characterized by a number of attributes weighted differently, which can be summed to an index as is performed by multiattribute decision analysis in which the utility function is a weighted sum of utilities for the selected attributes. Choice of an index is a value judgment in itself. To increase credibility and build confidence in the decision process, this index selection should be done with active participation of the stakeholders and involvement of the public.

Ortwin Renn, one of the contributors to the International Council on Risk Governance (IRGC) in Geneva, Switzerland, has been coauthor of several guidance papers. An introduction to the IRGC Risk Governance Framework[167] defined risk as: "Risk is an uncertain (generally adverse) consequence of an event or activity with respect to something that humans value. Risks are often accompanied by opportunities," while "Governance refers to the actions, processes, traditions and institutions by which authority is exercised and decisions are taken and implemented." Risk governance deals with the identification, assessment, uncertainties, management, and communication of risks in a broad context. The challenge of better risk governance is to enable societies to benefit from change while minimizing the negative consequences and impacts of the associated risks, which can be accomplished through a system approach to risk management.

The various stages and separate responsibilities to handle a risk problem, and the central role of communication with all parties involved, is best made clear in the drawing shown in Figure 3.35. The procedure of appraisal (scientific risk assessment and concern assessment) is crucial for understanding the risk, and risk management is key in decision making. The latter implies use of a predictive and quantitative decision model that recognizes the ranges, that is, distributions, of outcomes given a decision with associated uncertainty and therefore estimates the outcome level and the probability of outcome occurrence for each of the decision alternatives. This can be done on monetary value (cost–benefit) as we shall further discuss in Chapter 9. Responsibility for appraisal shall be independent and fully separate from the responsibility of risk management to maximize objectivity and

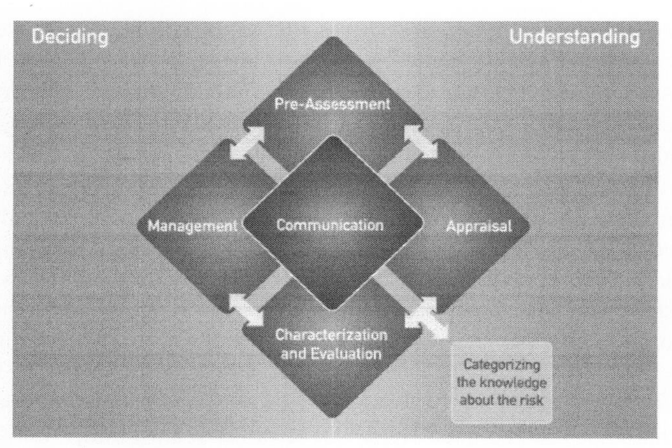

FIGURE 3.35

An introduction to the IRGC Risk Governance Framework's[167] representation of how to arrive at a best decision about a technology or a project that exposes people to certain risk. Further explanation is given in the text.

transparency of both activities. Preassessment, communication, characterizing, and evaluation are common activities involving all. The IRGC publication[167] lists the errors governance can make. It further emphasizes the crucial role and opportunities of two-way communication with stakeholders and the public. It lists what to avoid in risk communication so that trust in management can be increased and perception of risk can be more realistic. It also stresses the importance of knowledge because it is the resource that can diminish uncertainty, in particular for complex risks. When much uncertainty still remains "precaution-based" or "resilience-focused" strategies are recommended. In Chapter 12, we shall return to these subjects.

3.7 EVALUATION OF THE STATE OF RISK ANALYSIS METHODOLOGY

Results of a QRA calculation should be considered with caution because uncertainty is larger than an order of magnitude. Twice at the European Union level, a benchmark exercise was organized to evaluate risk analysis models; the second EU project ASSURANCE is most relevant. This risk analysis project, published by Lauridsen et al.,[99] was performed by seven highly experienced teams for an ammonia storage plant with loading/unloading operations. For this exercise, the plant was "virtually" located in Denmark. The teams were asked to perform a complete risk analysis so that all variability factors were included. Although the spread in results had decreased as compared to a similar exercise 10 years earlier in which only the dispersion model was a parameter, the root problems of spread had not been eradicated. Individual risk contours (for 10^{-5} per year, see Figure 3.36, left) differed by

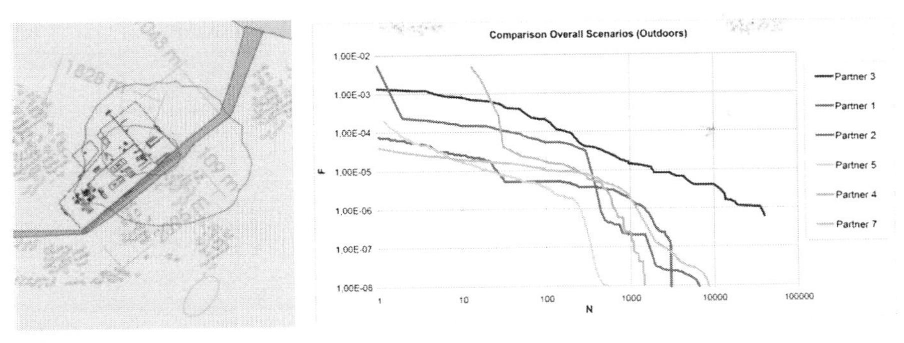

FIGURE 3.36

Results of EU Benchmark exercise ASSURANCE 2002 by Lauridsen et al.[99] Left: Maximum and minimum 10^{-5}/year risk contours found in the analysis; since the large one lies for a major part outside the plant's premises, this result will present a dilemma to decision makers. Right: Societal risk results expressed in *F–N* curves differing by two orders of magnitude. If realistic confidence limits were applied to these curves, the curve envelopes may overlap to varying degrees within the combined uncertainties.

at least a factor of three in radius, which is a very large spread in area. The two extremes, smallest and largest, result in either license granted or denied because the latter contour runs outside the plant premises and across parts of a nearby township. GR (*F–N* curves, see Figure 3.36, right) differed by at least two orders of magnitude. In conclusion, one can be truly skeptical about accuracy of predictive risk analysis results.

As a result of experience gained, a much better job was done this time in identifying the sources of variability. The relative importance of various contributing sources of uncertainty is summarized in Table 3.9. The largest contribution to uncertainty appeared to be the variety in the scenario definition, where analyst judgment and selection of release modes leads to large differences in effects. This scenario uncertainty had already been made clear in 1990 by Kaplan,[168] who illustrated this by Figure 3.37 in which is shown how from one initiating failure event many scenarios can evolve, not only branching and separating but even coinciding in pinch points with subsequent rebranching. This web of scenarios can only be modeled in a multiple of event trees but could more simply be modeled in a BN. Other significant contributions to uncertainty are the failure frequency assessments, which after what has been mentioned in Section 3.6.4 can be expected. Not explicitly mentioned in Table 3.9 are the uncertainties in determining the probability of fatality and injury given toxic loads, radiant heat or blast, and structural damage. Usually it appears that diminishing uncertainty by additional knowledge, estimated probability values increase, sometimes even by one or two orders of magnitude.

Summarizing: It is the task of the risk analyst to predict what can go wrong, its likelihood, and consequences. A systematic probabilistic approach is the only way open. However, analyst choices, case complexity, available computing time,

Table 3.9 Contributing Factors to the Variance in Risk Analysis Results

Uncertainty Factor	Importance
Differences in the qualitative analysis	**
Factors relating to frequency assessment:	
Frequency assessments of pipeline failures	***
Frequency assessments of loading arm failures	****
Frequency assessments of pressurized tank failures	****
Frequency assessments of cryogenic tank failures	***
Factors relating to consequence assessment:	
Definition of the scenario	*****
Modeling of release rate from long pipeline	***
Modeling of release rate from short pipeline	*
Release time (i.e., operator or shut-down system reaction time)	***
Choice of light, neutral, or heavy gas model for dispersion	****
Differences in dispersion calculation codes	***
"Analyst conservatism" or judgment	***

After Lauridsen et al.[99]

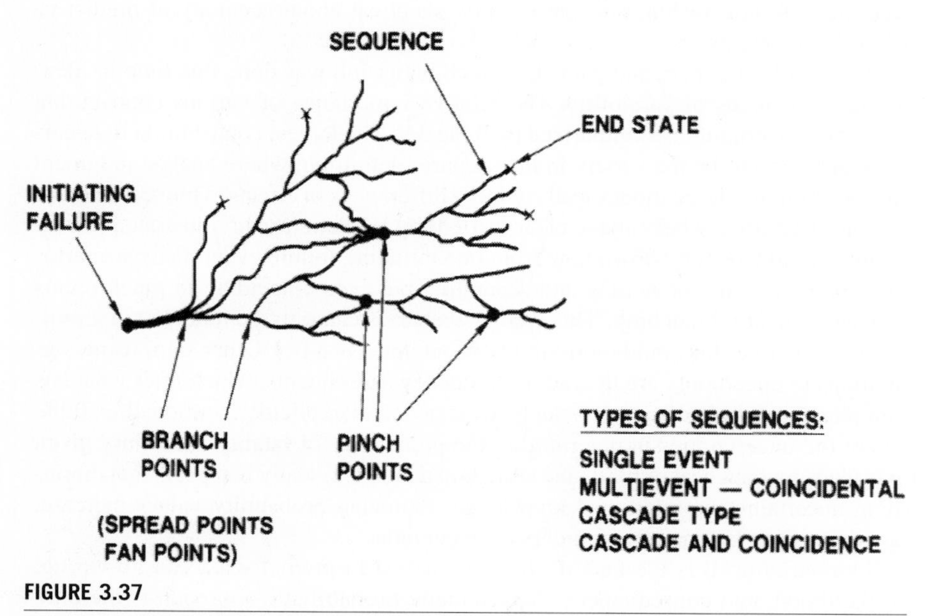

FIGURE 3.37

Scenarios emanating from an initiating event as described by Kaplan.[168]

and limited knowledge and experience will all contribute to unavoidable spread in risk analysis results and incomplete application of the method. It is therefore no surprise that accident scenarios that have actually occurred usually had not been predicted by risk analysis. John Murphy and Jim Conner[169] give some striking examples and warn to beware of the fully unexpected event that escaped the attention of the PHA team. Unforeseen mechanisms and domino effects play a role in this. It can be even a Nassim Taleb's[164] Black Swan of a catastrophic event never seen before (the unknown "unknowns" mentioned in Chapters 9 and 12).

Just a simplified QRA of potential consequences of inventories given three grades of leaks—rupture with an instantaneous release of whole content, a fracture with emptying in 10 min, and a smaller continuous leak—has the disadvantage of a large uncertainty of probability of occurrence. This is because it will be based on standard component failure rates assuming adequate construction and maintenance, and not accounting for operational errors and domino effects. LOPA as an order-of-magnitude QRA following a HAZOP helps greatly to choose the right protection layers in addition to the classical safeguards but is focused on reducing occurrence probability and does not provide the full picture of consequences, certainly not for more complex cases.

For detailed analysis of risks with the purpose of consideration of additional safety measures or optimal layout, a more elaborate and thus more expensive QRA is definitely a useful tool on a comparative basis. For example, Ulrich Hauptmanns,[170] emeritus professor at Otto-von-Guericke Universität Magdeburg, Germany, showed how despite reliability data set differences between various sources still useful results can be obtained to identify weak points in the safety of a design. George Apostolakis,[170] for certain a godfather of probabilistic risk analysis in the nuclear community, analyzed the various criticisms and presented a clear opinion about the great usefulness of QRA. Apart from various arguments already mentioned above, he stresses two more points: QRA does not enable risk-based, but risk-informed decision making, and a good QRA needs rigorous peer review. It will be clear, however, that in case of land use planning or licensing, the disagreements in model outcomes will cause in critical cases much debate and friction among planners from both private and public parties. As would be expected, there will be different interests, hence providing fertile grounds for lawyers, while competent authorities under pressure will be uncertain and will try to delay decision and realization or eliminate the activity altogether. Therefore, improvements must be made.

One of the aspects that can be improved is the following: physical models and system models used in risk analysis are embedded in computer software programs. For a reliable and reproducible answer, the software must be transparent, verifiable, and robust. "Transparent" means it shall be more than just a "black box." Insight into model assumptions and limitations, which inputs and equations are used at each step, and other information shall be easily obtainable. "Verifiable" means sources of input values (references) shall be traceable, as also the choices that were made, the reasons for those choices, and the information used. "Robustness" has to do with reproducibility. The outcome shall not be dependent on the team performing the calculation. Reliability of software forms a sector of science in itself. These requirements are simple but not easily satisfied.

Some suggestions to improve the prediction accuracy are the following:

- In identification of hazards and the definition of scenarios a new promising approach that is a synthesis of function-based HAZOP and component-based FMEA has been developed. This method, the Blended Hazid, will be explained in Chapter 7.
- BNs will enable more flexible modeling of scenario cause–consequence chains and interdependencies than fault and event trees can. BNs will also enable inclusion of full distributions, discrete and continuous, of uncertain parameter values, even fuzzy ones, and changes of values over time (dynamics). We shall explain, elucidate, and show examples of BNs in Chapter 7.
- Consequence analysis improvements by enhanced computational fluid dynamic models will be a slow but continuous progress. LES for gas dispersion and inclusion of some rough kinetics will help. Finite element methods are available for effect calculations on structures. Validating predictions by comparison with results of large-scale field test will be required. In the 1980s and 1990s, dispersion and other tests have been performed. In EU project SMEDIS, experts of the Model Evaluation Group have selected gas dispersion tests that are judged suitable for validation. For example, in 2006–2007 the Health and Safety Laboratory[171] in the UK, under contract by NFPA in the US, applied the selection to evaluate predictive LNG vapor dispersion models. This was followed by a study on LNG spill source terms.[172] New tests would certainly be welcome but will require large-scale cooperation and funding that undoubtedly will present an obstacle.
- With respect to failure rates, there is still a pressing need to obtain better figures; it depends on availability of historical data. HSE in the UK and RIVM in the Netherlands hope to make progress in a cooperative effort. RBI firms will have much data at their disposal. These data are proprietary, but they could be made anonymous and available at a cost. Delving deeper into the field highlights a number of other problems such as in what mode is the failure, under what environmental conditions did the failure occur, and what is the human influence. Depending on the number of cases and data considered, confidence limits must be specified. Another "hot" problem is cloud ignition probabilities. Pesce et al.[174] presented an overview and a framework to make further progress. An even more recent, detailed, and elaborate study of ignition probability has been made by Stack, Sepeda, and Moosemiller[175] and published as a CCPS Guideline.
- New approaches to tackle domino effect prediction will be mentioned in Chapter 7.
- A significant effect on failure rates and human error is influence of management and other organizational factors (the preconditions and latent failures). Quality of management is still an undetermined factor in (operational) risk analysis.

Trials so far in, for example, the ARAMIS project on barrier effectiveness have not been very successful. However, there are new attempts and also new insights on human factors and decision making, which we shall address in the next few chapters.

- Risk fluctuations on the short term due to, for example, weather, sudden vibrations, and personal exposure fluctuations are not yet taken into account. For longer-term risk-level changes, management influences must first be analyzed. However, accounting in risk analysis for monitored process safety metrics and indicators on integrity degradation trends over the longer term will offer new possibilities.
- Risk levels in nonroutine operations—start-ups, shut-downs, and turnarounds (maintenance operations)—are relatively high but as yet not accounted for.
- Handling and presenting uncertainties represented by distributions will increase the effectiveness of risk communication and in turn will help to convince stakeholders and the public to tolerate acceptable risk levels that are inherent to industrial production activity that benefit us all. Decision-making strategies, including cost–benefit analysis, need more attention. Also, these subjects will reemerge in later chapters.

Finally, Rae, Alexander, and McDermid[176] in their 2014 paper produced a very extensive and useful analysis and classification of the numerous (76) flaws a QRA can suffer of and developed a multilevel maturity model that will assess a QRA and will guide to a better practice. They start out considering accuracy and, for example, quote a statement that "the total error must be smaller than the margin between the estimated risk and the risk limit," which for several orders of magnitude does not seem easy to accomplish! But they go much further and deeper identifying flaws and show direction how to conduct best the QRA process and which research will improve quality and reduce errors most. To their analysis may be added, there is "no such thing as a quick fix."

3.8 CONCLUSIONS

Books such as the ones by Trevor Kletz[177] on "what went wrong" illustrate the point that possible mishap scenarios are almost endless. However, unless consequences have been really disastrous, lessons from mishaps learnt by survivors or investigators from such accidents do not spread far and after some time are usually virtually forgotten. To make progress, experience has to be formulated into systems, models, databases, and training. The goal shall be not only to predict risk a priori in design but to control risk level while processes are running. Qualitative analysis is not sufficient; in any science, measuring an effect quantitatively provides a key to improved controllability, so everything that affects the system must be measured to be managed. This chapter has given an overview of the current state of the art of QRA, its strong points but also its deficiencies and weaknesses despite the large body of process safety knowledge that has already been built.

Process safety knowledge is rather diverse and extensive. It ranges from properties of substances, via system safety and process technology and organization to risk assessment methodology. Several methods of experimental, theoretical modeling, and simulation have been established, and software for various applications is available. Yet, there is still much to investigate to aid making more accurate predictions. As shown, risk analysis results can easily differ over one to two orders of magnitude. Also, important data are lacking, for example, the probability of injury data on blast and toxicology of interest, for example, to emergency planning.

One can question whether this spread in outcomes is really a problem. When used for comparing risks by adding measures in a certain scenario, it may not be limiting. Moreover, it seems that the safety level in industrialized countries is not that much different from one to the other, and in general the safety level is high, at least in the globally operating companies. And, by the way, to span a large range of human signal processing of, for example, eye and ear, of light and sound, our mind converts the output to a logarithmic scale, and while looking at consequences and probabilities on a logarithmic scale dispersion of results does not look that serious! But of course, for predictive purposes and decision making about projects with potentially high consequences, we must improve the situation. Various suggestions have been made, and in Chapter 7 we shall see several promising new concepts that will help to accomplish this goal.

REFERENCES

1. Capt. Edward A. Murphy was an engineer working in 1949 on an US Air Force project to determine how much deceleration a person can stand in a crash. In fact, the volunteer in the rocket test sled vehicle, Dr Stapp, in press accounts after a successful experiment, attributed the test's good safety record to "Murphy's law".
2. Kidam K, Hurme M, Hassim MH. Technical analysis of accidents in chemical process industry and lessons learnt. *Chem Eng Trans* 2010;**19**:451–6.
3. Kidam K, Hurme M. Analysis of equipment failures as contributors to chemical process accidents. *Process Saf Environ Prot* 2013;**91**:61–78.
4. Broadribb MP. It's people, stupid!—human factors in incident investigation. *Process Saf Prog* 2012;**31**:152–8.
5. De Rademaeker E, Suter G, Pasman HJ, Fabiano B. A review on past, present and future of the European loss prevention and safety promotion in the process industries. *Process Saf Environ Prot* 2014;**92**:280–91.
6. a. Reason J. The contribution of latent human failures to the breakdown of complex systems. *Philos Trans R Soc Lond Ser B, Biol Sci* 1990;**327**(1241):475–84. http://dx.doi.org/10.1098/rstb.1990.0090;
   b. Reason JT. *Managing the risks of organizational accidents*. Aldershot (UK): Ashgate Publishing Limited; 1997.
7. TRIPOD BETA: Guidance on using Tripod Beta in the investigation and analysis of incidents, accidents and business losses. http://www.energypublishing.org/tripod/beta; Tripod Delta, http://www.energypublishing.org/tripod/delta.

8. CCPS, Center for Chemical Process Safety (AIChE). *Guidelines for technical management of chemical process safety*. 1989, 345 East 47th Street, New York, NY 10017; ISBN 0-8169-0423-5.
9. CCPS, Center for Chemical Process Safety. *Guidelines for risk based process safety*. Hoboken (NJ): John Wiley & Sons; 2007, ISBN 978-0-470-16569-0.
10. Cameron IT, Raman R. *Process systems risk management, in process systems engineering*, vol. 6. Elsevier Academic Press; 2005, ISBN 0-12-156932-2.
11. European Process Safety Centre, EPSC. *Safety management systems, sharing experiences in process safety*. Copyright © Institution of Chemical Engineers; 1994, ISBN 0 85295 356 9.
12. Visser JP. Managing safety in the oil industry—the way ahead. In: *Proceedings 8th international symposium loss prevention and safety promotion in the process industries*, vol. III. Antwerp, Belgium; June 6–9, 1995. p. 171–220.
13. International Association of Oil & Gas Producers. *Safety performance indicators—2013 data*. Report no. 2013s. July 2014. OGP Data Series, www.ogp.org.uk/pubs/2013s.pdf (2001–3 data from 2010 version added).
14. Fabiano B, Pasman HJ. Trends, problems and outlook in process industry risk assessment and aspects of personal and process safety management. In: Nota G, editor. *Advances in risk management*. Sciyo; August 2010, ISBN 978-953-307-138-1. p. 59–91. http://www.intechopen.com/books/advances-in-risk-management.
15. Baker JA, Leveson N, Bowman FL, Priest S, Erwin G, Rosenthal I, et al. *The report of the BP US refineries independent safety review panel*. 2007. http://www.bp.com/liveassets/bp_internet/globalbp/globalbp_uk_english/SP/STAGING/local_assets/assets/pdfs/Baker_panel_report.pdf.
16. Hopkins A. Thinking about process safety indicators. *Saf Sci* 2009;**47**:460–5 and following articles in the same Special Issue 4 on process safety indicators, p. 459–510.
17. CCPS, Center for Chemical Process Safety. *Guidelines for process safety metrics*. Wiley; 2010, ISBN 978-0-470-57212-2.
18. Guldenmund FW. *Understanding and Exploring safety culture* [Ph.D. Dissertation]. Delft University of Technology; 2010, ISBN 978 90 8891 138 5.
19. Zohar D. Safety climate in industrial organizations: theoretical and applied implications. *J Appl Psychol* 1980;**65**(1):96–102.
20. Zohar D. Thirty years of safety climate research: reflections and future directions. *Accid Anal Prev* 2010;**42**(5):1517–22.
21. a. Hudson P, Safety Management and Safety Culture The Long, Hard and Winding Road, 2001, http://www.skybrary.aero/bookshelf/books/2417.pdf.
    b. Hudson P, Parker D, Lawrie M, Van der Graaf G, Bryden R. How to win hearts and minds: the theory behind the program, the seventh SPE (society of petroleum engineers) international conference on health, safety and environment. In: Hudson P, editor. *Oil and gas exploration and production, Calgary, Alberta, Canada, March, 2004. Implementing a safety culture in a major multi-national, Safety Science*, vol. 45; 2007. p. 697–722.
22. http://www.energyinst.org.uk/heartsandminds/.
23. ISO. *Risk management—principles and guidelines*. 2009. ISO 31000:2009.
24. ISO. *Risk management—vocabulary*. 2009. Guide 73:2009.
25. Aven T. On the new ISO guide on risk management terminology. *Reliab Eng Syst Saf* 2011;**96**:719–26.

26. Heinrich H. *Industrial accident prevention: a scientific approach.* 1st ed. London: McGraw-Hill Book Company; 1931.
27. Johnson WG. MORT: the management oversight and risk tree. *J Saf Res* 1975;**7**(1): 4–15. Downloadable versions of MORT Chart and User's Manual are made available by NRI Foundation, http://www.nri.eu.com/mort.html.
28. HSE, Health & Safety Executive, UK. *Root causes analysis: literature review.* Contract Research Report 325/2001, ©Crown Copyright. 2001. WS Atkins Consultants Ltd; ISBN 0 7176 1966 4, www.hse.gov.uk/research/crr_pdf/2001/crr01325.pdf.
29. Johnson Ch. *A handbook of incident and accident reporting*. Glasgow University Press; 2003, ISBN 0-85261-784-4.
30. Saleh JH, Marais KB, Bakolas E, Cowlagi RV. Highlights from the literature on accident causation and system safety: review of major ideas, recent contributions, and challenges. *Reliab Eng Syst Saf* 2010;**95**:1105–16.
31. Saleh JH, Saltmarsh EA, Favarò FM, Brevault L. Accident precursors, near misses, and warning signs: critical review and formal definitions within the framework of discrete event systems. *Reliab Eng Syst Saf* 2013;**114**:148–54.
32. Ferjencik M. IPICA_Lite or how to improve root cause analysis. *Reliab Eng Syst Saf* 2014;**131**:1–13.
33. *Figure concept first published in Dutch advisory council on hazardous substances advice on knowledge infrastructure, 2008, and later in the process safety research agenda for the 21st century, A policy document developed by a representation of the global process safety academia*. College Station, Texas: ©Mary Kay O'Connor Process Safety Center; October 21–22, 2011, ISBN 978-0-9851357-0-6. www.process-safety.tamu.edu/.
34. A listing of NFPA ratings can be found on http://safety.nmsu.edu/programs/chem_safety/NFPA-ratingJ-R.htm.
35. Havens J, Walker H, Spicer T. Bhopal atmospheric dispersion revisited. *J Hazard Mater* 2012;**233–234**:33–40.
36. Duijm NJ, Carissimo B, Mercer A, Bartholome C, Giesbrecht H. Development & test of evaluation protocol for heavy gas dispersion models. *J Hazard Mater* 1997;**56**:273–85.
37. Plasmans J, Donnat L, de Carvalho E, Debelle Tr, Marechal B, Baillou F. Challenges with the use of CFD for major accident dispersion modeling. *Process Saf Prog* 2013; **32**:207–11.
38. Sam Mannan, editor. *Lees' loss prevention in the process industries. Hazard identification, assessment and control.* 4th ed., vols. 1–3. Butterworth-Heinemann; 2012, ISBN 978-0-12-397189-0; there also exists a shortened more convenient version: Mannan, S., *Lees' Process Safety Essentials: Hazard Identification, Assessment and Control*, 1st ed. Butterworth-Heinemann/Elsevier; 2014, ISBN 978-1-85617-776-4.
39. Crowl DA, Louvar JF. *Chemical process safety: fundamentals with applications*. International Series in the Physical and Chemical Engineering Sciences. 3rd ed. Prentice Hall; 2011. ISBN-13 978-0-13-138226-8.
40. Schmidt J, editor. *Process and plant safety.* Wiley-VCH Verlag GmbH & Co. KGaA, ISBN:978-3-527-33027-0.
41. RIVM. *Reference manual Bevi risk assessments version 3.2.* http://infonorma.gencat.cat/pdf/AG_AQR_2_Bevi_V3_2_01-07-2009.pdf.
42. Hansler RJ, Gooijer L, Wolting BG. Acute inhalation toxicity in quantitative risk assessment—methods and procedures. *Chem Eng Trans* 2013;**31**:751–6. http://dx.doi.org/10.3303/CET1331126.

43. EMI SIG, Emergency management issues, special interest group, ERPG definitions and background information. http://orise.orau.gov/emi/scapa/chem-pacs-teels/erpg-definitions.htm, see for values http://www.aiha.org/get-involved/AIHAGuidelineFoundation/EmergencyResponsePlanningGuidelines/Documents/2013ERPGValues.pdf.
44. US Environmental Protection Agency, Acute exposure guideline levels (AEGLs). http://www.epa.gov/oppt/aegl/pubs/define.htm. This website also provide data.
45. TEEL, AEGL and ERPG data can be found at http://www.atlintl.com/DOE/teels/teel.html.
46. 2012 ACGIH® threshold limit values (TLVs®) and biological exposure indices (BEIs®), Appendix D, http://www.nsc.org/news_resources/facultyportal/AnswerKeys/FIH%206e%20Appendix%20B.pdf.
47. European Commission Employment. Social affairs & inclusion health and safety at work—the scientific committee on occupational exposure limits (SCOEL), http://www.google.com/url?sa=t&rct=j&q=&esrc=s&source=web&cd=2&ved=0CC0QFjAB&url=http%3A%2F%2Fec.europa.eu%2Fsocial%2FBlobServlet%3FdocId%3D3803%26langId%3Den&ei=_mlAVJWAHs3KggSZ0YDQAg&usg=AFQjCNH2ozmtlyXKP5ySZYYOP_3fursgDA&bvm=bv.77648437,d.eXY.
48. Wrixon AD. New ICRP recommendations. *J Radiol Prot* 2008;**28**:161−8.
49. Abbasi T, Pasman HJ, Abbasi SA. A scheme for the classification of explosions in the chemical process industry. *J Hazard Mater* 2010;**174**:270−80.
50. Recommendations on the Transport of Dangerous Goods. *Manual of tests and criteria*. Fifth revised edition, 2009. ST/SG/AC.10/11/Rev.5, Copyright © United Nations.
51. HarsNet: http://www.harsnet.net/. (For S2S does not use the link on the HarsNet website, but the one below).
52. S2S. The european web Portal for process safety. http://www.safety-s2s.eu/index.php.
53. DECHEMA ChemSafe database. http://www.dechema.de/en/chemsafe.html.
54. Pasman HJ, Pekalski AA, Braithwaite M, Griffiths JF, Schroeder V, Battin-Leclerc F. Playing with fire: safety and reaction efficiency research on gas phase hydrocarbon oxidation processes: project SAFEKINEX. *Process Saf Environ Prot* 2005;**83**(B4):317−23. http://www.morechemistry.com/SAFEKINEX/.
55. Lemkowitz SM, Pasman HJ. A review of the fire and explosion hazards of particulates. *Kona Powder Part J* 2014;**31**:53−81. http://dx.doi.org/10.4356/Kona.014010. Review Paper, ISSN 02884-532.
56. a. Oppenheim AK, Laderman AJ, Urtiev PA. The onset of retonation. *Combust Flame* 1962;**6**:193−7;
    b. Urtiev PA, Oppenheim AK. The onset of detonation. *Combust Flame* 1965;**9**:405−7.
57. S2S, Safety to Safety. Training—evaluation chemical reaction hazards, the stoessel diagram. http://www.safety-s2s.eu/modules.php?name=s2s_wp4&idpart=2&op=v&idp=755.
58. Casson V. Integrated Calorimetric techniques applied to runaway reactions analysis [Ph.D. Dissertation]. University of Pisa, Leonardo da Vinci Engineering School; 2012.
59. Urben P, editor. *Bretherick's handbook of reactive chemical hazards*. 7th ed. Oxford: Academic Press, Elsevier; 2007. two volumes, set ISBN-13 978-0-12-372563-9.
60. University of South Alabama. CHETAH the computer program for chemical thermodynamics and energy release evaluation. http://www.southalabama.edu/engineering/chemical/chetah/intro.html.

61. Lawrence Livermore National Laboratory. Cheetah 7.0 thermochemical code. https://www-pls.llnl.gov/?url=science_and_technology-chemistry-cheetah.
62. Oran AS, Boris JP. *Numerical simulation of reactive flow*. 2nd ed. Cambridge University Press, © Naval Research Laboratory; 2001, ISBN 0-521-58175-3.
63. Di Sarli V, Di Benedetto A, Russo G. Using large Eddy simulation for understanding vented gas explosions in the presence of obstacles. *J Hazard Mater* 2009;**169**:435–42.
64. Eckhoff RK, Olsen W. A new method for generation of synchronized capacitive sparks of low energy. Reconsideration of previously published findings. *J Electrost* 2010;**68**:73–8.
65. Eckhoff RK, Ngo M, Olsen W. On the minimum ignition energy (MIE) for propane/air. *J Hazard Mater* 2010;**175**:293–7.
66. Glor M. Electrostatic ignition hazards in the process industry. *J Electrost* 2005;**63**:447–53.
67. Glor M. Modelling of electrostatic ignition hazards in industry: too complicated, not meaningful or only of academic interest? *Chem Eng Trans* 2013;**31**:583–8.
68. Choi KS, Mogami T, Suzuki T, Kim SCh, Yamaguma M. Charge reduction on polypropylene granules and suppression of incendiary electrostatic discharges by using a novel AC electrostatic ionizer. *J Loss Prev Process Ind* 2013;**26**:255–60.
69. McGrattan K, McDermott R, Floyd J, Hostikka S, Forney G, Baum H. Computational fluid dynamics modelling of fire. *Int J Comput Fluid Dyn* 2012;**26**(6–8):349–61.
70. Galea E. Several publications of University of Greenwich. http://fseg2.gre.ac.uk/HEED/publications/.
71. Baker WE. *Explosions in air.* Austin, TX: University of Texas Press; 1973.
72. Strehlow RA, Baker WE. The characterization and evaluation of accidental explosions. *Prog Energy Combust Sci* 1976;**2**(1):27–60.
73. Van den Berg AC. The multi-energy-method, a framework for vapour cloud explosion blast prediction. *J Hazard Mater* 1985;**12**:1–10.
74. Sari A. Comparison of TNO multi-energy and Baker–Strehlow–Tang models. *Process Saf Prog* 2011;**30**(1):23–6.
75. Gexcon website. http://www.gexcon.com/.
76. Bjerketvedt D, Bakke JR, Van Wingerden K. Gas explosion handbook. *J Hazard Mater* 1997;**52**:1–150. Also available as pdf from the Gexcon website, http://www.gexcon.com/.
77. HSE, Methods of approximation and determination of human vulnerability for offshore major accident hazard assessment. www.hse.gov.uk/foi/internalops/hid_circs/technical_osd/spc_tech_osd_30/spctecosd30.pdf.
78. Alonso FD, Ferradas EG, Sanchez Perez JF, Aznar AM, Gimeno JR, Doval Minarro M. Consequence analysis to determine damage to buildings from vapour cloud explosions using characteristic curves. *J Hazard Mater* 2008;**159**:264–70. (Also from the same researchers: *J Loss Prev Process Ind* 2008;**20**:264–70).
79. Zhang M, Jiang J. An improved probit method for assessment of domino effect to chemical process equipment caused by overpressure. *J Hazard Mater* 2008;**158**:280–6.
80. *Methods for the Calculation of Physical effects, Yellow Book*. 2005. PGS 2, Dutch government, Ministry VROM (meanwhile replaced by Ministry I&M), http://www.publicatiereeksgevaarlijkestoffen.nl/publicaties/PGS2.html.
81. Tugnoli A, Gubinelli G, Landucci G, Cozzani V. Assessment of fragment projection hazard: probability distributions for the initial direction of fragments. *J Hazard Mater* 2014;**92**:714–22.

82. 'Green book', Committee for the prevention of disasters, methods for the determination of possible damage to people and objects resulting from releases of hazardous materials, Rep, CPR 16E (Voorburg) 1992, in Dutch.
83. HSE, Daycock, J.H., Rew, P.J., WS Atkins Consultants Ltd. Thermal radiation criteria for vulnerable populations, © Crown Copyright 2000, ISBN 0 7176 1837 4, www.hse.gov.uk/research/crr_pdf/2000/crr00285.pdf; and HSL, O'Sullivan, S, Jagger, S., Human vulnerability to thermal radiation offshore. HSL/2004/04, © Crown Copyright 2004, http://www.hse.gov.uk/research/hsl_pdf/2004/hsl04-04.pdf.
84. Casal J. *Evaluation of the effects and consequences of major accidents in industrial plants. Industrial safety series.* 1st ed., vol. 8. Elsevier; 2008, ISBN 978-0-444-53081-3.
85. Raj Ph K. Field tests on human tolerance to (LNG) fire radiant heat exposure, and attenuation effects of clothing and other objects. *J Hazard Mater* 2008;**157**:247–59.
86. Möller N, Hansson SO. Principles of engineering safety: risk and uncertainty reduction. *Reliab Eng Syst Saf* 2008;**93**:776–83, ISBN-13: 978-1439804551 ISBN-10: 1439804559.
87. Kletz T, Amyotte PR. *Process plants: a handbook for inherently safer design*. 2nd ed. Boca Raton (FL): CRC Press; 2010.
88. Kletz T. *Process plants: a handbook for inherently safer design*. Philadelphia, PA: Taylor & Francis; 1998.
89. Pasman HJ, Rogers WJ, Mannan MS. How to consolidate Trevor's experience and knowledge? Impossibilities, reflections on possibilities, and call to action. *J Loss Prev Process Ind* 2012;**25**:870–5.
90. Khan FI, Amyotte PR. Integrated inherent safety Index (I2SI): a tool for inherent safety evaluation. *Process Saf Process* 2004;**23**:136.
91. Modarres M, Kaminskiy, M., Krivtsov V. *Reliability engineering and risk analysis, A Practical Guide*. 2nd ed. CRC Press, Taylor & Francis Group; ISBN 978-0-8493-9247-4.
92. Khan FI, Sadiq R, Husain T. Risk-based process safety assessment and control measures design for offshore process facilities. *J Hazard Mater* 2002;**A94**:1–36.
93. Lewis S, Hurst S. Bow-tie an elegant solution? *Strategic Risk* November 2005:8–10.
94. De Dianous V, Fiévez C. ARAMIS project: a more explicit demonstration of risk control through the use of bow-tie diagrams and the evaluation of safety barrier performance. *J Hazard Mater* 2006;**130**:220–33.
95. Ebeling Ch E. *An introduction to reliability and maintainability engineering*. 2nd ed. Long Grove, IL: Waveland Press Inc; 2010. ISBN 978-1-57766-625-7.7.
96. Kaplan S, Garrick BJ. On the quantitative definition of risk. *Risk Anal* 1981;**1**(1):11–27.
97. Publication series hazardous materials 1-4 (2-4 in English). http://www.publicatiereeksgevaarlijkestoffen.nl/.
98. International standard IEC/ISO 31010 risk management—risk assessment techniques, Edition 1.0, 2009-11.
99. Lauridsen K, Kozine I, Markert F, Amendola A, Christou M, Fiori M. *Assessment of uncertainties in risk analysis of chemical establishments*. The ASSURANCE project Final summary report. Roskilde, Denmark: Risø National Laboratory; May 2002. 52p, http://www.risoe.dk/rispubl/sys/syspdf/ris-r-1344.pdf.
100. Salvi O, Debray B. A global view on ARAMIS, a risk assessment methodology for industries in the framework of the SEVESO II directive. *J Hazard Mater* 2006;**130**:187–99.

101. Pasman H, Reniers G. Past, present and future of quantitative risk assessment (QRA) and the incentive it obtained from land-use planning (LUP). *J Loss Prev Process Ind* 2014;**28**:2–9.
102. Ross Gemini Centre, NTNU/SINTEF, Accident databases, http://www.ntnu.edu/ross/info/acc-data.
103. American Institute of Chemical Engineers (AIChE). *Dow's fire & explosion index hazard classification guide.* 7th ed. June 1994. ISBN 978-0-8169-0623-9.
104. American Institute of Chemical Engineers (AIChE). *Dow's chemical exposure Index guide.* 1st ed. 1994. ISBN 0-8169-0647-5.
105. Lawley HG. Operability studies and hazard analysis. *Chem Eng Prog* 1974;**70**:45–57.
106. Crawley F, Preston M, Tyler B. *HAZOP: guide to best practice.* Institution of Chemical Engineers, © 2000 European Process Safety Centre; ISBN 0 85295 427 1.
107. Delvosalle Ch, Fievez C, Pipart A, Debray B. ARAMIS project: a comprehensive methodology for the identification of reference accident scenarios in process industries. *J Hazard Mater* 2006;**130**:200–19.
108. Pitblado R, Turney R. *Risk assessment in the process industries.* 2nd ed. Institution of Chemical Engineers; Copyright © 1996, ISBN 0 85295 323 2.
109. Cozzani V, Gubinelli G, Salzano E. Escalation thresholds in the assessment of domino accidental events. *J Hazard Mater* 2006;**A129**:1–21.
110. Uit de Haag PAM, Ale BJM. *Guidelines for quantitative risk assessment, purple book.* CPR18E. 2005. RIVM, http://www.publicatiereeksgevaarlijkestoffen.nl/publicaties/PGS3.html.
111. Beerens HI, Post JG, Uit de Haag PAM. The use of generic failure frequencies in QRA: the quality and use of failure frequencies and how to bring them up-to-date. *J Hazard Mater* 2006;**130**:265–70.
112. Pasman HJ. History of Dutch process equipment failure frequencies and the Purple Book. *J Loss Prev Process Ind* 2011;**24**:208–13.
113. OREDA. *Offshore reliability data handbook 5th edition, volume 1—topside equipment, volume 2—subsea equipment.* Norway: SINTEF; 2009. http://www.oreda.com/handbook.html.
114. Taylor JR. *Taylor associates ApS, hazardous materials release and accident frequencies for process plant.* volume. I. version 2, Issue 1. May 2006. and Volume II, 1st version, Issue 7, Sept. 2006, personal communication.
115. a. Taylor JR. Understanding and combating design error in process plant design. *Saf Sci* 2007;**45**:75–105;
see also Taylor JR. Statistics of design error in the process industries. *Saf Sci* 2007;**45**:61–73.
116. HSE, UK. *Failure rate and event data for use within risk assessments.* June 28, 2012. www.hse.gov.uk/landuseplanning/failure-rates.pdf (with certain parts still in preparation). The data are also to be found in Int'l Ass Oil & Gas Producers, OGP Risk Assessment Data Directory, *Process Release Frequencies*, Report No. 434-1, March 2010.
117. Vilchez JA, Espejo V, Casal J. Generic event trees and probabilities for the release of different types of hazardous materials. *J Loss Prev Process Ind* 2011;**24**:281–7.
118. Guldenmund F, Hale A, Goossens L, Betten J, Duijm NJ. The development of an audit technique to assess the quality of safety barrier management. *J Hazard Mater* 2006;**130**:234–41.

119. Duijm NJ, Goossens L. Quantifying the influence of safety management on the reliability of safety barriers. *J Hazard Mater* 2006;**130**:284–92.
120. Pitblado R, Bain B, Falck A, Litland Kj, Spitzenberger C. Frequency data and modification factors used in QRA studies. *J Loss Prev Process Ind* 2011;**24**:249–58.
121. Pasman HJ, Rogers WJ. Risk assessment by means of Bayesian networks: a comparative study of compressed and liquefied H_2 transportation and tank station risks. *Int J Hydrog Energy* 2012;**37**:17415–25, and Erratum in 38 (2013) 1662.
122. Wiersma T, Boot H, Gooijer L. Societal risk on a map. In: *12th symposium loss prevention and safety promotion in the process industries. IChemE symposium series,* Edinburgh, UK **No. 153**; 22–24 May, 2007.
123. Hirst IL, Carter DA. A "worst case" methodology for obtaining a rough but rapid indication of the societal risk from a major accident hazard installation. *J Hazard Mater* 2002;**A92**:223–37.
124. *CCPS, guidelines for chemical transportation safety, security, and risk managemen*t, NY: Center for Chemical Process Safety, John Wiley& Sons; © 2008 AIChE, Chapter 7, ISBN 978-0471-78242-1.
125. CCPS. *Center for process safety of AIChE in New York, layer of protection analysis, simplified process risk assessment.* 2001. ISBN 0-8169-0811-7 (In 2014 is expected: Guidelines for Independent Protection Layers and Initiating Events.).
126. Dowell III AM. Layer of Protection analysis: a new PHA tool after hazop, before fault tree analysis. In: *Int'l conference and workshop on risk analysis in process safety.* Atlanta, Ga: CCPS/AIChE; October 21–24, 1997. pp 13–28.
127. Huff AM, Montgomery ML. *A risk assessment methodology for evaluating the effectiveness of safeguards and determining safety instrumented systems requirements, Int'l conference and workshop on risk analysis in process safety.* Atlanta, Georgia: CCPS/AIChE; October 21–24, 1997. pp 111–126.
128. IEC 61511. Functional safety—safety instrumented systems for the process industry sector-, parts 1–3. *International Electro-technical commission:* 1st ed. Geneva, Switzerland; 2003.
129. ANSI/ISA-84.00.01-2004 Parts 1-3 (IEC 61511 Mod), Functional safety: safety instrumented systems for the process industry sector.
130. Vansina P, Biermans K. *PLANOP, een methode voor het uitvoeren van procesveiligheidsstudies (In English: PLANOP, method for conducting process safety studies),* version 3. May 2012. www.planop.be.
131. Wang Y, Rausand M. Reliability analysis of safety-instrumented systems operated in high-demand mode. *J Loss Prev Process Ind* 2014;**32**:254–64.
132. Jin H, Lundteigen MA, Rausand M. Reliability performance of safety instrumented systems: a common approach for both low- and high-demand mode of operation. *Reliab Eng Syst Saf* 2011;**96**:365–73.
133. UK. HSE. *A review of layers of protection analysis (LOPA) analyses of overfill of fuel storage tanks.* Prepared by Health and Safety Laboratory for the Health and Safety Executives, Buxton, UK; 2009. RR716.
134. SINTEF, PDS reliability of safety instrumented systems. http://www.sintef.no/Projectweb/PDS-Main-Page/.
135. Bukowski JV, Goble WM. Analysis of pressure relief valve proof test data. *Process Saf Prog* 2009;**28**(1):24–9.

136. Lucchini S. Safety instrumented function reliability and the art of confusion. In: *14th annual symposium, Mary Kay O'Connor process safety Center.* College Station, TX: Texas A&M University; October 25–27, 2011. p. 582–95.
137. Jin H, Rausand M. Reliability of safety-instrumented systems subject to partial testing and common-cause failures. *Reliab Eng Syst Saf* 2014;**121**:146–51.
138. Bridges WG, Clark T. Key issues with implementing LOPA (Layer of Protection Analysis)—perspective from one of the originators of LOPA. In: *Proceedings 5th global congress on process safety, 11th plant process safety symposium.* New York: CCPS/AIChE; 2009.
139. Bridges WG. LOPA and human reliability—human errors and human IPLs. In: *2010 Spring meeting, 6th global Congress on process safety.* San Antonio, TX: CCPS/AIChE; March 22–24, 2010.
140. Schmidt MS. Villains, victims, and heroes: accounting for the roles of human activity plays in LOPA scenarios. In: *15th annual symposium.* College Station, TX: Mary Kay O'Connor Process safety Center, Texas A&M University; October 23–25, 2012. p. 454–68.
141. Kumamoto H, Henley EJ. *Probabilistic risk assessment and management for engineers and Scientists.* 2nd ed. IEEE Press; 1996. ISBN-13 978-0-7803-1004-9, ISBN 0-7803-1004-7; 2nd Edition, Wiley IEEE Press, 2000, ISBN-13 978-0-7803-6017-4, ISBN 0-7803-6017-6.
142. Pasman HJ, Knegtering B, Rogers WJ. A holistic approach to control process safety risks: possible ways forward, reliability engineering and system safety. *Reliab Eng Syst Saf* 2013;**117**:21–9.
143. Rahimi M, Rausand M. Monitoring human and organizational factors influencing common-cause failures of safety-instrumented system during the operational phase. *Reliab Eng Syst Saf* 2013;**120**:10–7.
144. US Chemical Safety and Hazard Investigation Board, Chevron refinery fire, Current investigation. http://www.csb.gov/.
145. US Chemical Safety and Hazard Investigation Board. *Catastrophic rupture of heat exchanger (seven fatalities), Tesoro anacortes refinery, April 2, 2010.* May 2014. Report 2010-08-I-WA.
146. SAE JA1011. *Evaluation criteria for reliability-centered maintenance (RCM) processes.* Society of Automotive Engineers; August 1, 1998;
SAE JA1012. *A guide to the reliability-centered maintenance (RCM) standard.* Society of Automotive Engineers; January 1, 2002.
147. National Fire Protection Agency. *NFPA 30: flammable and combustible liquids code.* 2012 Edition.
148. *FM global property loss prevention data sheets*, 7–88, October 2011.
149. Di Padova A, Tugnoli A, Cozzani V, Barbaresi T, Tallone F. Identification of fireproofing zones in oil & gas facilities by a risk based procedure. *J Hazard Mater* 2011;**191**:83–93.
150. Pekalski AA, Zevenbergen JF, Lemkowitz SM, Pasman HJ. A review of explosion prevention and protection systems suitable as ultimate layer of protection in chemical process installations. *J Trans IChemE, Part B Process Saf Environ Prot* January 2005;**83**(B1):1–17.
151. Bernechea EJ, Vilchez JA, Arnaldos J. A model for estimating the impact of the domino effect on accident frequencies in quantitative risk assessments of storage facilities. *Process Saf Environ Prot* 2013;**91**:423–37.

152. Rad A, Abdolhamidzadeh B, Abbasi T, Rashtchian D. FREEDOM II: an improved methodology to assess domino effect frequency using simulation techniques. *Process Saf Environ Prot* 2014;**92**:714–22.
153. Reniers G, Cuypers S, Pavlova Y. A game-theory based multi-plant collaboration model (MCM) for cross-plant prevention in a chemical cluster. *J Hazard Mater* 2012; **209–210**:164–76.
154. Krausmann E, Cozzani V, Salzano E, Renni E. Industrial accidents triggered by natural hazards: an emerging risk issue. *Nat Hazards Earth Syst Sci* 2011;**11**:921–9.
155. Antonioni G, Spadoni G, Cozzani V. A methodology for the quantitative risk assessment of major accidents triggered by seismic events. *J Hazard Mater* 2007;**147**:48–59.
156. Landucci G, Antonioni G, Tugnoli A, Cozzani V. Release of hazardous substances in flood events: damage model for atmospheric storage tanks. *Reliab Eng Syst Saf* 2012;**106**:200–16.
157. Necci A, Antonioni G, Cozzani V, Krausmann E, Borghetti A, Nucci CA. A model for process equipment damage probability assessment due to lightning. *Reliab Eng Syst Saf* 2013;**115**:91–9.
158. US Environmental Protection Agency's CAMEO website for downloading. http://www2.epa.gov/cameo.
159. Woodcock B, Au Z. Human factors issues in the management of emergency response at high hazard installations. *J Loss Prev Process Ind* 2013;**26**:547–57.
160. Le Guen J. Reducing risks, protecting people, HSE UK, discussion document, DDE11 C150 5/99. 1999.
161. ISO 2394. 1998. *General principles on reliability for structures*.
162. Boot H. The use of risk criteria in comparing transportation alternatives. *Chem Eng Trans* 2013;**31**:199–204.
163. Prem KP, Ng D, Pasman HJ, Sawyer M, Guo Y, Mannan MS. Risk measures constituting a risk metrics which enables improved decision making: value-at-risk. *J Loss Prev Process Ind* 2010;**23**:211–9.
164. Taleb NN. *The black swan—the impact of the highly improbable*. New York: Random House Trade Paperbacks; 2010, ISBN 978-0-8129-7381-5.
165. Slovic P, Weber EU. *Perception of risk posed by extreme events, conference "Risk management strategies in an uncertain world"*. New York: Palisades; April 12–13, 2002. http://cursos.campusvirtualsp.org/pluginfile.php/7062/mod_page/content/1/modulo2/content/perception-of-risk-posed-by-extreme-events.pdf.
166. Fischhoff B, Watson SR, Hope C. Defining risk. *Policy Sci* 1984;**17**:123–39.
167. International Risk Governance Council. *An introduction to the IRGC risk governance framework*. Geneva. 2008; www.irgc.org.
168. Kaplan S. On the inclusion of precursor and near miss events in quantitative risk assessments: a Bayesian point of view and a space shuttle example. *Reliab Eng Syst Saf* 1990; **27**:103–15.
169. Murphy JF, Conner J. Beware of the black swan: the limitations of risk analysis for predicting the extreme impact of rare process safety incidents. *Process Saf Prog* 2012; **31**(4):330–3.
170. Hauptmanns U. The impact of reliability data on probabilistic safety calculations. *J Loss Prev Process Ind* 2008;**21**:38–49.
171. Apostolakis GE. How useful is quantitative risk assessment? *Risk Anal* 2004;**24**(3): 515–20.

172. Ivings MJ, Jagger SF, Lea CJ, Webber DM. *Evaluating vapor dispersion models for safety analysis of LNG facilities research project.* Technical report. Quincy (MA, USA): Health & Safety Laboratory for The Fire Protection Research Foundation; April 2007.
173. Webber DM, Gant SE, Ivings MJ, Jagger SF. *LNG source term models for hazard analysis: a review of the state-of-the-art and an approach to model assessment.* Final Report. Quincy (MA, USA): Health & Safety Laboratory for The Fire Protection Research Foundation; March 2009.
174. Pesce M, Paci P, Garrone S, Pastorino R, Fabiano B. Modelling ignition probabilities within the framework of quantitative risk assessments. *Chem Eng Trans* 2012;**24**:141–6.
175. Stack B, Sepeda A, Moosemiller M. Guidelines for determining the probability of ignition of a released flammable mass. *Process Saf Prog* 2014;**33**(1):19–25.
176. Rae A, Alexander R, McDermid J. Fixing the cracks in the crystal ball: a maturity model for quantitative risk assessment. *Reliab Eng Syst Saf* 2014;**125**:67–81.
177. Kletz K. *Still going wrong, sequel to the classic "what went wrong": case histories of plant disasters and how they could have been avoided*, Butterworth-Heinemann, Reed Elsevier group, ISBN 13:978-0-7506-7709-7.

CHAPTER

Trends in Society and Characteristics of Recent Industrial Disasters

4

Culture means control over nature.

Johan Huizinga (1872–1945)

SUMMARY

This summary starts as an introduction and continues with a brief sketch of long-term dynamics in our societies. By some mystical mechanism, cultural trends in societies around the world change and at different locations often even in the same direction, albeit with local delays. Societies influence each other, more so after we obtained mass communication means such as television and worldwide travel, which give us insight into how others are behaving and surviving. As a result of the technology development and the introduction of personal electronic products such as laptops and smartphones over the last 50 years, on the one hand, materialistic requirements and dependence on income went up. It also brought more focus on "me," that is to say, a more egocentric attitude of individuals became favored. Societal ties became looser; the individualistic way of life with changing partners became more common. On the other hand, as a result of liability practice, damage supposedly sustained as a consequence of others to "the person" or "on the property" became more heavily sued and compensated. Societal respect for people working in science and arts, technology, and particularly in technical and governmental sectors declined, while those amassing personal wealth such as top company executive officers, entrepreneurs, sports stars, and bankers obtained greater admiration. Commerce and finance rose; differentials in income between poor and rich increased. With all that, our dressing habits changed tremendously. From ancient Greek and Roman authors, we know that older people have looked upon youth as being not eager to follow the pattern of life of the older generation and learn the "hard" lessons that their generation had to go through. But, so far over the centuries, beside negative outcomes there have been always very positive developments too. Progress will be possible, but also change always poses new challenges.

This chapter focuses first on trends in the process industry over the last 50 years. The diversification and enlargement in scale of operations, later the increase in competition, higher costs, and globalization all affected the objective priorities of industrial managers throughout and hence also on the work floor. Cost cutting

Risk Analysis and Control for Industrial Processes - Gas, Oil and Chemicals.

became a major drive and stimulated the introduction of more efficient technology and automation and put more emphasis on energy economics but delayed in many cases investment in safeguards. Plants became more complex. Reorganizations, outsourcing, and continuous change, also due to splits and mergers, added to the complexity. Oil and gas exploration gained importance.

Societal trends had an influence on the attitude of people toward risk. Tolerability to so-called involuntary risk caused by industrial activity decreased, but the trends had also affected the safety attitudes of people. Because the level of safety in society had increased, young people became less sensitive to the pain accidents can bring; they are exposed to fires and explosions in videos and games without the physical experiences, so to learn the importance of safe working takes more effort and puts more emphasis on company training. Performance pressures in work have increased. To accomplish a task within a certain time received priority, while on the other hand quality management systems, which under the same umbrella include environmental and safety management systems, shall guarantee a high level of work quality. As we shall see in upcoming chapters, the high reliability and resilient organization became an ideal.

In this chapter we shall also analyze the causes of two disastrous accidents that have recently attracted much attention and the consequences of which are still in part affecting us: the demise of the Deepwater Horizon offshore drilling rig and the associated oil spill in the Gulf of Mexico and, secondly, the Japanese Fukushima-Daiichi nuclear power plant catastrophe due to earthquake and consequential tsunami. In contrast to earlier process industry accidents for which causal mechanisms were still unknown, the experience and knowledge were present to indicate that both of these events could happen. Despite the precautions that were taken, the accidents still occurred. Therefore, this raises the question of how control of highly complex socio-technical systems, which are important for our well-being, can be improved.

4.1 BUSINESS, INDUSTRY, AND ENERGY TRENDS

Process industries are focused on energy and material products. Before World War II, coal was the most important source of energy and chemicals, but coal has been replaced by oil and gas, in particular with respect to materials. The innovations that led to the manufacture of various polymers as intermediates of extremely useful and very practical plastic construction materials, textile fibers, coatings, and many more, covering a large variety of applications, made the petrochemical industry great. Also, energy consumption per capita grew. Although over time the variety of products manufactured from materials available in the market continually increased and created a further expanding demand, profitability generally decreased overall in each category of product. This is quite a normal phenomenon for any product, since high profits will attract other manufacturers to start working in the sector that in the end will dilute profitability. Hence after World War II, a large expansion and scale increase of process industry occurred in the so-called

Western industrialized countries, Russia, and Japan, and this expansion spread later to regions with lower personnel cost such as the Middle East, Asia, and Brazil.

After the end of the Cold War in the early 1990s, globalization became a buzzword standing for lasting peace and a flourishing economy, thanks to a widening market. It became a reality due to the relatively low transport costs enabling the mining of raw materials and manufacture of intermediates and end products on completely different continents. This globalization changed the landscape of the process industries in the second half of the 1990s and later. Previously, enterprises had diversified their portfolio of products to make themselves as strong as possible in satisfying the local demand, or rather the regional demand within a geographical area within which transport cost did not become a factor in the competition for price. Thus, a company such as Imperial Chemical Industries (ICI) in the UK had a very large assortment of products manufactured in specialized plants, which in addition were to be found in many places around the world, in particular in the Commonwealth countries. Today, ICI has all but vanished; following, we shall see what kinds of factors played a role.

Besides enhanced trade, globalization also brought enhanced competition, because more suppliers entered the market such as those based in China and the oil-producing countries in the Middle East, that is, places where labor or feedstock costs are comparatively low. Meanwhile, the position of shareholders (often major pension funds) also became stronger, and the share values became an increasingly important performance indicator. Many companies therefore embarked on a strategy of specializing as opposed to diversifying as in the 1980s. A coherent, branded product range and market segment became important. This strategy was realized by selling off divisions or large parts of a company and buying from others plants that made similar products to the remaining parts of their own company with the objective of obtaining a larger market share in that particular product. So, the time of changes in ownership and building larger units by mergers and takeovers wiped away some great names, such as Rhône-Poulenc, Hoechst, Union Carbide, and Ciba–Geigy, and introduced or strengthened others.

The process outlined above was accompanied by cost cuts, which in the 1990s was denoted by "downsizing." This meant reduction in labor per mass of unit of product. It started first in the United States (US) but soon also became a trend in Europe. At the same time, it was enabled by the technical automation possibilities introduced during that period, as mentioned in the previous chapter where layers of protection analysis were treated. This put more pressure on output of workers and a higher priority on short-term gains. However, it did not diminish competition nor pressure on product price. This pressure made companies move out of their traditional areas (e.g., in man-made fibers or other commodities) and change to producing fine chemicals with a higher profit margin; later, fashionable areas included biochemicals, materials for the electronics industry, and nanomaterials.

Mergers and cost cuts brought reorganizations and mixing of cultures, which did not help to foster or retain process knowledge and competency. There have been very good exceptions and companies that kept their identity and a steady course, but many others that lost just part of their corporate core knowledge started to rely on outsourcing and on consulting agencies. Outsourcing can be helpful where specialist tasks must be performed for which a company does not have full-time employment of a critical mass of high-level specialized workers. It is, however, more vulnerable to lack of communication or misunderstanding among parties. There are quite a few examples of catastrophic accidents as a result of contractor misunderstanding or lack of specific knowledge of the local situation with respect to the contractor's assignment. This communication deficiency can become exacerbated when the contractor is in turn subcontracting part of his tasks.

Reducing energy consumption costs per unit of product has become an element in the competition, partly because of pressure to reduce use of fossil fuels, partly because of increase of energy prices, and just to outperform a competitor on product price. Since the turn of the century, sustainability and the energy economy have become higher priorities. Reduction of emission of carbon dioxide as a greenhouse gas became an item on the political wish list and the Kyoto Protocol, agreed to in 1997 and in effect since 2005, which committed endorsing nations within the UN Framework Convention on Climate Change to reduce their emissions. This protocol thus put an obligation on companies and countries but built in so-called flexibility mechanisms, which consisted of, for example, trading emission rights or compensating emissions with investments in clean technology (project mechanisms). Political instabilities in the Middle East, Africa, and elsewhere not only affected hydrocarbon production rates but also transportation, resulting in at least temporary price increases. To reduce the energy price tag in the overall product cost as far as possible, this led to optimizing processes for energy (usable energy) and even to heat integration between individual plants (using pinch technology). This caused processes to become more complex and plants more interdependent, so more sophisticated management would be required to maintain overview and oversight.

Besides tapping fossil energy, resources generating renewable energy became a significant industrial activity. Although it had little effect on the energy price, producing biofuels became a new branch of process industry. What will make a difference is the exploration and production of shale gas; in particular, this has had a major effect in the US. Just after the turn of the century, liquefied natural gas (LNG) import terminals were planned quite near to several major cities to satisfy their energy appetite. However, five years later these projects have been turned into export terminals for LNG. Because of its large molecular ratio of hydrogen compared to carbon, natural gas is an attractive fuel within an intermediate period for the development of renewable energy sources to help curb carbon dioxide emissions. Also, the conversion in the gas-to-liquids (GTL) processes, for example, of Shell and Sasol of natural gas to diesel and kerosene, is a relatively new viable possibility. (Latest development in this field is now use of the term XTL, where X indicates any potential fuel feedstock, i.e., gas, coal, or biomass.) In the transport sector these fuels will remain in high demand. In the GTL process, syngas is first made by partial oxidation, while in a second step over a Fischer—Tropsch catalyst, longer carbon chains are synthesized. The process provides at least technically, but maybe not yet economically, full interconnectivity into the traditional petrochemical industry.

Over an extensive period of time, the exploration of oil and gas also went through a development process, which has still not ended. Drilling for oil is the oldest, and during the 1960s and 1970s it progressed from onshore to offshore in order to exploit the rich subsea reservoirs. Drilling went into deeper waters and progressively down into deeper rock formations. The Gulf of Mexico was the theater for these competitive endeavors of various oil companies. Later in this chapter we shall describe the disaster that occurred with the Deepwater Horizon drilling platform and analyze the causes of this unfortunate accident. On the other hand, when the oil price is sufficiently high, improved technology also allows the huge reserves in oil sands, for example, in Alberta, Canada, to be exploited profitably and to do so in an environmentally responsible way.

Gas exploration developed in a similar way. Cavity gas had been running low, but Texas A&M petroleum engineering graduate George P. Mitchell (1919–2013) started an oil company, which worked in the 1980s and 1990s to develop horizontal drilling and the hydraulic rock fracturing or "fracking" technique to exploit the gas and oil-containing shale layers. In the US around 2010 this technology took off on a large scale and led to the production of an abundance of natural gas. There are quite a few other places in the world where this technology could be applied, but, for example, in Europe there is a strong environmental lobby and a fear that aquifers used for drinking water could become contaminated with the fracking liquids. Moreover, the open pit wastewaters release volatiles into the atmosphere. Experience in the US with half a million drill holes already made is so far not bad, although certain long-term effects of, for example, relatively weak, but irritant earthquakes may not become apparent immediately. The total length of gas pipelines has increased tremendously. As a result of this exploration, the price of natural gas has started gradually to decrease. Meanwhile, fracking also produces new oil, and in 2014 the price of oil started falling.

Because of the predictions of the Intergovernmental Panel on Climate Change (IPCC) and the Kyoto Protocol at the start of the twenty-first century, activities for constructing new nuclear power reactors were renewed after these had been reduced following the Chernobyl disaster in 1986. Several countries, China amongst others, initiated an active construction program. However, the events of the earthquake in Japan followed by a tsunami causing core melting and destruction of the Fukushima-Daiichi nuclear power plants with widespread radioactive contamination in the region, evacuation of hundreds of thousands and even permanent loss of property for thousands of people in the vicinity of the plant, has dampened the enthusiasm for nuclear power being an element of the solution to climate change. It led to abolition of nuclear power generation in Germany, including closing of older operational plants with an immediate effect and newer ones in the longer term. Germany will now focus on renewable energy and is making large investments in R&D for new technology, production, and construction of equipment and electricity transportation. Nuclear fusion is still a promise but is too far in the future to be considered as a factor of influence today. The Fukushima-Daiichi catastrophe and particularly its background causes in organization and management will be also discussed in this chapter.

Mankind must cut back further on burning fossil fuels so that as an energy carrier other than electricity, with its problem of storability, as mentioned in Chapter 1, hydrogen may become the option of choice. Fuel cell technology is becoming mature. In an intermediate period for a number of applications, hydrogen can be mixed with natural gas.

4.2 SOCIETAL TRENDS

With a rising standard of living enabled by a thriving economy as a result of expanding industry and exports, people changed in life styles and in appreciation of values. Attention to environmental and collective, involuntary safety risks have increased over the years, reflected in the development of regulations that we have seen in Chapter 2. Usually, this was preceded by activist group activities, such as demonstrations against plans for establishing another industrial production facility or after discovery of an environmental scandal. Regulation almost always came about following an accident. At the same time, prosperity and an expanding population also caused a large increase in housing areas around cities, which quite often advanced in the direction of industry premises, so that land use planning became a necessity to save some industrial parks from "suffocation." This is in the sense that their license to operate became very restricted as a result of risks imposed on neighbors and residents. Moreover, one serious accident with effects outside the premises was often sufficient to close down the operation. In contrast, on an individual basis, voluntary risk taking in terms of car driving or in adventure sports became more accepted.

On the downside, though, came a number of human factor developments. First, there was the downsizing in the 1990s with many early retirements, which resulted in a loss of knowledge and experience in the industry and an increase in workload of the remaining personnel. Also, attitudes and work ethos degraded, and the general level of education and ability to express thoughts in oral or written form did not develop favorably. E-mail can be efficient but can also hamper communication. Where safety depends on knowledge, communication, commitment, and discipline this trend was not very helpful. Prosperity had to some extent loosened the reins of discipline. Because a job for life could no longer be taken for granted, dedication and company loyalty decreased. Workers became more assertive with respect to career chances elsewhere and job hopping became common, and even a requirement for being well respected. The frequent job exchanges further reduced a company's memory horizon. At the same time, because of the pressure and progressive automation, process operators had to work on a higher level and are now surveying and managing the operation rather than controlling process variables directly by manipulating manual actuators.

In sum, because of factors such as higher work pressures, increased complexity of plant, changes in the organization, cost pressures and savings tied to bonuses, outsourcing on a large scale with the effect of higher chance of miscommunication, delayed investment in plant renewal, loss of competence due to early retirement and job hopping, and reduced dedication, safety will be under pressure and can readily become compromised. On the other hand, we have gained much from improved process hazard analysis and risk assessment methods because of more reliable hardware, much improved process instrumentation, better suited control rooms, versatile computing and information technology, effective and highly reliable layers of protection, and other resources. Hence, there are pluses and minuses and the resulting effect is not clear, as disastrous accidents still do occur. In the next section, we analyze two accidents in complex installations/plants both owned by large organizations and in the chapters that follow will dig even deeper and investigate new, recent approaches, concepts, and methods that may help us further to control risks of industrial operations adequately. Apart from safety risks, since the September 2001 New York World Trade Center attacks security risks have become an issue to take serious too.

4.3 TWO EXAMPLE ACCIDENTS ANALYZED

For both accidents described below quite extensive reports have been published; so readers interested in details, which are not relevant to the reason why we are looking at these two examples, are referred to these reports. This chapter's reference list contains particulars of most important ones.

4.3.1 DEEPWATER HORIZON PLATFORM DISASTER, APRIL 20, 2010

The BP Macondo, oil well drilling operation by means of the Transocean Deepwater Horizon drilling rig, took place in the Gulf of Mexico some 150 (statute) miles southeast of New Orleans, LA, and 50 miles from the nearest shore. Just before finishing the job and temporarily closing the drilled well, the huge, stabilized floating platform (see Figure 4.5) suffered a so-called "kick" of hydrocarbon gas and oil resulting in an explosion and fire that killed 11, destroyed the rig, and polluted the Gulf with an unprecedented, massive oil spill going on for weeks. The rig was owned by Transocean and leased by BP. A host of reports appeared on the causes and background of this disastrous event. Only the most important ones will be mentioned. The first one was the BP Investigation report[1] (incl. appendices, 760 pages) in September 2010, which is accompanied by an excellent slide presentation. The main line presented in this report shall set the scene. Second was the Report to the President of the National Commission on the BP Deepwater Horizon Offshore Oil Spill and Offshore Drilling[2] (398 pages) in early January 2011. The commission extended its investigation to the role of the governmental regulatory bodies involved and the environmental aspects of offshore drilling in general. It starts with an impressive and dramatic narrative of the episode during the explosion, fire, and rescue. Chapter 4 of this report (48 pages) provides a narrative of occurrences leading up the blow-out event and reveals technical details on causes. In fact, the chief counsel's report[3] of the National Commission (371 pages), which appeared just weeks later, contains even much more technical detail of designs, operations and postevent well investigation illustrated with many graphics, as well as details of deliberations and decisions of all parties involved in drilling the well. This report explains the drilling terms, clearly shows the timeline, the problems, and the various events preceding the disaster, and includes comments on organization and safety management. In addition, the National Commission published a report with recommendations.[4]

In March 2011, the report (126 pages) was published by the Deepwater Horizon Study Group[5] formed in May 2010 on the initiative of the Center for Catastrophic Risk Management (CCRM), led by Professor Robert Bea at the University of California, Berkeley. This group had broad international participation. The report analyzes and explains with graphics many technical aspects of the Macondo well drilling operation, the risks involved, and deficiencies in risk management. In April 2011, the US Coast Guard[6] produced Volume I of a report of an investigation on the event jointly performed with BOEMRE, the then-newly formed Bureau of Ocean Energy Management, Regulation and Enforcement. Volume 1 of the report focused on the maritime aspects, in particular the emergency operation; Volume 2 (not obtainable via Internet) that came out in September, described findings on the causes of the blowout and made recommendations. In September 2011, the US Department of the Interior[7] issued a report, which went into details of well design, cementing, and possible failure. Finally, in June 2014, the first two volumes of the report of the event from the Chemical Safety and Hazards Investigation Board (CSB), were published.[8] Volume 1 described in a compact way the parties involved, the backgrounds, and the development of the disaster event. Volume 2 focused on the blowout barriers, in particular the blowout preventer (BOP), and why it did not work. Volume 3 will go into the role of the regulator and the limitations of the regulatory regime, and Volume 4 will analyze organizational safety issues, such as corporate governance, safety performance indicators, organizational behavior, and safety culture. CSB will also make recommendations.

The Report to the President of the National Commission[2] mentions at the outset that "BP and its corporate partners on the well, Anadarko Petroleum and MOEX USA, had, according to government reports, budgeted $96.2 million and 51 days of work to drill the Macondo well in Mississippi Canyon Block 252. They discovered a large reservoir of oil and gas, but drilling had been challenging. As of April 20, 2010, BP and the Macondo well were almost six weeks behind schedule and more than $58 million over budget." The overall cost played a role.

For a brief description of in particular safety-related technical facts and events, we shall follow rather closely, often verbatim, relevant parts of the reported observations made by the CCRM report,[5] although even finer details with some more worrisome facts are presented in the reports of the National Commission. In particular, the chief counsel's report[3] is very detailed.

The main parties involved were BP, its lease partners Anadarko and MOEX, rig owner Transocean, and for all cementing jobs, Halliburton. Schlumberger was the well logging contractor, collecting data about the properties of the oil and gas-containing layer: the pay zone or pay sands, and drilling-mud subcontractor, M-I SWACO. The engineering design for the well had been prepared a year in advance.

The Deepwater Horizon rig was a huge, very advanced, semisubmersible, dynamically positioned drilling rig. Figure 4.5 near the end of the section gives an impression of what it was like. The rig was nine years old and had drilled tens of wells. It had on the day of the disaster 126 people on board working for 13 different companies. A different rig, the Marianas, had started drilling the Macondo well, but this rig had been damaged in a hurricane, and the regulator, then still the Mineral Management Service (MMS) of the US Department of the Interior, approved Deepwater Horizon taking over, which actually occurred in early February 2010. The sea floor was at 5000 ft (about 1500 m) depth and as shown in Figure 4.1, the plan had been to drill various sections of progressively smaller-diameter-size casing strings down to over 18,000 ft (5500 m). But when it drilled a 16.5 in × 20 in hole section down, planned to a depth of 12,500 ft (3810 m), the first signs of high pore pressure appeared and it achieved only 11,585 ft (3531 m) for the 16 in casing.

Quite soon after starting the 14.75 in × 16 in hole section, a well control event occurred and the drill pipe got stuck. An obstruction had been hit, and a sidetrack

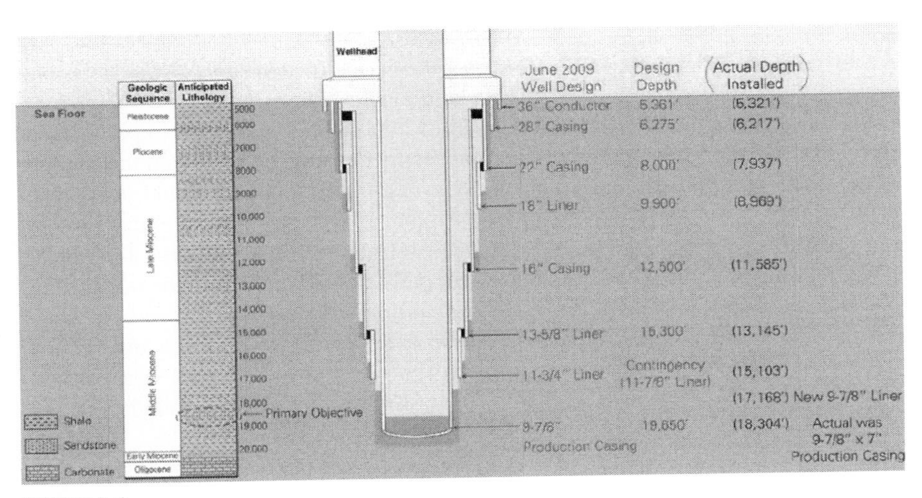

FIGURE 4.1

Schematically, the design plan as in the BP investigation report[1] of drilling the Macondo well with, at the left side the geology, and in the middle the drill hole casing sizes in inches, and at right the planned depth of 19,600 ft. The actual oil-producing achieved depth was 18,304 ft. The casing is the tubing lining the drill hole, and each part of the string at the bottom end is cemented to the rock wall. The smaller drill pipe (roughly 6 in diameter) running in the center is not shown. The diameter of hole becoming smaller toward the bottom results from balancing pressure at the formation with a column of dense mud, while not surpassing the strength of the rock and hence not fracturing and penetrating it with mud, but on the other hand keeping the pressure high enough so that no oil or gas will enter the bore hole (The figure is taken from the CCRM report[5]).

bore hole had to be made. In view of the high formation pressure, the design was modified and approved by MMS, and further sections were drilled. Near the end, a lost circulation event (mud penetrated into the formation, which may result in well pressure and a subsequent kick) was solved by applying "pill" material, and by April 9 drilling was finalized. The National Commission report[2] notes that drilling the last part was done cautiously because of the high pressure and the risk of fracturing the formation. Therefore, the planned final depth was not achieved. Subsequently, the well was logged and found to be stable (no gas in the mud). This would be the end of the drilling job, because a production rig would take over. On April 16, 2010, the MMS approved temporary abandonment of the well. A production casing was run (9.875 in $\times$ 7 in) to a depth of 18,304 ft (5579 m). Checking and cementing started on April 19 and was completed on April 20 at 00:36 a.m., followed by further testing and deliberations between BP and its service providers of finalizing the abandonment.

The following pages will be a somewhat abridged but still rather detailed version of part of the well-documented CCRM report[5] (only a few selected graphics from the report will be reproduced here, while references will not be copied) presenting the most important timeline facts with some additional explanation from this author for a clear understanding of terms. It will show the complexity of the operation, the special tools and attachments needed, and the series of decisions made. In some instances the chief counsel's report[3] will be quoted.

On April 12, 2010, BP asked the cementing contractor Halliburton to perform an analysis of a long string tapered production casing design reaching from the bottom of the well to the subsea wellhead at the sea floor. This analysis would utilize the performance of a lighter-weight nitrogen-foamed cement to reduce the bottom hole pressures to avoid losses into the weak formation found near the bottom of the well. An alternative option, a liner (lower part) and tieback (from the larger liner edge diameter upwards), could provide an additional barrier to hydrocarbon intrusion into the well by means of seals with the outer casing where liner and tieback meet. This would take about 3 days more time and cost about \$7–\$10 million more than the single long string. BP felt both designs "provided a sound basis for design" and chose the first option.

Halliburton performed tests on the proposed foamed cement mix that showed it would take 48 h at a temperature of 180 °F (82 °C) to achieve a compressive strength of 1590 pounds per square inch (psi), or 110 bar and that the foam was stable (in other words the nitrogen bubbles did not coalesce). Despite the use of lighter foamed cement, slow pumping rates to minimize bottom hole pressures, and 10 centralizers to center the long string casing in the open hole, the Halliburton report to BP indicated the well was likely to have a "moderate" (nitrogen) gas flow problem.

On April 15, BP applied to the MMS for a change from an exploration well to a production well and a completion design consisting of a single long string of tapered casing, which was approved the same day. BP requested Halliburton to analyze the use of seven casing centralizers on the long string. If the casing is not well centralized in the bore hole, cement pumped in later will not adequately fill all of the space between the bore hole wall and casing outer wall and hence will not close off perfectly. BP's original plan required the use of 21 centralizers, six of which were available at the rig and the extra 15 centralizers were located onshore.

On April 16 and 17, the additional 15 centralizers were delivered to the Deepwater Horizon. According to BP, the engineers erroneously believed that the additional centralizers they had received were of the wrong type, and they decided not to use them. (This decision added to the series of decisions decreasing the margin of safety.) The drilling mud at the bottom of the well was circulated out, down the drill pipe and up the casing, to clean out all of the hydrocarbons and settled cuttings. Then a wiper trip was made to ensure that the well bore was clear before the long string casing was installed. BP set the casing shoe 56 ft above the bottom of the hole, leaving an uncased area called a rat hole at the bottom of the well to allow debris to fall out of the way during the completion work.

April 18: The long string casing was installed with six centralizers by first installing a shoe track assembly with the shoe at the bottom and the float collar 190 ft above the shoe, as shown in Figure 4.2. The shoe track serves to ensure that the shoe becomes surrounded in high-quality cement. Above that assembly,

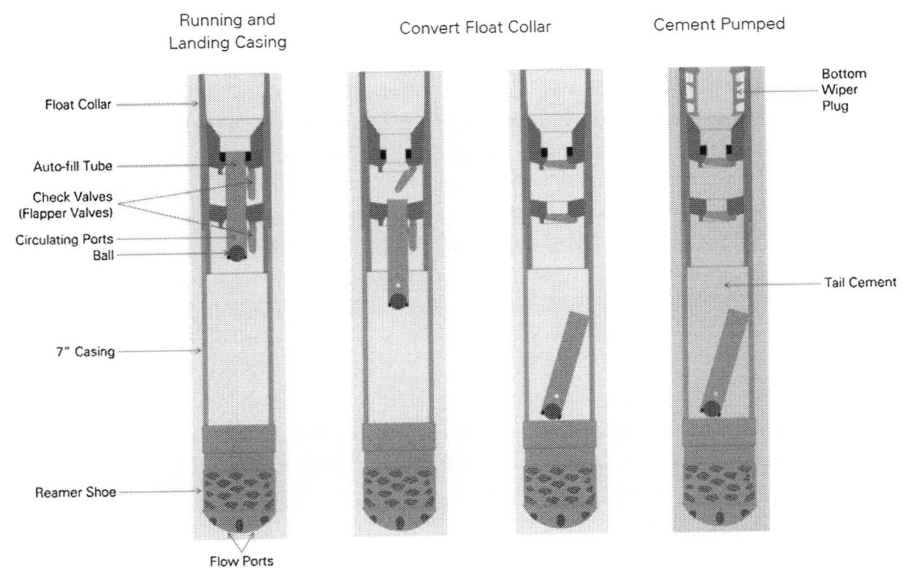

FIGURE 4.2

Bottom part of the casing assembly above the rat hole showing the reamer shoe and the conversion of the float collar with the autofill tube allowing mud to pass upwards to fill the production casing when the shoe was lowered and placed while lifting the ball. Next, the casing mud is pressurized pushing down the tube, upon which the flapper valves close. Then, cement is pumped in, while backflow is prevented by the closed valves.

After the CCRM report.[5]

the 7 in diameter casing was connected to a crossover to flare out to 9.875 in diameter for connecting to the casing above. Halliburton issued a report to their staff and BP on their recommended procedures for the cementing of the Macondo well. Halliburton reported to BP that the well with seven centralizers was likely to have "severe" gas flow problems. According to BP, the complete lab test results for the cement that was designed to be used had not been provided by Halliburton before cementing began.

April 19: The final production casing string was run segment by segment down to 18,304 ft. After the long string casing with the shoe track and six centralizers in place reached the bottom, mud was pumped to create the pressure difference needed to push down the autofill tube and unlock (convert) the float collar flapper valves. In the tube there is a small hole allowing at the same time some circulation of mud. An attempt was made (pumping mud from the inside through openings in the shoe and along the outside annulus up), but according to BP either the float collar or reamer shoe was plugged. The establishment of circulation and float collar conversion should have taken one attempt at approximately 400–700 psi (28–48 bar). Instead, the circulation/conversion took nine attempts with gradually increasing pressure until a final pressure of 3142 psi (217 bar) was developed. The high pressures required for the float collar conversion raised concerns about blockage in the reamer shoe at the bottom, breakdown of the weak formations at the bottom of the well, and that the float collar might not have converted. It is unclear as to whether the 3142 psi of pressure actually converted the float collar or if it just cleared a plugged shoe. There were additional concerns with the lower-than-expected circulating pressures following the conversion. It was decided that the standpipe pressure gauge reading was inaccurate.

BP decided not to do a complete bottoms-up circulation of the well before placing the bottom/tail cement plug. Such mud circulation would remove hydrocarbons and debris in the bottom of the well and provide a check on the condition. Due to BP engineer concerns for washout in the weak formation at the bottom of the hole, accompanied by disturbance of the lost circulation pill material that had been placed in this part of the well, only about half of the well-casing volume was circulated out. The incomplete bottoms-up circulation did not meet Halliburton's best practices. This incomplete circulation would have left any residing hydrocarbons in the well bottom, lingering in the upper part of the well.

Based on the plan proposed by Halliburton, a spacer fluid consisting of oil and mud followed by a wiper plug was placed at the bottom of the well to clean the well bore. The plug was special because of the variable diameter of the long string casing. This was followed with a cement sandwich consisting of regular cement followed by the foamed cement followed by regular cement and again a spacer. (The cement is pumped in, passes the shoe flow ports, see Figure 4.2, and flows up the annulus forming in the end a column of about 1000 ft cement. Top and tail part are regular cement, for the largest part of the annulus it is foamed.) BP reported the "job pumped per plan—no cement losses observed."

This report was later challenged during the formal hearings on the disaster and remains to be confirmed. It should be recognized that the foamed cement volume starts out at the mix unit at almost 100 barrels (of 159 L each) of mix volume. This is reduced by almost half (by pressure increase and collapse of the nitrogen bubbles) as it is pumped downhole making volume control for detecting losses difficult.

4.3.1.1 April 20, the day of the blowout

The Deepwater Horizon had an outstanding record of preventing lost-time incidents. In 2008, the Deepwater Horizon had received an award for its safety record, and on the day of the explosion there was a ceremony on board the rig celebrating seven years without a lost-time incident. There was also a tour organized for BP and Transocean managers to see the rig.

All of the cement of the tail plug was in place by 00:36 a.m. Pressure monitoring confirmed the floats were holding and the well was static. Next, the casing hanger seal at the seafloor wellhead was set between the side of the casing hanger and the wellhead walls to seal off the annulus. Two pressure tests on the seal were successfully completed. The plan developed by BP for temporary abandonment of the well consisted further altogether of seven major steps:

1. perform a positive test on the casing to check and confirm that the casing and wellhead seal assembly can contain the pressure inside the well to assure no fluid was capable of flowing from the well,
2. perform a negative test to confirm no fluid was capable of flowing from the producing formation into the well, by testing the integrity of the casing shoe track, the casing, and the wellhead assembly to withstand the formation pressure,
3. displace the drilling mud in the riser above the location of the cement surface plug (to be placed 3300 ft below the seafloor) with lighter sea water,
4. set a 300-ft-long cement surface plug,
5. test the integrity of the surface plug,
6. perform an impression test (with a lead block) to assure that the casing hanger was seated properly, and
7. install a lockdown sleeve to secure the casing hanger and seal.

At 7:00 a.m. after a discussion with the Macondo well contractors, as per the well plan decision tree, BP concluded that a cement bond log—hence a check on the quality—was not required. The Schlumberger crew departed.

At 10:55 a.m. a positive test was performed on the well to determine if there was any outflow from the well. For the second stage, the casing was pressured up to 2700 psi (186 bar) and held for 30 min. This test was performed about 10.5 h after the cement placement was completed, well before the 48 h that the Halliburton lab tests indicated to be necessary for the foamed portion of the cement to develop sufficient strength. It did not test further the integrity of the bottom assembly and cement due to the presence of the wiper plug at the top of the assembly.

At 11:30 a.m. the placement of the surface plug after displacement of the upper portion of the drill mud column was reviewed with the Transocean drill crew. Significant concerns for displacement of the drill mud before placement and testing of the surface plug and installation of a lockdown sleeve were expressed. After an agreement was reached, a drill pipe was run in the hole down to 8367 ft. In preparation for the mud displacement and the negative-pressure test, the displacement procedure was reviewed. At 1:28 p.m. the Deepwater Horizon started offloading mud to the supply vessel. Concerns were expressed by the mud loggers that given the simultaneous operations, the mud pit levels could not be accurately monitored—so, there was no good check on the difference between volumes in and out.

At 3:04 p.m. the blowout preventer (see Figure 4.4), connecting the riser to the rig with the long string casing in the well, was flushed with seawater to displace the mud as well as the choke, boost, and kill lines, while the kill line remained pressurized at 1200 psi. The choke and kill lines (see Figure 4.3) are piping and valves on the BOP that allow access to the well when the BOP is either fully or partially closed. The purpose of the choke line is to release pressure from the well, while the purpose of the kill line is to pump mud to stop the backflow in the well. The BOP devices and the choke and kill lines are also often used to conduct pressure tests on casing, set seals, activate tools, et cetera.

At 3:56 p.m. as part of the displacement operations, 424 bbls (barrels of 159 L) of 16 ppg (density of mud in pounds per gallon) spacer mud, followed by 30 bbls of freshwater were pumped into the well. (The National Commission report[2] mentions that usually water-based spacer mud is used. However, BP directed M-I SWACO to prepare the spacer from two leftover lost circulation pills: heavy viscous drilling fluids, to avoid obligatory waste disposal later.) A total of 352 bbls of seawater was used to complete the displacement to 8367 ft. This placed the top of spacer fluid in the riser 12 ft above the BOP (the estimate is optimistic, according to the chief counsel's report[3]). The pumps were then shut down, and the drill pipe pressure was at 2325 psi, while the pressure in the kill line was 1200 psi (as mentioned before). Next, the BOP's annular preventer (a heavy rubber ring at the top edge tapered—details not visible in Figure 4.4—pressed for closing upwards against a conical steel chamber top, hence forced moving toward the center around the drill pipe) was closed for the negative-pressure test.

For this critical part of the negative-pressure tests (not required by MMS), besides the CCRM report we shall also closely follow the chief counsel's report[3] because of its detail and clarity, so the text will be mixed. The latter report makes several critical comments with regard to the abandonment procedure of BP containing a few last-minute changes and unusual practices.

Shortly before 5:00 p.m. pressure was reduced in the drill string by bleeding water off at the top. From approximately 4:00 p.m. to 5:50 p.m., the trip tank for measuring the volume of displaced fluid was being cleaned. Because of the simultaneous offloading and cleaning operations, the fluid levels in the tank were changing, making it difficult to monitor how much fluid was bled off. Other simultaneous operations such as preparing for the setting of the surface cement plug in the casing and bleeding

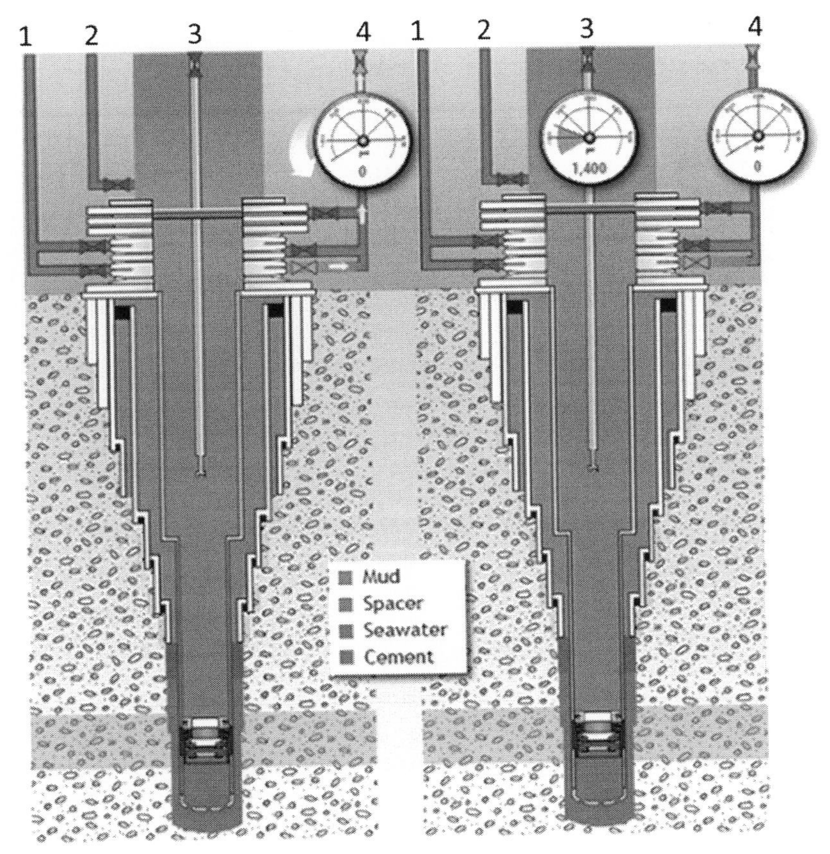

FIGURE 4.3

Chief Counsel's report[3] figure 4.6.18, page 158, showing at two moments in time in a longitudinally compressed drawing fashion the drill pipe within the riser above the sea floor and below that floor the long string casing reaching to the float collar and the shoe string at the well bottom. Numbers 1–4 refer to the choke line, boost line, drill pipe, and kill line respectively. The tip of the drill pipe is at an indicated depth of 8367 ft. The bottom of the well is still about 10,000 ft deeper, with the top of the shoestring at 18,304 ft. The figure shows the situation shortly after 6:40 p.m., after bleeding down to 0 psi (left), and again monitoring pressures 30 minutes later (right). Spacer fluid is indicated at the level of the kill line.

off the riser tensioners were occurring at the same time, which may have distracted the crew and mud loggers from accurately monitoring the well and observing whether the well was flowing. Nevertheless, some witnesses estimate the bleed-off between 23 and 25 bbls, but it may have been more or less. Then, at 5:00 p.m. the well for the first time was underbalanced: pressure at the bottom of the well (by the column of mud in the riser from the top downwards, then spacer fluid, water, and again mud at the

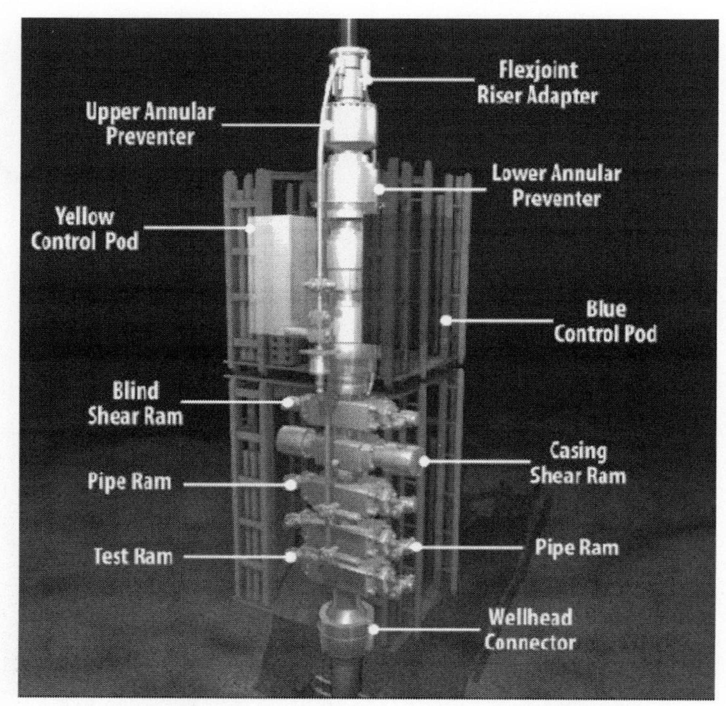

FIGURE 4.4

Blowout preventer (BOP), schematically, as shown in the Chief Counsel's report[3] figure 3.7, page 30. The height of the 300 tons weighing BOP is 56 ft, hence about six times the length of man. See for details also the CSB reports.[8] The Upper Annular Preventer closes the space between casing and drill pipe by compressing a heavy rubber ring, preventing flow from the well into the riser. LMRP is the lower marine riser package with upper and lower annular preventer; The blind shear ram squeezes and cuts the drill pipe and seals the annulus. The Casing Shear Ram cuts casing and even thicker drill pipe at tool joints but does not seal.[8] Upper and middle Pipe Rams are variable bore rams; hence devices closing annular space around variable bore tube with several concentric semicircular rubber-lined steel pieces. All rams are hydraulically operated. Not shown are the connecting cables to the blue and yellow control pods.

bottom) was less than the formation pressure. If the well was properly sealed, only enough fluid would backflow to account for the pressure release, the pressure at sea surface in both the drill string and the blowout preventer kill line would drop to zero (atmospheric), and the well would remain static with no backflow. Results from the test were not positive. There was (much) more backflow than anticipated (but no calculation had been made; from compressibility change 3.5 bbls can be calculated) and the drill pipe pressure never dropped to zero, but stayed during the bleed-off around 260 psi. After closing the drill line, the pressure went up over a period of 6 min to 1250 psi (or 1262 psi according to report[3]). Around 5:10 p.m. it was also noticed that the mud level in the riser had dropped, an indication that the annular preventer had leaked. At that time the night shift was coming up, a group of officials visited, and a leading manager noticed some confusion.

Around 6:00 p.m. after deliberation how to proceed, the hydraulic preventer pressure was increased to improve the seal, the riser topped-up again with 20–25 bbls mud (but the spacer was not recirculated to above the BOP, which is of interest for the final test later) and the test repeated. This time the crew was directed to bleed-off through the kill line instead of the drill pipe (actually this is according to the procedure submitted to the MMS, but the crew saw no difference in using the drill pipe, with both connected to the production casing). Pressure of the stagnant fluid in the drill pipe went up to 773 psi and would have gone higher if bleeding-off had not started. An estimated 3–15 bbls water came out of the kill line when the pressure in the drill pipe reached 0 psi. At that time the crew closed the kill line, while still water was flowing out. In the following 30–40 min the pressure rose again to 1400 psi (which is about the pressure in the formation).

The Transocean tool pusher and BP company man had different interpretations of the negative test results. The tool pusher asserted the evidence was indicative of the "bladder" effect or "annular compressibility" transmitting the pressure to the drill line. The BP company man asserted that the anomalous results were caused by the riser leak. It was decided to conduct another test.

Sometime after 6:40 p.m. the crew first pumped a small amount of water in the kill line to be sure it was full and then began by bleeding the kill pipe pressure to zero. This time they were measuring the amount of water bleeding-off but it remained less than a barrel until the pressure in the kill line was zero. For the next 30 min this situation stayed the same. Although the flow from the kill line had stopped, the drill pipe pressure remained constant at 1400 psi (!) as illustrated in Figure 4.3 (right).

At 7:55 p.m. the negative pressure test result was accepted as a good result despite the high pressure observed in the drill pipe. The first negative pressure tests were clearly not positive. Yet, they were accepted as positive—false positive. In the last test it is likely that viscous spacer mud, that had entered the BOP at previous tests, now had flowed into the kill line during the bleeding and had blocked the kill line, giving a false result of no flow. As a next step, the internal blowout preventer (IBOP) and the annular preventer were opened and seawater pumping was continued for about the next hour down the drill pipe to displace all mud out of the riser. Unfortunately, because seawater was being pumped into the well and the mud pits were only receiving mud coming from the well, it was not possible for the mud loggers to positively determine influx.

At 8:50 p.m. the pumps were slowed to monitor the water-based mud spacer's arrival so that it could be tested in order to be sure it had not been contaminated with the oil-based drilling mud (sheen test: a sample in water should not give a "sheen"). A decrease in the flow of the spacer was expected as the pumps were slowed; however, real-time data indicated that the flow actually increased.

According to BP, the first indication of flow into the well would likely have been at approximately 8:58 p.m. However, inaccuracies in the trip tank reading (due to the emptying) could have made this influx difficult to observe and detect. Shortly after 9:00 p.m. pressure on the drill pipe increased from 1250 to 1350 psi. The drill pipe pressure should have decreased due to the removal of the heavier mud (14.7 ppg) and its replacement with the lighter-weight seawater (8.6 ppg). This abnormality could have been detected from monitoring the drill pipe pressure data and could have been the first clear indication of the flow of hydrocarbons into the well, visible to the crew. Spacer fluid was then observed at the surface.

At around 9:08 p.m. the pumps were shut down to allow the sheen test to be performed. The records indicate that with the pumps off, the drill pipe pressure continued to increase, indicating that there was flow into the well. However, it is possible that due to overboard discharge during the sheen test, the inflow could not be directly observed. Following the successful sheen test, at about 9:14 p.m. the pumps were restarted to continue displacement of the spacer fluid overboard, and a spike in pressure occurred when pump #2 was started. The available records of drill pipe pressure show that it continually increased during this period of time. Because of the pumping overboard, there was no method to accurately record the outflow volume. At approximately 9:20 p.m., the tool pusher told the senior tool pusher that the results of the negative test were "good" and that the displacement was "going fine." Overall, the crew seemed unaware of the situation at hand that the well was flowing.

So far, we have roughly cited the description of events and observations made in the CCRM report[5] with some information from the chief counsel's report[3] added for improving clarity. Details about events that followed are for the purpose of this chapter less important. In particular, Chapter 1 of the National Commission report[2] and the chief counsel's report[3] provide many details of the functioning of the BOP after the kick and the tumultuous events of the developing disaster. Main subevents are as follows: After some further pressure irregularities at 9:40 p.m. seawater, subsequently mud and then gas was spewed from the riser into the air. Once it is in the riser, a flow of hot oil mixed with gas is continuously accelerating because due to the decreasing pressure more expanding gas is formed. The crew tried to deflect the flow away from the rig, but earlier that day the diverter was set to the mud–gas separator and not to exiting overboard. This worsened the situation due to the large amounts arriving and gas being blown out of the separator vents. The gas was then partly sucked into the generator engine air inlets, which revved up the engines. Power was lost, explosions followed, and fire engulfed the rig (Figure 4.5). Meanwhile the crew activated the various closing devices of the BOP, first the upper annular (postmortem found that the rubber ring of the upper annular had eroded and leaked and that the lower annular was open), and then the variable bore/pipe rams. From the panel indicators the crew thought the preventers worked, but the measures failed. After the first explosions, the crew activated the emergency disconnect system, which should have closed the blind shear ram, but there was no

FIGURE 4.5

The Deepwater Horizon semisubmersible rig before and at the night of the blow-out.

indication it worked. Finally, the automatic mode function (AMF)/deadman should have triggered as last resort. However, the flow of oil continued.

There have been quite a few studies on the reliability of these complex BOP devices. DNV[9] performed a forensic study on the Deepwater Horizon device for the US Department of the Interior. On the Internet one can find videos[10] of BOP functioning. The CSB Volume 2 report[8] goes into many BOP malfunctioning details. A new fact that CSB identified is that the drill pipe due to forces on the wall resulting from the pressure difference over the wall and deviations from perfect roundness must have buckled so that the blind shear ram could not function as intended and could not seal completely. Emergency response and rescue operations started. Eleven workers lost their lives in the inferno and 17 were injured.

All eight barriers mentioned in the BP report[1] had been breached (annulus cement leaked, shoe track cement failed, negative pressure test wrongly interpreted, influx unnoticed from real-time data until it had passed the BOP, activating the BOP preventers failed, flow diversion went wrong, gas dispersed beyond classified explosion zoning ignited, BOP AMF/deadman did not seal). Appendix 2A of the CSB reports[8] contains a detailed postmortem failure analysis of the recovered BOP. A BOP contains single failure points, for example, the shuttle valve supplying hydraulic fluid operating the blind shear ram. Failure was, however, not the result of failing ram operating systems but rather failure to seal the passage of the hydrocarbon flow: the upper annular due to rubber eroded packer (may have eroded in pulling out casing pipe at an earlier stage), the upper variable bore pipe ram had probably not been activated by the crew, the middle one functioned but the elastomer sealing ceased due to high temperature of the reservoir fluids, and the blind shear ram failed to squeeze the drill pipe due to its off-center location. The latter was probably the result of buckling of the pipe by pressure differences inside and outside the drill pipe and vertical loading of it after the explosion (see Volume 2 of the reports[8]). Apart from these main deficiencies of the BOP action, some smaller ones were discovered.

4.3.1.2 Subconclusions and aftermath implications

Looking back, we can observe that the complex operation required, in particular on the last day, a series of decisions by BP management that demonstrated a lack of hazard awareness, of situation awareness, of evidence-based reasoning, and of predictive system thinking leading to mistakes and misjudgments. Each of these decisions was on its own not unsafe in the sense that an event was unavoidable given the defenses, but all narrowed the safety margin, one more than another, and so the risk was increased. The go-ahead following the negative-pressure test was undoubtedly bad decision making. Precursor signals and expert opinions were ignored. An unmistakable background to these decisions was the drive for efficiency fed by cost reduction and time pressure, while complexity obscured an overview. The mindset may have been too much that "today the operation will get finished." Process safety thinking was underdeveloped, and training/retraining was insufficient. It appears that operation leaders had no clear mental image of how the process situation down in the well unfolded. Instead, the physically unrealistic "bladder" effect became the focus of "group think" or "tunnel vision" in which no alternative causes are considered. The failing safety management system of the organization showed itself in the failure of the eight layers of protection. Communication errors also played a role. In hindsight, Smith et al.[11] identified a series of 25 human errors in eight different categories, most of which were latent and a few of which had already been made in the design. In addition, it is clear that the barriers that should safeguard against nasty surprises had not been tested or maintained adequately, hence the barriers lost degrees of independence because the system lost its resilience. The crew had probably never trained for an overall failure scenario with a developing kick. It is obvious that a safety management system must be built, tested, and implemented with rigor, its performance monitored with indicators, and that priorities must be rearranged to prevent similar reoccurrences.

The environmental damage from the oil spill and damage to businesses of a maritime nature in neighboring states was colossal. Total financial losses have been record breaking and fines and compensation payments threatened the continued existence of BP. It also damaged the interest of many other oil and service companies involved in drilling in the Gulf or those who were dependent on such activities because a 6-month moratorium was imposed on related activities in the Gulf. Even after the moratorium was lifted, activities only slowly returned to their previous levels.

The US National Commission report[2] devoted much attention to the role of the MMS, the performance of the industry in general, the environmental disaster in the Gulf area, and it added a second volume with recommendations. The commission concluded that the *loss could have been prevented* and that it can be attributed to a series of mistakes made by the companies involved revealing systematic failures of risk management. A crucial problematic aspect was the independence of MMS, since besides a regulating, oversight, and enforcement task with respect to safety of the operations and environmental protection, it also had responsibility for the functions of leasing exploration, permitting operations, and collecting revenues. Already in June 2010 the Secretary of the Interior had announced that MMS would be reorganized into three separate entities with different missions: a Bureau of Ocean Energy Management (BOEM); a Bureau of Safety and Environmental Enforcement (BSEE); and an Office of Natural Resources Revenue. As regards BSEE, it would obtain stronger regulatory instruments, for example, to prescribe a safety and environmental management system and the possible development of a safety case—such measures had not been required previously. In the intervening years all this has been realized, but in the interim years MMS regulatory and enforcement functions were in the short term transferred to the Bureau of Ocean Energy Management Regulation and Enforcement (BOEMRE), which was subsequently split into BOEM and BSEE. Meanwhile, in 2013 the latter assigned Texas A&M Engineering Experiment Station's Mary Kay O'Connor Process Safety Center, in partnership with Texas A&M University, University of Texas, and University of Houston to establish and run the Ocean Energy Safety Institute.

In July 2012, the CSB board organized a public hearing in Houston on the possible introduction of process safety performance indicators for offshore drilling by the US oil and gas community.

4.3.2 FUKUSHIMA-DAIICHI CATASTROPHE, MARCH 11, 2011

On Friday, March 11, 2011, an exceptionally strong earthquake with a magnitude of 9 on the (logarithmic) Richter scale hit Japan's east coast, followed by high amplitude tsunami waves. Because of its lack of natural resources, Japan generates about 30% of its electrical energy from 54 nuclear power plants.[12] The sequence of events will be summarized from the report by the International Atomic Energy Agency[13] (IAEA) and the initial report by Professor Magdi Ragheb[14] (University of Illinois at Urbana–Champaign). Meanwhile also, Wikipedia featured detailed, but probably nonpeer-reviewed articles on the subject. Along the Japanese east coast, quite a few reactors had been in operation. At Fukushima there are two sites both owned by Tokyo Electric Power Company (TEPCO): Daiichi and Daini. As designed, as soon as earthquake shocks are detected, all operational plants shut down automatically by inserting the control rods (such action is called SCRAM or Safety Control Rod Axe Man, a term attributed to Enrico Fermi who built and operated the first reactor in the early 1940s). All sites were hit after the shocks by tsunami waves and sustained various degrees of damage by water, but only at Fukushima-Daiichi did it lead to disaster. An overview of the site is shown in Figure 4.6.

The six General Electric–designed boiling water reactors (BWRs) were built and commissioned in the period 1970–1979. For a highly simplified schematic of a plant, see Figure 4.7. Unit 1 at Daiichi is a Mark I type and capable of producing 460 MW(e), Units 2–5 are also of Mark I type but with a larger capacity of

FIGURE 4.6

The Fukushima-Daiichi site with units 1–6.

IAEA mission report.[13]

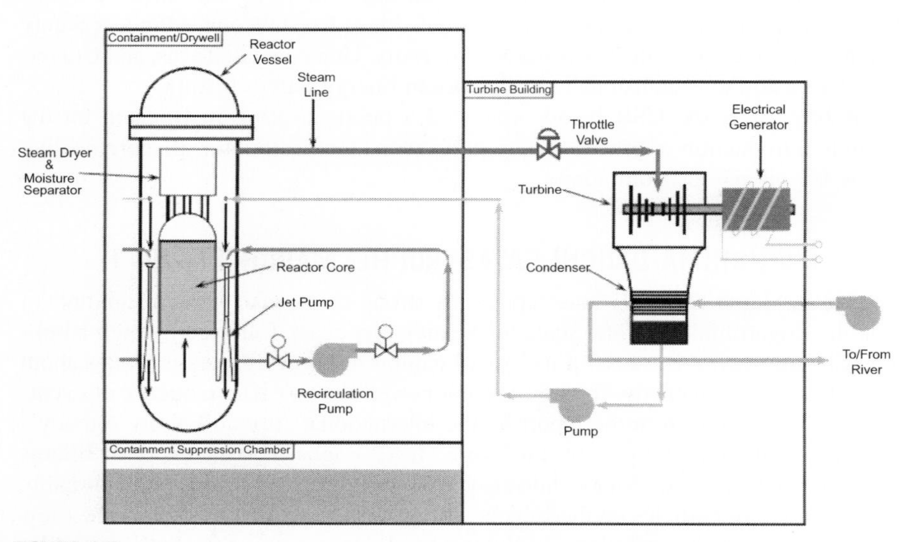

FIGURE 4.7

Simplified flow scheme of a generic boiling water reactor and the turbine power generator.

Reactor concept manual of the USNRC technical training center.[15]

784 MW(e), and Unit 6 is Mark II with 1100 MW(e) output. (Daini has four Mark II reactors of 1100 MW(e)). At the time of the earthquake, Daiichi units 1–3 were operational and 4–6 were in maintenance; 6000 workers were present at the site.

SCRAMming also stops the turbines, and because the earthquake damaged an offsite transformer station connecting to the grid, all electricity supply ceased and the emergency core cooling systems possessing redundancy had to take over. First, 12 of 13 large emergency diesel generators (one was in maintenance) started delivering power to the pumps for keeping the reactors and the spent fuel stores cool. Eight of them had been placed in a basement of the turbine hall some 150 m from the shore and the others in various places on ground level. The site was 10 m above sea level. Forty-six minutes after the earthquake, giant 10—15-m-high tsunami waves hit the coastline and flooded the area of the site penetrating various buildings and rendering the diesel generators inoperative (except one at Unit 6) by destroying the electrical switching boxes due to the short-circuiting action of the seawater. In principle, there should still be battery power sufficient for driving the pumps for about 12 h, but in Units 1 and 2 the batteries were flooded, and also the seawater pumps were destroyed. For fear of aftershocks most personnel had been evacuated, and communication between the On-site Emergency Control Center and personnel was only possible by one wired connection within each control room. Moreover, the tsunami inundated rooms, washed away vehicles, storage tanks, and machinery, and left the site with much rubble and debris.

Due to the radiation activity that must die down after inserting the modifying control rods, residual heat in a nuclear reactor after shutdown decays slowly in an exponential fashion. This heat must be extracted. In case of a loss of cooling accident (LOCA), all reactors use the principle of pressure suppression of the (radioactive) steam, for which in the Mark I the torus below the reactor serves with a pool of water (Figure 4.8). Further, various systems are available to activate pressure suppression. In Unit 1 the isolation condenser (IC) is a system of cooling by applying natural convection with a heat exchanger in a large tank of water at an elevation above the core. The IC initially functioned but was temporarily interrupted

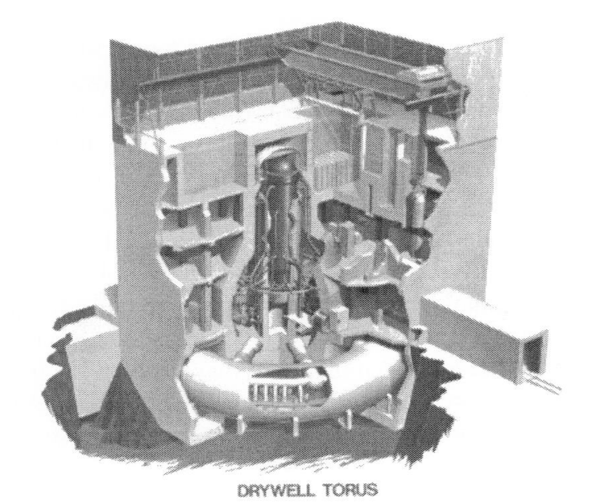

FIGURE 4.8

The General Electric Mark I BWR after which design Fukushima-Daiichi Unit 1 was built; the lower circular torus is the pressure suppression pool.[15]

because, according to TEPCO, the temperature rate of decrease was too large. However, due to an unexplained cause the system failed later. In the hours following, first a very low pressure was observed, and then some hours later workers succeeded in supplying 80 m^3 of water until available water ran out.

If there is no circulation, water at the core is turned into steam and pressure in the reactor vessel increases. Overheating and uncovering of the core leads to the reaction of the zirconium-based (zircaloy) cladding of the fuel rods with steam to produce zirconium oxide and hydrogen. To reduce pressure in the reactor vessel, which otherwise may burst, after much effort safety relief valves could be opened, although apparently hydrogen escaped with the steam. Venting should be through the stack but somehow it also penetrated the reactor building, mixed with air and exploded, blowing the roof off the building with dispersion of radioactive fission products into the atmosphere. The core came to partial meltdown. The explosion occurred on March 12, about 24 h after the tsunami flooding. The explosion and its aftereffects caused emission to the environment of large amounts of radioactive material. Some hours later evacuation was ordered of the population within a radius of 3 km to take shelter from 3 to 10 km, and the next morning by direction of the prime minister evacuation distance was increased to 10 km. Workers succeeded in injecting seawater into the core, which could be continued until the freshwater supply was reestablished. If the core became dry it would severely overheat and meltdown would take place. In that case the hot melt would burn through its containments causing widespread radioactive contamination. Because at all units in the building a basin was present in which spent fuel rods were stored, it was feared that if these pools become dry, meltdown could also occur there.

Units 2 and 3 applied the reactor core isolation cooling system: by a pump driven by steam from the reactor pressure vessel, water is circulated to the reactor from the suppression pool. This system went undisturbed for a longer time than in Unit 1 but eventually also stopped. After that, improvised means of seawater supply had to be used to cool the reactors, while steam was bled to the pressure suppression pool below the reactor pressure vessel. Working conditions were dreadful: there was no light, instruments did not function and could not be read, motor-operated valves did not work, and there was exposure to radiation. It took hours to open the safety relief valves to enable water injection. Eventually Unit 3 also exploded on March 14, while on March 15 a hydrogen explosion took place in Unit 4 and a fire broke out. At about the same time due to insufficient cooling and subsequent explosion the primary containment of Unit 2 was breached and another discharge of radioactive material occurred. The hydrogen in Unit 4 may have had as its source Unit 3, because both units shared the same stack header. The nuclear fuel pools at all six units had to be supplied with water under dreadful conditions to keep the stored rods cool.

Fortunately, local leadership was strong and persistence and morale of the workers was high. Three workers were killed by the direct effects of the earthquake and tsunami and some were injured; 167 workers were exposed to radiation dose levels of more than 100 mSv[16] of which about 30 were between 100 and 250 mSv[13] (milliSievert; 1 Sv = 1 J/kg radiation energy). According to the 2008 recommendation a dose of 20 mSv on average per year is allowed for nuclear power plant workers over a period of five years (and for the public 1 mSv/yr), see Chapter 3, Section 3.5.1.1. In an emergency, the maximum reference level is 100 mSv. Three workers stood in heavily contaminated water; their feet sustained radiation burns; they were treated for a few days in a hospital but seemed to recover. In general, TEPCO was blamed for not sufficiently preparing its workers for emergency situations.

The lack of preparedness with respect to evacuation caused confusion that was further amplified by shifting the borderline in steps from 3 km from the site to 10 km, and finally to 20-km radius and on a voluntary basis to 30 km. Even in some areas at that distance radiation levels became high. Evacuation, certainly when it is unprepared, is upsetting to people and creates victims due to the stress alone. The area around the site including animals was contaminated by airborne dust containing iodine-131, caesium-134, and caesium-137 isotopes. Part of the area will remain closed for many years to come, so many residents cannot return to their property. There have been complaints about lack of communication. Besides airborne dispersion and fallout, seawater was also contaminated by cooling water and leaks from storage basins, affecting fish and other foodstuffs. In 2013, the Japanese government ordered TEPCO to close and scrap the site altogether. This effort will take an estimated 30–40 years to accomplish.

At the Fukushima-Daini site, and other sites, electricity production was not interrupted and reactors could be saved.

4.3.2.1 The deeper causes

In this case, the main problem also arose from decisions by management and governing bodies in the precursor years. The Japanese Parliament, the Diet, created an independent commission to investigate the case. The report[16] by the commission revealed a number of human and organizational factors, which are important and are an essential part of a system approach to risk management. The main conclusion is that the accident was "human made," and therefore *it was preventable*. The main parties involved were the operator TEPCO, Nuclear and Industrial Safety Agency (NISA), Nuclear Safety Commission (NSC), and Ministry of Economy, Trade & Industry (METI). Interactions had been such that NISA and NSC did not act according to their missions.

The Diet report expresses doubts about TEPCO's opinion that all damage was caused by the tsunami. In Unit 1, a small-scale LOCA could have occurred as a consequence of the tremors. The commission states that it believes that the risks of earthquake and tsunami were known or should have been known. Several components were up to standard earthquake resistant strength, which in hindsight has been confirmed by TEPCO and NISA. Over the years, warnings have been made that tsunamis would reach higher levels than had been assumed at the time of construction. Since 2006, regulatory bodies and TEPCO were aware of the possibility of a total power failure in the case that a tsunami should reach the site. This was also true for failure of seawater pumps and therefore damage of reactor cores.

NISA did not inform or prepare the public about these warnings and did not keep any records of the information it obtained. NISA did not validate the method to determine the relative low height value of tsunamis, which was used in constructing the site, while there are many stone monuments right around the site marking the height of tsunamis of the past (see Figure 4.9).

The Diet report also claims that TEPCO had an inappropriate attitude with respect to risk management. It delayed taking countermeasures by commissioning further studies and ironically or paradoxically used the argument of uncertainty to justify not taking measures. NISA was aware of these tactics but did not itself act. It was clear that in trying to vent Unit 1 reactor, costly time was lost, because

FIGURE 4.9

Atsuo Kishimoto of the National Institute of Advanced Industrial Science and Technology (AIST), Research Institute of Science for Safety and Sustainability (RISS), Tsukuba, Japan, in a presentation about risk governance deficits shows some pictures of stone monuments commemorating the occurrence of high tsunamis. The texts on the stones pronounce a hearkening warning, e.g., "Remember the calamity of the great tsunamis. Do not build any homes below this point." But after a number of years and no events being observed, people and organizations ignored the advice.

workers had no procedure to work from and had not been trained to cope with this kind of situation. The Diet commission even doubted, based on hearing from workers involved, that the reason TEPCO gave for interrupting the isolation cooling of Unit 1 is true. In contrast to Units 2 and 3, there is no evidence that the safety relief valve of Unit 1 was opened.

TEPCO's top management was slow in reacting to the emergency, while the command line between NISA and the local TEPCO management, foreseen in case of emergency, did not work because the emergency control center was without power. In addition, the prime minister's decision to go to the site resulted in further confusion and needless time loss, which could have been prevented if the TEPCO board had adequately communicated about the situation. Similar confused situations occurred with other emergency actions, communication with the prefecture and emergency responders, which were totally unprepared for a nuclear accident. In total 150,000 people had to be evacuated, but an official message did not come or came to them slowly and without explaining the circumstances and directions of where to go.

The NSC had not updated internationally accepted provisions for community protective measures because of not wanting to arouse anxiousness in the population, which also is ironic given the misery caused by the disaster. In the Fukushima prefecture an area of 1800 km^2 is contaminated at a cumulative level of 5 mSv per year or higher. Only below this threshold value is the population at large considered safe. (Normal background radiation by natural sources is on average about 2 mSv per year). Decay of the radiation level to acceptable levels will take decades. Decontamination has been started but requires substantial funding and effort. The Nuclear Emergency Response Headquarters Government and TEPCO Mid- and Long-term Response Council[17] developed a road map for clearing the site and the environment covering a time period of at least 30 years. Economic losses will vastly exceed the ones of the Macondo well disaster.

4.4 CONCLUSIONS

National commissions investigating the two largest industrial accidents in the 2010–2011 time span, which in terms of financial losses (not human loss) may be considered as the largest ever, concluded that these accidents could have been prevented. The main causes are found in decision making by a lack of effective risk management based on a system approach. Imagination of decision makers of (uncertain) likelihood of a risk event with known large consequences (which when the event actually occurs turn out to be even more dramatic than anticipated) falls tragically short of what must be expected. Cost considerations and feeling that "it will not happen to us" kicks in too early due to a lack of a forecasting based on current information. Regulatory and oversight bodies appear far too accommodating. In later chapters we shall see what can be done to alleviate these problems.

REFERENCES

1. BP, Deepwater Horizon, accident investigation report. September 8, 2010. http://www.bp.com/liveassets/bp_internet/globalbp/globalbp_uk_english/incident_response/STAGING/local_assets/downloads_pdfs/Deepwater_Horizon_Accident_Investigation_Report.pdf.
2. National Commission on the BP Deepwater Horizon Offshore Oil Spill and Offshore Drilling. Deep water, the gulf oil disaster and the future of offshore drilling. January 2011. http://cybercemetery.unt.edu/archive/oilspill/20121210200431/http:/www.oilspillcommission.gov/final-report.
3. Macondo – The Gulf Oil Disaster, Chief Counsel's Report. National commission on the BP Deepwater Horizon oil spill and offshore drilling. 2011. http://www.eoearth.org/files/164401_164500/164423/full.pdf.
4. National Commission on the BP Deepwater Horizon Offshore Oil Spill and Offshore Drilling. Deep water, the gulf oil disaster and the future of offshore drilling, recommendations. January 2011. http://www.eoearth.org/files/164401_164500/164420/osc_deep_water_summary_recommendations_final.pdf.
5. Deepwater Horizon Study Group. Final report on the investigation of the Macondo well blowout. March 1, 2011. http://ccrm.berkeley.edu/pdfs_papers/bea_pdfs/DHSGFinalReport-March2011-tag.pdf.
6. U.S. Coast Guard. Report of investigation into the circumstances surrounding the explosion, fire, sinking and loss of eleven crew members aboard the mobile offshore drilling unit Deepwater Horizon in the Gulf of Mexico. April 20–22, 2010. http://media.nola.com/2010_gulf_oil_spill/other/FINAL%20REDACTED%20VERSION%20DWH.pdf.
7. Bureau of Ocean Energy Management Regulation and Enforcement, U.S. Dept. of Interior. Report regarding the causes of the April 20, 2010 Macondo well blowout. September 14, 2011.
8. CSB, U.S. Chemical Safety and Hazard Investigation Board. Investigation report overview – explosion and fire at the Macondo well, Deepwater Horizon Rig, Mississippi Canyon block #252, Gulf of Mexico. April 20, 2010. Report No. 2010-10-I-OS, June 2014; Volume 1 Macondo-specific incident events: Relevant background on deepwater drilling and temporary abandonment; Volume 2 Technical findings on the Deepwater Horizon blowout preventer (BOP) with an emphasis on the effective management of safety critical elements; Volume 3 and Volume 4 will follow, www.csb.gov.
9. DNV, Det Norske Veritas, Forensic Examination of Deepwater Horizon Blowout Preventer. Final report for the U.S. Dept. of Interior, Bureau of ocean energy management regulation and enforcement. contract award No. M10PX00335, Report No. EP030842. March 20, 2011.
10. See e.g., http://www.nytimes.com/interactive/2010/06/21/us/20100621-bop.html?_r=0 or http://vimeo.com/30391513.
11. Smith P, Kincannon H, Lehnert R, Wang Q, Larrañaga MD. Human error analysis of the Macondo well blowout. *Process Saf Prog* 2013;**32**:217–21.
12. For a general introduction on nuclear power, see Ferguson ChD. Nuclear energy: what everyone needs to know. Oxford University Press, Inc.; 2011. ISBN:978-0-19-9759-45-3. For Dutch readers is recommended: Bogtstra, F.R., Kernenergie: Hoe zit dat?, Betatext, Bergen NH, Netherlands, 2013, ISBN:9789075541137.

13. IAEA Mission Report. The Great East Japan earthquake expert mission, IAEA international fact finding expert mission of the Fukushima Dai-ichi NPP accident following the Great East Japan earthquake and tsunami. May 24—June 2, 2011. Tokyo, Fukushima Dai-ichi NPP, Fukushima Dai-ni NPP and Tokai Dai-ni NPP, Japan.
14. Ragheb M. Fukushima earthquake and tsunami blackout station accident, department of nuclear, plasma and radiological engineering Illinois. April 2011. http://mragheb.com/NPRE 402 ME 405 Nuclear Power Engineering/Fukushima Earthquake and Tsunami Station Blackout Accident.pdf.
15. USNRC Technical Training Center, Reactor concepts manual, boiling water reactor systems, Rev 0200, http://www.nrc.gov/reading-rm/basic-ref/teachers/03.pdf.
16. The National Diet of Japan, The official report of the Fukushima nuclear accident independent investigation commission, © 2012, The National Diet of Japan.
17. Nuclear Emergency Response Headquarters Government-TEPCO Mid-and-long Term Response Council. Mid-and-long-term roadmap towards the decommissioning of Fukushima Daiichi nuclear power station units 1—4, TEPCO. December 21, 2011.

CHAPTER

5 Sociotechnical Systems, System Safety, Resilience Engineering, and Deeper Accident Analysis

"Chaos umpire sits,
And by decision more embroils the fray
by which he reigns: next him high arbiter
Chance governs all."
John Milton, 1608–1674: *Paradise Lost*

SUMMARY

In the previous chapter we have seen how complexity and pressure from competitors translating into cost and time pressure on the work floor, are conditions that may hamper safe working. Even in large commercial organizations this can lead to decision making in which assessed risk margins prove to be too small or in which one is even crossing into unsafe conditions of operation. This can be enhanced by governmental, regulatory, inspection, and enforcement entities becoming too lenient and relinquishing their authority under various pressures. These pressures include economic, political, or other influences or budget restrictions and having insufficient in-house expert capability. Warnings are often ignored and/or not well analyzed; whistleblowers can be ignored or even persecuted; audits are not completed or their recommendations ignored. As a result, damage to society can become orders of magnitude larger than the potential gain if the operation had been managed successfully. Before moving on to subsequent chapters in which some new promising preventive methods will be explained, we shall first delve deeper into the problem field of evolving risks in large process industry organizations, commonly characterized as a category of sociotechnical systems, while in the next chapter human factors will be explored in more detail.

In this summary, we shall first introduce a few of the pioneers who attempted to discover patterns in the emergence of catastrophic accidents of large and complex sociotechnical systems. The more recent of these scholars applied a real systems approach, each developing their own version of an accident analysis method. We shall look at these from a certain distance and at the end of the chapter make a brief comparison.

Risk Analysis and Control for Industrial Processes - Gas, Oil and Chemicals.

In *Man-Made Disasters, The Failure of Foresight*, first published in 1978 with a second edition in 1997, Barry Turner,[1] organizational sociologist and professor in organizational behavior at Middlesex Business School London, characterized disasters as (misplaced or misdirected) energy plus misinformation. There are preconditions and factors that may combine to produce its onset. He portrayed the occurrence of a disaster and its aftermath as a process of various stages starting with violation of rules, followed by an incubation period in which events pass unnoticed and are misunderstood. Examples are disregard of complaints by "outsiders," lack of communication, and reluctance to fear the worst outcome.

Charles Perrow,[2] professor in sociology at Yale and Stanford, USA, departed further from a system's point of view. In his 1984 book *Normal Accidents: Living with High-Risk Technologies*, he was one of the first to warn about accidents resulting from "tightly coupled, non-linear, complexly interactive systems." Accidents were qualified as normal by Perrow because he believes mishaps are unavoidable, even when all conditions are right and all work safely.

Jens Rasmussen[3] started his career in the late 1960s as a reliability and control engineer at Risø in Denmark. This institute was then still on nuclear power research but focuses meanwhile on sustainable energy. Rasmussen was later in the 1990s also a professor in cognitive systems at the Technical University of Denmark and is considered a pioneer in industrial systems safety and control.[4] Rasmussen came to study complexly interactive systems at about the same time as Perrow. What Perrow and Rasmussen both concluded in the 1980s was that as a result of feedback loops and nontransparent connections among (automated) subsystems, components, human operators, supervisors, and management, all causing complexity, full control of such systems is extremely difficult, if not impossible. We must keep in mind, however, that although with current practices much greater control of complex systems is achievable, at the same time complexity has increased.

In the mid-1980s organizational researchers mainly from the University of California, Berkeley, such as Karl Weick, Gene Rochlin, Karlene Roberts, and others studying what makes an organization successful proposed the concept of the high reliability organization. Such organization showed among other things the property of being resilient. In other words, such organization is able to absorb unexpected shocks. It is able to understand warning signals, minimize damage by an unforeseen threat, and to restore quickly from damage sustained.

Erik Hollnagel, a psychologist specializing in human error and decision making, who started in Rasmussen's time at Risø, later was a professor at universities in Sweden, Norway, France, and Denmark. He did a great deal of work in the fields of nuclear and aviation safety. Besides his own human error model, he contributed to the safety field early in this century with some remarkable concepts, such as the efficiency-thoroughness trade-off principle and resilience engineering. He propelled the latter concept forward together with David Woods, professor in ergonomics at Ohio State University, and Nancy Leveson, specialist in computer science and professor of aeronautics and astronautics at Massachusetts Institute of Technology by organizing a symposium in 2004. In this context resilience is the capability to absorb and overcome unexpected, unforeseen, and unknown threatening disturbances that could otherwise result in a catastrophe. The purpose of resilience engineering is to enshrine such a capability within the process and plant design.

This chapter will first analyze in more detail the problem field outlined above and will then describe the system approach Leveson conceived to tackle risk control. She developed a new accident investigation model by analyzing the system on identifying and finding the control loops required to keep the system in a safe state, in other words, for it to remain within its envelope of safe operating conditions. The model added new causal factors to the ones found by traditional hazard analysis methods that we reviewed in Chapter 3. This model was followed by development of a hazard and operability (HAZOP)-like tool called system-theoretic process analysis (STPA). This tool enables a team guided by questions to identify as a first step those weaknesses or deficiencies in the functioning of control loops that allow the system to depart from its safe operational envelope. A subsequent step then determines what component failures in the control loop could cause such a flaw. In turn, this identification would enable corrections to be made and take appropriate preventive measures. The method can be applied at any stage in the system process life cycle so that, for example, degradation over time can be traced and repaired.

Finally, the third part of the chapter is dedicated to what resilience engineering is and why Hollnagel, Woods, and Leveson generated momentum to communicate the concept to the engineering community at large. From the outset it addressed organizational resilience, what factors influence it, and how it can change over time. But resilience in an industrial process context also has a technical component. Margins, flexibility, and controllability must already be considered in design. Several other factors that reinforce resilience are identified. Detecting and understanding early, mostly weak warning signals proves to be crucial, which emphasizes the importance of system monitoring and measuring. Some existing tools to measure the extent of resilience are mentioned, but much improvement and extension of these tools is still needed.

5.1 SOCIOTECHNICAL SYSTEMS AND SAFETY

The disastrous sequence of events in 2011 at Fukushima-Daiichi was for Charles Perrow[5] an opportunity for renewed comment on the highly significant risks that are still present in many large sociotechnical systems, despite many years of research, technical improvements, training, and regulation. And it is undeniable that the partial meltdowns in Japan occurred just at a time after the Chernobyl catastrophe when the world was starting to regain confidence in nuclear power. In various countries new nuclear power plants were planned or being built, which was welcome in view of climate change and the pressure to reduce carbon dioxide emissions. As we have seen in the previous chapter, the latent causes allowing this tragedy to occur are exactly those that Perrow had been mentioning in his various publications: "failed regulation, ignored warnings, inept disaster response, and commonplace human error." In his 2011 article he gives some details regarding warnings with respect to nuclear power plants in Japan and the risks imposed by earthquakes and tsunamis in the "Ring of Fire"; this is the rift near the Japanese East Coast, where the tectonic plates meet and move relative to each other. He also mentions reasons why warnings are so often ignored—there is a fear that they may turn out to be a false alarm, or similar ones previously have proven to be false, or they are not precise enough for a decision maker to justify acting upon them. Nonetheless, the Fukushima event has been quite discouraging for those supporting the trend of larger contributions from nuclear power, certainly in Germany, which decided in due course to ban all nuclear power stations. With new, safer fourth-generation nuclear plants in sight, this has to many been a significant disappointment.

Now, Perrow as a sociologist has the rather dim view in that he is doubtful about us, mankind, ever being able to control the risks that a sociotechnical system will expose people to. Similarly, Jens Rasmussen[6] in his seemingly last published paper that he wrote with Inge Svedung, is not optimistic either, but they do lay out a way for improvement. Rasmussen and Svedung start out with the observation that there are many levels of politicians, managers, safety officers, and work planners that are involved in risk control of an industrial activity with associated hazards by means of regulation, rules, and instructions. All of these levels try to maintain safety by motivating, educating, and guiding workers, and operators trying to adhere to the rules, as shown in the sociotechnical system of Figure 5.1. In analyzing industrial accidents they included this observation by identifying decision-making actors and showing the findings in a cause–consequence chart. An example is shown in Figure 5.2. This chart also shows the flow of events (in rectangular boxes) and prevailing conditions (in rounded boxes), which results in a critical event. In part the idea stemmed from Johnson's MORT (see Chapter 3, Section 3.4). The chart can explicitly include barriers, fault tree elements as "AND" and "OR" gates, while to the right of the critical event it can branch out to further consequences as in an event tree. The chart is called "Accimap," which clearly represents decisions made at the various levels of the sociotechnical system.

Rasmussen and Svedung's concern is the dynamics within society with fast changes in technology, the steadily increasing scale of operations, the higher degree of integration by logistic networks, information technology, and the "just-in-time" (JIT) philosophy leading to tighter coupling. As a consequence of that mechanism, a relatively small disturbance could lead to a catastrophic deviation. Rasmussen and Svedung refer to an example in the financial world, viz, the Wall Street stock exchange upset in 1986 reinforced by computerized trading. Indeed, to this can be added a more recent example with even a much larger impact, namely, the 2007–2008 credit crunch. In an Organisation for Economic Co-operation and Development (OECD) report, Stefan Thurner[7] shows, supported by a mathematical simulation model of financial markets, that the high leverage situation allowed to financial institutions at that time created an unstable potential that reduced the resilience of the market. This potential developed with banks and institutions such as hedge funds in the years before the crisis. The model shows that because of a

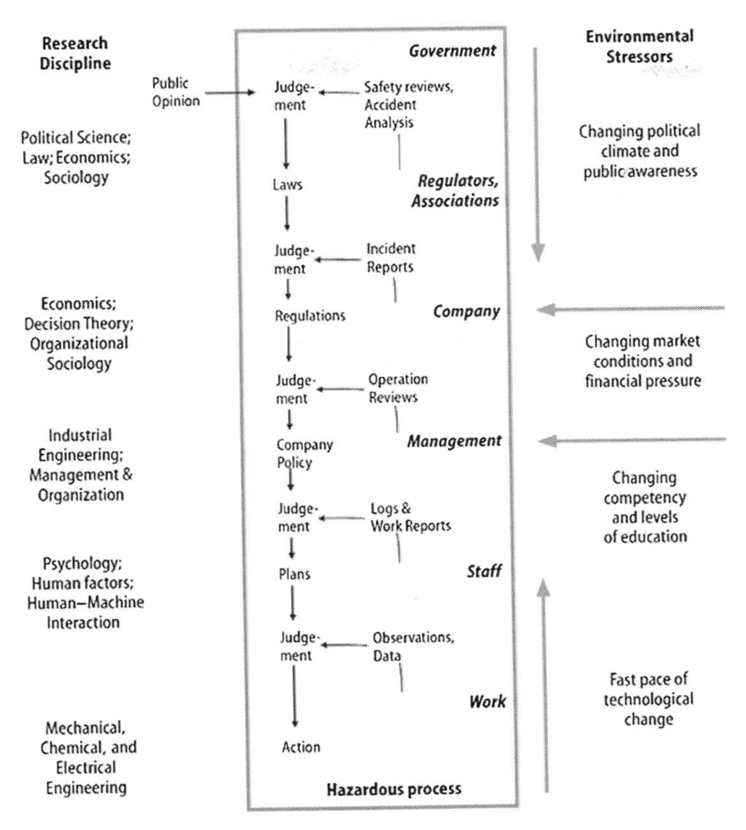

FIGURE 5.1

The various levels involved in risk management and control of a hazardous process and the feedback at each level, which at evaluation lead to renewed action in a sociotechnical system, as sketched by Rasmussen and Svedung[5] in 2000 and by Rasmussen in earlier papers.

single default by a small investor the entire market, given low resilience, can suddenly collapse.

According to Rasmussen and Svedung,[5] various influences, called environmental stressors, are acting on this sociotechnical system to make "*companies today live in a very aggressive and competitive environment that will focus the incentives of decision makers on short term financial criteria during economic crises rather than on long term criteria concerning welfare, safety, and environmental impact.*" This statement was written near the end of the 1990s but became more evident in the years after the financial breakdown in 2007–2008 and the shrinking economies later. Striving for commercial success tends to drive a company to the edge of

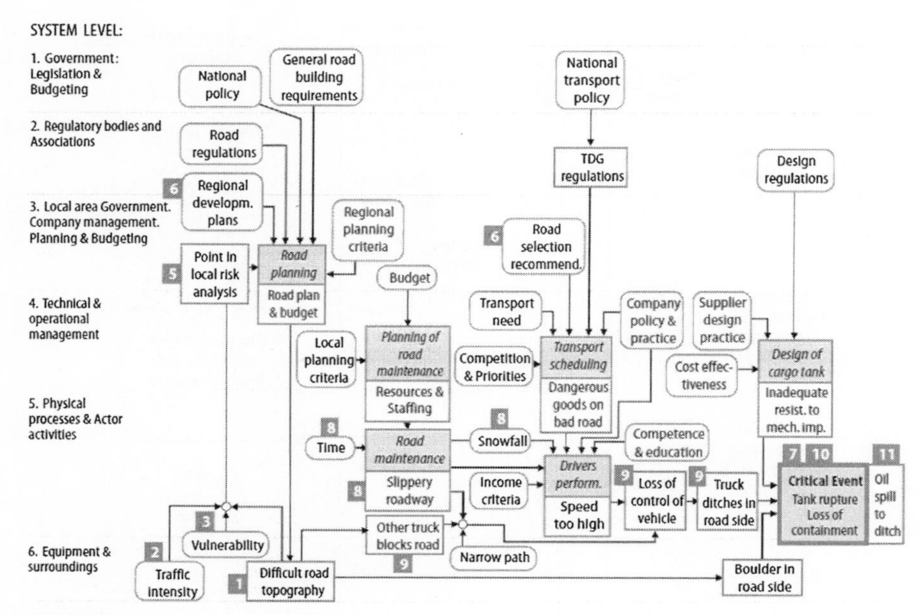

FIGURE 5.2

Rasmussen and Svedung's Accimap representation of the causal structure with conditions and events of an accident of a truck transporting hazardous material.[6] It shows at *left* levels of the sociotechnical system and at *right* the causal structure with the critical event. Numbers in the structure refer to annotations in the original article. (Additionally, one could think of a pillar "Planning of truck maintenance" under Technical & Operational Management.) In a subsequent graph in Rasmussen and Svedung's article, the accident events following the critical event have been detailed (together it would form a bow-tie-like structure).

the possible and allowable. Following Le Coze,[4] discussing the cause of major accidents Rasmussen stated already in 1995: "*This is the case partly because technical knowledge is not maintained during normal management activities at higher levels of the organisation, partly because high level managers often are law and business school graduates with a general financial background, not technically competent people promoted from the technical staff.*"

Rasmussen and Svedung[6] continue by arguing that the best way for proactive risk management to tackle this development is not by trying to predict scenarios better and to cater for that. This is not even by including the improbable ones and all the human errors possible—they use the word *exotic*—but to ensure that the process remains within the envelope already fixed in the design stage. This approach requires that at the various levels decision makers obtain adequate information about the state of progress of the process so that on their level they can keep conditions right for safe advancement. This objective will require a cross-disciplinary research effort and a study of the present information environment and what it should be. They then continue by emphasizing the difference between procedural control, which is preplanned and in fact open loop, and, in contrast, closed-loop control in which feedback provides the stimulus enabling correction of the course to the target. Safety is a *control problem* and *proactive*, so an optimal risk management strategy requires closed-loop control.

Nancy Leveson[8] follows the same line of thinking, but she makes the systems approach even more explicit. In her book *Engineering a Safer World, Systems Thinking Applied to Safety*, she starts out with the following observations, part of which we have seen before:

- fast pace of technological change,
- reduced ability to learn from experience,
- changing nature of accidents,
- new types of hazards (such as new man-made chemical effects in food and the environment),
- increasing complexity and coupling,
- decreasing tolerance for single accidents,
- difficulty in selecting priorities and making trade-offs (companies are exposed to harsh competition and governments face budget restrictions. This is all forcing cost cuts and increased temptation to accept shortcuts.),
- more complex relations between humans and automation,
- changing regulatory and public views on safety (individuals have a growing problem to control the risks around them and look for the government to take the responsibility for their safety).

N.B. This *author's* note: The last development took place in Europe a few decades ago.

These changes require a different type of safety engineering. The system that she envisages is from an engineering point of view more nearly complete than in Figure 5.1, as it puts the system development/design activity upfront of operations, as depicted in Figure 5.3. Both hierarchical pillars have regulatory government as the top level. However, design enshrines features for the entire life cycle of a technical installation/construction (the manufacturing branch pertains to prototypes and parts), and it is therefore crucial that it be done right. Also, the design lays down boundary conditions for maintenance, while operations complete the burden for maintenance, as shown in the figure. It goes without saying that communication between levels is of crucial importance. In her articles and book, she presents examples of how in these structures an accident can develop. In the next section, we shall describe one of those examples in more detail.

The majority of the changes mentioned by Leveson have their root in economic drive and competition. There are also changes that have to do with complexity and aversion from uncertainty. We shall dwell somewhat longer on the latter subjects based on a recent article by Johansen and Rausand.[9] These authors treat complexity and uncertainty in relation to risk assessment of sociotechnical systems. The larger the complexity of a system, the larger the uncertainty with which risks can be

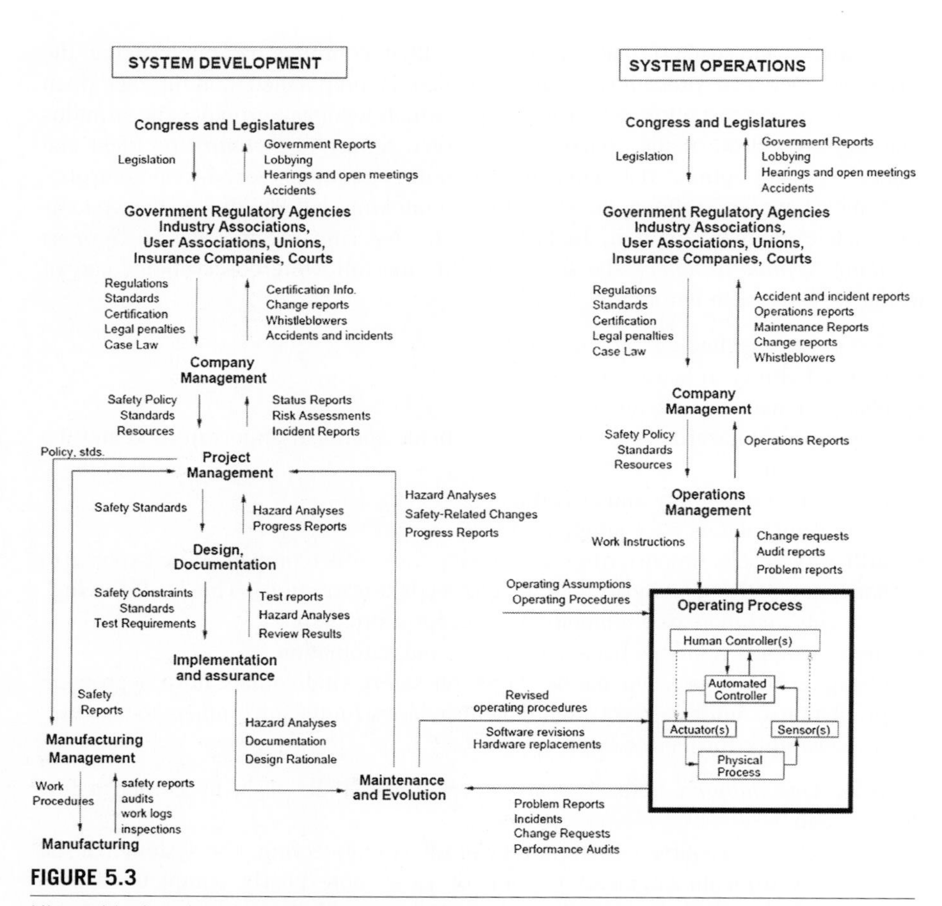

FIGURE 5.3

Hierarchical sociotechnical system of a safety-critical industry in the US with levels connected by information control flows and feedbacks, according to Leveson.[8]

determined and the lower the confidence one can have in the results of a risk assessment. Their main point, though, is that complexity is not a property of a system. Complexity is a consequence of the limitation of human understanding of all mechanisms of interaction within a system and the failure to predict how a system will behave given stimuli. The interactions are what Perrow has designated to be tightly coupled and nonlinear. So, one can characterize a system as being complex.

In nonlinear interaction, cause and effect are linked in an indirect way. A small cause can have an effect that is growing, while propagating through the system, but it can also be the other way around. Feedback loops generate circular causality resulting in amplification or damping of a disturbance. Johansen and Rausand[9] put against each other a simple system with linear causality that comes with symmetry, rationality, singularity, and determinism, and in contrast to what we perceive as a complex system with circular causality, bounded rationality, contextuality, self-organization, and emergence. Self-organization of a system is the kind of ordered behavior that we can observe in the twisting and turning of a school of fish. Emergence means that system properties cannot all be derived from the properties of its components alone, which is expressed by the old adage: "the whole is more than the sum of parts." Leveson calls safety typically an emergent system property. All components can be fully reliable and not at fault but still the system can be unsafe due to dysfunctional interactions of the components, leading to disaster. All this makes a risk assessment not easier, to say the least, but also the more important to perform at the system level! Reductionism, that is linear decomposability of a system in cause—effect scenarios as we have done in binary fault trees, event trees, and bow-ties, all of which model only two states, therefore has limited value.

To improve understanding of complexity and shed more light on it, Johansen and Rausand[9] ask themselves the following questions:

1. Can a system be complex to one analyst but not to another?
2. Can one system be more complex than another system?
3. Can a system be more complex by the choice of one model than another?
4. Can a system be complex in one assessment context but not in another?
5. Can a system be complex today but not tomorrow?
6. Can a system be complex tomorrow but not today?

The answer to all of these questions is an affirmative "yes." The first is rather obvious that it depends on the analyst. The second does not need any explanation. The third is inherent to a model that is a simplification of reality; so, one model, such as a two-state fault tree, is simpler than another, such as a multistate or continuous Bayesian network. The answer to the fourth question depends on the depth of the assessment that in turn depends on the type and nature of the decision to be made on the basis of the assessment. An example is that following a sensitivity analysis, variables that contribute significantly to the overall risk are represented in a complex context by distributions, and the other variables are set at their nominal values, whereas in a simple, point-value context all variables are set at their nominal values. Increasing knowledge about the system can reduce its complexity in our mind, so this explains the answer to the fifth question, while the other way around by acquiring more knowledge about the system, we can find it to be more complex than we thought earlier. Johansen and Rausand then carry on by developing a detailed system of indicators characterizing complexity. After applying weighting, the indicators are aggregated to an index with the purpose of presenting it to the decision maker together with the outcome of a risk assessment. The index is meant as a measure of confidence in the risk assessment result or, rather, the lack of confidence.

In the chemical engineering community, Venkat Venkatasubramanian,[10] formerly at Purdue but now at Columbia University, drew attention to systemic risks. Earlier, Venkatasubramanian looked at automating HAZOP for improving predictively identification of risks adhering to a system (work that we shall summarize in Chapter 7,

Section 7.5.3). But he also spent much effort on real-time detecting and diagnosing operational process risks in view of process control and an operator being placed to identify an abnormal situation. To advance in beating systemic failures due to internal or external factors, the latter being, for example, unstable goals by changes in supply or demand, a prognostic tool is needed through applying complexity science. This requires not only a decomposing "reductionist" approach as the present risk analysis techniques but also an "integrative" one of smaller elements to a self-aware organism. So far, the latter approach has been much less successful. He further recommended multiperspective modeling, that is, looking at a system from different perspectives, such as its structure and connectivity (as the approach in Blended Hazid, Section 7.3.1, and models of the process installations' behavior under different conditions, such as multilevel flow modeling developed by Lind, Sections 7.3.3 and 7.9.1). Combining the above should then lead to intelligent real-time decision support systems alerting on an anomaly (*prognostic*) with information about its cause (*diagnostic*).

5.2 SYSTEM APPROACH TO RISK CONTROL

The arguments explained in the previous section have been reasonably conclusive. As Leveson argues in various articles,[11] safety is a system control problem. In her book,[8] she describes the accident model she developed, based on her approach, and the tool to identify risk control flaws in a system. The accident model is called system-theoretic accident model and processes (STAMP) and the tool is STPA. STAMP integrates technical and human failure aspects by focusing on the whole, hence a holistic approach. The following paragraph will describe in brief the model development as it is to be applied.

After defining the system boundary and the system hierarchical levels with their information flows as depicted in Figure 5.3, *unacceptable losses* to the system must be defined, for example, a release of a certain quantity of hazardous material with a chance of injury and fatality and a business interruption of a certain duration. Next, the *hazardous states* of the system that can cause these losses must be identified, which enables by negation the identification of the safe states of the system with their borderlines that must not be breached. This identification produces system-level *safety constraints* that must not be violated. These constraints appear instead of component failures and human error in the decomposition or reductionism approach. But violation of the constraints may result from component failure, errors in software and decisions, improper component interactions, conditions, or external disturbances. The most obvious constraints from a technical process point of view are physical, defined by combinations of process variable values, such as those of temperature, pressure, concentration, and flow rate. Many other constraints are procedural and organizational and can be very diverse in their specific purpose, as we got a flavor of in Chapter 3 looking at the safety management system, but there are also constraints to the activity posed by the community. These social constraints can have the form of not to exceed certain health impact threshold levels, as we have seen in Chapter 2 in EPA RMP's rule or certain individual risk levels as in the UK, France, and the Netherlands. These constraints can be politically influenced.

The final step of the model will be to identify the risk controls in the system that prevent violation of the safety constraints. An accident investigation will then search for the risk controls that failed. For this, Leveson developed the new STPA tool to identify causal factors. In principle, STPA replaces HAZOP and FMEA, fault tree, and event tree analysis, all explained in Chapter 3, and will include all further possible types of errors, faults, and flaws, such as in design and software, procedures, organization, management, and decision making. One should be able to use STPA at all life cycle stages of a system.

STPA is performed in two major steps. The first step guides the analyst by queries similar to HAZOP. One by one the identified safety constraints shall be examined and queried for potential violation by a failing risk control. Guideword queries are only four, in which time is an important element:

1. Is a control action *not* provided?
2. Is the control action unsafe?
3. Is the control action too early, too late, or out of sequence?
4. Is the control action stopped too soon or applied too long?

If any of these questions must be answered positively, in a second step the possible causal scenarios leading to a hazardous state are identified. This is carried out by considering one by one the various elements in the control loop whether they can cause or contribute to failure of the risk control. The step is rather similar to what is done with the traditional methods of hazard analysis, except that not only failures are examined but also dysfunctional interaction of healthy components is considered by various kinds of information mismatches, wrong design, missed possibilities, and the like, possibly leading to an accident. This is illustrated in Figure 5.4. If required, for each component in the loop, a level-deeper STPA procedure can be applied.

Subsequently, effects of degradation of system components over time must be considered. Besides loss of integrity, degradation encompasses failure of management of change procedures by neglecting safety constraints. Performance audits shall be held to investigate whether or not unplanned changes violate safety constraints, while incidents shall be investigated to learn what has been overlooked and must be corrected. Yes, one could even go a step further and for proactive risk management use indicators to monitor the risk controls. Indeed, Khawaji[12] proposed to develop leading indicators that can monitor the risk controls at various levels of the hierarchy.

The previous discussion considered the safety structure of the system, but a system as a whole can degrade or upgrade in performance, also with respect to its safety records and culture. So, apart from the static picture, it is of interest to look at the dynamics and to identify trends. Indicators are a means to do that. As Leveson has presented in various papers, the actors and loops can be made visible in a system dynamics model. Once developed by Forrester of MIT in Boston in

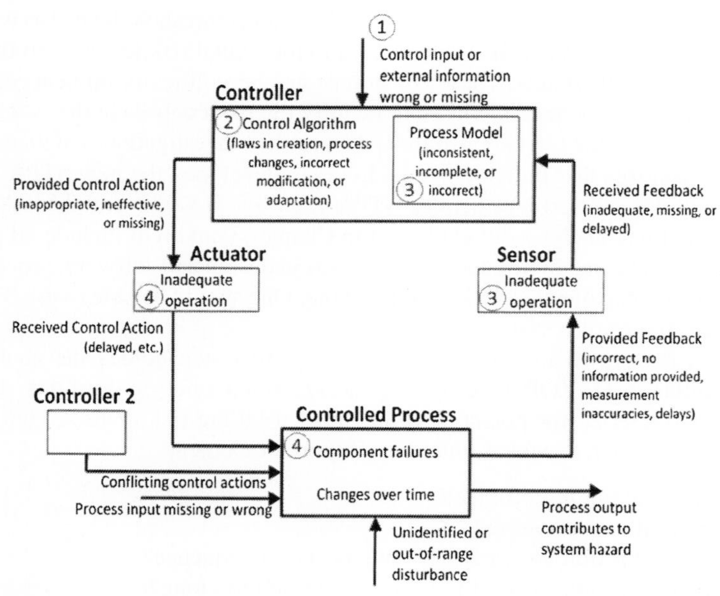

FIGURE 5.4

Step 2 of an system-theoretic process analysis considers all parts of the control loop, here shown as a generic one, to identify the scenarios that can cause or contribute to failure of risk control, after Leveson.[8] The circled figures refer to the query type number.

the late 1950s to model business processes, the models are a graphical representation of a system structure. They consist of causal links and feedback loops (positive or negative), connecting and controlling valves, which are restricting flow to or from resources or stocks (a state variable). Links can contain delays, which lowers the transparency and stability of system behavior. For an even modest size system, the number of loops can easily grow so large that the model becomes impractical. Khawaji developed a model for indicators derived from incident and near-miss investigation, which is reproduced in Figure 5.5. Although the diagram is quite busy, one can notice indicators as "incidents under investigation," "overdue engineering studies," and many more. It can easily be seen that if management decreases funds for incident investigation, the risk level will go up, although there may be a certain delay. These pictures help to understand qualitatively the feedback and the pattern of behavior of a system. In principle, system dynamic models can be quantified given known loop time constants and other quantities. But, data are difficult to obtain.

Summarizing, the system approach to analyze systemic failures offers perspective, certainly when it is applied in a top-down approach. However, further research could show how to deal with the myriad of details when one would apply the STPA tool to an actual process at the lower, more detailed levels of components

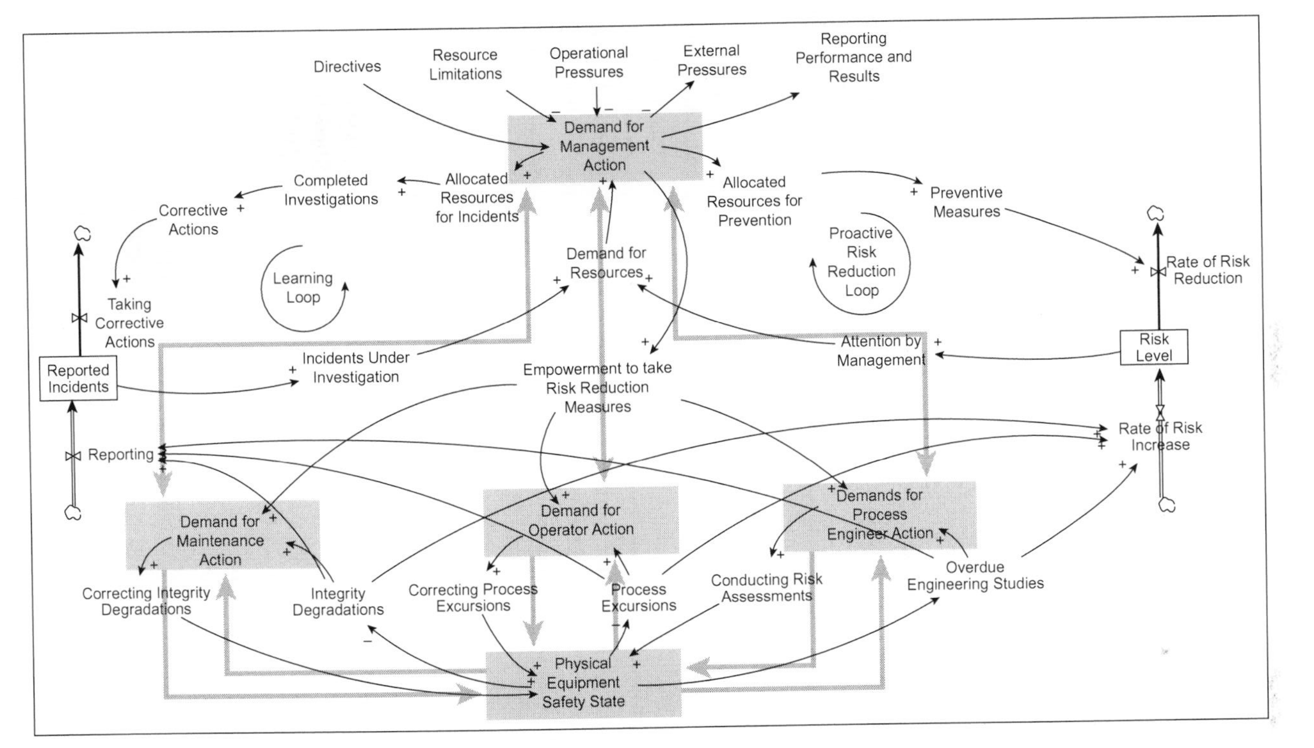

FIGURE 5.5

System dynamics model of trends from investigated and reported incidents (stock, at *left*). These feed a learning loop for which management has made funding available and to which maintenance, operators, and process engineers contribute with information. To the *right* is the stock of risk level, which is decreasing by preventive measures in a proactive risk-reduction loop, but which increases by slack in finishing engineering studies, integrity degradation, and process excursions.

Reproduced from Khawaji's thesis.[12]

and parts. In Chapter 7 we shall return to this matter. To close this chapter, we shall briefly describe an example accident investigated by Leveson and coworkers in 2003 applying STAMP, which shows its features. The investigation is published as an Appendix in Leveson's book[8] to which we will refer for details.

5.2.1 AN STAMP ACCIDENT INVESTIGATION EXAMPLE

In Walkerton, Ontario, Canada, in May 2000, about half of the town population of 4800 fell ill, in particular children, some seriously, and seven people died. There was a suspicion that the source was the drinking water, but this was repeatedly denied by the manager of the plant, although the water was contaminated by *Escherichia coli* bacteria from manure spread during an unusual wet rainy season on a relatively thin layer of soil on rather porous bedrock near well 5, from which drinking water was extracted. At the time, the manager was on leave. The manager's brother was the plant's foreman and he was supposed to check chlorine residuals, which, if low, is a sure sign of contamination, but he did not do this.

The day the manager returned, he had well 5 shut and well 7 opened, however, after this change, he learned that the chlorinator on well 7 was in maintenance and for a few days water went unchlorinated. Due to the delay in manifestation of the illness, at the time the victims fell ill, the chlorinator was working again. Upon local inquiries about the water, the manager assured callers that the water was safe. A routine water analysis result still from the water of well 5, coming in from an independent, private laboratory with two days delay, showed an *E. coli* problem, which the plant manager hid. Current well 7 water he superchlorinated. A local pediatrician contacted the governmental Health Unit because of the *E. coli* suspicion. The Health Unit started an investigation, *E. coli* infection tested positively (in well 7 samples just before chlorination started), but the Health Unit was assured by the plant manager that the water was all right, a message that was relayed to the population. The Health Unit sent water samples for microbiological analysis. Two days later, *E. coli* contamination of these samples was confirmed. Meanwhile, a plant employee anonymously called the Ministry of the Environment (MoE) about the earlier analysis results. MoE called the manager but was at first fobbed off. Later, the manager had to produce all records and analysis results, but had his brother fake the ones of well 7 for the days the chlorinator had not been installed. After handing over the documentation, the Health Unit immediately had a boil-water advisory radioed. An official investigation was carried out.

Now, the interesting question is, how could it happen? Clearly, the hazard is water contamination. The (abridged) safety constraints are:

1. Drinking water quality shall stay within certain norms;
2. Public health measures shall protect the public.

The various entities and procedural information flows of the rather complex sociotechnical system as it had been set up originally are reproduced in Figure 5.6. We can notice many control elements. If the control loops of the two constraints are

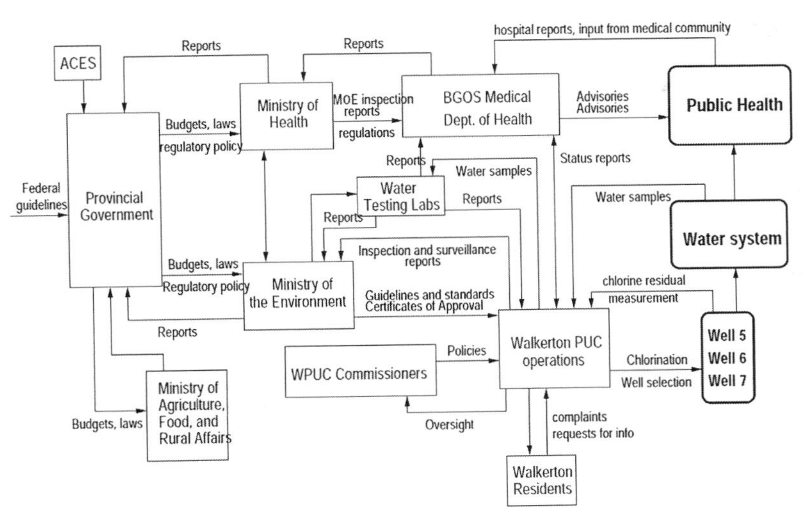

FIGURE 5.6

The sociotechnical system of the Walkerton, Ontario, Canada, public water service with governmental and regulatory entities, a water quality testing laboratory and the Walkerton Public Utilities Commission (PUC) Operations plant, after Leveson.[8] Opposite to previous sociotechnical schemes (e.g., Figure 5.3) this one is depicted with its top rotated left over 90°. Shown is the setup with the various *control elements as it was originally structured.* ACES stands for Advisory Committee on Environmental Standards.

examined, many deficiencies, called inadequate control actions and mental model flaws, surface. It will not be possible to reproduce here all the observations made by Leveson and coworkers, but we shall select some salient ones. Important is, for example, the brothers' lack of knowledge about potential health effects of untreated water. They had grown in the job without formal qualification. They personally had not noticed any adverse effects of drinking directly water from a well. Moreover, the chlorine was perceived by many as unnecessary and spoiling taste.

Another point is the change of the sociotechnical system of Figure 5.6 over time in Figure 5.7. First of all, conditions changed, for example, the number of cattle and hogs strongly increased. Then, a few politically motivated changes occurred, mainly in the second half of the 1990s. In 1995, a conservative government was elected, and, as Leveson put it, bias against more environmental regulation and "red tape" resulted in taking away a number of controls. In addition, the water-testing laboratory was privatized and had only to report to the plant. The reporting occurred formally "over the top," hence slower than before. The Advisory Committee on Environmental Standards (ACES) was disbanded. The budget of the Ministry of Environment was reduced 42%. So, alertness had diminished. The Ministry of Health complained to local health officers about bad reporting on water quality by municipal plants operators and private laboratories. The Provincial Drinking Water Surveillance program was stopped.

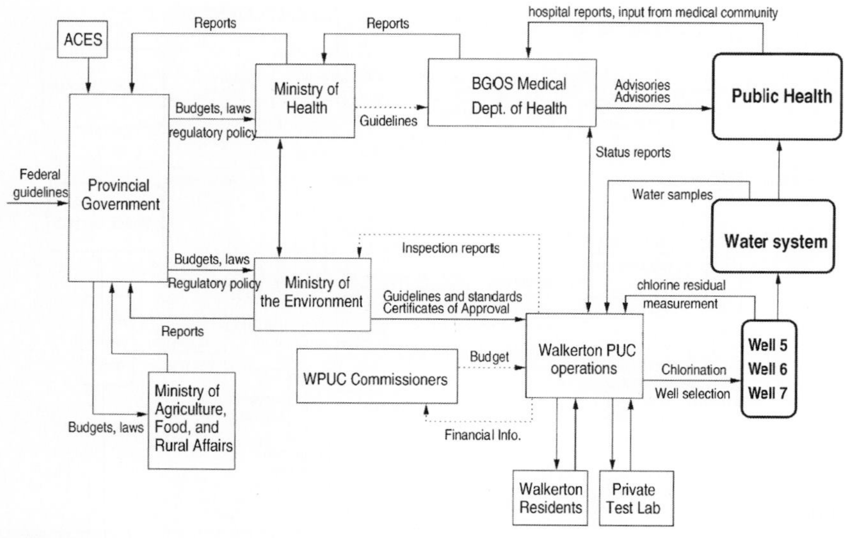

FIGURE 5.7

Sociotechnical system with *control loops at the time of the Walkerton contamination event.* In comparison with the original setup, numerous control loops disappeared. The water-testing laboratory with a rather central position has been privatized. (ACES is still indicated in the scheme, but was disbanded in 1996.)

In 1998, the local Walkerton council became concerned to the extent that they sent a letter to the premier, however, they received no reply. From various other sides, complaints and warnings were made. Three months before the accident, the MoE staff sent a letter to the Provincial Government warning that "Not monitoring drinking water quality is a serious concern for the Ministry in view of its mandate to protect public health." In 1996 in a town near Walkerton, a less serious "precursor" had even occurred. But all of that did not trigger action from the side of government. On the contrary, all blame was put on the brothers, typical for showing lack of insight in accident causation. Three years after the event, the brothers were arrested.

Leveson applied a system dynamics model for this case. The investigation showed that STAMP with its analysis of control loop deficiencies is quite helpful to reveal not only "root causes" but to obtain the grand picture of the sociotechnical system with all the players and all possible missing or inadequate control actions including government action.

Although the approach by Leveson is system based and is also meant to tackle the sheer unpredictable, tightly coupled, and nonlinear interaction scenarios, it is control oriented. It is therefore different from mathematical system theories also studying nonlinearity, such as catastrophe theory describing how a small change can cause an equilibrium to collapse, for example, a beam buckling under a load or chaos theory (of Poincaré and Mandelbrot). The latter describes dynamic systems that for a while are predictable but later appear increasingly random albeit scalable, for example, a weather system. In Chapter 7 we shall discuss practical limitations of STPA and the direction in which these shall be solved.

5.3 RESILIENCE ENGINEERING OF SOCIOTECHNICAL SYSTEMS

From a mechanical engineering point of view, resilience means the property of a material that enables it to resume its original shape or position after being loaded by a force. So, one would require a spring to be perfectly resilient. Resilience in a psychological sense is the ability to restore quickly after a mental blow, hence for an organization to restore from a disastrous setback. In a figurative sense, resilience was used by Weick et al.[13] in 1999 as a quality of a high reliability organization (HRO) to discover and manage unexpected events. Later, in 2007 in their second edition of *Managing the Unexpected*, Weick and Sutcliffe[14] mention a series of attributes of HROs, such as "a broad repertoire of action and experience, ability to recombine fragments of past experience into novel response, emotional control, skill in respectful interaction, and knowledge how the system functions." The authors further state that "the best HROs know that they have not experienced all the ways their system can fail. They (the HROs) also know that they have not deduced all possible failure modes. And they have a deep appreciation for the liabilities of overconfidence." Weick and Sutcliffe stress consistent *mindfulness* as a property of an HRO. By mindful, they mean that HROs organize themselves such that they are better able to discover an unexpected threat emerging and take precautions. We shall see that again when discussing resilience engineering in the sections following.

5.3.1 THE "SOCIO" SIDE OF RESILIENCE ENGINEERING: PSYCHOLOGICAL AND ORGANIZATIONAL

Erik Hollnagel in his publications over the course of years increasingly stressed the large variability of a person's performance and the intractability of systems due to nonlinearity. In response, he developed the functional resonance accident model (FRAM),[15] which after some preliminary papers he published as a book in 2012. Previous accident/hazard analysis models that we reviewed before in Chapter 3, such as fault tree analysis, Hollnagel characterizes as a sequential type. These are structurally decomposable, with linear independence in its dynamics, resulting in probability values. The Swiss cheese barrier slices he describes as an epidemiological model, structurally decomposable, but with linear dependence showing likelihoods of weaknesses in defenses. In contrast, FRAM is functionally decomposable, the dynamics are determined by nonlinear dependencies (couplings), and the representation is through organizational functional modules, see Figure 5.8 *left*.

The modules or organizational functional units Hollnagel proposes are characterized by six parameters: *inputs* (I), forming the link to other modules; *resource* (R), what is needed by function to process the input, for example, energy; *controls* (C), which supervise the function via, for example, procedures; *time* (T), which can be a constraint or a resource; *precondition* (P), with which is meant the system conditions required to perform the function, and *output* (O). Modules of various work stages are linked via parameters.

FRAM follows four principles: (1) Equivalence of successes and failures, meaning human variability in successfully performing tasks according to rule or procedure means that it sometimes goes wrong; (2) Approximate adjustments: nothing can ever be done exactly according to plan; (3) Emergence: "the whole is more than the sum of parts"; (4) Functional resonance, with which Hollnagel means the occurrence of unexpected peaks in a continually varying performance, see Figure 5.8 *right*. Such peaks can mean extremely successful achievement or failure. The analysis then goes through the following steps:

1. Identify essential system functions; characterize each function by the six basic parameters.
2. Characterize the (context dependent) potential variability using a checklist.
3. Define functional resonance based on possible dependencies (couplings) among functions.
4. Identify barriers for variability (damping factors), and specify required performance monitoring.

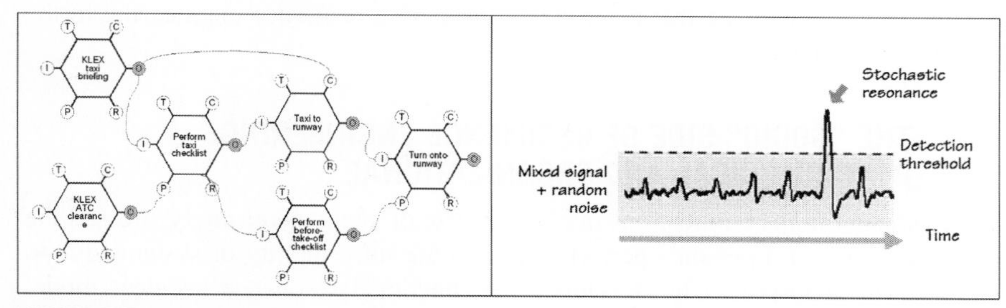

FIGURE 5.8

Left: Part of a functional resonance accident model applied on the Conair flight 5191 crash described by Hollnagel et al.[16] (I) is *inputs*, (R) *resource*, (C) *controls*, (T) *time*, (P) *precondition*, and (O) *output*; see also the text. The submodel represents the taxiway process forming an essential part in the overall model, which revealed some facts not discovered before. *Right*: Nonlinear accident model of organizational/human performance variation possibly leading to failure by resonance due to random disturbances.

After Hollnagel.[17]

Here is not the place to explain the model in detail but just to get a flavor of the thinking. And why does *resilience* come around the corner? It is because of the unexpected dips or error/disturbance peaks an organization should be able to absorb to remain safe. Human and hence organizational performance variation is not only a fact of life but even a necessity for success. Robots are different, and even those are difficult to program without application of fuzzy set theory. Hollnagel notes that "working to rule" and procedure is an efficient way to sabotage a work process. Human creativity and the ability to learn, interpret, adjust, and improvise are a blessing and needed for safety but leave us, on the other hand, with undetermined risk.

Hollnagel took the initiative of organizing with David Woods and Nancy Leveson a symposium in Söderköping, Sweden, in 2004, and the three edited a first book on resilience engineering based on the symposium papers.[18] The "engineering" in this bundle is mainly to understand how from the point of view of lack of resilience of organizations that accidents have occurred, emphasizing the need to obtain methods to measure and monitor organizational resilience and ways to improve where necessary. The concept made an impact, and further meetings and papers followed. It has become clear that traditional risk assessment methods may determine with some accuracy a hazard potential and possible consequence effects, but the probabilities of a failure event on the basis of historical data can give a false impression of safety. This is because of the complexity we fail hopelessly in identifying all possible scenarios leading to a rare, but disastrous event and quantifying its likelihood. Moreover, traditional risk assessment is almost all static and deterioration of a situation, and an increase of risk level that would justify decision making, is not measured. In Chapter 7 we shall see a tentative start of dynamics in risk assessment. In short, resilience engineering does not put its focus on human limitation and avoidance of human error as in safety endeavors, but on a positively contributing attitude at all levels to anticipate and to be best prepared.

5.3.1.1 Dynamics of change and drift to failure

In the mentioned book on resilience engineering,[18] Leveson gives a clear example how change in an organization induced by external factors can lead to looping feedback processes that in the end degrade safety thinking and cause accidents. Her example is the NASA space shuttle Columbia disaster over Texas in 2003 from a hole punctured in its wing by a part of the insulating foam of an external fuel tank that loosened at launch. The mechanism was not unknown and had shown up in previous flights to a lesser extent. Following up and learning from the previous near misses would have greatly reduced the likelihood of this disaster, but the urge for launch performance had become dominate.

In the system dynamics model in Figure 5.9, one can notice four loops. The primary one is "Pushing the limit." Fed by external pressure and the resource launch success, performance pressure goes up, but this first loop is part of a larger loop named "Limits to success." One limitation comes from budget cuts, which affect system safety efforts in the "Do more with less loop," diminishing the rate of safety increase. At first, this is not given much weight, because the stock of complacency is

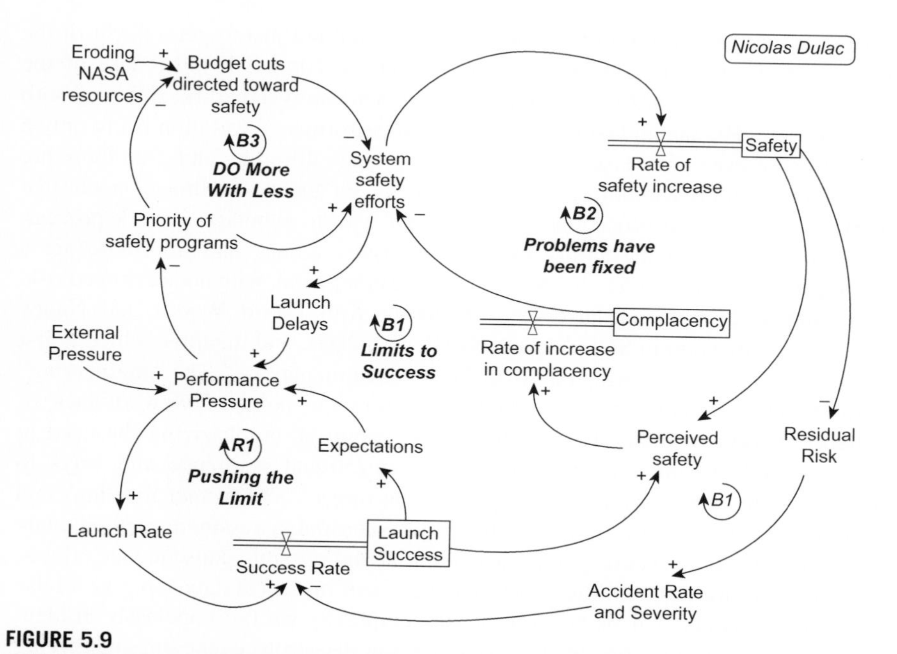

FIGURE 5.9

Simplified system dynamics model of drift to failure exemplified by the processes in NASA behind the Space Shuttle Columbia loss in 2003, according to Leveson and Dulac[19] and used by Leveson again in the bundle on resilience engineering.[22]

high. "Problems have been fixed," and we had no accident for a long time, so why worry? This attitude will dominate the scene, and so time becomes ripe for an accident to occur because risk has been increasing. Because these changes take place over a period of months or years, people involved will not usually notice them. The only way would be to define indicators and to measure and monitor their values over prolonged periods of time to identify trends of increasing risk.

M. Gajdosz, Tim Bedford, and S. Howick[20] (University of Strathclyde, Scotland) went a step in the direction of further exploring the possibility of making system dynamics a source of information and analysis of organizational factors in the context of risk analysis applied on a maintenance process.

5.3.1.2 Comparison of accident analysis models

In this chapter we briefly reviewed Accimap of Rasmussen, STAMP of Leveson, and FRAM of Hollnagel, while in Chapter 3 we described Reason's Swiss Cheese and others. Although related, there are fundamental differences in approach. Accimap describes a causal structure of decisions and events while taking account of conditions; STAMP maps an organizational hierarchy and its control loops while it identifies where control fails; FRAM shows the coupled system functionalities and their propensity to human performance variability; and the Swiss Cheese with in its wake Tripod Beta the direct and latent causes, as well as preconditions leading to the accident event, in fact no different from Turner's incubation period.[1] This is apart from tools mentioned in Chapter 3 such as MORT and IPICA, which go into depth of identifying management failures, where IPICA Lite root cause analysis is focusing on failing safety management system elements. One can ask whether an attempt to make a full comparison makes any sense, because it also depends on the nature of the accident which model would apply best, for example, whether human factor or management influence is very important or not. In case of the Tesoro Anacortes refinery explosion and fire in 2010 investigated by the US Chemical Safety and Hazard Investigation Board,[21] in which technical aspects dominated, Accimap produced a good overview. In principle, STAMP should work well in any case. Salmon et al.[22] compared the applicability of Accimap, Human Factors Analysis and Classification System (HFACS) inspired by Reason's Swiss Cheese model and mentioned in Chapter 6, and STAMP on a sports accident. Accimap and STAMP result in recommendations to modify the system. However, these miss the human factor taxonomy of HFACS. Therefore, a light preference was concluded for Accimap of which the infrastructure allows inclusion of such taxonomies at all levels of the sociotechnical system.

5.3.2 THE TECHNICAL SIDE OF RESILIENCE ENGINEERING AND THE RISK MANAGEMENT VIEWPOINT

Resilience in relation to safety has a technical side. In all parts of an installation, safety margins are present. As in other aspects, competition will drive the safety margins down unless there are clear threats to be countered. As a "super inherent safety," resilience shall safeguard us for unexpected, unforeseen, and even unknown threats. Steen and Aven[23] made this very clear by formulating it in a more exact way, including the problem of uncertainty inseparably connected with risk probability. Their reasoning, going to what resilience ultimately should be, can be summarized as follows:

The public at large is interested only in consequences that affect them and uncertainty makes them worry. Hence, risk, R, for them is a function of the threat/attack, A, severity of consequences, C, uncertainty of likelihood and strength/effectiveness of barriers, B, or:

> $R = (A, C, B, U)$, in which, of course, risk decreases if barriers increase in effectiveness.

To this shall be added the element of uncertainty, U, which depends on conditions. Risk analysts, though, like to express risk as the product of consequences, C, and probability, P, but as we have seen at the end of Chapter 3 it cannot be denied that the result of an analysis contains considerable uncertainty:

$$R = (A, C, B, P, U)$$

Uncertainty diminishes as knowledge, K, about the threat event, its likelihood and its impact, increases (e.g., uncertainty reduces if the variable can be expressed as a distribution), so that:

$$R = (A, C, B, P, U, K)$$

Given an attack, A, the vulnerability, V, of a receptor can be written as a function of

$$V = (B, C, P, U, K|A), \text{ while robustness is the complement of } V = (1 - \text{Robustness}).$$

Now, resilience, $\mathcal{R}$, can be considered as robustness against any attack or

$$\mathcal{R} = 1 - (B, C, P, U, K|\mathit{any}\ A), \text{ including new types of threat, which puts it } \textit{ad extremum}.$$

Measures that can be taken to increase resilience can be preplanned and activated if first signs of a threat appear, but as the very nature of the threat can be unknown, signs may be missed. Resilience will also depend on improvisation, hence adaptability and flexibility of the organization and its installations are critical. This assumes sufficient time is available to install measures, which again depends on detection of early warning signals, understanding those, and taking action. But it goes a step further—an organization can also be considered resilient to the extent that after receiving a damaging blow, it can quickly restore its capabilities. This restoration capability requires containment of cascading or escalating aftereffects and a well-organized emergency response organization.

A preliminary study by Dinh et al.[24] attempted to extend the concept of resilience as a forward and proactive defense to technical process plant aspects and the organization around it. This was done by defining characterizing key variables and by quantifying the degree of resilience by determining index values. Besides the normal operation state, Dinh et al. distinguish upsets and the catastrophic state. In case of the latter two, the goal will be to minimize the *extent of failure* due to an upset or, worse, to detect and recognize *early warning signals*, and to design for higher *flexibility* and higher *controllability* of the process and functioning of the organization, while minimizing *effects of failure* and optimizing *administrative controls*, *predictive methods*, and *procedures*.

The extent of failure can be minimized by applying the principles of inherent safety, as summarized in Chapter 3 in the section on system safety, while in general increasing safety margins will help, although their effect by themselves will be limited. The cost effectiveness of increased safety margins to reduce risk can be predicted within a risk assessment. At the same time, proper protective equipment shall be installed, and appropriate safety management shall be performed to the maximum extent. Strengthening *situational awareness* is of utmost importance in many respects, but early warning signals will be very diverse in nature and often not simple to interpret. A primary source will be the process control system, which we shall treat in more detail in Chapter 8, but signals will also be arriving from upcoming external events of which violent weather conditions, flooding, and earthquake are rather obvious. More intricate are signals on technical wear of components before failure. Vibration can be detected. Maintenance workers can trace unexpected deviations potentially causing serious future trouble. Also, cyber hacking or other malevolent action shall be intercepted. Also important, but with effect more on the longer term, are signs by lagging and leading indicators that discipline and safety culture of the organization are changing. As mentioned in the previous chapter in connection to the Fukushima disaster, warnings are very often ignored when brought to top management, certainly if there is no emotion or urgency connected to them or they are not sufficiently precise or perceived to be credible.

Flexibility is the ability of the process to cope with an input change such that output will not go beyond the set limit conditions. It assumes that people and equipment are available to respond. Controllability, on the other hand, is the ability or the ease with which process control can bring the process to a target state. It is clear that when these properties are better, resilience is higher. In a metaphorical sense, the same will be true for the functioning of the organization. It is quite obvious that by minimizing effects of failure, hence failure with fewer and less severe consequences and damage, resilience increases. Finally, administrative controls and procedures is one-to-one translatable into a safety management system. Effective emergency response and crisis management will be part of it, certainly when critical infrastructure is involved.

5.3.3 MANAGEMENT FOR RESILIENCE AND TOOLS TO PROBE RESILIENCE

Managing distribution of one's resources over measures to obtain an optimized result starts with assessment, which in turn first needs analysis. As follows from the equation for resilience $\mathcal{R}$, mentioned in the previous section, in principle, for resilience the same parameters must be determined as for risk, albeit for any threat. This already brings a first limitation, because a number of still realistic but uncommon threats can be thought of, but there remains the uncertain and unforeseen. A second limitation is as previously shown that traditional risk analysis does not present a complete risk picture—Leveson underscored the dysfunctional system component interactions and Hollnagel the human/organizational performance variability. Hence, it is clear that traditional methods of scenario identification are deficient, not so much with respect to consequences but in particular regarding likelihood. This deficiency calls for new methods such as STPA, which we will discuss further in Chapter 7, while we shall mention below a method based on early warning signals.

Just as a design should be scrutinized for resilience, detailed process simulation can be used for study of process flexibility and controllability to determine cost-effective safety margins. The quality of simulation has made vast improvements. A simulation model is well reusable in later stages, for example, operator training, although the simulated environment and results shall be sufficiently realistic to fulfill minimum requirements for learning the right responses to deviations and disturbances. More information about methods of simulation as well as about extended process control will be given in Chapters 7 and 8.

Organizational resilience depends in part on structure and command and control, and for another part it is linked to safety culture. This type of resilience depends on the competence and commitment of people involved, but most of all on the quality of leadership. Tools to measure the extent of organizational resilience are audits, interviews, and safety climate questionnaires, staff-training exercises supported, for example, by virtual reality playing events and assessing crisis management and emergency response results. Recording indicators on precursors and near-miss incidents and monitoring their follow-up investigation and intervention response may reveal trends. We shall in part address these issues in Chapter 6 discussing safety culture and performance indicators of the safety management system; for another part we shall return to resilience engineering aspects in Chapter 7.

It remains to assess effectiveness of early warning and administrative controls/procedures. Early warning depends heavily on detection and reliable interpretation of signals, formulating a clear message, and communicating this convincingly within an organization to the leadership. Reliability is important, as false alarms are effective killers of action on true alarms. Joël Luyk[25] identified four main groups of underlying factors affecting the ability of signal detection. In a negative sense these are: human factor with a cognitive bias not recognizing a signal as important, indifferent attitude toward risk and stress level; nonreceptive internal environment with its system technology, structure, culture, and strategy; lax external stakeholder engagement and external communication; and exogenous factors such as an overwhelming information input. It is clearly not a simple task to evaluate an organization on its ability of receiving and understanding weak signals. SWOT (Strengths, Weaknesses, Opportunities, and Threats) analysis and auditing appear to be the most effective ways.

Administrative controls/procedures shall go through a permanent feedback and correction process, as depicted in Chapter 3, Figure 3.4.

5.4 CONCLUSIONS

- In this chapter we briefly reviewed what challenges sociotechnical systems pose to maintaining safety. Following the pioneering work of Rasmussen, first the structure of a sociotechnical system was explained and the effects of complexity making its functioning opaque and intractable. Johansen and Rausand[9] explained the nature of complexity and its causes. Complexity is not a system property but a consequence of the limitations of the human mind.
- Next, Leveson's approach was explained to control the risks that a socio-technical system can present, and the progress offered by the STAMP accident model and STPA tool she developed. The latter tool enables grasping in

principle all possible scenarios including the category of dysfunctional component interactions that has escaped traditional risk assessment.

- Finally, we considered the advent of the concept of organizational resilience engineering in the world of safety with the initiative of Hollnagel, Woods, and Leveson in 2004. Weick and Sutcliffe explained what organizational resilience means, while Hollnagel et al. explained how resilience can help to improve safety. It was shown how resilience can be extended to the technical domain, and although evaluation tools are in part available, further development is needed.

REFERENCES

1. Turner BA, Pidgeon NF. Man-made-disasters. 2nd ed. Butterworth-Heinemann, Reed Elsevier; 1997, ISBN 0-7506-2087-0 [first edition in 1978 was by Turner, BA, with the sub-title: *The failure of foresight*].
2. Perrow C. Normal accidents: living with high-risk technologies (First published by Basic Books 1994). Princeton (NJ): Princeton University Press; 1999, ISBN 0-691-00412-9.
3. Rasmussen J. Risk and information processing. DK-4000 Roskilde (Denmark): Risø National Laboratory; August 1985. RISØ-M-2518.
4. Le Coze J-C. Reflecting on Jens Rasmussen's legacy. A strong program for a hard problem. Saf Sci 2015;71(Part B):123–41.
5. Perrow Ch. Fukushima and the inevitability of accidents. Bull At Sci 2011;67(6):44–52. http://dx.doi.org/10.1177/0096340211426395.
6. Rasmussen J, Svedung I. Proactive risk management in a dynamic society. 1st ed. Karlstad (Sweden): Swedish Rescue Services Agency; 2000, ISBN 91-7253-084-7. (The Swedish Rescue Services Agency is meanwhile renamed as Swedish Civil Contingencies Agency (MSB)).
7. Thurner S. Systemic financial risk. In: OECD reviews of risk management policies. OECD Publishing; 2012, ISBN 978-92-64-11272-8. http://dx.doi.org/10.1787/9789264167711-en.
8. Leveson NG. Engineering a safer world, systems thinking applied to safety. The MIT Press; 2011. 608 pp., ISBN-10:0–262-01662-1, ISBN-13:978-0-262-01662-9 [For the Walkerton case, see Appendix C, page 429].
9. Johansen IL, Rausand M. Defining complexity for risk assessment of sociotechnical systems: a conceptual framework. J Risk Reliab 2014;228(3):272–90.
10. Venkatasubramanian V. Systemic failures: challenges and opportunities in risk management in complex systems. AIChE J 2011;57:2–9.
11. Leveson NG. Applying systems thinking to analyze and learn from events. Saf Sci 2011; 49:55–64.
12. Khawaji IA. *Developing system-based leading indicators for proactive risk management in the chemical processing industry* [Master of Science Thesis]. Massachusetts Institute of Technology, 2012. http://sunnyday.mit.edu/safer-world/Khawaji-thesis.pdf.
13. Weick KE, Sutcliffe KM, Obstfeld D. Organizing for high reliability: processes of collective mindfulness. In: Sutton RS, Staw BM, editors. Research in organizational behavior, vol. 1. Stanford: Jai Press; 1999. p. 81–123.
14. Weick KE, Sutcliffe KM. Managing the unexpected, resilient performance in an age of uncertainty. 2nd ed. Jossey-Bass; 2007, ISBN 978-0-7879-9649-9.

15. Hollnagel E. FRAM: the functional resonance analysis method, modelling complex socio-technical systems. Aldershot (UK): Ashgate Publ. Ltd; 2012, ISBN 978-1-4094-4551-7.
16. Hollnagel E, Pruchnicki S, Woltjer R, Etcher S. Analysis of Comair flight 5191 with the functional resonance accident model. In: 8th Int'l symposium of the Australian aviation psychology association, Sydney, Australia; 2008. http://hal.archives-ouvertes.fr/docs/00/61/42/54/PDF/Hollnagel-et-al—FRAM-analysis-flight-5191.pdf.
17. Hollnagel E. Modelling of failures: from chains to coincidences, keynote symposium Budapest, 2007. http://www.resist-noe.org/DOC/Budapest/Keynote-Hollnagel.pdf.
18. Hollnagel E, Woods DD, Leveson N, editors. Resilience engineering, concepts and precepts. Aldershot (UK): Ashgate Publ. Ltd; 2006, ISBN 0-7546-4641-6.
19. Leveson NG, Dulac N. Safety and risk-driven design in complex systems-of-systems. In: 1st space exploration conference: continuing the voyage of discovery; Orlando, FL, USA; 30 January—1 February 2005; 2005. p. 1—25.
20. Gajdosz M, Bedford T, Howick S. Analysis of organizational factors for probabilistic risk analyses (PRAs). In: Steenbergen, et al., editors. ESREL 2013, safety, reliability and risk analysis: beyond the horizon. London: ©2014 Taylor & Francis Group; 2013, ISBN 978-1-138-00123-7. p. 317—25. Amsterdam, 29 September—2 October.
21. U.S. CSB. Catastrophic rupture of heat exchanger (seven fatalities). Tesoro Anacortes Refinery; 2 April 2010. Investigation Report 2010-08-I-WA, May 2014, www.csb.org.
22. Salmon PM, Cornelissen M, Trotter MJ. Systems-based accident analysis methods: a comparison of Accimap, HFACS, and STAMP. Saf Sci 2012;50:1158—70.
23. Steen R, Aven T. A risk perspective suitable for resilience engineering. Saf Sci 2011;49: 292—7.
24. Dinh LTT, Pasman HJ, Gao X, Mannan MS. Resilience engineering of industrial processes: principles and contributing factors. J Loss Prev Process Ind 2011;25:233—41.
25. Luyk J. *Towards improving detection of early warning signals within organizations, an approach to the identification and utilization of underlying factors from an organizational perspective* [Ph.D. dissertation]. The Netherlands: Beta Research School for Operations Management and Logistics, Eindhoven University of Technology; 2011, ISBN 978-90-386-2544-7.

CHAPTER

Human Factors, Safety Culture, Management Influences, Pressures, and More

Ah ne'er so dire a Thirst of Glory boast,
Nor in the Critick let the Man be lost!
Good-Nature and Good-Sense must ever join;
To err is Humane; to Forgive, Divine.
—Alexander Pope, An Essay on Criticism, Part II, 1711

SUMMARY

From the previous chapters, it is rather evident that human behavior, such as human action in response to assignments, and human decision making for achieving a goal, is a central issue in process safety and risk control. Therefore, it might have been even better if we would have started with this chapter on human aspects and then shown where it is leading with respect to process safety. But alas, it is also typical human weakness to first watch the consequences of action and then to care about causes. Much has been already written about the human performing in his/her team and operating a technical system within an organization. Knowledge about human factors, behavioral science, and ergonomics has enabled great advances in productivity and safety. Obviously, in process safety the main interest is not productivity but safe operation (although the reverse is often true, i.e., an operation performed safely also proves productive). Yet, task analysis and behavioral science have been applied to analyze what and how an error can occur. If something goes wrong, one of the first questions is, was it caused by human error or human failure? It appears to be a rather senseless question, because any accident can in the end be attributed to human shortcomings when one goes deep enough into the causes, although that may not always be immediately clear. Even a "spontaneous" fracture of a pressure vessel or the accidental onset of an unstoppable "runaway" reaction can be attributed to wrong decisions somewhere down the line, perhaps in the process or plant design phase. The exception is an event due to a real "unknown—unknown," but those are rare. Reason's Swiss cheese model (Chapter 3, Figure 3.2.) shows that the erroneous decision can have been made long before the accident occurs. The question, "what was the

Risk Analysis and Control for Industrial Processes - Gas, Oil and Chemicals.

error?" can evoke a large variety of answers: a wrong decision by an estimation error, a calculated risk, a moment of distraction, tiredness, a memory lapse, a violation of a rule, an incorrect model, a calculation error, a wrong assumption, a miscommunication, a misunderstanding, a misperception, et cetera.

At the lowest steps of the culture ladder (Chapter 3, Figure 3.9), the first question of management after an accident is often, who made the error, who is to blame? In a sociotechnical system it is the deadliest question to ask for the survival of a good safety culture. It has been demonstrated over and over again that management, or rather leadership, creates the atmosphere, the safety climate in an organization and hence the safety attitude of its workers. Indeed, the safety management system (SMS) in combination with an indicator monitoring system has been conceived to assure that conditions for safe work are right, but it is not sufficient. Safety culture and measures to maintain a high level of it are equally important. Safety culture will include a "just culture" in which a good balance exists between attributing to individuals unnecessary blame and at the same time imposing disciplinary measures when unacceptable behavior occurs. Indeed, in stepping up the culture ladder, safety thinking must become "second nature." This shall be in the first place with the top decision maker in the enterprise and this shall be "beyond regulatory compliance" (to quote mottos used by the Mary Kay O'Connor Process Safety Center of Texas A&M University)! Apart from other board activities, during the last decade proactive risk management in general has become an important agenda item. How can risks be determined and risk control be realized? This includes safety risks.

Human factors (HFs) and ergonomics concern understanding and taking account in design of human capability limitations to achieve optimal human–system interaction. Knowledge of human failure, its taxonomy, and likelihood in different conditions and ways to prevent failure is important in all stages of engineering and processing and at all levels of the hierarchy. It will be most helpful in training and writing procedures, and it will be useful input for design decisions with respect to, for example, tasks, the work environment and what to automate. We shall therefore begin by briefly reviewing various aspects of the broad concept of "human factor" with its implications for design in a wide sense, as well as its relation with occupational safety and health (OSH) and the attempts to achieve adequate influence and control on HF requirements and efficacy in practice. Next we discuss occupational risk modeling, which has made great progress in the last decade. We then will review past work on human error with the aim to obtain human action reliability figures. It is one of the lacking factors in detailed scenarios of risk assessments. Much about human error has to be learnt from past accidents. Human error models are numerous and their bases can be rather different, so various generations of models are distinguished. This turns out to be a complex topic.

Next, Erik Hollnagel's vision on decision making will be summarized. Hollnagel is known for his work as a psychologist in the field of human–machine interaction and safety. Decision making is crucial at all levels of the sociotechnical hierarchy, but when one analyzes how humans make decisions, it is not all that systematic and thought out. Recent work by Hollnagel culminated in his efficiency-thoroughness trade-off (ETTO) principle and human variability in decision making that turns much upside down again.

In the late 1990s, rather suddenly safety culture became a buzz word in the process industry, although at least a decade earlier the lack of it had been recognized as contributing significantly to accident causation. The essays of Dov Zohar, James Reason, and others on safety culture as a derivative of organizational culture have had much effect on safety thinking. We shall learn the distinction between safety culture, safety climate, and safety attitude and how to measure and influence these.

If there is something going wrong then quite often the management level has been failing. Leadership is crucial, and not only have we become aware of it, but we shall see initiatives to educate those in command of the organizations to such an extent that they at least can ask the right process safety questions before decisions are made. Note that most often these individuals are nontechnical people. We shall also encounter some new features as risk registers, the ISO 55000 standard, and involvement of top management—in other words, aspects of corporate governance for effective process safety. Attention will be given to rules and procedures, compliance, and work discipline. However, in the fervor to reach a state of zero accidents, the mistake can be made to avoid complacency by more rules and more frequently checking off items at screen displays or paper. This may easily become counterproductive. An atmosphere of safety thinking and trust on the work floor, freedom to take action and improvise if necessary, motivates. It also might be more productive being addressed without arousing emotion when a colleague points to one's error, and the right to stop the work if feeling unsafe. However, everything has its limits and keeping the right balance is management's challenge.

We shall conclude the chapter with a brief history of the evolution of key performance indicators (KPIs) adapted to process safety. Explanations will be given of why KPIs for human factors (HFs) and organizational safety culture indicators help so much, how the industry looks at them, and some practicalities of their introduction. How we can further use performance indicators as weak risk signals once we collect them, besides just watching trends, will be a topic in Chapter 7.

6.1 HUMAN FACTORS AND OCCUPATIONAL SAFETY AND HEALTH

Process safety and risk management must give HFs much attention and must devote much effort to treating HFs with great care. HFs is a much broader concept than human errors. In the limited sense of ergonomics it is a multidisciplinary field of study of designing equipment and devices optimized for the human physical and cognitive abilities. Hence, adaptation of the machine with its screen display to human senses, body measures, and time needed to make decisions (See Figure 6.1 for the reverse situation of human adaptation to the machine). It can be interpreted even in a still wider sense as human factors engineering (HFE), which comprises besides the

FIGURE 6.1

In the old days, an operator had to be knowledgeable and also athletic!

above, task analysis, writing procedures, design of personal protection equipment including protocols for use, training aids, simulators, communication, enhancing team work, and system ergonomics, here all with reference to safety. System ergonomics in man–machine interaction has a strong cognitive accent, as a human operator must make the right decisions based on information generated from information technology systems and offered to him by the system in the right colors and lighting for optimum observation. Hence, achieving a good state of affairs with HFs is a matter of optimizing demands and requirements on the one hand against human capacities and system performance characteristics on the other, and so minimizing human error. Good examples can also be found in design and layout of equipment such that effective and safe maintenance can be performed. Sound procedures for work permits (e.g., for hot work) and confined space entry are other important items. One distinguishes even organizational ergonomics as optimizing the socio-technical system for sustainable high performance. To support work on methods to optimize workplace safety and health many countries have dedicated institutes such as NIOSH (National Institute of Occupational Safety and Health) in the US.

Viewed from the negative, inadequacies in HFs lead to errors and unsafe acts, which can result in injury or death, both by acute or by chronic exposure, and to loss of productivity and efficiency. Therefore, HFs are also heavily connected to OSH, entailing an impressive body of relevant regulation and (risk-based) standards on amongst others machinery safety. There is also a large volume of literature and text-books available on OSH topics and perhaps even more material, at least in volume, from consultants. A known book by Daniel Della-Giustina[1] contains chapters such as Hazard Communication and Hazardous Materials Handling, Job Safety Programs, Safety Committees, Lockout/Tagout, Confined Space Entry, Personal Protective Equipment, Occupational Noise and Ventilation, Bloodborne Pathogen Standards, and chapters on emergency response issues. The last few years there have been quite a few laboratory accidents. *CRC Handbook on Laboratory Safety* by A. Keith Furr[2] contains a wealth of information that would help to prevent these. For process industry practice in preventing human error API 770[3] provides a useful guide.

A historical review of the interesting evolution of OSH is available from Paul Swuste et al.[4,5] In the 1920s−1940s (the time of Herbert W. Heinrich), recounted in these reviews, the worker was blamed for causing the accident. In contrast to this opinion, in the present day "just culture" came up. Just culture was a phrase first coined by James Reason[6] as one of the components of safety culture (to be treated in Section 6.5). Pushed forward more recently by amongst others Sydney Dekker,[7] just culture attempts to propagate the creation of justice inside an organization by learning from accidents and fair accountability. It endeavors to build an atmosphere of trust within an organization so that safety-related information is not withheld or swept under the carpet ("Keep your mouth shut"). Criminalization and civil liability of an unsafe act (resulting from error, negligence, recklessness, or violation) may keep a person from providing information and is a concern, while personal culpa-bility in organizational accidents shall only be apportioned with care. However, the borderline between acceptable and nonacceptable behavior must always be clear. Just culture applied to aviation and health-care organizations is well described in respective guidelines.[8,9]

In explaining the value of just culture, John Bond[10] summarized Reason's distinction of three potential accident causation types, resulting from HF inade-quacies as

- the *personal factor*: the main emphasis here is on inadequate capability, lack of knowledge, lack of skill, stress and improper motivation, fatigue, sleep deprivation;
- the *workplace factor*: this includes inadequate supervision, engineering, purchasing, maintenance, inadequate training, and work standards;
- the *organization factor*: this views human error more as a consequence rather than a cause and is indicative of latent inadequacies in the leadership or management system.

Around the year 2000, a consortium guided by the European Process Safety Centre accomplished a European Union (EU) project called PRISM[11] dedicated to the HF.

The project was reported at the 11th International Symposium on Loss Prevention and Safety Promotion in 2004 and covered the following topics spread over 16 papers:

- organizational and cultural issues, such as team working, and *behavior-based safety*;
- optimizing human performance: task design, procedures, ergonomics, man-machine, and human–computer interface;
- HFs in high demand situations: diagnosis of process upsets, cognitive (alarm) overload, emergency response, control room layout, *abnormal situation management*;
- HFs as part of the engineering design process: an application guide.

Behavior-based safety is an approach to reduce unsafe behavior in the working place. It is a process to help perform routine tasks safely by, for example, behavioral observation, intervention, and feedback, and further by setting improvement goals, training and education, motivation, and coaching. UK's Health and Safety Executive (HSE) issued a useful guide.[12] A separate activity but with the same goal is the issuing of the Napo safety promotion videos,[13] prepared by a consortium supported by the European Commission and distributed by the EU-OSHA. Napo is the name of the hero in the cartoons and is a normal, willing worker who finds himself in all kinds of situations. Hopkins[14] points, though, to the weakness of safe behavior programs, as these tend to shift causation to the first link in the chain—the worker (the sharp end)—and not to the last one—management (the blunt end), where for the sake of general prevention more gain can be obtained. Also, observations may be biased and attitude may be more important than observed behavior.

Abnormal situation management is the operator response to deviation from the normal course of a process, such as unexpected process parameter excursions, runaways, leaks with emissions possibly followed by explosion or fire, unplanned shutdowns, et cetera. Besides technical causes, it has quite a few HF aspects. We shall return to this topic in Chapter 8, where we shall also discuss the activities of the Abnormal Situation Management Consortium led by Honeywell in the US, which since 1994 has been working on research to prevent and mitigate.

As the PRISM papers show, there are many detailed recommendations to be made with regard to HFs. UK's HSE issued two very practical HF advisory notes: the HSE Human Factor Briefing Notes[15] and the Inspectors' Toolkit on Human Factors[16] in the management of major accident hazards at lower-tier COMAH sites. These contain suggestions for questions to be asked by inspectors on the various HF aspects. Further, Bridges and Tew[17] provided an overview of how HFs are addressed in the US OSHA Process Safety Management standard, CCPS Risk Based Process Safety guideline, and the Responsible Care® recommendations. They published a list of practical HF items that were missing in the OSHA PSM standard. More recently, following the 2005 BP Texas City refinery accident, for personnel in refining and petrochemical industries in the US, the API Recommended Practice 755 on Fatigue Risk Management has been introduced. (In 2005 at the Texas City refinery shifts worked 12 h.)

In the 2005 time frame, another initiative of HSE's Hazardous Installations Directorate led to the effort of Bellamy, Geyer, and Wilkinson[18] to develop a functional model integrating HFs, SMS, and the organization. The ultimate goal was to find in practice patterns identifying weaknesses that could initiate major accident scenarios. Hence, the identification would generate warning signals. This work was specifically aimed at safety in the process industries and to support inspectors and companies. The idea has been to break down the three separate areas into a taxonomy of smaller components (they number nearly a thousand) and use these as building elements to reconstruct an integrated model that can serve during both audits and inspections. The product is what the authors called an archetypal warning triangle, of which an example is reproduced here in Figure 6.2, and a sector-specific taxonomy to be used for defining safety constraints. Four warning triangles themes were proposed: Understanding Major Accident Prevention; Competence for Tasks; Priorities, attention, and conflict resolution; and Assurance. The latter means making sure operations are safe by monitoring and evaluating. The four triangles can be combined to a regular tetrahedron (also called triangular pyramid). The triangles can be made sector specific. Case histories of eight major accidents that had occurred in various countries were used to check how warning signs were ignored. Inputs are issues of concern/interest; outputs from the taxonomy are "definitions of meaningful integrations in the knowledge space" that can act as guide to a user. The user himself must specify criteria and should have his own knowledge resources. The paper announced that before validation of the proposed model in practice, a workshop would be held.

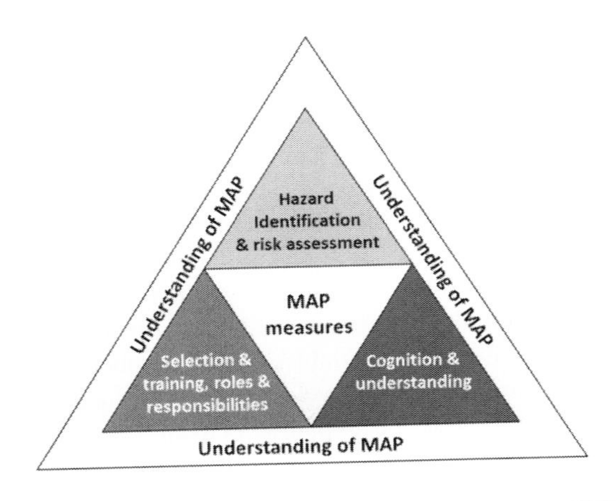

FIGURE 6.2

Example of warning triangles embedded in a large one representing the theme, as developed by Bellamy, Geyer, and Wilkinson.[18] The triangle here is for Understanding Major Accident Prevention (MAP), within the center what is necessary in MAP for Risk Control, above what is relevant from the SMS, below left from organization, and to the right from HFs.

In line with the international standards ISO 9001 quality management system requirements, and ISO 14001, Environmental Management System, American ANSI Z10-2012 (secretariat at the American Society of Safety Engineers, ASSE), and similar ones in other fields, standard OHSAS 18001 (Occupational Health and Safety Assessment Series) provides requirements for an occupational safety and health management system. The standard is widely accepted internationally, although it is not an ISO standard. It gives an organization guidance on how to build, maintain, and improve a management system that is directed to minimize hazards and risks to employees and to others working for the organization. It is all based on Deming's Plan—Do—Check—Act management cycle. Among others, the standard requires compliance with applicable law. By working with an independent, unbiased evaluator, implementation of the standard can be certified and the organization registered. Based on the standard, an extensive guide to managing for health and safety for all involved, leaders to workers and safety staff, is given in HSE's publication[19] HSG 65. The guide also addresses the issue of contractors' health and safety in outsourced work. The EU's project Total Operations Management for Safety Critical Activities (TOSCA) running from 2013 to 2016 intends to integrate the various management systems.

6.2 OCCUPATIONAL RISK MODELING

If an effort is made to collect large quantities of data and make the database publicly available, much can be achieved to improve personal safety. This was demonstrated in a project run by the Dutch Ministry of Social Affairs and Employment (Min SZW), among others responsible for occupational safety. The project is described by Ale et al.[20] and Linda Bellamy et al.[21] and formally reported by RIVM,[22] the Dutch National Institute for Public Health and the Environment. The project aim was quantification of occupational risk, and the project team took off in 2003 with the development of the Workplace Occupational Risk Model (WORM). Data on "slips, trips, and falls" had already been gathered from 1998 onward in the database GISAI. An occupational accident is reportable if it results in a serious physical or mental injury or death within one year. The accident is serious if the victim has to be hospitalized within 24 h and for at least 24 h or when the injury is permanent but hospitalization is not required. Reporting must be done within 24 h. At the start of the project about 20,000 cases were available, and the number continually increases.

The first part of the project was to develop a list of accident types, for example, "fall from height." Next is to describe all pathways to an accident type by means of a bow-tie structure (as in Chapter 3, Section 3.5.2 and illustrated here in Figure 6.3) with the accident type in the center and the various possible consequence events to the right, while to the left incidental factors (conditions) are added such as wet, slippery surface. The exercise resulted in 64 bow-ties. Additionally, 60,000 worker exposure data (among others Danish data) were collected enabling statistics on determining accident frequency data (these are point averages; there are no

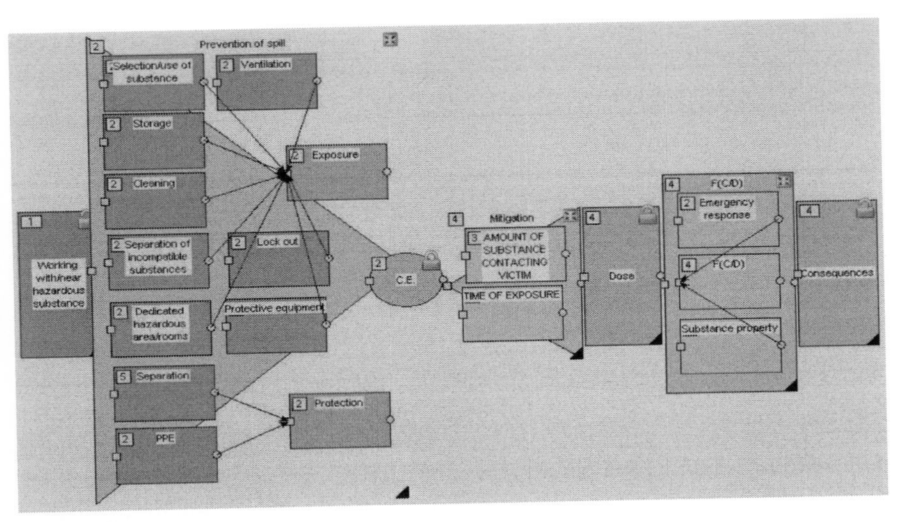

FIGURE 6.3

Example bow-tie of working with a chemical (here without loss of containment), with far left the work activity, left preventive barriers that can be breached connecting to the center event of the injury occurrence, and to the right variables determining consequences.[20]

distributions yet). The data set contained more than 40,000 barrier breaches, which could also be analyzed statistically. Three main types of barriers are distinguished: safeguarding, operational control, and danger zone control, which are elaborated in a later paper[23] in an attempt to extract effective leading indicators. Barrier effectiveness is affected by probability influencing entities. All this was built into a quantified occupational risk model (ORM). Regarding barriers and other measures, a cost model is included in terms of capital and operational expenditures, thus enabling optimization against a set risk or a cost level. In real life a worker is not exposed to a single hazard but to a multitude, therefore ORM contains a composite model calculating the risk of exposure to multiple hazards either simultaneously or sequentially. ORM software has been used to derive an occupational risk calculator (ORCA), which is publicly available as a webORCA for an employer to assess his situation. The plan is further to expand ORM with a health impact model for chronic occupational diseases.

Following the idea of Heinrich's accident pyramid, Bellamy[24] succeeded in using the bow-tie models too for relating per type of accident injury frequencies with much less frequent fatal accidents and even with rare major hazard ones. So, the more frequent injury accidents would as a kind of "precursor" provide a warning for the relatively rare, more lethal mishaps. In the next chapter we shall see this kind of signaling effect back in other forms and being used for quantitative prediction of the major hazard frequency. Bellamy et al.[25] also used the accident data, webORCA, and exposure data to calculate worker risk rates for different sectors of industry. They obtained a risk profile of, for example, the top 10 accident types, for example, in construction with "fall from height" as highest. Such a profile was called a "horoscope," and it enables setting best risk reduction priorities. Due to the relatively large numbers, these rates have predictive power.

6.3 METHODS TO ASSESS HUMAN ERROR, OR RATHER HUMAN RELIABILITY

A primary concern of having any hazardous activity done safely is avoiding errors. So, it is about human reliability and the human failure event (HFE; note that this acronym is also used for human factor engineering). As we saw earlier in Chapter 3 (see Figure 3.2), James Reason classified errors into *slips* (of the mind; an error in executing an intention), *lapses* (of the memory), *mistakes* (intention was right, but the decision/action was in error because of lack of knowledge or bad procedure), and *violation of rule* (routine or exceptional). In Denmark at Risø during the 1970s—1980s, from reliability perspective and with the aim of failure prediction, Jens Rasmussen[26] developed the *skill—rule—knowledge model*, the SRK model, with different decision time intervals and error mechanisms at the three levels of decision tasks.

The SRK model, as schematically shown in Figure 6.4, is in the context of an operator and his mental model interacting with a highly automated technical

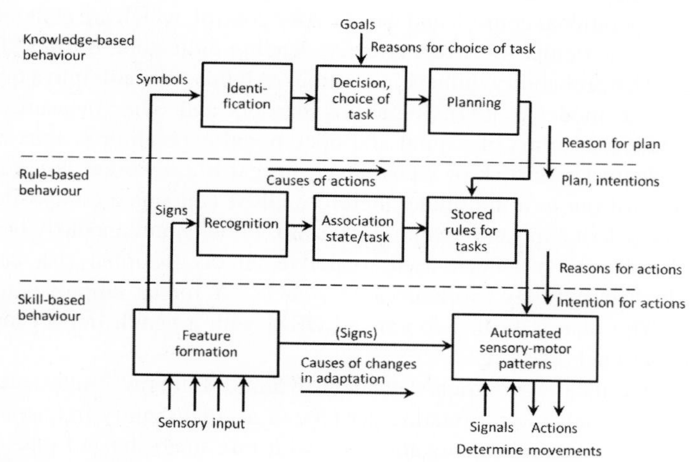

FIGURE 6.4

The three levels at which operator task decisions are performed according to the skill—rule—knowledge-based model proposed by Jens Rasmussen.[26] At skill level, action is taken without conscious attention or control; the difference with rule based is not always clear. When confronted with an unusual process upset, the operator working at knowledge-based level has first to make a plan.

installation, hence a man—machine interaction. In qualitative terms, skill can be described as reacting to a pattern that is known as abnormal and requires action, without conscious attention or control. While at the level of rule-based behavior an operator recognizes the state the process is in, and selects from stored rules for tasks (procedures/"cookbook") the appropriate one(s) for an action, upon which he gets feedback. At the level of knowledge-based behavior, the operator encounters an unfamiliar situation, hence he has to develop a plan based on his mental model of the process in view of the goal to be achieved, while getting feedback on the effects of his action on the goal. Rasmussen draws a parallel of SRK with signals, signs, and symbols. The level distinction provides cues for causes and error classification, making Reason to consider slips to be at a skill-based level, while mistakes can be either at the rule- or knowledge-based levels.

Of course, many examples of knowledge-based errors can be found. A striking one appears from the report of the US Chemical Safety and Hazard Investigation Board[27] about the 2005 Texas City refinery vapor cloud explosion. This explosion occurred in the so-called ISOM plant after overfilling a distillation column (raffinate splitter). Apart from many other errors and deficiencies such as a series of miscommunications, a clear knowledge-based error can be observed. At start-up, according to procedure the operator should fill the bottom of the column to a 50% reading of the level transmitter and then start pumping and heating the reboiler. The level concerns only the bottom part of the column to achieve a well-running distillation process. However, routinely they filled the column higher (routine violation) because otherwise during continuation of start-up the level may drop too much. The tower side glass was dirty at the inside and could not be read. The operator who had taken over without having obtained instruction from the previous shift continued filling while he had closed the level control valve. From the two high level alarms one had triggered at 72% during the previous shift, and the other did not sound even at 99% of the level transmitter range. The operator was further not aware that the instrument reading was at fault. At his start, the operator observed a level of 97% full. While filling the tower for a further 3 h the indicator level even gradually dropped to 73%. He did not realize that based on elapsed time or flow rate measurements, he was severely overfilling and the liquid level came near the top of the tower. Other observations, such as increasing pressure, he tried to explain based on a wrong mental image of what occurred inside the installation. Indeed, inconsistence in observations did not occur to him. A knowledge-based error occurs if the operator lacks process safety expertise and makes a wrong diagnosis of an abnormal state. Operators in such situations, trying to correct a deviation, in fact, often make it worse as was the case here.

In the US during the late 1990s, Wiegmann and Shappell[28] expanded Reason's idea for the purpose of aviation safety to a Human Factors Analysis and Classification System (HFACS), which proved useful as framework of human error causes in accident investigation. HFACS is now generally applied in human performance analysis. Twenty years earlier, also in the US, where the nuclear community was, and still is, much engaged in human reliability, classification is different. There one distinguishes *errors of commission* (actively doing something wrong), *errors of omission* (failing to do something), and *cognitive task errors*, such as making an incorrect diagnosis or taking an unsubstantiated decision.

Knowledge by analysis of human reliability would help in many ways, for example, for drafting better procedures, for designing better man—machine interactive workstations, for designing better risk controls, and for performing better risk assessments. The literature on human reliability analysis (HRA) is vast and the field is rather complex. The UK HSE[29] has listed and discussed 35 qualitative and quantitative models from 72 that have been identified.

The oldest quantitative model, Technique for Human Error Rate Prediction (THERP), which originated in the American nuclear power plant community in the late 1970s—early 1980s, considered man as a machine component and gave it a reliability figure. Developers and describers were Swain and Guttmann[31]; the manual provides a listing of human error probability values (HEP = number of errors occurred/number of opportunities for error) that are the result of task decomposition (see for an example Figure 6.5) and error estimation based on data. Local influences were accounted for by plant-specific performance shaping factors (PSFs). It all hinges on historical data, extensive, rather resource intensive task decomposition event trees, and HRA specialists. THERP is a first-generation preprocessed model focusing on skill- and rule-based operator action but not taking into

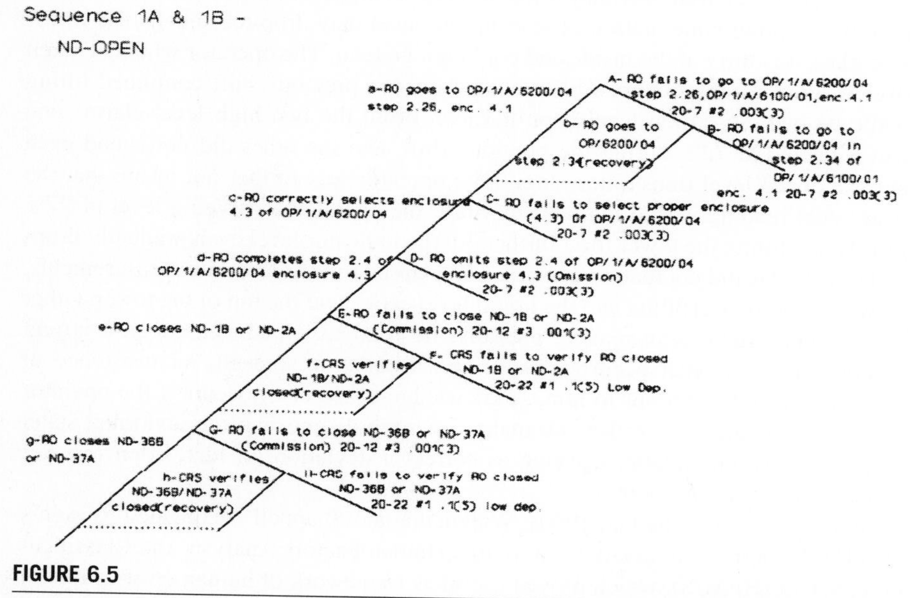

FIGURE 6.5

Example HRA task event tree, illustrating THERP decomposition into subtasks of failure to close nuclear reactor residual heat removal system suction isolation valves.

After Kelly.[30]

account cognition and organizational context. In a later stage, time–reliability correlations (TRCs) were added of the type we have seen in Section 3.6.6.3 discussing operator response to a rare initiating event—the shorter the available decision time, the larger the probability for human error. These TRCs in the case of THERP are expert judgment based, while later models were based on simulator data. There also exist first generation models that are fully based on expert judgment.

Because of skepticism toward human error modeling, in 1995 HSE in the UK sponsored a project led and reported by Barry Kirwan[32] to compare and validate three first-generation models and to determine their predictive accuracy and precision. The models were THERP, HEART, and JHEDI. Human Error Assessment and Reduction Technique (HEART) was developed by Williams,[33] 1985; Justification of Human Error Data Information (JHEDI) was a proprietary model developed by Kirwan, 1990. HEART was the only one that has been developed for application in engineering and process industry risk assessments. JHEDI specifically addresses nuclear reprocessing operations. As described by Kirwan,[32] HEART is simpler to use than THERP. It is based on a database of nominal errors and distinguishes eight generic task categories (and a ninth being none of the others) and a wide range of 32 error producing conditions (EPCs) such as task unfamiliarity and time pressure. The task categories are described in Table 6.1. The EPCs are multipliers ranked in decreasing order, where, for example, a shortage of time available for error detection and correction multiplies unreliability with an order of magnitude and operator inexperience with a factor of three. The lowest 10 or more EPCs affect HEP values only slightly. For multiple influences, a composed factor must be derived according to assessed weight factors and a prescribed formula. (A rather successful derivative of HEART in use by the nuclear community for less-detailed assessment is Simplified Plant Analysis Risk Human Reliability Assessment (SPAR-H).

In JHEDI for determining an error factor, first one of the nine in Table 6.1 mentioned task categories is selected in which the actual task fits. For each category a dozen or more error descriptors exist from which a choice must be made. Then, PSFs questions (time available, place of action, frequency, et cetera) have to be answered, and then finally a computer returns HEPs, which for composed tasks through fault tree cut sets yield human performance limiting values.

Further objectives of the project of Kirwan were to test for consistency of usage, such as how to improve model prediction and what error reduction measures could be based on HRA. The project was meant to apply to all high-risk industries but focused on human error in the nuclear power and reprocessing industries. Errors were mainly slips and lapses in skill- and rule-based tasks and not so much cognitive (knowledge-based) mistakes such as misdiagnosis. A validation occurred against human error data from the Computerized Operators Reliability and Error Data Base (CORE-DATA). The first output from the project was the analysis of all the data: 30 assessors each determined 30 HEPs. HEP examples are operator sets an incorrect calibration pressure, HEP $= 0.03$; or gasket not fitted correctly, HEP $= 0.05$, et cetera. Kirwan et al.[32] show a significant correlation (at $p < 0.01$) between estimates and their corresponding true values. Most errors produce conservative, hence

Table 6.1 HEART's Generic Task Categories and Corresponding Proposed Nominal Human Unreliability (Incl. 5th–95th Percentile Bounds) of Williams[33] and Nine JHEDI Categories of Kirwan[32]

	HEART Task Category	Unreliability, HEP (90% Conf. Interval)		JHEDI Action Category
A	Totally unfamiliar, performed at speed with no real idea of likely consequences.	0.55 (0.35–0.97)	1	Response to alarms (no diagnosis required)
B	Shift or restore system to a new or original state on a single attempt without supervision or procedures.	0.26 (0.14–0.42)	2	Response to alarms (diagnosis required)
C	Complex task requiring high level of comprehension and skill.	0.16 (0.12–0.28)	3	Checking and monitoring
D	Fairly simple task performed rapidly or given scant attention.	0.09 (0.06–0.13)	4	Control actions
E	Routine, highly practiced, rapid task involving relatively low level of skill.	0.02 (0.007–0.045)	5	VDU (visual display unit) input actions and calculations
F	Restore or shift a system to original or new state following procedures, with some checking.	0.003 (0.008–0.007)	6	Sampling and analyzing
G	Completely familiar, well-designed, highly practiced, routine task occurring several times per hour, performed to highest possible standards by highly motivated, highly trained and experienced person, totally aware of implications of failure, with time to correct potential error, but without the benefit of significant job aids.	0.0004 (0.00008–0.009)	7	Calibration and maintenance
H	Respond correctly to system command even when there is an augmented or automated supervisory system providing accurate interpretation of system stage.	0.00002 (0.00000–0.0009)	8	Communications
			9	Evacuation (not used by any subject throughout the validation)

pessimistic results. JHEDI tended to perform slightly better than the other two methods, which can be explained by its bias toward nuclear reprocessing. Most results (60–87%, or 72% on average) are accurate within a factor of 10, which was judged reasonable, but there are outliers. Weaknesses and differences in consistency of the three techniques were analyzed. The conclusion is that no method convincingly outperformed the others. Recommendations for improvements were made.

The second generation, US-developed model, A Technique for Human Event Analysis (ATHEANA), attempts to make up for lack of cognitive error in the first generation and identifies "error forcing conditions" that could lead to accidents. The main reason for development is to model operator action and also because advances in behavioral science and psychology enabled model improvements. ATHEANA also includes more realism by representing spread in results (aleatory uncertainty, defined in Chapter 7).

In 2006 Forester et al.[34] reported to the US Nuclear Regulatory Commission a comparative study of the performance of eight American HRA models, including THERP and ATHEANA. As regards quantification of HEPs, three methods were distinguished: (1) adjusting basic HEPs by a list of influencing factors and TRCs based on expert judgment; (2) applying a context-defined set of factors and more expert judgment; and (3) using information derived from plant simulator output of nuclear power plant accidents. None of the methods emerged as being the best in all contexts; all had their specific strengths and weaknesses. An example of a simulator used is the one in the OECD Halden Reactor Project in Norway in an exercise in 2009. Interestingly, the latter study showed a relatively large variability in performance of experienced crews coping with a simulated accident.[29]

More recently, based on the PhD work of Katrina Groth, Groth and Mosleh[35] proposed an improvement by giving degrading PSFs such as, for example, stress, a much needed, better foundation. Depending on conditions, PSFs can be upgrading as well. PSFs, also called performance influencing factors (PIFs), are used in many models. Groth and Mosleh developed a standard set of PIFs. For that they started with an information-decision-action cognitive model and a human events database (HERA) of recorded operator actions. After a laborious effort, the result is a hierarchical set of PIFs, collapsible according to need, which can serve a range of HRA purposes. We shall see the way it can be used in Chapter 7.

Published results of successful use of the techniques described above in risk analysis studies for the process industry are very limited. We already mentioned Williams's HEART method[33] as having received attention. Robert Taylor,[36] as Rasmussen and Hollnagel originally from Risø, performed quantitative risk analysis of process plants in various stages of their life cycle.

For more than 30 years, Taylor collected relatively reliable data on failure rates. One of the categories was human error data. In the years 1978–1980 the *Action Error Analysis* method was developed, which combines HAZOP with the SRK human error model of Rasmussen.[26] Application of the model does not require HRA specialists, but it can be applied by project engineers carrying out the risk analysis. The method follows a written procedure and the same documentation as HAZOP:

Table 6.2 Standard Error Modes for Action Error Analysis According to Taylor[36]

Omission	In Wrong Sequence	Wrong Materials
Too early/too late	Repetition	Wrong tool
Too fast/too slow	Wrong object	Wrong value
Too much/too little	Wrong substance	Wrong action
Too hard/too slight	Extraneous action (unrelated to the task but interfering with it)	
In wrong direction	Other similar wrong choices for each aspect of the task step, e.g., wrong label placed on a product package	

a piping and instrumentation diagram (P&ID) and cause-and-effect matrices for control description. Error modes are as listed in Table 6.2. While passing through the P&ID at each HAZOP node, consequences of erroneous actions are estimated and, when necessary, measures are proposed. Supervisory activity, such as alarms checking, can be added as additional nodes. It is important to include latent failures. They can be tracked by detailing the event sequence upon an action. If error causes can be determined, it will be possible to take preventive measures. For finding error causes, Rasmussen's SRK model was used to derive a checklist that was validated against case histories. Quantification in the form of the failure frequencies required for QRA requires historical data of errors made and the conditions under which these errors occurred. Due to the low frequencies of error and the varying conditions, data collecting requires either a vast cooperating network or decades of time. Taylor[37] applied Kirwan's[32] "anchor point" method using the observed frequency ratio of two types of errors made under similar conditions. If one error frequency can be established, that of the other one derives from the ratio. So far, Taylor has succeeded in determining ca. 60 anchor points and in quantifying more than 200 typical accident types.

In order to identify specifically latent failures, Kožuh and Peklenik[38] developed similar to Taylor a HAZOP-like method applying guide words specific for this purpose. They showed the usefulness of the method with a case study based on a master logic diagram instead of a P&ID.

A different approach by Noroozi et al.[39] following Kirwan's[32] application of the Success Likelihood Index Method on expert judgment, derived HEPs for risk assessment of detailed activities in offshore maintenance. These showed a spread between lower- and upper-bound values of at least three orders of magnitude.

Erik Hollnagel, whose work we encountered in the previous chapter when treating resilience engineering, earlier in his career also developed a second-generation human error model, Cognitive Reliability and Error Analysis Method (CREAM), explained in the HSE review.[30] In a note[40] about the method, he mentions how technology changed the character of work from much "doing" to much "thinking" and knowledge intensive tasks so that conventional ergonomics had to change into cognitive ergonomics. Second-generation HRA models address cognitive ergonomics by identifying those tasks/actions affected by cognitive reliability, conditions in which this reliability may be reduced, whether this has a consequence for system safety, and designing measures to reduce the risk. In 2005, Hollnagel[41] expressed in clear arguments serious doubts about the foundation of HRA, which, he argues, as an input to QRA will be of limited value. In a 2012 disclaimer,[40] he distances himself from CREAM, as he latterly became convinced of larger variability of human performance than expressed in the model, rather than the human circumstances in the work environment or context, which cannot be well specified, cause the main part of the variability. A system approach may be the answer, but not without considering the resilience of the entire sociotechnical system, although for the time being this too requires a better definition.

Despite Hollnagel's disclaimer, work on and with the CREAM model continues with satisfaction of the researchers involved.[42,43] The model distinguishes competence and control, and classifies "genotypes" (causes) and "phenotypes" (manifestations; error modes). The former has three categories: the first is linked to behavior, for example, emotional state and personality; the second to man—machine interface/interaction (MMI); and the third to organization and environment (light, noise, temperature). Phenotypes can be distinguished by action at the wrong time, of the wrong type, on the wrong object, or in the wrong place. The genotypes and phenotypes are further classified into general ones (consequents) with each again in general antecedents, and the latter in further specific antecedents. An example mentioned is the genotype "communication," the consequent "communication failure," the general antecedent "distraction," and the specific one "noise." Prior to an HRA, a task analysis is made and the levels of nine common performance conditions (CPCs) are estimated. CPCs are adequacy of organization, working conditions, MMI and operational support, availability of procedures/plans, number of simultaneous goals, available time, time of day, adequacy of training and experience, and crew collaboration quality.

Following HSE's review,[30] as a subconclusion it can be stated that although there is not much uniformity in terminology and validation databases are rather specific, with quite a few models reasonable successes have been obtained. Some models behaved well when applied even in different branches of industry. Yet, with the complexity in human decision making in general as Hollnagel describes it and treated in the next section, for sure human error modeling will undergo further development.

6.4 HUMAN MECHANISMS FOR DECISION MAKING AND THE ETTO PRINCIPLE

Hollnagel[44] in proposing his ETTO principle of people deciding and acting according to an *efficiency-thoroughness trade-off* gave a summary of the different views on how decision making is performed. In general there is a time to think (evaluate; select) and a time to do (execute), but these are often mixed together, and there is at least some overlap zone. He cites Kahneman and Tversky[45] who distinguish two rational stages: *elimination by aspect* and *prospecting*, meaning selection among risky alternatives. However, brain experts are convinced that rationality comes usually after the decision already has been taken subconsciously. In decision making, people normally rely on ETTOs by applying *heuristics*, which are experience-based techniques for problem solving. Hollnagel mentions the following collection:

- *Similarity matching/frequency gambling* (think it is the same thing and just take a gamble on it);
- *Representativeness*: in case of uncertainty, resort to how it looks from the outside, compare it with a known example, and make adjustments;
- *Focus gambling*: (opportunistically changing one hypothesis for another)—*conservative focusing* (slowly and incrementally building a hypothesis)—*simultaneous scanning* (trying out several hypotheses at the same time);
- *Satisficing*: (contraction of satisfying–sufficing) emanating from inability to be rational resulting from limited human cognitive capacity, also to be explained as *sacrificing* due to intractability of work environment or to difficulty of coping with complexity;
- *Muddling through*: defining principal objectives, outlining a few alternatives that pop up, select one that is a reasonable compromise between means and values (Lindblom).
- *Recognition primed decisions* or naturalistic decision making (initiated at a workshop in Dayton, Ohio, 1989, and described by G. A. Klein and others): constraints imposed by time pressure, ambiguous information, ill-defined goals, changing priorities; relevant in cases, for example, of firefighters or surgeons in rapidly changing situations. Decision on action is made based on comparison with generic situation, otherwise reject and think of new possibility.
- *Schemata*: mental representation of some aspect of the world, recognizing a situation. (Era of humans as info processors; *It looks like Y, so it is Y*). It may lead to tragic errors.

At least in the first three, the uncertainty can be modeled by probabilistic methods as we shall see later in Chapters 7 and 9. We have further to reckon with: *speed-accuracy trade-off*; *time–reliability correlation* (TRC); and *information input overload*. Coping with the load can cause escape by omission, reduced precision, queuing, filtering, cutting categories, and decentralization. However, in safety-critical situations, accuracy is more important than speed.

In summary, ETTO-ing is a cause of human performance variability producing both positive and disastrous results. Sources of ETTO-ing for humans and organization are distinct: Humans are subject to physiologically and psychologically induced variability due to physical or mental fatigue. We dislike monotony, and variability may appear as ingenuity/creativity fluctuations or trying to conserve resources. Socially induced variability emanates from trying to meet expectations or complying with work standards. Organizationally induced variability arises from meeting demands, stretching resources, and resolving ambiguity. Contextually induced variability can arise from ambient working conditions such as noise, humidity, temperature, vibration, et cetera.

Another classification of performance variability (mixed organizational and human) is by differentiating according to level:

- Teleological (regarding final causes): goals are unstable due to external factors (supply or demand). This is variability on strategic level;
- Contextual/situational: seeking best possible outcome, which is tactical level;
- Compensatory/incidental: when something is missing (tool or resource), for example, procedure cannot be remembered. This is variability on operational level.

Adjustments of tasks and activities can be approximate and incomplete due to lack of information and/or time. Drift in the organization may lead to success but also to failure.

Risk assessors must understand the nature of and measure the variability of normal performance and use it to identify conditions that can be expected to lead to both positive and adverse outcomes (hence, this is another plea for the necessity to replace point value outcomes by distributions). Some aspects of variability should be predictable, but for others that will not be the case. (Certainly not at prima facie, but as we shall see in Chapter 7, Section 7.7.4 with just some observations and a Bayesian approach much can be achieved.) There will be differences in ETTO at the sharp end (those involved in operations) and at the blunt end (e.g., designers with lack of information and work dispersed). One should find out what conditions lead to ETTO-ing and what extent of ETTO-ing may lead to variability of multiple functions, which then by combining in unexpected ways lead to disproportionally large consequences. Numerous smaller ETTOs of various team members in less easily tractable systems can lead to disaster. The latter statement of Hollnagel looks exactly like what happened in the case of the Deepwater Horizon demise, as seen in Chapter 4. The same tragic event led Hopkins[46] to comment on the contributing human and organizational factors playing a significant role. He tried to find out how the decision makers perceived their own situation and how their organizational culture influenced them to think and act as they did. Moreover, this comment concerned a large corporation of which one may expect to have sufficient capability and knowledge to perform a self-reflection about their state of culture—but what about the small companies that have to toil to keep afloat? The next section will dwell on culture, how to characterize and measure it, and also how to influence it.

6.5 SAFETY CULTURE, SAFETY CLIMATE, SAFETY ATTITUDE

Culture is an abstract concept, a construct, with different aspects, and it expresses itself in learned practices, in the way people forming a group behave, things they like and dislike, dress codes, shared values, unified objectives, and many others. Safety culture is a derivation of organizational culture for which Frank Guldenmund[47] presented an extensive overview in his dissertation. It starts with Edgar Schein's 1980s layered model of organizational culture with a *core of basic assumptions* (true values setting priorities), around it a *layer of* "espoused values" (an organization's stated values and rules of behavior) and on top, or as an *outer layer, artifacts.* These artifacts are all visible and tangible elements of a culture such as posters, slogans, and dress codes. Culture is relatively stable, it is shared by members of a group that agree on what they think is important and how they believe things should be done, and it is also functional in that it provides a "frame of reference for behavior." It is the product of externally and internally influenced processes in which a leader sets the preferences: "*The way we do things around here*," a concise definition used by many scholars in the field.

Safety culture as a concept emerged in the International Nuclear Safety Advisory Group (INSAG) of the International Atomic Energy Agency (IAEA) Chernobyl accident reports of 1986 and in 1991. Since 2002–2003, the number of publications on the topics in the heading of this section has markedly increased. Earlier, "climate" was used where one would now write "culture." Over the years the concept of organizational climate, and hence also *safety climate*, narrowed slightly to the manifestations of the culture of an organization. Culture expresses itself through a climate. So, if one wants to characterize the culture, one "measures" the climate. This is largely done by psychometric questionnaire surveys. We shall return to this below, because measuring is a first step to diagnosis and, if necessary, improvement. Lack of good safety culture leads to noncompliance with safety rules.

Guldenmund[47] distinguishes an analytic (psychological), a pragmatic (engineering), and an academic (anthropological) approach to the study of occupational safety culture. His review about *safety attitude* states that there are processes preceding its formation. Attitude sprouts from the basic core, which then expresses itself in hardware as safety measures and protection equipment, in software as safety procedures, training and knowledge, attitudes toward people at the various levels in an organization, and finally, toward behavior including responsibility, safe working, skepticism, and communication about safety.

Dov Zohar[48] (Technion, Haifa) is an analytic pioneer in the field. His publications go back to 1980. Although we all intuitively know what safety culture means, and when entering a facility we sense almost immediately its local level, it is hard to explain exactly what it is. There is one thing Zohar is certain about—it is the leadership that determines it. Various paraphernalia (artifacts) such as "safety first" signs or the advice "to hold the handrail when going up the stairs" do not reflect much about the true state of affairs in an organization. But, the core of culture is hard to fathom.

Zohar also notes that there are competing demands in an organization between safety and economic continuity of the enterprise, the latter depending on healthy financial conditions as turnover, cash flow, profit margin, share value, et cetera. This competition translates into conflicting internal demands at the level of "artifacts" between safety and cost of measures, and externally at the level of "espoused values" in conflicts with legal responsibility and social norms. The former expresses itself in "misalignments," where there is much superficial talk about safety but few inspections and audits, or mitigation is preferred over prevention. Maintenance is neglected, and large differences in safety performance of similar units are accepted. The latter, the so-called "espousal-enactment gaps," one can observe as talk about safety but overtime and overload are enforced, production goals are more concretely specified than safety goals, and there is talk about process safety but one measures only personal safety as manifested in the "safety first" signs. (On the contrary, throughout this book it appears that investment in safety is a strong contributor to long-term economic continuity and viability of a company.) By surveying employees' opinions about the priority supervisors/superiors really place on safety, aspects of safety climate can be quantified.

Dov Zohar has led the development of safety climate measurement (or safety climate scale) since 1980 and developed Zohar's Safety Climate Questionnaire (ZSCQ). Regarding questionnaires on safety climate, Stephen Johnson[49] of the University of Phoenix, following Zohar, notes that these have evolved as a tool, because of dissatisfaction with the traditional ways of trying to explain safety performance variations in terms of engineering solutions or by HFs, for example, attitudes, behavior, or compliance. In fact, Johnson refers to Hale and Hovden,[50] who turned for explanation and prediction to the so-called "third age of safety research." In this third age organizational constructs such as culture, safety climate, organizational commitment, and leadership were tested for predictive capability. In the first age, all the focus was on engineering, enforcement, and training, while in the second age, behavioral sciences were called in to make predictions. According to Johnson, of the third age organizational constructs, safety climate appeared to be the easiest and most successful contender to explain changes in safety performance because questionnaires can produce quantitative results. Moreover, the outcomes appeared to have good capability in predicting safety trends.

After some modification of the definition of safety climate over the years, Zohar concluded, as we have seen above, that safety climate is the employees' shared perception of the priority that an organization, or rather their direct supervisors, place on safety and hence reward. A measurement should uncover the relative priority of safety compared with competing demands of productivity and efficiency, the gaps between words and deeds, and possible inconsistencies in policies and procedures from the top and how these are put into practice by supervisors. While today this kind of measurement is common practice, as described in the EU-OSHA Culture Assessment document,[51] in the following only an illustrative example will be shown of a study Stephen Johnson[49] performed. The objective of the study was to establish to what degree safety climate could predict, for example, injury rates. A sample of 292 employees of a heavy-manufacturing industry spread over three locations filled the ZSCQ reproduced in Table 6.3.

Analysis of variance on the outcomes produced results that showed significant differences with respect to employees' Time in Area (6 months−1 year, or 1 year or longer) and Time at the Company (less than 2 years and longer than 5 years, the seasoned ones), while age and gender did not make a difference. After further statistical treatment of the data and application of confirmatory factor analysis, a

Table 6.3 Zohar's 16-item Safety Climate Survey (ZSCQ)[49]

My Direct Supervisor
1. Makes sure we receive all the equipment needed to do the job safely.
2. Frequently checks to see if we are all obeying the safety rules.
3. Discusses how to improve safety with us.
4. Uses explanations (not just compliance) to get us to act safely.
5. Emphasizes safety procedures when we are working under pressure.
6. Frequently tells us about the hazards in our work.
7. Refuses to ignore safety rules when work falls behind schedule.
8. Is strict about working safely when we are tired or stressed.
9. Reminds workers who need reminders to work safely.
10. Makes sure we follow all the safety rules (not just the most important ones).
11. Insists that we obey safety rules when fixing equipment or machines.
12. Says a "good word" to workers who pay special attention to safety.
13. Is strict about safety at the end of the shift, when we want to go home.
14. Spends time helping us learn to see problems before they arise.
15. Frequently talks about safety issues throughout the work week.
16. Insists we wear our protective equipment even if it is uncomfortable.

All statements rated on a scale from 1 (strongly disagree) to 7 (strongly agree).

three-factor model of caring, compliance, and coaching appeared to fit the data well, although due to mutual correlation of the three elements, a one-factor model would fit too. However, the three-factor model was found to be more descriptive. In addition, a direct relationship between safety climate as a predictor and safe behavior could be observed. The safety climate predicted injury frequency trends too, although this conclusion was less evident for the case investigated. Morrow et al.[52] used surveys to "measure" safety culture and reported some success in correlating the results with safety performance. (As mentioned, Zohar has doubts whether a survey exposes the deepest core of culture.)

These results seem to be rather sound, but the field is still developing and not free of fuzziness. Zohar, for example, in Zohar and Luria,[53] added a multilevel model approach to safety climate to grasp more effectively the differences in an organization that may arise in implementing in various groups or suborganizations policies and procedures emanating from the top. For example, consider an order that during preparation sustained a delay but still has to be readied and delivered to a customer on time. In spite of procedures a supervisor can allow certain shortcuts to be taken. If this occurs more frequently, it will undermine the safety climate, at least locally in that unit. An alternative would be to authorize the shortcut in this unique situation but to introduce some other new activity or action so that, despite the shortcut, safety performance is not undermined or degraded. Their investigation in this direction on a two-level example has shown promising results. Extension to multilevels for larger organizations is an obvious goal, and the need for further research, also with respect to the relation between leadership and climate, is clear.

The Shell-sponsored development in the 1990s of the Tripod Delta tool has been successful in measuring on a comparative basis safety-related organizational performance. As explained in Chapter 3, this tool was derived from Reason's idea of the Swiss Cheese. Tripod refers to the three elements: general failure types, unsafe acts, and incidents causing losses. Tripod Delta[54] is finding out about latent failures that could enable or promote unsafe acts by asking 275 questions randomly selected from a set of 1500 ones to team members concerning performance of 11 basic risk factors. Results are evaluated against those stored in a database of replies. Basic risk factors are designs leading to errors, hardware (quality and availability of tools and equipment), maintenance management, housekeeping, error enforcing conditions, procedures, training, communication, incompatible goals, organizational deficiencies and defenses (weaknesses in detection, warnings, protection, recovery, containment escape, and rescue). The method is proprietary.

The EU-OSHA Culture Assessment document mentioned above provides an overview of the various safety climate measuring tools, which meanwhile have become operational. It also provides a list of pragmatic safety culture maturity assessment and improvement tools that can provide an image of the culture due to the organizational triangle coherence represented by Figure 6.7. A nonexhaustive list of those follows

- the Hearts & Minds program/tool kit, developed for Shell Exploration and Production by Leiden and Manchester universities and available from the Energy Institute in the UK, with the culture ladder shown in Chapter 3, Figure 3.9;
- the Safety Culture Maturity Model of the Keil Centre in Edinburgh, UK, as with the Shell program developed from its original application in offshore oil and gas safety;
- the Safety Culture Indicator Scale Measurement System, developed for the Federal Aviation Agency (FAA) by the University of Illinois.

Most popular is the Hearts & Minds program.[55]

- The program starts with understanding the present culture in the company, or in other words, on which step of the HSE culture ladder are we currently standing?
- The second step is "seeing yourself," in which a questionnaire shall be used for answers to the question of how others see you as a manager. The questionnaire has four chapters: Walk the talk, Informedness, Trust, and Priorities.
- The third step is "Making change last," which as a task is similar to quitting smoking and maintaining the abstinence. The stage model of change provides a framework for dealing with this.
- Subsequently, as a fourth step a simple risk matrix shall be made by the employees to assess the HSE risks in the area where their work occurs.
- The fifth step is learning to apply the traffic light stopping rule, also called the "Rule of Three" (Patrick Hudson et al.[56]). Suppose an activity will start up or runs as it should be, no problem in sight, the light is *green* and it is a "go." If a condition appears to clearly surpass a threshold of risk, the light turns *red*, and to proceed is definitely a "no go." But what about *amber* or *orange*, the cases in which there is doubt, although not enough to stop the activity? The rule now

states that with *one* amber or orange condition in an operation, one can proceed normally; with *two*, one shall proceed with caution; but when simultaneously amber or orange conditions or factors occur *three* times, it becomes a "no go." So, for example, when an important tool is missing, the assistant is there but rather sick, and it looks as if a thunderstorm is coming up, then stop the operation.

- The sixth step is "Managing rule breaking";
- the seventh step is "Improving supervision" and leadership styles;
- the eighth step is "Work Safely" (look, speak, and listen), and finally the ninth step is about (car) "Driving for Excellence" as a sign of taking safety seriously and making it second nature.

Patrick Hudson[57] (Emeritus Professor Human Factor in Safety, Delft University of Technology, the Netherlands) reported his experiences in implementing the program. The background about the development and the link of safety management system and safety culture at Shell can be found in Hudson.[58]

The Keil Centre's Safety Culture Maturity® model[59] is inspired by a software development model and is a stepwise ascent of five levels of maturity, from "safety is not perceived as a business risk" all the way to "the prevention of all injuries or harm to employees is a core value of the company." It defined 10 elements (reproduced from their Web site): visible management commitment, safety communication, productivity versus safety, learning organization, supervision, health and safety resources, participation in safety, risk-taking behavior, contractor management, and competency. Before participating in training, a company should have its technical safety in order and it should be compliant with regulation.

Dov Zohar has his own method to improve a culture because in his opinion only the informal daily conversations between employee and supervisor can reveal what is in the core (see Figure 6.6), and for change one has to adjust the core. In case leadership invites him to investigate and improve if necessary, he performs first a safety

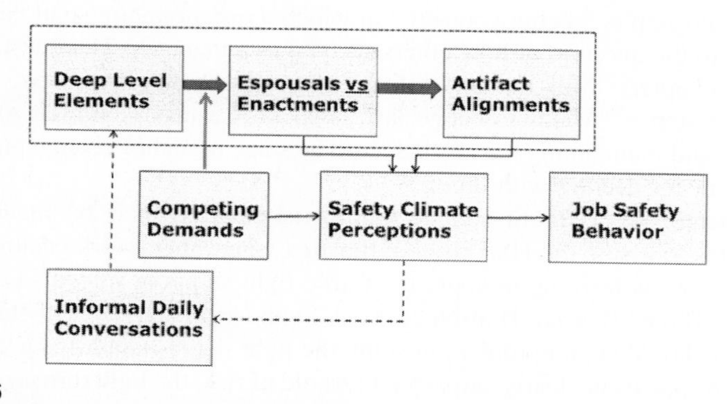

FIGURE 6.6

Culture—Climate—Behavior model.

After Zohar, Lecture at MKOPSC, Texas A&M University, October 2010.

climate measurement. Then, for improvement he calls a meeting of all employees, including management. At the meeting he announces that each employee can expect at random to be called, and he or she will be asked by Zohar or one of his coworkers to provide details about their last conversation with their supervisor/superior. After collecting all the interview results and interpreting the "subtexts," he returns to the employees, but now in their role of supervisors/superiors, and based on his expertise shows them what are the thorny points or what went wrong and how to improve. In addition, an action checklist for the organization is developed in view of its strategic goals and objectives. Repeated safety climate measurement will reveal progress made.

Since OSHA in the US for industry in general, and the new Bureau of Safety and Environmental Enforcement (BSEE) for oil and gas exploration and production, started emphasizing the need for a high level of safety culture, companies in the US are looking for ways to improve. As we have seen in the 10 elements of the Keil Centre, after management makes its commitment clear, one needs safety communication. For personal safety, tools are available, but for process safety one is at a loss. In the past, process safety was not everywhere seen as relevant for their operations. Just as in the fourth step of Hearts and Minds in which a risk matrix has to be set up for the area for which one is responsible, Megan Weichel[60] of DNV US underlines the importance of being aware of the existing risks. One must be able to name them and talk about them in a common language. One needs risk "ownership," which can be obtained by implementing asset-specific "risk registers," and beside the nature of the risk and various particulars, registering the name of the "owner." This brings identification of risk to the people who are in the field closest to the assets and improves attention in inspection and maintenance. By aggregating registers upward to corporate level and integrating them with other business risk tools, the top management will obtain better insight. This will enable improved enterprise-wide risk management. At the same time, the frontline will note a keen interest of the "top brass" in hazards and risks, which will build credibility and stimulate the culture. Presumably, risk registers fit well with the new ISO 55000 standard explained in the next section.

6.6 ORGANIZATIONAL HIERARCHY, MANAGEMENT DILEMMAS AND RULES

Much has been written about organizations, their structures, and the theories behind them. This is not the place to go into any depth. As we have seen in the previous chapter, as part of a sociotechnical system in any organization there is a hierarchy of levels each with their specific responsibilities, tasks, and authority, and connected to each other by communication lines. The latter shall be equally effective in all directions, internally within a level and also between adjacent levels. From the previous section in this chapter it should be clear that in an organization there are performance-determining relationships between the "processes" going on, its "organizational culture," and its "structures and behaviors" as laid down by its

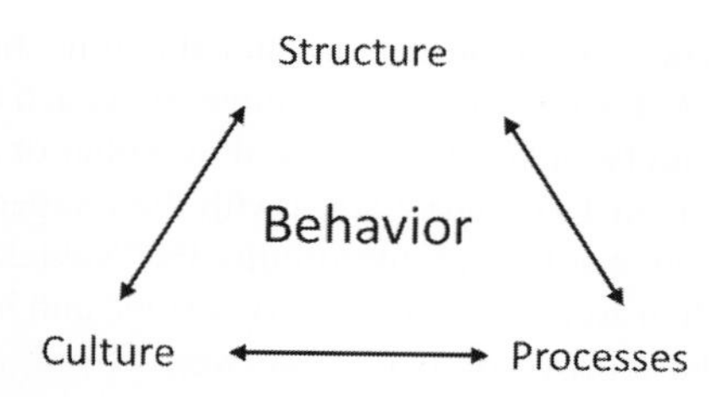

FIGURE 6.7

Organizational triangle of which many different representations are found. This particular one with "behavior" in the center is adapted from Guldenmund,[47] who referred to the older paper of Van Hoewijk.[61]

management system and recorded by performance indicators. These relations form the so-called organizational triangle as shown in Figure 6.7.

Rasmussen and Svedung[62] distinguish between the *functional* work organization with content communication and the *social* role configuration. The latter depends on the management style and the form of communication and social interaction. Allocation of work to individuals or teams occurs based on norms and practices, on load sharing, on functional decoupling to minimize exchanges, on competence, information access, and on safety and reliability when staff are exposed to hazards. These criteria can compete with one another. Rasmussen and Svedung quote as an example of how well an organization can function, the one of the well-known Rochlin et al.[63] study, namely the high reliability organization (HRO) of a US aircraft carrier crew. Given the freedom the crew forms a complex, dynamic structure of overlapping networks with different management modes well geared to perform critical tasks. Further below, the new Norwegian concept of integrated operation (IO) in the oil and gas industry will be mentioned, which has some similar characteristics, but first, attention will be paid to the lowest organizational cell, the team.

The team has been the subject of many studies with regard to leadership, differences in types of team members, and optimal composition. In a study by Skjerve et al.[64] of the Norwegian Institute for Energy Technology at Halden, teamwork competence requirements for nuclear power plant control rooms have been determined by actual surveys and group interviews of 20 American operators. For this study, nine teamwork dimensions have been defined: attitudes toward colleagues and the plant, backup behavior, communication, coordination, decision making, leadership, learning and refreshing of competencies, personality fits, and situation awareness. All of these have been provided with positive and negative teamwork quality indicators. It turned out that determination of required competencies in the normal operational state is not sufficient. In addition, abnormal situations (e.g., an emergency or incident) must be included.

In another study, Skjerve et al.[65] address effectiveness of collaboration between groups onshore and offshore for the Norwegian oil industry in the traditional way and applying the relatively new concept of IO through a so-called distributed collaboration structure with work arenas. The differences in the ways of working are

Table 6.4 Ways of Working in Traditional Operation and Integrated Operation (IO)[65]

Traditional Way of Working	IO Way of Working
Serial	Parallel
Single discipline	Multidiscipline
Dependence of physical location	Independence of physical location
Decisions are made based on historical data	Decisions are made based on real-time data
Reactive	Proactive
Continuous relationships with teammates	More fragmented relationships with teammates
Collaboration will imply a higher degree of informal exchange	Collaboration will be more formal
Lower degree of technology-mediated teamwork	Higher degree of technology-mediated teamwork

explained in Table 6.4. On first sight, such an arrangement puts higher requirements on the functioning of the people involved in the network. Collaboration must be resilient to maintain a high safety level. For this, effective communication among partners in a network, who even do not need to know each other, will be crucial. To be successful, the collaboration must be high quality and assessment criteria are being worked out, for which further research is needed.

In the same sphere, Albrechtsen et al.[66] argue that IO will improve proactive safety and risk management by real-time data capture, communication infrastructure and information access, visualization of risks by combining a risk picture with risk indicators, and taking account of uncertainties. (In the next chapters, we shall see elements of this return again). Further, the IO mindset and operational capability will enable better interpretation and decision making in the work arena, while finally a higher sensitization to unknowns will contribute. It can be expected that the IO model and its operational capability resulting from new ways of organizing work processes will have an impact on the traditional main- and subcontractor collaboration and on outsourcing to service companies.

6.6.1 MANAGEMENT DILEMMAS

In this chapter it is further of interest to make explicit how managements under pressure in the stress field of competition and with vision blurred by complexity are inclined to give priority to short-term time and effort gains, despite that such a trend may introduce larger risk. Safer operations may in the longer term be more cost effective, but short-term gains are of course tempting. Many aspects have already been described, but we shall focus in the last sections of this chapter on management influence on an organization's propensity. First, a simple straightforward example is presented.

Sonnemans et al.[68] investigated the causation of recurring process disruptions/deviations/disturbances/defects, which are safety critical or at least have a negative influence on safety, and can be considered as "precursor" signals. All of these can be attributed to some HF inadequacy in an organization's hierarchy as depicted in Figure 6.8 by either overlooking the disruption, deciding it is not important, not acting upon it despite action that was agreed to, or not giving it any attention from management. As one of the many examples a leaking reactor valve was found but the operational level did not correct it, as the tactical level had no eye for problems of maintenance because top management had moved the engineers away from the plant to the remote main office. It is also known from the conclusions of many other studies that in the end the majority of failures can be attributed to underlying management failures.

One can ask what drives and influences management. This will be in the first place continuity of the enterprise, which puts a multitude of requirements on them and exposure to many risks, in which safety and within that process safety are only one facet. There are many worries: a positive result this year; the outlook for next year; the competition (e.g., in China or the Middle East); quality and delivery time; the shareholders meeting next month; the bonuses, the accountants. Costs of personnel, maintenance, and energy go up next year! We may need to reorganize: leaner and meaner! And, oh yes, there is the environment and health and safety. But, we have specialists in the latter areas and the number of accidents decreases each year. Thus, the temptation is, "let's take that for granted!" Therefore, it is clear that without a safety program with effective leadership that can balance the short- and long-term issues and anticipate the future, there is an inherent bias to limit the scope and focus on present and short-term issues. However, it is the right balance that in the end should be successful.

Expectedly, the new ISO 55000 Standard for Asset Management (AM) will attract the attention of management to risks and thus will add to the weight for tipping the balance to the positive side. In fact, the ISO standard concerns a "trilogy": ISO 55000 provides overview, concepts, and terminology; ISO 55001 specifies requirements for good AM in the asset management system (AMS) and ISO 55002 provides interpretation and implementation guidance for an AMS. Just as for the other management systems mentioned at the end of Section 6.2, it complies with the ISO 72 Management Standard and it follows the plan–do–check–act model. Also, it is important to note that ISO 55000 is risk based. It means classifying the assets, identifying and analyzing risks during the life cycle (for which the risk registers mentioned in Section 6.5 will be a useful input), developing and setting a risk control strategy, and putting performance measurements in place with KPIs so as to be able to strive for continuous improvement. Similar risk-based activities are required for maintenance and other asset integrity measures as part of a system approach. Integrity in this context encompasses the whole of leadership/management, design, equipment, operations, and information/data.

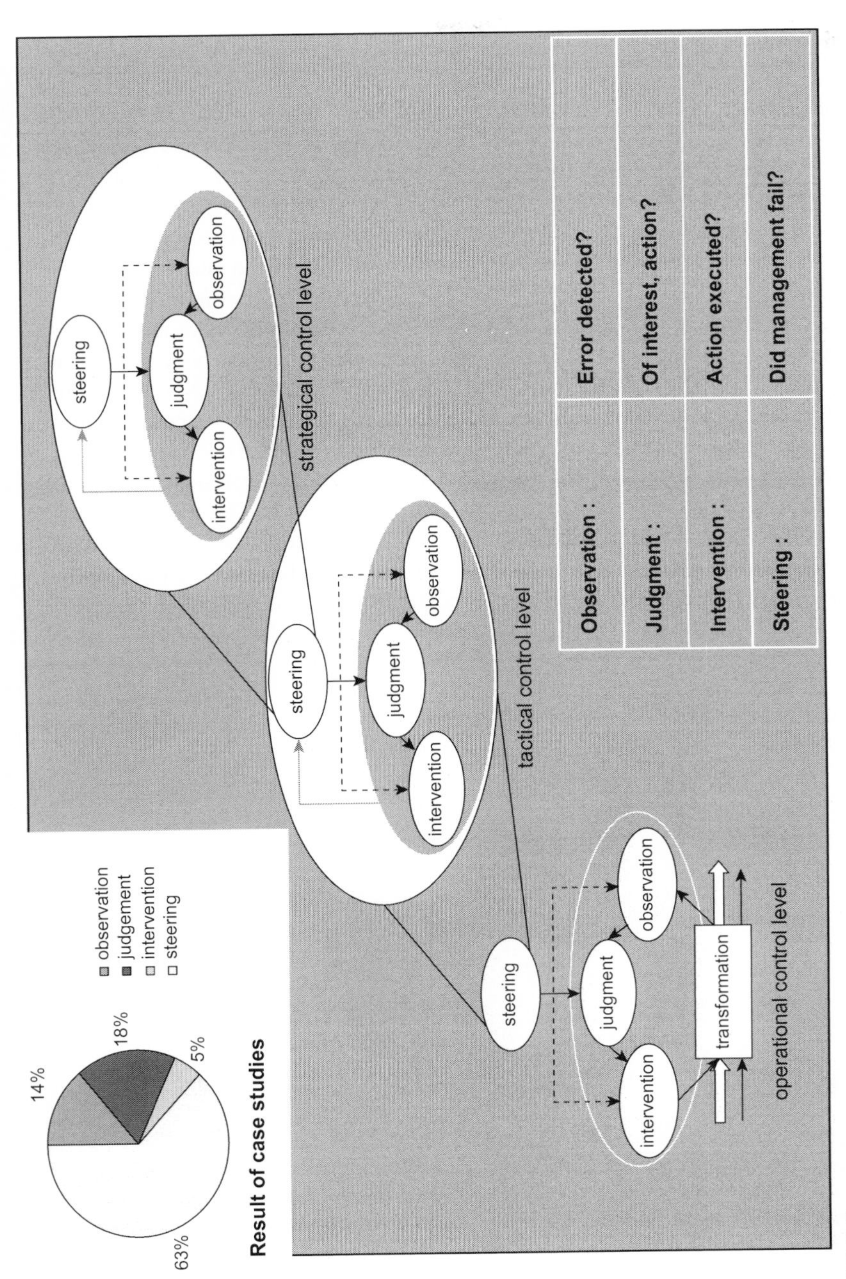

FIGURE 6.8

Failures at different levels in an organization of not detecting a disruption, not considering it important, not executing agreed action, or not steering the activity from a higher level (*after Sonnemans et al.*[68] *based on Körvers's PhD dissertation*[67]) showing as a result of the case studies performed that 63% of failures found is in the steering and hence is due to management failure.

Another weight on the balance that is even more specifically directed to process safety is the June 2012 Paris meeting organized by the OECD Environment, Health and Safety, Chemical Accidents Program that focused on process industry CEO participation. In preparation for this, in Brussels in January of that year, an earlier meeting of global industrial chiefs of safety had been organized (see Section 6.7 below) to discuss safety performance indicators, while also an OECD Guidance for Senior Leaders in High Hazard Industries document[69] had been drafted. This Guidance document started with a description of the business case for effective safety management. Leaders must recognize that major accidents are credible business risks. As part of anticipating the future, they also must recognize the integrated nature of many major hazard businesses and the risk of supply chain disruption. Management of process risks should have the same focus as for financial governance, markets, decisions on investments, et cetera. The document mentions high reliability organizations. Two characteristics are (1) a feeling of permanent unease, that is, an absence of complacency and a realization that there are risks in our operation that can manifest themselves today, and (2) detection of, and response to, weak signals. In other words, put the trigger level for intervention at a low level and accept that as a result there will be more false alarms. Boards also have to be aware of the regulatory requirements placed on them in the countries where they operate. The document depicts the essential elements for governance of process safety as in Figure 6.9. The document also contains a checklist for self-assessment, it refers to

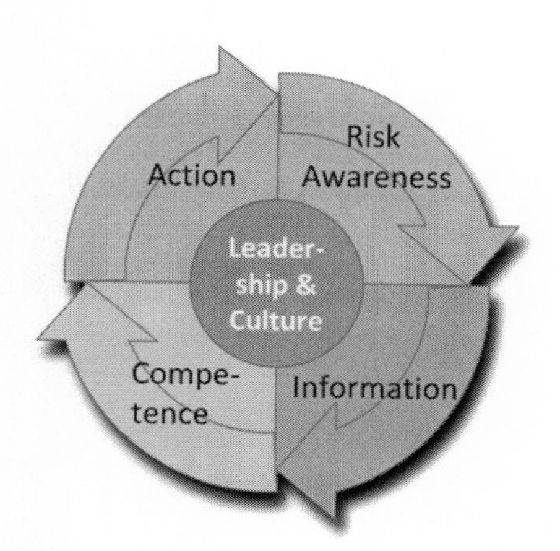

FIGURE 6.9

Essential elements of Corporate Governance for Process Safety, according to Guidance for Senior Leaders in High Hazard Industries, presented to the meeting organized by the OECD Environment, Health and Safety, Chemical Accidents Program, in Paris, June 2012.

several Web sites and to the "The Business Case for Process Safety" from the Center for Chemical Process Safety, AIChE in New York.

The document also contains a few quotations from CEOs of large chemical companies stressing the importance of safety for the industry, for example, one by Kurt Bock, CEO of BASF: "For us in the Chemical Industry, safety is key for our 'licence to operate.' At BASF, one of our core values is 'We never compromise on safety.' Process Safety is of particular importance, because of the severe consequences of major incidents. Through strong process safety performance we protect our employees and neighbors, our environment, and our reputation and our business success. We have implemented—and are further strengthening—strong programs to reduce process safety risks, ranging from safety conscious plant design to excellence in safe plant operation."

According to Judith Hackitt,[70] chair of UK's HSE and president of IChemE in the UK, and who previously held positions in industry and in Cefic, the European Chemical Industry Association, leadership commitment in the high-hazard industry is crucial. Causes of complacency are the assumption that problems are fixed, remoteness from the process, and too much faith in technology. Simultaneously, lessons from the past get lost and a culture develops that seeks reassurance rather than assurance, hence more leaving it to insurance than actively auditing, inspecting, and monitoring indicators that processes and operations are safe. The IChemE is setting up board member training in process safety in the UK, following the Corporate Governance for Process Safety guidance by OECD, with the purpose that board members must learn what questions to ask when projects are initiated or other key decisions must be made.

An interesting interview with Rex Tillerson, CEO of ExxonMobil, conducted by Zain Shauk on April 5, 2013, just before Tillerson was to receive a safety award from the National Safety Council in the US, was published in FuelFix.[71] In the interview, which also covered cyber security, human behavior, and a pipeline spill in Arkansas, Tillerson explained his motivation for safety, which was among others due to his personal experience with a fatality in an operation. He further explained that training people, giving them rules and procedures is not enough. Risk management ought to follow a holistic systems approach, identifying risks and preparing people with what they have to deal with, if something goes wrong. It is important that "intervention" is accepted in the culture, that is, when somebody notices something unsafe that a colleague is doing, he can mention that to the colleague without risking a harsh response. Indeed, that is exactly what ExxonMobil expects him to do. People sometimes make wrong decisions, although they know they take a greater risk, but they may perceive that as "it will not happen to me."

Another dilemma is human capital management. According to a report of AXA Investment Managers,[72] the Oil and Gas industry is going to lose due to retirement a large number of very skilled and experienced employees. This is true in general for the process industry. Because of the low interest in technology in the 1990s, salary level, recruitment policy, and the fact that fully mature professionals take 10–15 years to develop, a gap is arising[a]. The need is exacerbated by increasing complexity of conducting operations safely. We shall return to this in Chapter 11.

6.6.2 "SAFETY OBJECTIVE TREES"

In complex plants at remote places under harsh weather conditions, managing barrier failures is rather essential for maintaining safe operation (think of the Macondo drilling catastrophe described in Chapter 4). Robin Pitblado and William Nelson of DNV[73] proposed a quick and practical top-down procedure to support staff under pressure in barrier management. (Barrier, a variety of which is shown in Table 3.7, is a much broader applied concept than protection layer or safety instrumented system). After reiterating the need of including human and organizational aspects in risk assessments, they mentioned a methodology used in the nuclear industry called "safety objective trees" grafted on fault tree structure, here to be replaced by bow-tie enabling barriers to clearly appear. If barrier effectiveness degrades/gives way, as it always does, the overview contains important instructions: at top level the safety objective, for example, containment integrity of a specific content; at second level the critical control functions, for example, pressure control; at the third level the types of challenges that could threaten the function, for example, process upset; at the fourth level the types of mechanism causing the challenges, for example, defect valve; and at the fifth level the response strategy or "success path," for example, activate emergency cooling. The method will be tested for offshore application in connection with the Norwegian Integrated Operations (IO) Centre mentioned above. Key is also knowledge of barrier status. In practice this is a rather difficult problem because of the large number of barriers on a site and ways to detect possible degraded states casting doubt on barrier effectiveness. The authors recommend using incidents and applying their accident investigation tool specialized on barrier functioning, BSCAT, to acquire barrier health status information. In addition, analysis by applying Bayesian network (Chapter 7) helps.

6.6.3 RULES AND PROCEDURES

Indeed, rules and procedures are not enough, but an organization cannot do without those either. Andrew Hale and David Borys[74] made an extensive analysis underpinned with many references about following safety rules and procedures. At the start they noted the ambivalence, or what they called the Janus faces of following rules. Working exactly to the rules paralyzes—to express it in an extreme way.

[a]At the Mary Kay O'Connor Process Safety Center and elsewhere, graduated engineers with a Master of Science degree (hence five years of university education) to become educated in the concepts need four years for obtaining a PhD in process safety and risk management. They then still have to mature in the practice of operations and project management.

They focused on rules and procedures for those working at primary hazardous processes, they distinguished categories of rules (performance goal, process, and action rules) and quoted Sydney Dekker's[75] two rule models. Because of brevity of the rendering here we may oversimplify and do some injustice to the authors. Model 1 is described as a rationalist and prescriptive approach characterized as "the one and only best way to carry out activities," whereas Model 2 rules are "routines" or "patterns of behavior, socially constructed, emerging from experience." The latter serves high-hazard professions such as air pilots and health-care workers. Rule violation can be routine, situational, exceptional, or optimizing. The study then analyzes reasons for violation and other aspects, and ends with summarizing strengths and weaknesses of both models. What this essentially boils down to is that Model 1, the more dominant one, is providing clarity but is inflexible, and Model 2 yields good guidance but contains fuzziness, also with respect to responsibility. The second part of the article discusses the various items of a management framework for making and redesigning/scrapping rules and procedures. It concludes that the cyclical framework can be applied to both models. Strengths of both models should somehow be combined: fully top-down does not work and neither does fully bottom-up. The authors provide some thoughts for improvement and also pose some questions for further investigation. The articles pay a great deal of attention to reasons of *rule violation* and how to overcome them.

For making rules practical advice can be obtained from many sources, but a rather succinct summary of items to think of can be downloaded from the TopVes Web site.[76] Regarding rule violation, Phil La Duke[77] summarizes the reasons as Misinterpretation of the rule; Distraction (by other activity); It's worth it; The rules don't make sense; The organization fails to sell the rule; The rule seems temporarily unnecessary; The rule seems trivial or overly protective.

A known example of violating or rather ignoring instructions that saved quite a number of lives is the escape from the disastrous burning Piper Alpha offshore platform[78] in 1988 by diving from "10 stories" high in the North Sea. These survivors chose to ignore the incident and evacuation procedures/rules based on their rapid assessment of the manner in which the accident was unfolding. Their judgment was well founded. Such a case is of course exceptional.

Indeed, keeping the right balance is a real challenge to management, but what exactly is right? In a well-documented 2014 article, entitled "The Bureaucratization of Safety," Sydney Dekker[79] analyzed trends over the past 50 years. Dekker treats the proliferation of occupational safety regulation, responses of management in a competitive climate, practices on the work floor, bureaucratization of accountability with outsourcing of expertise, and changes in liability and insurance of harm and damage. Many well-intended rules may provoke secondary counterproductive effects. To be mentioned are demotivation, stifling innovation, "number games," "paper safety," hiding incidents, and workarounds, while weak signals that really matter are not picked up. Overregulation may become a menace. Hence, the challenge seems to find the "golden middle" between top-down regulated and controlled safety and bottom-up risk awareness. The latter relies on the right feel for what matters together with sufficient freedom to act and, of course, on effective communication and organizational transparency.

What can the company's chief safety officer do besides contributing to well thought through rules and procedures, providing training, giving advice, investigating near misses, auditing, and warning? Well, with his board's support he can at least install a system of process safety performance indicators, collect values, interpret those, and look for trends. The rationale behind indicator metrics has a link with the balance between remaining profitable as an enterprise and maintaining a good safety level, as we shall see in the next section.

6.7 PROCESS SAFETY PERFORMANCE INDICATORS

In Chapter 3, Section 3.2, we have already been exposed to the development of management systems in the process industry in the 1990s. Following the Deming cycle, management decision making in general is on the basis of KPIs. This was true for finance, productivity, marketing, and other facets, but it took a while before a follow-on step was made of conceiving process safety indicators. Installing and using indicators started in the larger companies but is still not introduced enthusiastically and adopted within industry overall. The skilled use of good metrics can be much more effective, as we shall see in the next chapter.

The use of the backwards looking, lagging indicators is rather obvious. It is just a matter of criteria to classify incidents, count, and compare. As mentioned in Chapter 3, the lost time injury frequency (LTIF) indicator for employees losing one or more days due to an accident at work has been in use in the process industry at least since the early 1960s. Near-miss investigation has been applied also for many years although it is not as widespread as LTIF. Near misses provide valuable information to improve preventive measures as shown in Figure 6.10, which was adopted but slightly modified from a 2001 project report, written for the nuclear power industry. Avoiding both bankruptcy by losing the competition or occurrence of a catastrophic accident (with another risk of bankruptcy), usually a downward drifting safety level could traditionally be kept sufficiently high by a kind of trial-and-error approach. In case an incident happened, management then decided for a one-time investment in preventive measures. Otherwise, investment was only in productivity improvement. However, in the learning process it becomes harder to have internal incidents or failures to learn from, yet the threat of a rare, catastrophic accident remains. Therefore, the project accomplished in assignment with the Electric Power Research Institute (EPRI)[81] was to find out, whether one can develop leading indicators of human performance with sufficient predictive ability. The result was positive.

Indeed, just a few years later, in 2003, the OECD Environment, Health and Safety, Chemical Accidents Program, which we discussed in the previous section regarding the 2012 document for leaders of industry, began work on a guidance document for indicators measuring the effectiveness of a safety management system; this was published as an interim report in 2005.[82] This was followed by a

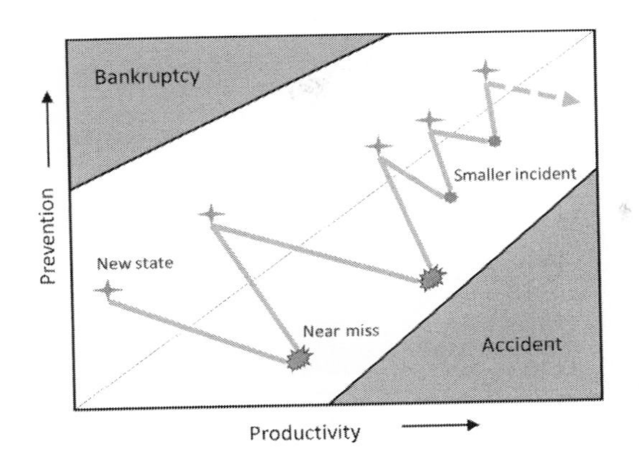

FIGURE 6.10

The traditional way of deciding to invest in preventive measures rather than in productivity was only when a near miss occurred. However, when the safety level rises, fewer incidents and failures will provide information for learning new ideas for improvement, while on the other hand the risk of disaster is still present. Safety levels tend to drift downward. The way forward appears to be defining leading indicators.

Adapted from final EPRI report[80].

wealth of other reports from various institutions on the importance of indicators and how to implement them. In 2006 in the UK, HSE published a guideline[83] on how to proceed in six steps to performance measurement of the safety management system elements. The performance measurement of risk controls by lagging indicators and process controls by leading indicators on plant change, inspection and maintenance, staff competence and permit to work will provide "dual assurance." It concluded with a worked example of a high-hazard liquid chemicals bulk storage and handling site. In 2007 after the CSB and the Baker reports came out on the Texas City isomerization plant vapor cloud explosion and the Baker review of BP's US refineries, CCPS issued a brochure-type document[84] urging introduction of indicating metrics. This was followed a few years later by the more extensive description of the risk-based SMS elements and a guideline on metrics (see Chapter 3, references 9 and 17). Other institutions issuing guidelines have been OGP, the International Association of Oil & Gas Producers[85] in 2008, a combined effort of ANSI/API[86] in 2010, Cefic, the European Chemical Industry Council,[87] and UK Oil & Gas.[88]

Leading indicators in particular will monitor changes in safety culture, so the metrics are a good instrument to accompany a culture maturity process. In 2009, Andrew Hopkins[89] initiated a broad discussion on indicators in the journal *Safety Science*, issue No. 47. Indeed, distinction between lagging (outcome) and leading (activity) indicators is not always clear and one can argue whether an indicator is a process or an occupational safety one, but does it really matter?

A global conference was held in Brussels in early 2012, organized by Cefic and the European Process Safety Centre (EPSC) under the auspices of the Responsible

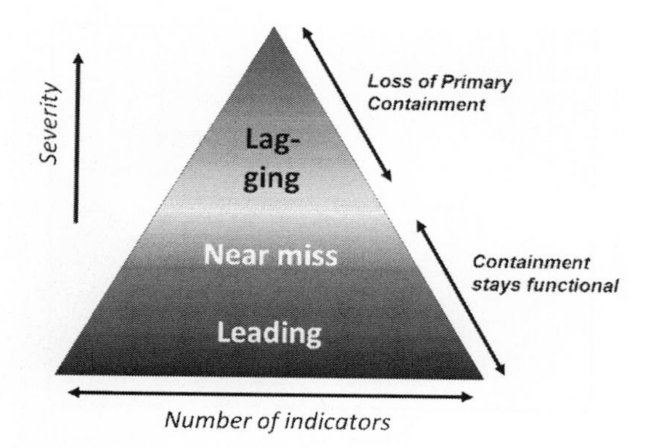

FIGURE 6.11

Indicator triangle as shown in William Garcia's introductory presentation, on behalf of the International Council of Chemical Associations, to the Brussels conference in January 2012. Besides near misses, other precursors shall also be monitored and interpreted as weak signals that can contain important information about the organizational "organism."

Care® program and the International Council of Chemical Associations. This was to discuss with a majority of industry attendees the importance of process safety performance indicators (Figure 6.11), how to set up a system, what the experiences are, and how to proceed. In brief, the conclusions were

- for obtaining a stable systems of metrics, one needs 3–5 years;
- lagging indicator criteria of loss of containment consequences shall be globally standardized; see Chapter 3, Section 3.2;
- leading indicators in abundance are suggested. Each user should make his own selection; see also Section 3.2.
- the basis should be broadened; and small and medium enterprises stimulated to participate;
- when the system is sufficiently reliable, one should go public with the outcomes to gain trust of the public at large.

Following the conference in Brussels, Knijff et al.[90] reported a European perspective on leading indicators based on contributions of members of the European Process Safety Centre. This common position will also help to formulate a minimum requirement in view of the Seveso III implementation in national law in the various EU countries. In July 2012, the US Chemical Safety and Hazard Investigation Board (CSB) organized a public hearing in Houston to discuss offshore indicator metrics with inputs from British, Norwegian, and Australian experiences.

In the next chapter, we shall learn what more can we do with indicator metrics than just looking at trends—because for convincing top management, more may be necessary than just observing with satisfaction a downtrend of some of the indicators. Once a system of metrics is established, problems of interpretation will emerge. In an extreme, ideal case all lagging indicators over the last period will be nil, so that over time leading indicators, of which selection is more difficult, become more important. Then, there will be fluctuations in outcomes so that apart from unusual changes raising the question of causation and learning from it, for trend analysis statistics must be applied. As we shall see in the next chapter, distribution parameters of a data set can be determined easily by applying Bayesian networks. A further problem is the sheer number of indicators, which can spoil the overview. As Figure 6.11 shows, this is particularly true for the leading indicators. For an overview, 5 indicators will be acceptable and 10 will be maximum. However, CCPS suggested almost 400 indicators based on the 21 principal ones in 4 groups, referred to in Chapter 3, Figure 3.3. Hence, for assessment at various levels in a company an aggregation method will be needed. Such a method has been presented by Hassan and Khan[91] for asset integrity indicators, in which they weighted the individual indicators with their relative importance for asset integrity based on expert judgment, and with a risk factor the indicator is associated with. Aggregation went in three steps, first from specific indicators to 35 key indicators, then to 13 activity ones, and finally at the top to three element indicators of, respectively, mechanical, operational, and personnel integrity. Pasman and Rogers[92] took this approach a step further and proposed to couple weighted indicators as risk factors to a risk assessment model of a process, so that the effect of the performance results of the safety management system on the overall plant risk level can be monitored. In Section 7.7.2 of the next chapter, it will be shown first how by applying Bayesian networks in a convenient way with periodical inputs, statistical treatment and aggregation can be performed, and subsequently for a simple example how an indicator-influenced risk-assessment model can be structured.

6.8 CONCLUSIONS

This chapter has presented a colorful palette with aspects of HFs influencing both process and personal safety. We have seen agreement and differences at the sharp end, the frontline workers, and at the blunt end, the management. Safety ought to be a line responsibility. Leadership definitely shapes the safety culture. Safety culture must be determined by measuring safety climate, and this proves to be a rather reliable predictor. We have seen many different approaches to achieve the same result, viz a safe process and a safe workplace. Although there is continual progress, nothing can deliver a panacea. In any case, by applying the bow-tie concept occupational risk assessment has made impressive progress.

For any prediction of risk caused by a sociotechnical system, knowledge of human error is crucial. To be able to make an estimate of failure rate one needs to know what decision/action can be taken wrongly, at what decision point, due to what cause, and in what context/set of conditions. Although with present tools some reasonable successes for operator action with regard to control of process installations have been claimed, there is no question of general application yet. Due to the large variability in human decision making and the large variety in abnormal conditions and what can go wrong, a general solution appears still a rather remote possibility. Present results are helpful though in design, such as of measures in operability of equipment to adapt to human capabilities and of training means and procedures. Not taking account of HFs will certainly lead to mishap.

We looked at organizational aspects and the daily field of tensions in which management must make decisions. For the greater part of time, safety as an issue plays only a small role. On the other hand, experienced companies such as Dupont know to value safety as an important factor. This company launched the Dupont Integrated Approach (DnA) to safety.[93] Indeed, as expressed in the organizational triangle, good safety culture, good safety management, and well taking account of HFs go together to the benefit of a company. It should be well integrated with business strategy and operations. A pillar is providing training for sustainable safety performance based on the latest in safety science with emphasis on behavior and competence development. Sam Mannan et al.[94] have summarized the attributes of a framework for a so-called "best-in-class" safety culture, in which items appear treated in this chapter and elsewhere in the book. But, even companies considered to be safety champions can be hit by disastrous accidents as the facts show. In striving to exclude fatal error, an optimal balance between rules and work floor risk awareness is key.

Bureaucratization of safety is looming. Yet, safety performance indicators can form a powerful tool to monitor risk level, but a few developments are needed to fully unleash their potential. In the next two chapters, we shall describe new methods based on systems approach, which can have a positive impact, also by the use of indicators. This can further improve process safety.

REFERENCES

1. Della-Giustina DE. *Developing a safety and health program.* Boca Raton, FL: Lewis Publishers; 1999. ISBN:1-56670-518-5.
2. Furr AK. *CRC handbook of laboratory safety.* 5th ed. Boca Raton, FL: CRC Press; 2000. ISBN:0-8493-2523-4.
3. American Petroleum Institute. *A manager's guide to reducing human error: improving human performance in the process industries.* Washington, DC: API Publication 770; 2001.
4. Swuste P, Van Gulijk C, Zwaard W. Safety metaphors and theories, a review of the occupational safety literature of the US, UK and The Netherlands, till the first part of the 20th century. *Saf Sci* 2010;**48**:1000–18.
5. Swuste P, Van Gulijk C, Zwaard W, Oostendorp Y. Occupational safety theories, models and metaphors in the three decades since World War II, in the United States, Britain and The Netherlands: a literature review. *Saf Sci* 2014;**62**:16–27.
6. Reason J. *Managing the risks of organisational accidents.* Aldershot, Hampshire, England: Ashgate Publishing Ltd; 1997. ISBN:1-84014-105-0.
7. Dekker S. *Just culture: balancing safety and accountability.* 2nd ed. Farnham, Surrey, U.K.: Ashgate Publishing Ltd; 2012. ISBN:978-1-4094-4060-4.

8. GAIN. *A roadmap to just culture: enhancing the safety environment*. 1st ed. September 2004. http://flightsafety.org/files/just_culture.pdf.
9. Marx D. *Patient safety and the "Just Culture:" a primer for health care executives*. April 2001. http://www.safer.healthcare.ucla.edu/safer/archive/ahrq/FinalPrimerDoc.pdf.
10. Bond J. A safety culture with justice: a way to improve safety performance. In: *12th international symposium on loss prevention and safety promotion in the process industries, IChemE symposium series no 153*; 2007. 67.1–6. Edinburgh, UK.
11. Turney R, Alford L. Improving human factors & safety in the process industries: "The Prism Project". In: *11th international symposium on loss prevention and safety promotion in the process industries, Praha*; June, 2004. p. 501–16.
12. HSE. *Reducing error and influencing behaviour*. 2nd ed. U.K.: Health and Safety Organization; 1999/reprinted 2009 ISBN:978-0-7176-2452-2 http://www.hse.gov.uk/pubns/books/hsg48.htm.
13. Napo videos. http://www.napofilm.net/en/napos-films.
14. Hopkins A. What are we to make of safe behaviour programs? *Saf Sci* 2006;**44**:583–97.
15. http://www.hse.gov.uk/humanfactors/topics/complete.pdf.
16. http://www.hse.gov.uk/humanfactors/topics/toolkit.pdf.
17. Bridges W, Tew R. Human factors elements missing from process safety management (PSM). In: *2010 Spring meeting, 6th global congress on process safety, and the 44th annual loss prevention symposium*. San Antonio, TX: American Institute of Chemical Engineers; March 22–24, 2010. http://www.piii.com.
18. Bellamy LJ, Geyer TAW, Wilkinson J. Development of a functional model which integrates human factors, safety management systems and wider organisational issues. *Saf Sci* 2008;**46**:461–92.
19. HSE. *Managing for health and safety*. 3rd ed. U.K.: Health and Safety Organization; 2013 ISBN:978-0-7176-6456-6 http://www.hse.gov.uk/pubns/books/hsg65.htm.
20. Ale BJM, Baksteen H, Bellamy LJ, Bloemhof A, Goossens L, Hale AR, et al. Quantifying occupational risk: the development of an occupational risk model. *Saf Sci* 2008;**46**:176–85.
21. Bellamy LJ, Ale BJM, Whiston JY, Mud ML, Baksteen H, Hale AR, et al. The software tool storybuilder and the analysis of the horrible stories of occupational accidents. *Saf Sci* 2008;**46**:186–97.
22. RIVM, The quantification of occupational risk; The development of a risk assessment model and software, WORM Metamorphosis consortium, Report No. 620801001/2008. http://www.rivm.nl/bibliotheek/rapporten/620801001.pdf.
23. Bellamy LJ, Aneziris ON, Papazoglou IA, Damen M, Manuel HJ, Mud M, et al. Safety management of safety performance-what factors to measure? In: Steenbergen, et al., editors. *ESREL 2013, safety, reliability and risk analysis: beyond the horizon*. London: © 2014 Taylor & Francis Group; 2013. p. 445–52. ISBN:978-1-138-00123-7.
24. Bellamy LJ. Exploring the relationship between major hazard, fatal and non-fatal accidents through outcomes and causes. *Saf Sci* 2015;**71**(Part B):93–103.
25. Bellamy LJ, Damen M, Manuel HJ, Aneziris ON, Papazoglou IA, Oh JIH. Risk horoscopes: predicting the number and type of serious occupational accidents in The Netherlands for sectors and jobs. *Reliab Eng Syst Saf* 2015;**133**:106–18.
26. Rasmussen J. Skills, rules and knowledge; signals, signs and symbols, and other distinctions in human performance models. *IEEE Trans Syst Man Cybern* 1983;**13**(3):257–66 (In Figure 1 of the paper at the lowest level this paper mentions "Knowledge based

behaviour", which as appears from earlier reports of Rasmussen: On the Structure of Knowledge - a Morphology of Mental Models in a Man- Machine System Context, RISØ-M-2192, November 1979, should be "Skill-based behaviour").

27. U.S. Chemical Safety and Hazard Investigation Board. *Investigation report, refinery explosion and fire (15 killed, 180 injured).* Texas city, TX: BP; March 23, 2005. Incident Description, Report No. 2005-04-I-TX, March 2007.
28. Wiegmann DA, Shappell SA. *A human error approach to aviation accident analysis. The human factors analysis and classification system.* Burlington, VT: Ashgate Publishing Ltd; 2003.
29. Bell J, Holroyd J. *Review of human reliability assessment methods.* Health and Safety Laboratory for Health and Safety Executive, U.K., © Crown copyright; 2009. Research Report RR 679.
30. Kelly DL. *Incorporating process mining into human reliability analysis* (PhD dissertation). Eindhoven University of Technology, the Netherlands; 2011. ISBN:978-90-386-2503-4.
31. Swain AD, Guttmann HE. *Handbook of human reliability analysis with emphasis on nuclear power plant applications.* Washington, DC: US Nuclear Regulatory Commission; 1983. NUREG/CR-1278.
32. Kirwan B. The validation of three human reliability quantification techniques - THERP, HEART and JHEDI: Part I - technique descriptions and validation issues. *Appl Ergon* 1996;**27**:359–73. Part II - Results of validation, Exercise, Applied Ergonomics, 28, 1997, 17–25; Part III - Practical aspects of the usage of the techniques, Applied Ergonomics, 28, 1997, 27–39.
33. Williams JC. Toward an improved evaluation analysis tool for users of HEART. In: *Proceedings of the international conference on hazard identification and risk analysis, human factors and human reliability in process safety.* Orlando, FL: AIChE/CCPS; January 1992. p. 15–7.
34. Forester J, Kolaczkowski A, Lois E, Kelly D. *Evaluation of human reliability analysis methods against good practices.* NUREG-1842. Sandia National Laboratories for U.S. Nuclear Regulatory Commission; 2006.
35. Groth KM, Mosleh A. A data-informed PIF hierarchy for model-based human reliability analysis. *Reliab Eng Syst Saf* 2012;**108**:154–74.
36. Taylor JR. *Incorporating human error analysis into process plant safety analysis. 14th international symposium on loss prevention and safety promotion in the process industries*, vol. 31. Chemical Engineering Transactions; 2013. http://dx.doi.org/10.3303/CET1331051. 301–306.
37. Taylor JR. *Human error in process plant operations.* Denmark: ITSA; 2012.
38. Kožuh M, Peklenik J. A method for identification and quantification of latent weaknesses in complex systems. *Cognit Technol Work* 1999;**1**:211–21.
39. Noroozi A, Khakzad N, Khan F, MacKinnon S, Abbassi R. The role of human error in risk analysis: application to pre- and post-maintenance procedures of process facilities. *Reliab Eng Syst Saf* 2013;**119**:251–8.
40. Hollnagel, E., CREAM - cognitive reliability and error analysis method, Disclaimer written 2012, http://erikhollnagel.com/ideas/cream.html.
41. Hollnagel E. Human reliability in context. *Nucl Eng Technol* 2005;**37**(2):159–66.
42. He X, Wang Y, Shen Z, Huang X. A simplified CREAM prospective quantification process and its application. *Reliab Eng Syst Saf* 2008;**93**:298–306.

43. Phillips RO, Sagberg F. What did you expect? CREAM analysis of hazardous incidents occurring on approach to rail signals. *Saf Sci* 2014;**66**:92–100.
44. Hollnagel E. *The ETTO principle: efficiency-thoroughness trade-off; why things that go right sometimes go wrong*. Farnham, Surrey, U.K.: Ashgate Publishing Ltd; 2009. ISBN: 978-0-7546-7678-2.
45. Kahneman D, Tversky A. Prospect theory: an analysis of decision under risk. *Econometrica* 1979;**47**(2):263–91.
46. Hopkins A. *Disastrous decisions: the human and organisational causes of the Gulf of Mexico Blowout, CCH*. 2012. ISBN:9781921948770.
47. Guldenmund FW. *Understanding and exploring safety culture* (Ph.D. dissertation). Delft University of Technology, Uitgeverij BOX Press; 2010. ISBN:978-90-8891-138-5.
48. Zohar D. Thirty years of safety climate research: reflections and future directions. *Accid Anal Prev* 2010;**42**:1517–22.
49. Johnson S. Predictive validity of safety climate. In: Li Sh, Wang Y, An Y, Sun X, Li X, editors. *Progress in safety science and technology*, vol. VII, Part A. Beijing, China: Science Press; 2008. p. 13–22. ISBN:978-7-03-022901-4.
50. Hale AR, Hovden J. Management and culture: the third age of safety. A review of approaches to organizational aspects of safety, health, and environment. In: Feyer AM, Williamson A, editors. *Occupational injury: risk, prevention and intervention*. London: Taylor-Francis; 1998. p. 129–65.
51. European Agency for Safety and Health at Work (EU-OSHA). Occupational safety and health culture assessment - a review of main approaches and selected tools. In: Taylor TN, editor. *Working paper*; 2011. TE-WE-11-005–EN-N, http://dx.doi.org/10.2802/53184, ISBN:978-92-9191-662-7.
52. Morrow SL, Koves GK, Barnes VE. Exploring the relationship between safety culture and safety, performance in U.S. nuclear power operations. *Saf Sci* 2014;**69**:37–47.
53. Zohar D, Luria G. A multilevel model of safety climate: cross-level relationships between organization and group-level climates,. *J Appl Psychol* 2005;**90**(4):616–28.
54. Tripod Delta. http://www.energypublishing.org/tripod/delta.
55 a. Hudson P, Parker D, Lawrie M, Van der Graaf G, Bryden R. How to win hearts and minds: the theory behind the program, the seventh SPE (society of petroleum engineers) international conference on health, safety and environment. In: *Oil and gas exploration and production*. Alberta, Canada: Calgary; March 2004;
   b. Hudson P. Implementing a safety culture in a major multi-national. *Saf Sci* 2007;**45**: 697–722.
56. Hudson PTW, Van der Graaf GC, Bryden R. The rule of three: situation awareness in hazardous situations. http://www.eimicrosites.org/heartsandminds/userfiles/file/ASA/ASA%20PDF%20rule%20of%20three%20paper,%20P%20Hudson,%20C%20vdGraaf.pdf.
57. Hudson PTW. Implementing a safety culture in a major multi-national. *Saf Sci* 2007;**45**: 697–722.
58. Hudson PTW. *Safety management and safety culture: the long, hard and winding road*. The Netherlands: Centre for Safety Research, Leiden University; 2001. http://www.skybrary.aero/bookshelf/books/2417.pdf.
59. Fleming M. *Safety culture maturity model, prepared by Keil Centre for the health and safety executive*. Offshore technology report 2000/049. © Crown copyright; 2001. ISBN:0-7176-1919-2.

60. Weichel M. Utilizing field-based risk registers to improve process safety culture and competence. In: *16th annual international symposium Mary Kay O'Connor process safety center, College Station, TX*; October 2013.
61. Hoewijk Van, van R. The meaning of organisational culture: an overview of the literature (in Dutch). *M&O Tijdschr Organ Soc beleid* 1988;**1**:4–46.
62. Rasmussen J, Svedung I. *Proactive risk management in a dynamic society.* 1st ed. Karlstad, Sweden: Swedish Rescue Services Agency; 2000. ISBN:91-7253-084-7.
63. Rochlin GI, La Porte TR, Roberts KH. The self designing high reliability organization: aircraft carrier flight operations at sea. *Nav War Coll Rev Autumn* 1987;**40**:76–90.
64. Skjerve AB, Kaarstad M, Holmgren L. Teamwork competence requirements in nuclear power plant control rooms. In: Steenbergen, et al., editors. *ESREL 2013, safety, reliability and risk analysis: beyond the horizon*. London: © 2014 Taylor & Francis Group; 2013. p. 401–8. ISBN:978-1-138-00123-7.
65. Skjerve AB, Nystad E, Rindahl G, Sarshar S. Assessing the quality of collaboration in an integrated operations organization. In: *ESREL 2013, ibidem*; 2013. p. 341–8. ISBN:978-1-138-00123-7.
66. Albrechtsen E, Grøtan TO, Haugen S. Improving proactive major accident prevention by new technology and work processes. In: *ESREL 2013, ibidem*; 2013. p. 2487–93. ISBN: 978-1-138-00123-7.
67. Körvers PMW. *Accident precursors: pro-active identification of safety risks in the chemical process industry* (Ph.D. dissertation). The Netherlands: Eindhoven University of Technology; 2004. ISBN:90-386-1868-9.
68. Sonnemans PJM, Körvers PMW, Pasman HJ. Accidents in "normal" operation - can you see them coming? *J Loss Prev Process Ind* 2010;**23**:351–66.
69. OECD C. Governance for process safety: guidance for senior leaders in high hazard industries. In: *OECD environment, health and safety chemical accidents programme, Paris*; June 2012. http://www.oecd.org/env/ehs/chemical-accidents/corporate%20governance%20for%20process%20safety-colour%20cover.pdf.
70. Hackitt J, Frank P. Lees memorial lecture: process safety-focussing on what really matters-leadership! In: *Mary Kay O'Connor process safety center, 16th annual international symposium, College Station, TX*; October 22, 2013.
71. http://fuelfix.com/blog/2013/04/05/exxon-ceo-on-death-it-makes-a-big-impression-on-you/.
72. Sagnier P, Le Floch M. *Mind the gap: experienced engineers wanted, ESG insight AXA investment managers responsible investment*. March 2012. http://www.fundresearch.de/sites/default/files/partnercenter/axa-investment-managers/news_2011/20120608_axa_esginside_brochure_en_.pdf.
73. Pitblado R, Nelson WR. Advanced safety barrier management with inclusion of human and organizational aspects. *Chem Eng Trans* 2013;**31**:331–6.
74. a. Hale A, Borys D. Working to rule, or working safely? Part 1: a state of the art review. *Saf Sci* 2013;**55**:207–21;

    b. Part 2: the management of safety rules and procedures. *Saf Sci* 2013;**55**:222–31.
75. Dekker S. Failure to adapt or adaptations that fail: contrasting models on procedures and safety. *Appl Ergon* 2003;**34**:233–8.
76. TopVes, How to Make Rules. http://www.topves.nl/PDF/How%20to%20make%20rules.pdf; [accessed August 2014].
77. La Duke P, Why We Violate the Rules. http://www.fabricatingandmetalworking.com/2011/05/why-we-violate-the-rules/; [accessed August 2014].

78. BBC video, Disaster, Stone City Films for BBC Education &Training MCMXCVI.
79. Dekker SWA. The bureaucratization of safety. *Saf Sci* 2014;**70**:348–57.
80. EPRI (Electric Power Research Institute). *Final report on leading indicators of human performance*. 1003033. October 2001.
81. EPRI (Electric Power Research Institute). *Predictive validity of leading indicators: human performance measures and organizational health*. 1004670. October 2001.
82. OECD environment, health and safety publications series on chemical accidents *OECD guidance on safety performance indicators, guidance for industry, public authorities and communities for developing SPI programmes related to chemical accident prevention, preparedness and response*. Interim Publication; 2011. Series No. 11, revised 2005; Series No. 10, 2nd ed.; Addendum.
83. HSE (Health and Safety Executive U.K.). Developing process safety indicators: a step-by-step guide for chemical and major hazard industries, HSG 254, ISBN:978-0-7176-6180-0; [downloadable from the web].
84. CCPS, Center for Process Safety AIChE. *Process safety leading and lagging metrics: you don't improve what you don't measure*. 2008.
85. OGP, *International association of oil & gas producers, asset integrity—the key to managing major incident risks*. Report No. 415. 2008.
86. API, American Petroleum Institute and ANSI, American National Standards Institute. *Process safety performance indicators for the refining and petrochemical industries*. ANSI/API Recommended Practice 754. 1st ed. 2010.
87. Cefic, The European Chemical Industry Council. *Guidance on process safety performance indicators*. May 2011.
88. U.K. Oil & Gas, Step change in safety, leading performance indicators: guidance for effective use. http://www.stepchangeinsafety.net/knowledgecentre/publications/publication.cfm/publicationid/26.
89. Hopkins A. Thinking about process safety indicators. *Saf Sci* 2009;**47**:460–5.
90. Knijff P, Allford L, Schmelzer P. Process safety leading indicators—a perspective from Europe. *Process Saf Prog* 2013;**32**(4):332–6.
91. Hassan J, Khan F. Risk based asset integrity indicators. *J Loss Prev Process Ind* 2012;**25**:544–54.
92. Pasman HJ, Rogers WJ. How can we use the information provided by process safety performance indicators? Possibilities and limitations. *J Loss Prev Process Ind* 2014;**30**:197–206.
93. Dupont Integrated Approach (DnA). http://www.dupont.com/content/dam/assets/products-and-services/consulting-services-process-technologies/articles/documents/DnA_USA_Brochure_06192012.pdf.
94. Mannan MS, Mentzer RA, Zhang J. Framework for creating a best-in-class safety culture. *J Loss Prev Process Ind* 2013;**26**:1423–32.

CHAPTER

7 New and Improved Process and Plant Risk and Resilience Analysis Tools

New brooms sweep clean.
Free translation of a Dutch saying

SUMMARY

As has been rather manifest in previous chapters, the deepest causes for rare industrial catastrophic events to occur are preoccupation of top management with the current economic situation of the enterprise (business perspectives, competition, and finance), little eye for safety risks, and complexity of the technology and the process. The latter results in lack of insight of what consequences of an event can be, also in an economic sense, how robust design of the defenses must include emergency response, lack of recognition of precursors or signals of an imminent event, and insufficient awareness of the trend in level of safety culture and commitment of the organization.

We have also seen that the instruments we have today are still far from perfect. Quantitative risk analysis needs further improvements in particular with scenario identification, hence possibilities that can go wrong with all ex- and internal causes, cascading and escalating effects, and human and organizational influences. One cannot overemphasize the importance of system monitoring to gather information useful for developing previously unidentified scenarios and also identifying emerging conditions for new scenarios. Monitoring must start in the design stage and continue in operations. Consequence calculations are still too inaccurate and have for complex situations too little resolution for knowing the consequence distribution and effect limits more precisely. On the other hand, consequences are the result of physical, mechanical, and chemical phenomena, which can be validated by experiment. In principle, failure rates of equipment can also be verified, but this is a rather laborious process. Risk-based inspection activity though can yield an invaluable wealth of information. Loss of containment probability distribution values for any particular plant will be for an unforeseeable time at best order of magnitude estimates and will often lack reliable confidence limits. Validation of rare catastrophic event probability values as a result of multiplying a number of uncertain precursor events is almost impossible, although we shall examine a new Bayesian method to make an improved estimate applying all available evidence. Undoubtedly, the best estimates of the risk level are achieved by using a system approach through estimations of the consequence and probability distributions for the specific process conditions. In this chapter, we shall review a few system approaches that will help to make improvements.

In Chapter 5, we became acquainted with the systems approach promoted by Nancy Leveson. The system-theoretic process analysis (STPA) tool is certainly a track to follow because it brings us knowledge of risks that a sociotechnical system poses that cannot be traced through simple component or communication failures, or failure to act on a signal. To be identified are safety flaws emerging from dysfunctional interaction of healthy components or justified acting, but of which the possibility has been overlooked in the system specification, the design, or in the software coding. Moreover, the system approach provides direct insight on the risk controls. So, the principles are there, and it proved to work for the large blocks as integral parts. But the sheer multiplicity of items and issues when going to the detail level from which an event usually evolves prohibits at this moment descending to the lower levels of process, machinery, and organization, applying the principles there, and subsequently putting it all together in a grand picture. This requires, at least in part, automation. In addition, it should be available not only in a static mode for a design or for a snapshot of an operational situation. Ideally, the approach should be available as a monitoring instrument, to follow the adequate functioning of the risk controls. For the time being, there has been a start in the direction of being able to cope with a larger number of control loops and components, which we shall review in this chapter.

Another development that is systematic and could, but in practice may not, cover all dysfunctional interactions, is Blended Hazid (BLHAZID). Thanks to a sound IT infrastructure, this tool is able to cope with a myriad of items. This effort is an initiative of Ian Cameron, of Queensland University, Australia, with contributions of a multidisciplinary team. It places plant, people, and procedures in a functional systems framework and applies the classical HAZOP and FMEA tools described in Chapter 3. So, it builds from detail level up and succeeds as a result of the computerization to link long strings of cause—effect information. These causal graphs can be produced in an eyewink, among others enabling an operator to identify quickly what causes an upset situation. The method also will save companies much effort having to repeat HAZOP-ing every 5 years. BLHAZID is still assuming the application of classical team-based HAZOP. This technique is quite time consuming and puts strain on the participants. There have been several attempts to at least semi-automate the process, and we shall review those. It is making progress.

The world is full of uncertainty—things we do not know precisely or only in a fuzzy fashion. Statistical science provides tools to delineate uncertainty and based on observations provide a probability distribution of the values a parameter can have. The more recent Bayesian statistics enable easy updating with the latest information, while making use of all types of background knowledge and both hard and soft information. Statistical correlations of parameters even of high strength are not proof of an existing cause–effect relation, but in time, effects result from causes. Bayesian networks (BNs) are an excellent tool to model cause–effect relations, perform diagnostics and inference, while taking account of the statistics. Their popularity has strongly increased the last few years because the difficulty of the associated arithmetic computation is absorbed by specifically developed algorithms and software. Their applications in economics, finance, medical diagnostics, psychology, and decision making have been spreading rapidly. We shall briefly review the method and applications in data treatment, risk analysis, analysis of factors influencing human performance, and relations among process safety performance indicator (PSPI) values, management effectiveness, and human factor effects on a risk level. This flexible and logically coherent method will likely replace the earlier fuzzy set method.

As a further step, dynamics of interactions in a process become a factor to be studied because wrong sequencing or untimely actions also introduce and add to risks. Dynamic BNs (DBNs) can be used, but there are other possibilities such as Petri nets and agent-based modeling. In addition, process simulation tools such as ASPEN and HYSYS will be of help. We shall briefly review perspectives that these methods will offer.

To be able to maintain a safe situation in a sustainable way, we shall also be prepared for unforeseen threats, hence plant and organization shall be resilient. Resilience means that after an unexpected disturbance threatening to cause, or already causing, an upset state, the normal operational state shall be back as soon as possible, so that a minimum of business interruption and damage will be sustained. Resilience analysis is however in its infancy, but its development is a necessity for the reason set out hereafter. Uncertainty was in the past compensated by introducing a safety margin or reserve capacity. (For that reason, the Romans standardized all their construction elements and methods.) Due to better current analysis methods and cost pressures, reserves have been reduced or even taken away. This makes the whole system more vulnerable, less robust and flexible, and less able to withstand or to cope with unexpected threats. A set of probabilistic tools of risk analysis and simulation in a setting of the systemic approach explained above must be developed to analyze what reserves shall be vital and what flexibility and room for improvisation shall be left.

Evidently, the chapter will have the character of a snapshot because fortunately continuing research is active in developing new concepts and elaborating existing ones.

7.1 INTRODUCTION

A number of promising new approaches and tools shall be described to better identify and measure risks and enable cost-effective risk control. The first will be the top-down systemic approach of Nancy Leveson's STPA tool, where the challenge is to make the tool operational and to what depth one can look with the tool into sociotechnical systems. Second will be the bottom-up approach of the Blended Hazid of Ian Cameron, starting at the component or even subcomponent level but including people and procedures in a functional system setting. Because HAZOP is needed and its associated effort large, HAZOP automation attempts will be briefly reviewed. Third will be applications of relatively new techniques. These will be mainly BN type, replacing fault and event trees, examples of optimizing layer of protection analysis (LOPA) designs, a full risk distribution analysis, aggregating and using indicator values, and human error or rather human reliability with effects of culture and management. In addition, a simple example with dynamic features will be shown. Dynamics can be simulated by a variety of techniques depending on the nature of the processes. Random changes of states in time can be modeled as a Markov process and also by a dynamic BN. Processes in which resources and logistics are important can be simulated by system dynamic models in a continuous fashion, or by Petri net if sent "packages" are discrete and of different type. Action of autonomous decision-making entities guided by certain rules and by assessing its situation is simulated in agent-based modeling (ABM). The chapter will close with a sketch of the directions analysis that shall be performed to probe the resilience a system possesses. In principle, the result will always be uncertain because as we have seen in Chapter 5, a resilient system shall resist unforeseen and even unknown threats.

7.2 SYSTEM-THEORETIC PROCESS ANALYSIS

The following is based on the PhD dissertation of John Thomas,[1] coworker of Nancy Leveson at MIT (Cambridge, MA). The characteristics and features of the accident model system-theoretic accident model and processes (STAMP) and of STPA are described in Chapter 5. It is about a hierarchy of "processes" and the control loops to steer these processes. The types of failures that can occur are shown in the generic control loop of Figure 5.3, for convenience here reproduced from Thomas's dissertation. The loops consist of a sensor(s), a controller responding on the information reaching him from sensors and from elsewhere in the organization, upper or lower level and externally, an actuator that reacts to action signals from the controller and upon which it performs its operational action in the process for which it is designed. The process can be subject to action from other control loops. It further depends on inputs, is subject to disturbances initiated by internal or external conditions, and it may run out of hand, hence generating a system risk. It should be noted that the controller box contains a *process model*. This model consists of all process variables and their interacting relations. This can take various forms, but if the controller is a human operator, the model is his mental image of the process.

The function of the process depicted in Figure 7.1 can also be controlling lower level control loops. The deeper one descends in the hierarchy the more detailed it becomes. So, in principle one can start at the top and go down to the lowest detail

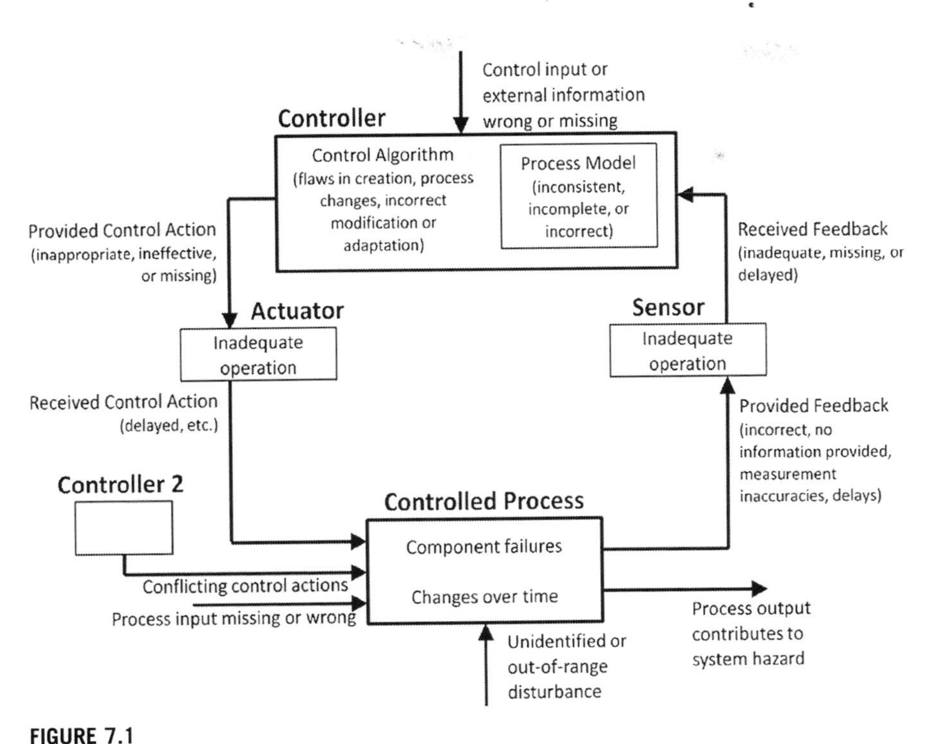

FIGURE 7.1

Generic control loop with possible failure causal factors from Thomas.[1]

of interest to control risks that the whole organizational structure will be subjected to. Thomas's assignment was to find ways to cope with detail because the information quickly becomes overwhelming. As mentioned in Chapter 5, the steps to perform an STPA are to determine successively the unacceptable *losses to the system*, the *hazardous states* that can cause these losses, and identification of the *safety constraints* preventing these hazardous states to arise. It is then crucial to identify which of the various forms of *hazardous control action*, shown in Figure 7.1 in a particular case, violate the constraints.

Thomas describes two example loops, a simpler one of a controller of train doors and a more complex one of operating a main steam isolation valve in the secondary circuit of a nuclear power reactor. To keep it simple at the start while introducing the conceptual approach, a simplistic example is preferable, but we shall do that here in a chemical semibatch process setting. Suppose an operator has to open the feed of reactant B while reactant A is in the reactor, warming up until the temperature is right and the stirrer is working. If the conditions are not fulfilled, reactant B accumulates and a hazardous state arises through which the system can run away and be destroyed. Hence, the safety constraint is that for adding reactant B the right conditions are present.

Next, Thomas distinguishes the *source controller* (*SC*), in this example the operator, providing or not, a *type* (*T*) of *control action* (*CA*), here activating dosing reactant B; this occurs in a certain *context* (*Co*), here reactant A present or not, temperature right or not, and stirrer on or not. In this particular example setting inaction will not occur, but clearly in others it will, inaction can be as hazardous as here untimely action. The context variables are safety critical and determine whether the hazard really arises. These variables shall appear in the process model.

So, the systematic procedure of identification boils down to describing the process by its variables and their functional relations, and identification of hazardous system states and given inputs, the control actions the controller enables. The process model variables come in a certain hierarchy determined by their relations, where the lower ones are safety critical, which at a certain value make a control action hazardous (see the STPA guide words in Chapter 5: CA not provided; unsafe; too early, too late, or out of sequence; stopped too soon or applied too long).

In a context table (Table 7.1), one can find out which possible combinations yield a hazardous state, so that the control action qualifies as hazardous. Hazardous state distributions can differ in severity and probability, so if opportune, different hazardous states can be defined. It is beyond the example applied here.

This very simple example with only bimodal variables is chosen at an intermediate level in a physical setting to make it better insightful, but STPA would be more valuable when applied at the management level controlling an organization in a command structure by means of actions (orders), following procedures, and with the aid of a management system with indicators and other feedback of how the process performs.

A necessary, subsequent step is identifying causal factors scenarios, namely to find out why on itself safe control actions are not be followed or executed, and what are the causes of hazardous control actions. The former could be the case in the example if reactant A due to contamination contains a small amount of inhibiting substance delaying reaction, which then at a certain point suddenly starts and causes runaway by overheating. Hence, the basic scenario for executed but unsuccessful control action is delayed reaction and runaway, and the safety-critical context is the contamination. This generates a safety-critical requirement with respect to the purity of reactant A. Another cause could have been a component failure (CF), for example, sticking of a valve disturbing a predetermined dosing rate. Hence, subdividing the basic scenarios can provide safety-critical design requirements for the control path.

The generic causal factors resulting in hazardous control actions have been shown in Figure 7.1. An inconsistent, incomplete, or incorrect process model is a source of much trouble. It can take many forms not all of which have to be hazardous. As Thomas reminds us, an incomplete or wrong process model in the mind of the operator, hence a failing mental image of the process (or typified in Chapter 6 as knowledge-based error), has been the cause of many accidents. We shall look at this again in the next chapter on control of industrial processes involving hazardous materials. These scenarios can also provide much useful information in the design stage

Table 7.1 STPA Context Variable Combinations Leading to Hazardous Control Actions

				Hazardous Control Action			
Control Action, CA	**Reactant A Present**	**Stirrer on**	**Temperature Right**	**Action Not Provided**	**Action Unsafe**	**Too Early Or Too Late**	**Stopped Too Soon Or Applied Too Long**
Dosing B	Yes	Yes	Yes	No	No	Too early, yes; too late, no	Stopped too soon, no; Applied too long depends on chemistry
Dosing B	No	Yes	Yes	No	No	No	No
Dosing B	Yes	No	Yes	No	Yes	Yes	Yes
Dosing B	Yes	Yes	No	No	Yes	Yes	Yes
Dosing B	Yes	No	No	No	No	Yes	Yes
Dosing B	No	No	No	No	No	No	No

of a system. Another type of cause will be feedbacks that are incomplete, incorrect, delayed, or missing, and further subdividing of the scenarios will again lead to formulating additional design requirements.

The last part of Thomas's thesis is on development of formal methods of analysis with the goal to cope with many and, when successful, with a myriad of details. It starts with formal specification of hazardous control action as a four-tuple (*SC*, *T*, *CA*, *Co*) and application of sets of each parameter. Context *Co* can be further decomposed into a number of variables. Once the system is specified, a list of control actions can be automatically generated at least for a considerable part. The same will be true for model-based design requirement specifications. The limitation will be that at the start of a project a complete specification will not be available, so that proceeding must go in an iterative way. The approach offers on the other hand the possibility of identifying conflicts and design deficiencies, while as a spin-off it can be applied to serve nonsafety functional requirement specification. The setup has further to be elaborated and the scalability investigated.

In general, the trend of continually more complex sociotechnical systems increases the share of nonlinear, dynamic causes of accident that cannot be captured with the traditional cause-finding techniques. Quite a few have to do with various kinds of human error, which cannot be modeled in a bimodal way, good or bad, with a certain probability for either one. It would also be valuable to apply STPA to conceptual systems before investments are made to carry out engineering development and production. As in other methods the quality of the identification process will be determined by the thoroughness and experience/imagination of the investigating person or team.

7.3 BLENDED HAZID: HAZOP AND FMEA IN A SYSTEM APPROACH

7.3.1 A SYSTEM VIEW

In a drive to improve results obtained by two classical, basic and bottom-up methods to identify possible failure scenarios: HAZOP and FMEA (respectively, hazard and operability study and failure mode and effect analysis as described in Chapter 3), Ian Cameron and coworkers (see e.g., Seligmann et al.[2]) realized that progress can only be made by taking the concept of a sociotechnical system as a starting point. Therefore, they considered not only process hardware but also the people and the procedures and by applying a system's approach. The functional systems framework they proposed is shown in Figure 7.2.

Identifying what can go wrong depends on the resourcefulness of people able to predict hazardous events by possible cause—consequence combinations. In general, people are not very strong in doing this (and this seems to be even more apparent when it concerns feared dreadful events). As we have seen before, a systems approach but also knowledge/experience and good knowledge management help.

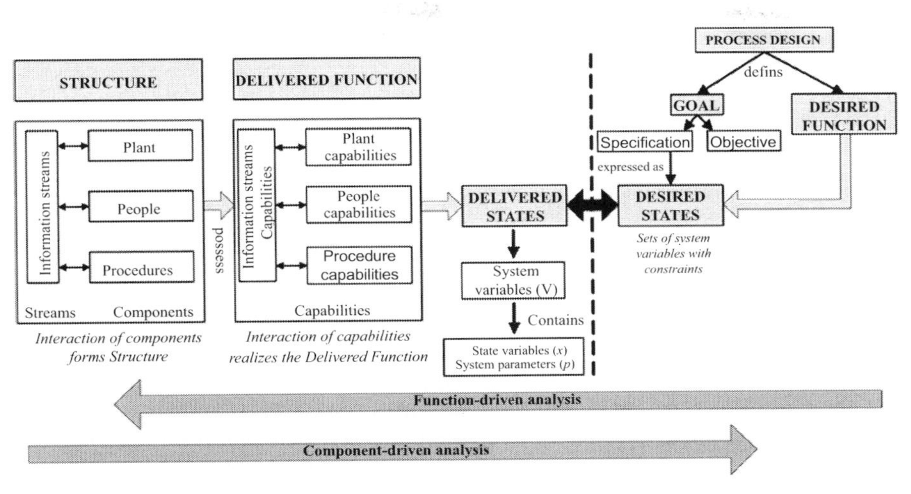

FIGURE 7.2

Functional systems framework of a process for the purpose of hazard identification by function-driven analysis (e.g., HAZOP) and component-driven analysis (e.g., FMEA), according to Seligmann et al.[2] The structure possesses capabilities to deliver the intended functions.

In this regard, Paltrinieri et al.[3] succeeded by applying a similarity algorithm to search incident databases most effectively for scenarios that would fit actual boundaries given by the situation at hand and that can be structured as a bow-tie, consisting of the sequence of causes, a critical event, and consequences. The new method is called dynamic procedure for atypical scenarios identification (DyPASI) and serves to generate supplementing scenarios for risk assessment. The method diminishes in this way the number of unknown "knowns." In Section 7.3.3, we shall discuss methods of HAZOP automation. Once an installation is operational, process supervisors will also be able to provide additional scenario information.

The basic idea of developing Blended Hazid (BLHAZID) is to perform combinations of types of hazard analysis and therefore to improve identification completeness. At the same time, causes and consequences shall be better defined to prevent inconsistency and improve diagnosing capability, while also a number of other improvements shall be made. Hazard analysis starts in the design stage but will have to be continued in subsequent stages and periodically in operation to verify whether new hazards have appeared. Also, management of change may require renewed (partial) systematic hazard analysis. The method will become more efficient if by improved structuring and computerization results of previous exercises are more easily retrievable. Results can then be reused in later, more detailed stages of HAZID and also during operation further checked, extended, and modified.

As Ben Seligmann[4] states in his PhD thesis, the objective of the BLHAZID development was to put hazard identification on an improved system theoretical basis and further, to "blend" results of different identification methods that identify function failures (FFs) and CFs as, respectively, HAZOP and FMEA do, although evidently partly overlapping. Apart from a certain overlap, the blending is complementary and generates synergy. A final objective was to apply a structured computer language (semantics) for storing and retrieving results. The latter will prove useful for fast fault diagnosis in case of process upset linking symptoms to causes to support abnormal condition/situation management (ACM/ASM).

However, the basis of BLHAZID is the functional systems framework describing the relations among the structure, the functions, and their goals. The structure consists of components—plant, people, and procedures—and further of streams—information (communication to and from people, and signals to and from controls) and material streams. Materials are characterized by properties, of which the values are also providing information. Hence, streams are generalized to information streams, although material streams can have properties very different from information streams. Plant consists of components subdivided in parts, which work together through a mechanism. Operating procedures can be in written form, software, or embedded in the control system.

Components have (a set of) *capabilities* for delivering a function, hence for exerting an *action* on a *property*, so, a pump can <increase> <pressure>, and a tank can <hold> <mass>. By connecting streams and components, capabilities are activated and this generates interaction, which in turn can change properties. A change in property can give rise to a capability and vice versa. One can think of many examples, but one given is about the capability of a heat exchanger. The transfer of heat (mechanism), hence interaction between two streams (carrying flow) and changing properties only activates when the temperatures (properties) are different, else the exchanger just holds mass. The set of activated capabilities produces the delivered function, which should realize the desired state. If the desired state is delivered, the system function meets its goals. Failure to deliver can occur in components, streams, and connections leading to a faulty state. The crux of hazard identification is therefore finding out causes and effects of differences between delivered and desired states. So, function-driven analysis is asking for the functions of subsystems and components and how function loss can occur, while component-driven analysis is searching for changes in activated capabilities.

The above can easily be seen for hardware components, with some limitation also applicable to procedures, but what about people? In the section on human error in Chapter 6 we have witnessed a proliferation of models and a clash of opinions. According to Erik Hollnagel,[5] identifying and predicting operation failures by modeling the operator "mechanistically" via task analysis and classifying a number of environments or contexts will not work because of the large variability in outcomes. Accident analyses show that variability in contexts make the human taking decisions, which in hindsight are designated as error. In other words, performance-shaping factors (PSFs), performance influencing factors (PIFs), are in his opinion too simplistic. So, failures shall be considered as a loss of control of the man–machine system. In the common performance conditions of Hollnagel's CREAM model (see Chapter 6) this is to a certain extent already the case, but in view of his 2012 disclaimer regarding the model clearly not sufficient.

Hollnagel's criticism focuses on cognitive tasks—the "K" of knowledge in the SRK rule. With the simpler skill and rule-based decisions, which are in daily practice the majority of action initiators, this discrepancy may be less emphatically present. Therefore HEART (see Chapter 6), because of its generic tasks and error-producing conditions, may be in the direction Hollnagel indicates and yield reasonable results. Taylor's Action Error Model (Chapter 6) and David Embrey's Systematic Human Error Reduction and Prediction Approach (SHERPA)[6], first published in 1986, have some characteristics in common. Both start with a task breakdown in subtasks and further task elements. This can be included in a HAZOP study, although it will require in that case the presence of human factor expertise in the team. SHERPA is more sophisticated than Taylor's model. It ranks task elements on the largest contribution to risk, identifies failure modes, and includes generic (time pressure, distraction, operator experience) and situation context-specific PIFs. Next, the PIFs are combined to a success likelihood index, of which the value has been calibrated to human error probability by a logarithmic relation as presented by Embrey and Zaed.[7] The future will reveal what model of human process interaction will feed BLHAZID best.

7.3.2 BLHAZID PRACTICAL WORKING OUT

For analyzing a system, the workflow followed by Seligmann[4] and summarized by Seligmann et al.[2] starts with defining the system boundary, after which first the function-driven analysis (HAZOP) is performed and subsequently the component-driven FMEA. As in conventional HAZOP, the first step is decomposition of the system into subsystems, which are divided in (major and minor) mass inventories, transfer subsystems and control subsystems, and in addition input, inlet, output, outlet, and environment ports, and direct and indirect process line connections (all with individual system graph icons and definitions given in Seligmann[4]). Then, for each subsystem the set of characterizing variables (*c-vars*) is identified. These variables represent the goal of the subsystem, they are or are associated with, internal subsystem variables, and for correct functioning of the subsystem to stay within certain tolerance limits. Examples of *c-vars* are temperature, pressure, or flow, but *c-vars* should be relevant for functioning of the subsystem and, therefore, for example, of a reactor, catalyst activity is more determining and hence more relevant than flow. Next, generate *deviations* (outside the specified *c-var* limits and consisting of a <guide word> and a <c-var>, for example, <high> <temperature>), elicit (usually as a team exercise) *causes* for a specific deviation and determine its *implications*. The causal structure of <cause> <deviation> <implication> forms a *triplet*. The structure is qualitative. Some implications may result in hazardous effects, which can lead to damaging consequences. In Table 7.2 an example for a benzene hydrogenation reactor (#940D), taken from the work of Seligmann et al.,[2] is reproduced.

Table 7.2 Example of Functional Triplets for a Benzene Hydrogenation Reactor

Characterizing Variable	Guidewords
Temperature	High, low
Flow	High, low, no
Benzene concentration	High
H_2 concentration	High, low, no
Extent of reaction	Low
Hydrogenolysis extent of reaction	High
Catalyst activity	Low
Vapor–liquid ratio	High, low
Impurity concentration	High
Exotherm	High, low
Pressure	High, low
Pressure drop	High, low

Subsystem	Causes	Deviation	Implications
940D: Reactor	High inlet temperature from TR1	High temperature	High outlet temperature to TR2
940D: Reactor	High benzene concentration	High temperature	High outlet temperature to TR2
940D: Reactor	940D maldistribution	High temperature	940D external leak
940D: Reactor	High inlet temperature from TR1	High temperature	940D external leak
940D: Reactor	High benzene concentration	High temperature	940D external leak
940D: Reactor	High inlet temperature from TR1	High temperature	940D rupture
940D: Reactor	High benzene concentration	High temperature	940D rupture
940D: Reactor	940D maldistribution	High temperature	940D rupture
940D: Reactor	High air impurity concentration	High temperature	High outlet temperature to TR2
940D: Reactor	High air impurity concentration	High temperature	940D external leak
940D: Reactor	High air impurity concentration	High temperature	940D rupture
940D: Reactor	High inlet temperature from TR1	High temperature	High vapor–liquid ratio
940D: Reactor	High benzene concentration	High temperature	High vapor–liquid ratio
940D: Reactor	High air impurity concentration	High temperature	High vapor–liquid ratio
940D: Reactor	High inlet temperature from TR1	High temperature	High outlet temperature to TR2
940D: Reactor	High benzene concentration	High temperature	High outlet temperature to TR2
940D: Reactor	940D maldistribution	High temperature	High outlet temperature to TR2

Table 7.2 Example of Functional Triplets for a Benzene Hydrogenation Reactor *Continued*

Subsystem	Causes	Deviation	Implications
940D: Reactor	High air impurity concentration	High temperature	High outlet temperature to TR2
940D: Reactor	High inlet temperature from TR1	High temperature	High exotherm
940D: Reactor	High inlet temperature from TR1	High temperature	High hydrogenolysis extent of reaction
940D: Reactor	High hydrogenolysis extent of reaction	High temperature	High outlet temperature to TR2
940D: Reactor	High hydrogenolysis extent of reaction	High temperature	940D external leak
940D: Reactor	High hydrogenolysis extent of reaction	High temperature	940D rupture
940D: Reactor	High hydrogenolysis extent of reaction	High temperature	High vapor–liquid ratio
940D: Reactor	High hydrogenolysis extent of reaction	High temperature	High outlet temperature to TR2
940D: Reactor	High hydrogenolysis extent of reaction	High temperature	High exotherm

After Seligmann et al.[2]

Apart from human inputs, basic knowledge sources for the HAZOP analysis are process flow sheets, engineering CAD sheets, and measured variables found on piping & instrumentation diagrams (P&IDs). Currently the software is making use of Bentley AutoPLANT intelligent P&ID, but there are other providers of intelligent P&ID such as AVEVA, which is offering an AutoCAD-based one. Németh et al.[8] explain the (semi-)automated data extraction from the P&ID. Further sources are the capability sets of components affecting <c-var> values mentioned in the previous section, physical and chemical processes in a subsystem, design documents, and importantly, failures (implications) passed from connected subsystems. Cycling through all selected deviations in all subsystems will yield a large number of triplets.

A further step is the FMEA analyzing failure modes of equipment, their causes and implications. A failure mode is defined as an observable failure, for example, a leak, affecting the function of the component. Failure modes can be listed for a component before further analysis takes place. In the component reliability database OREDA, a link is made between failure mode and operational mode, while here the latter specifies the capabilities that should be activated for the function to be properly executed. For example, operational modes of a pump are ON or OFF. In the OFF mode, the capabilities <hold> <mass> and <permit> <flow> are active, while in the ON mode, the capability <increase> <pressure> is added. Failure modes affect activated capabilities, so a <leak> affects <hold> <mass>.

Once the list of failure modes has been prepared, one by one causes of the mode are elicited. In part these are root causes due to slow degradation of parts, or they are causes by disturbing implications transmitted by streams. Thus, again triplets can be formed: <cause> <failure mode> <implication>. An example is shown in Table 7.3.

Table 7.3 Component Failure Triplets According to the Benzene Saturation Example in Seligmann et al.[2] In the Lower Half of the Table No Causes are Mentioned As the Failure Modes Are the Result of Slow Performance Degradation and, Hence, Failures Are Considered to Be Autonomous

Subsystem	Causes	Deviation	Implications
940D: Reactor	Failure of inlet PEOPLE from environment	940D maldistribution	Low extent of reaction High exotherm High temperature High flow Low pressure drop
940D: Reactor	High exotherm High corrosion products impurity concentration High temperature High pressure	940D external leak	Low pressure drop Low flow Low pressure drop High outlet flow to environment
	High exotherm High temperature High pressure	940D rupture	High pressure High pressure drop No flow
940D: Reactor		940D catalyst bed blockage	High pressure High pressure drop No flow
940D: Reactor		940D outlet blockage	High pressure High pressure drop No flow
940D: Reactor		940D inlet blockage	High pressure drop Low pressure drop No flow
940D: Reactor		940D inlet partial blockage	High pressure drop Low flow
940D: Reactor		940D catalyst bed partial blockage	Low flow High pressure drop
940D: Reactor		940D outlet partial blockage	High pressure drop Low flow

In contrast to the function-driven triplets, in the component-driven ones the implication is only dependent on the failure mode and not also on the cause. Obviously there is considerable overlap between the two approaches, which should be sorted out by a computer. Also, there is much repetition in causal pathways for similar functions and components in an installation. Semiautomatic generation of a triplet and instantiation in outcomes is therefore a must, although expert checks on outcomes are imperative. The main structured language elements developed by Seligmann et al.[2] are collected in Table 7.4. As one may notice, triplets can be composed of five different combinations of FF and CF.

A software tool has been developed that can also generate the causal graphs, see Figure 7.3. Erzsébet Németh and Ian Cameron[9] presented a more expanded report about these causal graphs or rather cause–implications diagrams. These diagrams can be expected to become very useful for consistency checking of BLHAZID outcomes, auditing hazard identification, supporting design decisions, online diagnostics for operator decision support, and operator training. The importance of the latter will be further discussed in the next chapter.

The human and procedural components are still missing in all of the above, but work is also continuing on those parts. It looks therefore as if BLHAZID will become an important contributor to process safety by considerably improving the grip on the hazard identification step. Reuse of identification results in later safety reviews should yield a significant increase in work efficiency.

The foregoing has been based on the assumption that use is made of HAZOP and FMEA methods as these are conducted in the conventional way. However, because of the intensive use of HAZOP in the process industry, and its drawback of considerable investment in time and effort to accomplish, several attempts have been made to significantly increase the efficiency of conducting it. These attempts will be briefly discussed in the next section.

Table 7.4 Structured Language Elements Summary (VA and TA Are Component Designations) According to Seligmann et al.[2]

Concept	Syntax	Example
Capability	<action> <property>	<increase> <pressure>
Component	<component>	<VA>
Failure (CF)	<failure mode>	<blockage>
Functional failure (FF)	<guideword> <c-var>	<high> <temperature>
BLHAZID triplet	(FF, FF, FF) or (CF, FF, FF) or (CF, FF, CF)	(<high> <inlet temperature>, <high> <temperature>, <high> <pressure>)
(Cause, deviation, implication)	Or (FF, CF, FF) or (CF, CF, FF)	(<high> <pressure>, <TA> <rupture>, <high> <flow to environment>)

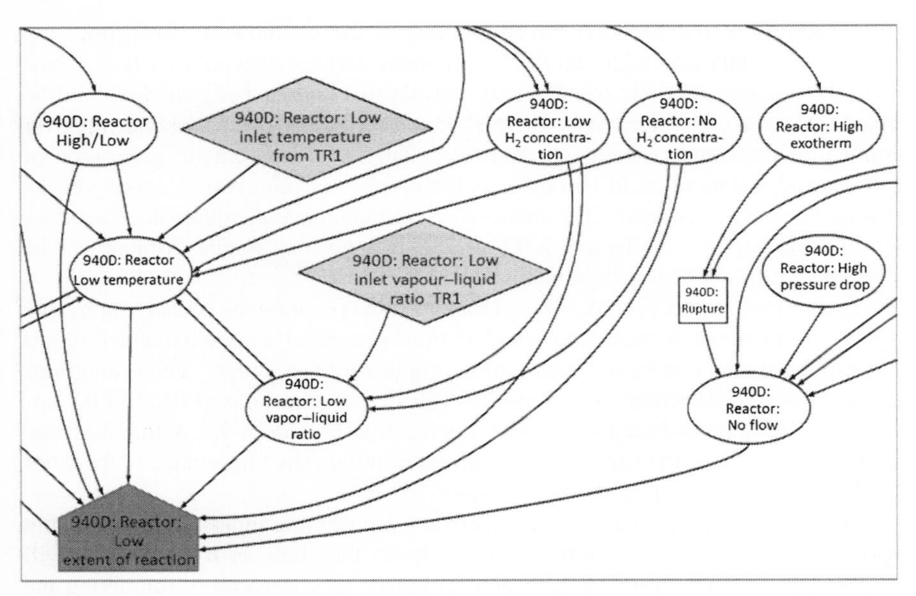

FIGURE 7.3

Part of a causal graph of possible causes of low extent of reaction generated by the BLHAZID software tool *(after Seligmann et al.[2])*. A full graph will also show the root causes (failures and faults) and implicated effects. Given all information stored by the BLHAZID about all deviations that can occur in the system, an automated diagnosis of an upset will appear on the horizon.

7.3.3 HAZOP AUTOMATION ATTEMPTS

HAZOP is an indispensable method of process hazard identification with a worldwide use. In fact, it became applied in many other fields to uncover undesired scenarios. In 1997 Trevor Kletz,[10] describing the past and future of HAZOP, foresaw an important role of the computer in HAZOP and in presenting HAZOP results in case of upset to diagnose the cause. In 1999 McCoy et al.[11] presented an overview account of work on computer aiding HAZOPs performed thus far, besides describing their own knowledge-based system HAZID. In 2010 Dunjo et al.[12] published a general literature review on HAZOP. Computer support of HAZOP began in the 1980s in an administrative sense but at various places attempts were soon made to expedite the identification steps proper, in fact to automate or at least to computer assist the procedure. These attempts tried also to mitigate incompleteness or human bias by lack of knowledge, experience, or creative thinking. Some of these attempts are based on a digraph model of the equipment and an expert system. Digraph will be briefly described in Section 8.3.2; an expert system is based on "if-then-else" rules and contributes by suggesting cause–consequence relations as HAZOP emulators. Other attempts consisted of making use of different graphical tools to assist the reasoning.

Venkat Venkatasubramanian of Purdue University, who was the lead author of a summary paper[13] on intelligent HAZOP, initiated some important advances. An overview is given of typical examples of the more recent approaches in Table 7.5. For the time being, none of the approaches has emerged as a champion. The literature contains more articles on HAZOP automation than are referenced here.

Looking at the developed methods in Table 7.5, one can distinguish two main lines of approach: (1) automation by applying qualitative modeling and reasoning by expert system and (2) HAZOP support by process simulation enabling revealing up- and downstream causes and consequences. Comparison of the various approaches yields the following. According to Rodriguez and De la Mata,[27] setting up the rule-based expert system costs more effort than that of a D-higraph model because much function and goal information about the process is already incorporated in the model. The same is true for the PetroHAZOP learning expert system of Zhao et al.,[23] although it has the advantage of being able to absorb new cases and it is learning. MFM versus D-higraph yields pluses and minuses. D-higraph is more intuitive because it reflects the equipment directly. It also can be used for other purposes such as curbing alarm floods and tracking and tracing faults. Integration of D-higraph with Aspen Plus process simulation has been attempted. MFM is more abstract but can still be used for further, more quantitative analysis of identified scenarios with respect to likelihood and severity. Likewise, MFM may be useful for HAZOP-ing start-ups, which is a topic that will come back in the last section of Chapter 8. Thunem[30] developed a Java-based MFM editor for model design.

In the case of continuous processes, a DBN as proposed by Hu, Zhang and Liang[28] is suited to assist in conducting a preoperational HAZOP to reveal and track the fault propagation paths and possible coinciding of faults and associated consequences. Also, based on historic data, probabilities can be quantified. Then, later in operation by monitoring observed process variables, development of initially hidden defects can be predicted enabling timely maintenance. Also, in view of the development of BLHAZID this approach is quite promising.

Few detailed reports are available on comparisons of automated HAZOP results with those of experienced HAZOP teams. Rather poor results of the SDG automated HAZOP system HAZID are reported by McCoy et al.[18] Roughly 10–30% of scenarios identified by the system were judged correct and useful. A more systematic and a coordinated effort over various groups would be needed to sort out the most effective and efficient approach, preferably in conjunction with BLHAZID as a follow-on for comparison.

A separate aspect is extracting information needed for a HAZOP from material safety data sheets, piping and instrumentation diagrams (P&ID), flow sheets, procedures, and other sources. As already mentioned in Section 7.3.1, an intelligent P&ID used as input to BLHAZID already contains much process data. As reported by Németh et al.,[8] it provides input as a result of functional layering of equipment (mass/energy inventories, process lines, instrumentation and control, and sublayers). Upon decomposition of the system (semi-) automatically component function classification and design intent is obtained. This enables grouping in sections with

Table 7.5 Various, Relatively Recent Approaches to Support and to (Semi-)automate HAZOP

Year	Authors	Approach
1995	Vaidhyanathan and Venkatasubramanian[14]	*Digraph*-based HAZOP. Digraph is explained in Chapter 8, Section 8.3.2. Basically it is modeling material and information flows initiating or undergoing changes.
1996	Vaidhyanathan and Venkatasubramanian[15,16]	HAZOPExpert is a *DiGraph* (HDG) model-based, object-oriented, intelligent system for automating HAZOP. The *expert system* is provided with semiquantitative reasoning that is checking whether, in case of loss of containment, conditions surpass the autoignition threshold and whether a spill presents a toxicity risk. It further checks the adequacy of protective devices and ranks consequences.
1998	Srinivasan and Venkatasubramanian[17]	HAZOP applied to batch processes (with additional challenges compared to continuous process of operator procedures and actions, and discrete process steps). This is realized by *Timed Petri Net* (see Section 7.8) representation of the batch process and DiGraph to represent causal relations between process variables in sub-tasks. In combination with the earlier *expert system* work it constituted the batch HAZOPExpert.
1999–2000	McCoy et al.[11,18]	Software system HAZID consisting of several modules: AutoHAZID is the heart of the system. The description is quite detailed. It has at the start a configuration checker after the program read in the plant description and built a Signed DiGraph (SDG) of the process units. It further has a qualitative effects module. The HAZOP emulation module was developed in the earlier STOPHAZ project. It is *a rule-based inference engine* generating scenarios. VTT (Finland) contributed with a fluid library and fluid rules distinguishing feasible from infeasible scenarios. Fault propagation was modeled by means of SDG. The output is filtered to remove redundant information.
1997–2000	Khan and Abbasi[19,20]	Development of a procedure to speed up “HAZOP-ing”: optHAZOP, followed by development of a *knowledge-based inference engine* software tool enabling automation. The tool consists of a general and a process specific part. It generates deviations and contains rule-based trees linking process specific attributes, via process parameters, and deviations to causes and consequences. Renamed from TOPHAZOP to EXPERTOP.
2005	Zhao et al.[21]	Software system PHASuite: Instead of DiGraph the process is now represented by *colored Petri net* (see Section 7.8), but the Petri net also represents the methodology to perform the HAZOP. The process is abstracted to two levels: operation and equipment, which have been functionally linked. Knowledge is externally stored in layered operation and equipment models in a structured database that can be approached by a user via knowledge builder. A *case-based reasoning engine* (CBR, stories containing knowledge/experience from experts) operating on two levels and two layers performs the automated HAZOP. Application is again to pharmaceutical batch processes, more difficult to HAZOP than continuous ones. Gain in time spent is about 50%.

2008–2009	Cui et al.[22]; Zhao et al.[23]	As an extension of the digraph method of Vaidhyanathan and Venkatasubramanian,[15,16] a *Layered DiGraph* (LDG) expert system was proposed. The digraph is now three-dimensional, which enlarged the flexibility and knowledge storage capability. Each layer or workspace is associated with a guideword. The workspaces contain nodes representing variables interconnected by (unsigned directed) arcs, implying that the deviation in the "parent" node determines the direction of deviation in the "child." Linked nodes can also be in different workspaces. The authors claim that a higher degree of completeness of HAZOP scenarios is achieved. Later the same group developed *PetroHAZOP*, an expert system but this time it is learning by *case-based reasoning* (see PHASuite above) making use of CAPE ontology (explicit specification of conceptualization) for process systems. A case consists of problem/situation, solution, and outcome description. A new problem is judged on similarity by an algorithm based on predefined indexes. It is thus highly domain dependent. It functions in the Chinese petrochemical industry with 900+ cases. A future effort was announced to combine the two approaches.
2009	Rahman et al.[24]	Further development of Khan and Abbasi's EXPERTOP to ExpHAZOP+ with some added features as an enhanced graphical user interface (GUI) and a selection method for an equipment node. It also added an update possibility of the knowledge base and introduced a unique *fault propagation algorithm*, identifying downstream causes and consequences from an identified upstream event.
2010 2014	Rossing et al.[25] Wu et al.[26]	*Multilevel flow modeling* (MFM), developed by coauthor Lind in the early 1990s, is applied to describe the plant goal-function structure. MFM can be used at various abstraction levels, applies symbols (of which a few resemble Petri net ones) for objectives (source, transport, storage) and functions (sink, barrier, balance), and it describes the interactions of mass, energy, and information flows, combined to flow structures. Also, symbols are available for functions as management, decision, and actor action. Further, a set of means-to-ends relations (with symbols for produce, producer product, maintain, and mediate) and causal roles (with condition, agent, participant symbols) describe dependencies between functions. The interconnected flow structures to achieve a goal are represented graphically. Combined with a rule-based causal reasoning engine, and quantitative dynamic simulation (with e.g., HYSYS), MFM can generate fault/cause and consequence trees/paths for a given deviation in a system function, and with a goal reasoning engine goal trees. The different trees can be used in reasoning to develop counteraction plans. The whole is called MFM workbench. After process variable deviations have been specified, the workbench facilitates HAZOP as a functional assistant by diagnosing the causes of abnormal situations. It does not have the aim of automating HAZOP. The concept is further elaborated, extensively described, and demonstrated on an offshore three-phase separator case by Wu et al.[26]

Continued

Table 7.5 Various, Relatively Recent Approaches to Support and to (Semi-)automate HAZOP *Continued*

Year	Authors	Approach
2012	Rodriguez and De la Mata[27]	*D-higraphs* are another way of modeling a process including controls. Developed in the late 1980s, D-higraphs represent in yet another way states (blobs, being a function effected by an actor—a machine—with an optional condition as a Boolean variable) and transitions (edges). Hence, D-higraphs combine in their representation function and equipment/structure, so there is more direct correlation with the real installation than with MFM. Distinction is made between mass, energy, and information edges. There are process (green), control (orange), and mixed (blue) blobs. The edges can be triggered or fired resulting in state changes. A blob can contain other (sub-) blobs and can also be partitioned to represent an OR-statement. Causal rules have been established. The system description is in three layers: structural, behavioral, and functional. Deviations are coded and the reasoning engine is constructing cause and consequence trees. For comparison the same distillation unit was "HAZOP-ed" as Rossing et al. did. The D-higraph HAZOP assistant results were not different.
2012	Hu et al.[28] Hu et al.[29]	Having in mind prognosis for enabling predictive maintenance to prevent process upset a HAZOP method was developed assisted by a *dynamic Bayesian network* (DBN), see Section 7.5. Application was for a gas turbine plant where wear, fouling, and corrosion lead to faults. A DBN was chosen because process faults often have multiple propagation paths to different effects, some of which propagate to adjacent parts. This may lead to fault coupling and disaster. A DBN can represent these interactions in space and time by conditional probabilities. Variables are represented by nodes and the causal structure between variables by edges (arcs). Degradation of components is modeled by a distribution, for example, Weibull. Observable variable values can be obtained from the supervisory control and data acquisition (SCADA) system. Then, DBN-HAZOP can predict failure before it occurs.

distinct function. Also, the storage of the structured information will facilitate further computerized HAZOP. For extraction of information in a general sense process ontology is important. Ontology is a hierarchical framework of concept descriptions within a certain domain. It is usually more extensive than just taxonomy. Zhao et al.[31] discussed the role ontology can have in automated HAZOP as in PHASuite. Morbach et al.[32] developed ontology for process engineering purposes within computer-aided process engineering (CAPE). CAPE was initiated in the late 1980s as a project with many contributors and by stimulating inputs of among others Rafiqul Gani of the Technical University of Denmark. CAPE is established now as a working party of the European Federation of Chemical Engineering. Many papers can be found in ESCAPE symposia proceedings. The systems approach embodied in CAPE shall play a future role in process safety.

7.4 INNOVATION AND EXTENSION OF CLASSICAL RISK ASSESSMENT METHODS

In previous sections we have seen progress in the foundation of risk assessment, namely in hazard identification (HAZID). The HAZOP method is the main, classical hazard identification tool used in risk analysis, on which scenarios can be based. We also saw attempts by (semi-) automation to increase quality of the HAZOP effort and reduce manpower required. Blended HAZID can be considered as an innovation to the scope of HAZOP and reusability of its results.

There have been quite a few attempts to improve existing quantitative, also called probabilistic, risk assessment methodology. Improving consequence analysis is a steady activity in which many contribute; obtaining better event frequency data is another perpetual endeavor. A possible estimation method of the usual extremely low, so-called rare accident event frequency by making use of recorded data on the higher frequencies of near misses and other precursors will be discussed in Section 7.7.3 after having explained Bayesian statistics. Availability of equipment failure data is a matter of willingness to share data and organizing a database, which over the years proved to be quite a thorny path. A few proprietary component failure data bases have been established, but no sound public ones with exception of the North Sea offshore industry where UK's HSE and Norwegian SINTEF made much information available. With respect to methodological addition, adding temporal effects of degradation of equipment can be mentioned. Under the name dynamic (operational) risk analysis, concepts associated with reliability engineering in more extended form can be taken into account. Examples are availability, mean time to repair, and various failure models (e.g., Weibull), which can be identified to model the failure population as sampled through tests. Dynamic versions have been proposed for fault and event trees (combined in bow-tie, see Khakzad et al.[33]) and master logic diagrams (Hu and Modarres[34]).

Hence in a qualitative sense renewal is modest, but from a quantitative effort point of view there is much renewal in source terms, dispersion and explosion models, substance properties, etc. Nevertheless, uncertainties still dominate the scene. The issue of uncertainty is the topic of Section 7.6 and of Chapters 9 and 12 with respect to decision making. Hopefully, risk analysts finally will learn to estimate uncertainties and to reflect these in the risk figures they produce. Most fundamental for a system approach is to think, work, and seek information in distributions for both consequences and event probabilities, and not just in intervals or point values. Then, the results of risk assessment should be displayed in terms of distributions or risk profiles with estimated confidence limits. Why this is useful will be explained in Chapter12.

Progress will be achieved by making use of a probabilistic approach to include Bayesian data analysis and information updating. If one method has changed the risk assessment landscape in this century it is Bayesian statistics and BNs, thanks to the development of software produced by various sources. The reason why, and the way these methods work and why they are necessary for a system approach, is explained rather extensively in the following sections.

7.5 BAYESIAN STATISTICS AND BNs

The Reverend Thomas Bayes, after whom Bayesian statistics is named, lived between 1701 and 1761 and was for a great part of his life a minister in Tunbridge Wells, UK. He had studied theology and mathematics, and published a theological work, "Divine Benevolence, or an Attempt to Prove That the Principal End of the Divine Providence and Government is the Happiness of His Creatures (1731)," and a mathematical one on "fluxions," in which he defended the work of Isaac Newton. His work on probability was published only after his death, and the definition of probability he gave was worded: "The probability of any event is the ratio between the value at which an expectation depending on the happening of the event ought to be computed, and the value of the thing expected upon its happening." This somewhat puzzling expression aroused enthusiasm of later physicists and mathematicians such as Pierre-Simon Laplace trying to distil logic from measurements ridden of random error. The citations also suggest that Bayes was looking for proof of the existence of divine powers.

The crux of Bayesian data analysis, which has become increasingly popular the last few decades, is the statistical theorem telling how to update existing knowledge, A, with newly obtained observations, B. It is based on the "product rule," which according to the great defender of Bayesian approach, Edwin T. Jaynes,[35] was already recognized by others, such as Jacob (also named James or Jacques) Bernoulli, long before Bayes. This in its simplest form is: $P(A|B) = P(B|A)P(A)/P(B)$, telling us that the probability of an uncertain event A, given an observed event or evidence B, is equal to the probability of the evidence B given A occurred (or as a hypothesis is considered true), weighted by the prior probability of event A, independent of B, and normalized by the total probability of observing the evidence B. A decomposition according to the law of total probability of the total probability of observing B yields: $P(B) = P(B|A)P(A) + P(B|notA)P(notA)$, where the sum is over all values of the uncertain A weighted by the prior probabilities of A. The posterior probability of A (or of the hypothesis) given the evidence B, $P(A|B)$, containing the prior and observed information, is then calculated using the Bayes model expression above.

In case of Boolean expressions of A and B as true or false, this is immediately clear. The theorem fits well also with the subjectivist approach of probability, hence the degree of belief in a hypothesis or expert judgment is increased or decreased based on the evidence modeled in the function, $P(B|A)$. Prior belief in the probability of the uncertain A, $P(A)$, can be updated with new evidence B modeled by the likelihood $P(B|A)$, which upon forming a joint distribution with the prior, $P(B|A)P(A)$, is normalized by the total probability of the evidence B, $P(B)$, producing a posterior value of the hypothesis, $P(A|B)$. The theorem can be used in a predictive sense by making some assumptions or by simulating a likelihood, while having background knowledge (prior) and producing a predicted/future observable outcome (as a posterior). Also the posterior can be used as a weight factor to calculate the posterior predictive probability of an observed event or of a parameter. But the theorem is used also for inference or diagnostics by having the background (prior), obtaining new observations providing the likelihood, and hence inferring a posterior. Thereby, each variable can have its own type of probability distribution, discrete or continuous.

If applied to, for example, information on a failure rate of a component, λ, expressed as a continuous probability density function (pdf) of past experience, a prior, $f(\lambda)$, and E, newly observed evidence (likelihood), expressed as a likelihood function, also called sampling density, conditional on values of λ, $L(E|\lambda)$, we can obtain a posterior distribution. This distribution can be derived from the product of the newly obtained information, E, and the prior distribution of failure rate values normalized by the total probability of E, $P(E)$, according to:

$$p(\lambda|E) = \frac{L(E|\lambda)f(\lambda)}{\int_0^\infty L(E|\lambda)f(\lambda)\mathrm{d}\lambda} \tag{7.1}$$

As before, the denominator integral (TP or total probability of the observed evidence E, $P(E)$), a constant, serves to normalize the result (constraining probability between 0 and 1) and so making the nominator product integrate to 1 and consistent with the probability axiom that the mutually exclusive contributions that span the sample space must sum to 1. If the distributions are continuous, the updating calculation updates in one operation of the continuous Bayes model all values of the uncertain parameter, here the prior distribution of λ. This single update is instead of updating separately each discrete value of a prior probability mass function in the case of discrete distributions.

The distribution functions can in principle all be different, but as we shall see later there are preferred combinations, the so-called *conjugates*. Averaging by weighting with the posterior of λ yields the posterior mean: $\int_0^\infty p(\lambda|E)\lambda\mathrm{d}\lambda$. Also useful is the determination of a probability interval. For example, 90% of λ's probability mass is between the fractions of $a(E)$ and $b(E)$ when is satisfied that: $0.90 = \int_{a(E)}^{b(E)} p(\lambda|E)\mathrm{d}\lambda$. Many pairs of a and b will fulfill this requirement, but the shortest interval is called the highest posterior density. This notion of confidence limit shows a typical difference between the classical frequentist statistics and the Bayesian. In the former one could specify a mean value and a margin of error at, for example, the 90% confidence level, only based on the size of the test. As seen above in the Bayesian context one can state there is 90% probability that the value lies between specified limits. This is because one assumes prior information expressed as a probability distribution, for example, a uniform one or a prior based on generic data or expert judgment. For example, determining a failure rate from a limited sample of items of which over a certain time some fail clearly shows the Bayesian advantage.

A further feature is the Bayes factor, B, which is the Bayesian alternative to hypothesis testing and to Fisher's likelihood ratio. The factor is the quotient of the probabilities of data observed, D, given two alternative models M_1 and M_2 with model parameter vectors θ_1 and θ_2. Applying the law of total probability to marginalize over the model parameters weighted by the prior probabilities of the parameters, this can be expressed for a relative test of the two models as a likelihood ratio assuming the prior distributions for models M_1 and M_2 are equally probable:

$$B = \frac{P(D|M_1)}{P(D|M_2)} = \frac{\int P(\theta_1|M_1)P(D|\theta_1, M_1)\mathrm{d}\theta_1}{\int P(\theta_2|M_2)P(D|\theta_2, M_2)\mathrm{d}\theta_2}. \tag{7.2}$$

If $B > 1$, the data favor M_1, while if $B < 1$, the data favor M_2, whereas if $B \approx 1$, each model likelihood is equally probable to represent the observed data, D. In contrast with traditional statistics, with Bayesian methods, we seek the relative probabilities of the models or hypotheses given the evidence without any constraining or arbitrary threshold or cutoff point between acceptance and rejection. This can be further evaluated to Schwarz's Bayesian information criterion (BIC), enabling to select the most likely, but simplest, model with the least number of parameters given the data, an operation also known as Occam's (or Ockham's) razor.

Bayesian analysis can also be seen as providing the predictive probability density $f_p(\lambda^*|E)$ of future observations λ^*, conditionally independent of the "new" evidence, E, given the existing "old" information on λ. The predictive density using the posterior probability, $p(\lambda|E)$, as the weight factor then becomes

$$f_p(\lambda^*|E) = \int f_p(\lambda^*|\lambda)p(\lambda|E)\mathrm{d}\lambda \tag{7.3}$$

That this is true follows from the assumed conditional independence stated above (for which there are exceptions) and the law of total probability, see, for example, Ronald Christensen et al.[36] for fundamentals. When we discuss in Section 7.7.3 the hierarchical Bayesian analysis, the significance of the relation will appear to estimate very small probabilities of catastrophic events related to more frequently occurring "precursors" of such rare events.

Mohammad Modarres et al.[37] elaborated reliability and risk applications of the Bayes theorem. With little information to start with as a prior distribution, a uniform distribution of λ (state of total ignorance in which all values within a range are assumed equally likely) can be assumed, while as the likelihood function, in the case of a constant failure rate, the continuous exponential distribution or the discrete homogeneous Poisson distribution (λ is constant in time and hence the process is memory-less) is suited. The exponential or the Poisson as the likelihood function with a gamma distribution as the prior produces also a gamma distribution as the posterior. A much-used prior of the parameter, p, of the binomial distribution is a beta-distribution, which is the generalization of the binomial distribution. A beta prior with a binomial likelihood produces also a beta posterior. These prior–posterior combinations are applied in failure in time and in failure on demand problems. They are called *conjugate pairs* and can be evaluated analytically. In other, more complex cases, numerical integration by means of Markov Chain Monte Carlo (MCMC) as incorporated in software package WinBUGS[38] will be needed, or alternatively for less complex cases solving a BN of which software shall be described later in this section. WinBUGS is quite well known for Bayesian data analysis applying complex statistical models (BUGS stands for Bayesian inference using Gibbs sampling). The BioStatistics Unit of the Medical Research Council hosted by the University of Cambridge, UK, developed it, later together with the Imperial College, School of Medicine in London. The use and applications of this package are extensively explained by Ronald Christensen et al.[36] The rapid spread in this century of Bayesian inference to a large spectrum of fields (finance, health, medicine, behavioral science, et cetera) results from both the advent of the MCMC sampling method to solve equations, WinBUGS, and Bayesian network software. Phil Gregory[39] formulated the differences between Bayesian and frequentist approach very clearly and explained how to make best use of both in data analysis.

Central in risk analysis are cause–effect relations of probabilistic nature. Probability of effect B occurring depends on, and hence is conditional on, instantiation of cause A, with probability $P(B|A)$. Graphical notation helps to structure problems and logical structuring leads to BNs. Sometimes these are called Bayesian belief networks (BBNs) because of possible subjective use of probability by assigning a value to a degree of belief, as with expert judgment. BNs are acyclic, directed graphs (DAGs) consisting of parameter nodes or vertices shown as ovals, and dependencies among parameters represented by edges or links, often called arcs and shown as arrows. The nodes represent causes and effects by random variables. The net represents the structure of the dependencies. The graph is acyclic because an effect will have no influence on its cause, but an effect can cause other effects.

Among others, Judea Pearl[40] has constructed the foundation under causality with BNs. Because of this work and the creation of causal inference algorithms he received in 2011 the A.M. Turing award. An effect can have multiple causes that in turn can have more basic, root, or underlying causes. There can be (unobserved or hidden) confounders or common causes to effect nodes that are otherwise dependent on different, mutually independent causes. A statistical derived correlation between two random variables, even with a high correlation coefficient, is not necessarily a cause–effect relation. There is quite a good correlation of Ferrari sports cars and the color red, but there is no cause—effect relation. For sure is that measured in time an effect comes after a cause, but to express this mathematically is not simple. In a mathematical sense, the direction of the arc does not make a difference. Pearl explains why and provides a solution.

BN variable values can have discrete or continuous distributions. Given a discretely distributed probability of causes, the conditional distribution tables of the effects can be calculated by hand as long as the number of nodes is limited to a few. In Figure 7.4 a simple example of a discrete net is shown representing a batch reactor in which a runaway may be caused by two types of failures. To understand the calculations shown in the table, a bit of theory is needed. For this we shall follow Norman Fenton and Martin Neil's[42] book, which not only explain BNs in detail but also presents many practical, daily life examples of problem solutions by BNs.

It all spins around conditional probabilities, joint distributions, and determining marginal distributions. In risk assessments one must predict the outcome of many dependent variables. The *chain rule* is most important. If we have a joint probability distribution of variables A, B, and C, for example, we can write

$$P(A,B,C) = P(A|B,C) \cdot P(B|C) \cdot P(C) \tag{7.4}$$

By expansion of a system joint probability density function (pdf) into products of conditional and marginal probabilities, the pdf can be simplified by dropping dependencies that are not considered significant based on the current information. For example, if the evidence suggests that B is not dependent on or influenced significantly by C, $P(B|C)$ can be replaced or approximated by $P(B)$ as an example of conditional independence.

The above expression (7.4) is used in the table shown in Figure 7.4. It can be generalized as follows:

$$P(A_1, A_2, A_3, \ldots\ldots, A_n) = \prod_{i=1}^{n} P(A_i|A_{i+1}, \ldots\ldots, A_n) \tag{7.5}$$

In Figure 7.4 the cause (parent) nodes are represented by marginal distributions, because no influences on these "parents" are known or are explicitly accounted for, so they will often be represented by point values as in Figure 7.4, or also as uniform distributions (fully arbitrary, no preference). All the rest of the nodes are conditioned on one or more variables in joint distributions. The marginal probability of, for example, $P(Runaway = True)$ derives from the chain rule equation above. BNs also allow inserting new evidence following which the network is updated by calculations applying the Bayesian theorem to identify the cause with the highest probability and therefore the most likely explanation (MLE), given the evidence. So, both prediction and diagnostic inference are facilitated, which explains the popularity of BNs in artificial intelligence, finance, economics, in medical diagnosis, and other decision systems. For a more detailed scientific and mathematical approach to BNs, Adnan Darwiche's book[43] is recommended.

Do not think that setting up a network to solve a problem is that easy. It requires tight reasoning. Given data, a most probable network fitting the data can be derived

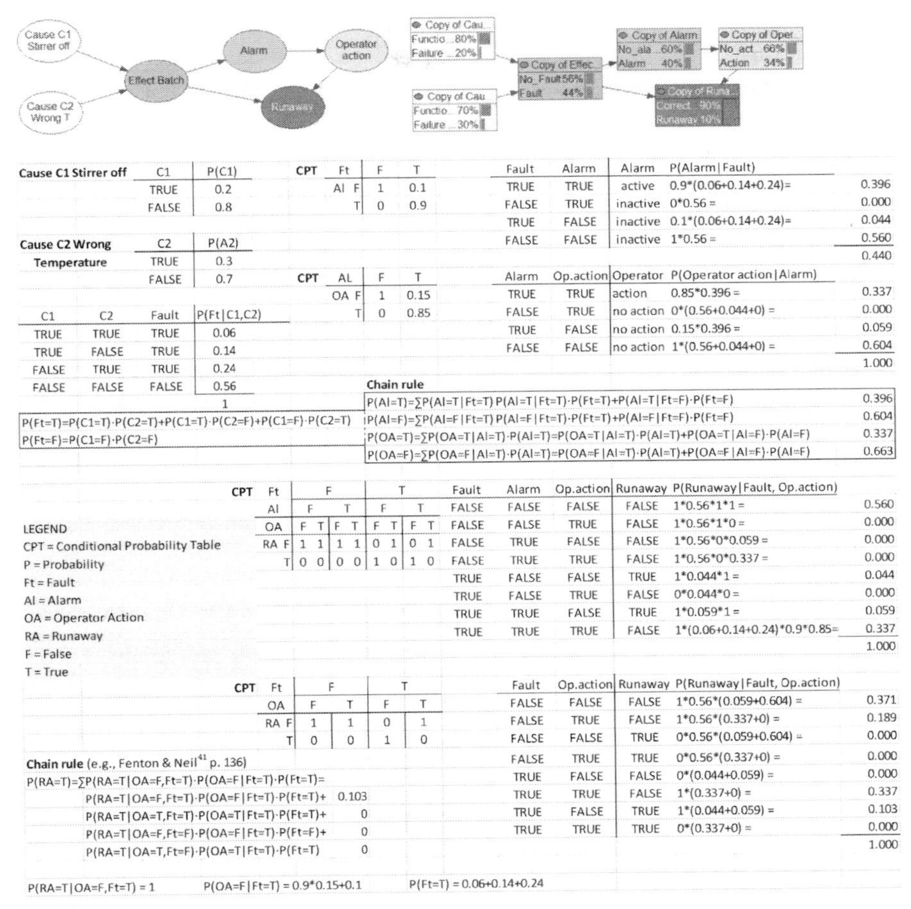

Cause C1 Stirrer off

C1	P(C1)
TRUE	0.2
FALSE	0.8

Cause C2 Wrong Temperature

C2	P(A2)
TRUE	0.3
FALSE	0.7

C1	C2	Fault	P(Ft|C1,C2)
TRUE	TRUE	TRUE	0.06
TRUE	FALSE	TRUE	0.14
FALSE	TRUE	TRUE	0.24
FALSE	FALSE	FALSE	0.56
			1

P(Ft=T)=P(C1=T)·P(C2=T)+P(C1=T)·P(C2=F)+P(C1=F)·P(C2=T)
P(Ft=F)=P(C1=F)·P(C2=F)

CPT	Ft	F	T
Al	F	1	0.1
	T	0	0.9

CPT	AL	F	T
OA	F	1	0.15
	T	0	0.85

Fault	Alarm	Alarm	P(Alarm|Fault)	
TRUE	TRUE	active	0.9*(0.06+0.14+0.24)=	0.396
FALSE	TRUE	inactive	0*0.56 =	0.000
TRUE	FALSE	inactive	0.1*(0.06+0.14+0.24)=	0.044
FALSE	FALSE	inactive	1*0.56 =	0.560
				0.440

Alarm	Op.action	Operator	P(Operator action|Alarm)	
TRUE	TRUE	action	0.85*0.396 =	0.337
FALSE	TRUE	no action	0*(0.56+0.044+0) =	0.000
TRUE	FALSE	no action	0.15*0.396 =	0.059
FALSE	FALSE	no action	1*(0.56+0.044+0) =	0.604
				1.000

Chain rule

P(Al=T)=∑P(Al=T|Ft=T) P(Al=T|Ft=T)·P(Ft=T)+P(Al=T|Ft=F)·P(Ft=F)	0.396
P(Al=F)=∑P(Al=F|Ft=T) P(Al=F|Ft=T)·P(Ft=T)+P(Al=F|Ft=F)·P(Ft=F)	0.604
P(OA=T)=∑P(OA=T|Al=T)·P(Al=T)=P(OA=T|Al=T)·P(Al=T)+P(OA=T|Al=F)·P(Al=F)	0.337
P(OA=F)=∑P(OA=F|Al=T)·P(Al=T)=P(OA=F|Al=T)·P(Al=T)+P(OA=F|Al=F)·P(Al=F)	0.663

CPT		F				T			
Ft		F				T			
Al		F		T		F		T	
OA		F	T	F	T	F	T	F	T
RA	F	1	1	1	1	0	1	0	1
	T	0	0	0	0	1	0	1	0

Fault	Alarm	Op.action	Runaway	P(Runaway|Fault, Op.action)	
FALSE	FALSE	FALSE	FALSE	1*0.56*1*1 =	0.560
FALSE	FALSE	TRUE	FALSE	1*0.56*1*0 =	0.000
FALSE	TRUE	FALSE	FALSE	1*0.56*0*0.059 =	0.000
FALSE	TRUE	TRUE	FALSE	1*0.56*0*0.337 =	0.000
TRUE	FALSE	FALSE	TRUE	1*0.044*1 =	0.044
TRUE	FALSE	TRUE	FALSE	0*0.044*0 =	0.000
TRUE	TRUE	FALSE	TRUE	1*0.059*1 =	0.059
TRUE	TRUE	TRUE	FALSE	1*(0.06+0.14+0.24)*0.9*0.85=	0.337
					1.000

LEGEND
CPT = Conditional Probability Table
P = Probability
Ft = Fault
Al = Alarm
OA = Operator Action
RA = Runaway
F = False
T = True

CPT		F		T	
Ft		F		T	
OA		F	T	F	T
RA	F	1	1	0	1
	T	0	0	1	0

Fault	Op.action	Runaway	P(Runaway|Fault, Op.action)	
FALSE	FALSE	FALSE	1*0.56*(0.059+0.604) =	0.371
FALSE	TRUE	FALSE	1*0.56*(0.337+0) =	0.189
FALSE	FALSE	TRUE	0*0.56*(0.059+0.604) =	0.000
FALSE	TRUE	TRUE	0*0.56*(0.337+0) =	0.000
TRUE	FALSE	FALSE	0*(0.044+0.059) =	0.000
TRUE	TRUE	FALSE	1*(0.337+0) =	0.337
TRUE	FALSE	TRUE	1*(0.044+0.059) =	0.103
TRUE	TRUE	TRUE	0*(0.337+0) =	0.000
				1.000

Chain rule (e.g., Fenton & Neil[41] p. 136)

P(RA=T)=∑P(RA=T|OA=F,Ft=T)·P(OA=F|Ft=T)·P(Ft=T)=	
P(RA=T|OA=F,Ft=T)·P(OA=F|Ft=T)·P(Ft=T)+	0.103
P(RA=T|OA=T,Ft=T)·P(OA=T|Ft=T)·P(Ft=T)+	0
P(RA=T|OA=F,Ft=F)·P(OA=F|Ft=T)·P(Ft=F)+	0
P(RA=T|OA=T,Ft=F)·P(OA=T|Ft=T)·P(Ft=T)	0

P(RA=T|OA=F,Ft=T) = 1   P(OA=F|Ft=T) = 0.9*0.15+0.1   P(Ft=T) = 0.06+0.14+0.24

FIGURE 7.4

Example of a simple Bayesian network applied to a batch reactor with two possible failure causes—one is failed stirrer (on average 2 times out of 10) and the other a wrong initial temperature (3 times out of 10)—both triggering an annunciator requiring the operator to take action to correct. The alarm (alert) has a probability of failure on demand of 0.1 and the operator taking action will fail with probability 0.15. (Failure probabilities are chosen high and many other upset causes are possible, but this example is just to illustrate how it works. No false alarms are taken into account.) *Top left* is shown the structure of the discrete variable Bayesian network with arcs and nodes, and *top right* is a copy showing the calculation results. The network is produced by the freely downloadable GeNIe software of DSL,[41] University of Pittsburgh. The *table below* the networks is to present the analysis in hand calculations through conditional probability tables (CPTs) via the alarm node and direct, and also through the chain rule. By hand, it starts to become puzzling, whereas GeNIe does it without any problem.

by applying the Bayes factor type of Eqn (7.2) repeatedly and use of a scoring criterion. Richard Neapolitan's book[44] about BN structure learning provides the method, initially developed by Cooper and Herskowits[45], in detail. Evaluating a small net can still be done by hand. However, if the number of nodes increases, it results in an exponential growth of elements to populate in the conditional probability tables. Hence, even with a modest number of nodes and the variable expressed as a discrete probability, evaluating each instantiation of a cause event by hand takes an intensive effort with a good chance of errors in the calculation. For continuous distributions it would be impossible to perform. Fortunately, from a variety of sources, both free and commercial software have become available to perform the computations. Kevin Murphy[46] listed an overview on the Internet.

Bayesian net is based on the Markov assumption, which means that they describe a stochastic process of linked events, but the transition to the next state only depends on the current state and not on states further in the past. This assumption applied to BNs is that child nodes are conditionally independent of their nondescendant nodes (the "higher-ups") given their parent nodes, which are the primary causes or influences of their child nodes.

For many problems a discrete BN suffices, but for engineering purposes often continuous distributions must be handled. For continuous or mixed discrete-continuous BNs, we shall compare three types of methods to solve the dependence equations, although more may exist:

1. *GeNIe* of Decision Systems Laboratory[41] of the University of Pittsburgh applies the Monte Carlo technique to solve the convolution of equations between nodes and is rather easy to work with. It is MS Excel compatible. For large continuous nets, Monte Carlo becomes slow and multiple runs are needed to get a mean result.
2. *AgenaRisk* by Norman Fenton and Martin Neil[42] applies an algorithm of dynamic discretization, which works well for not-too-large networks and is well described in Appendix D of the reference. Its basis is optimization of the discretization around the mode of density functions in an iteration process till an acceptable precision is reached.
3. *UniNet* developed by Roger Cooke and coworkers, among others Anca Hanea and Dorota Kurowicka[47] at the Delft University of Technology, follows a new, smart method to solve a net of mixed distributions. No joint distribution between the variables of the source and sink node is formed, which makes the BN nonparametric and therefore independent of particular distributions. Their line of thought will be briefly summarized here because it works well; for more detail, see the original work or for a short description, see Pasman and Rogers.[48] Because correlation of the data (interval coefficient $[1,-1]$) in a direct mode works only with normal distributed variables, a transformation is performed. To that end the probabilistic influence of the arcs is associated with the correlation of the *ranks* of the data and not with the correlation of the data itself. The rank values generating function for each node is the cumulative density function of

the ordered data. The conditional rank correlation is the product moment correlation (pmc) of the rank functions of two random variables given a third. From the rank functions transformed standard normal variables can be derived by inverting the standard normal univariate cumulative distribution. This operation enables construction of the net with the transformed variables. Because the transformed variable increases monotonously with the increasing rank function, the same rank correlation between pairs of transformed variables can be applied as between the original variables. Next, the rank correlation is realized using the joint normal copula having the property of independence at zero correlation. Finally, the joint normal distribution, which contains all of the dependency information, is sampled with the correlation matrix, and for each sample the original marginal distribution is calculated. This procedure yields the joint distribution of the initial variables with their specified dependence structure, which enables efficient analytical conditioning and updating of the net.
To include *discrete* distributions, Hanea and Kurowicka[47] wrote discrete ordinal variables as monotone transforms of uniform variables. The dependence structure in the BN is then to be defined with respect to the underlying uniform variables. Because the rank correlation of the discrete variables and that of their underlying uniform counterparts may not be equal, their relation was established. In the derivation a copula is involved. It turned out that these rank correlations over even a small number of states are quite close; with 10 states the difference is of the order of 10^{-3}. To execute all this, UniNet obtained two types of nodes: probability nodes to which a univariate distribution of some kind is assigned that fits the parameter values modeled and a functional node in which the operations between parameters are realized. The arcs between the probability nodes reflect the correlation among the modeled variables. For Cooke's specialism of elicitation of expert knowledge, special software, Excalibur, is available and a special procedure has been developed to determine rank correlations in that case. In addition, data mining can be performed. A good example of a study making use of combining Cooke's BN approach with expert knowledge is by Daniela Hanea[49] in fire safety, engineering and dimensioning a building for timely evacuation by calculating the required safety egress time, given a certain acceptable risk level, versus the available safety egress time.

In most risk analyses, time is not an explicit variable because usually only end states matter and the effect on transitions by change of conditions due to processes in the course of time can be neglected or are just ignored. However, there are many time effects such as equipment degradation, effect of spare components on overall system failure, effects of lasting fires in, for example, domino effects, and many more. There are various approaches, and Labeau et al.[50] present an overview with respect to dynamic reliability. One possibility is Markov chains (MC). In general, Markov chains describe state transitions according to a state transition probability matrix. The problem is solved by applying Monte Carlo (MC) sampling and applying statistics, hence the combination is indicated by the acronym MCMC.

However, most known are "memory-less" continuous-time Markov chains of Poisson processes. In that case, stochastic occurrence of state transition events is described by a Poisson distribution with a certain transition rate λ. In case the rate, for example, a failure rate, is constant, the process is time homogeneous. In principle, an analytic solution of the chain can be applied. However, calculation is manageable as long as the number of transitions is small. An example of how Markov analysis is used to solve more complex problems in safety instrumented systems as part of a layer of protection is shown by Innal et al.[51] Another possibility of modeling a time-dependent process is dynamic BN or DBN, not mentioned yet in the 2000 paper of Labeau et al.[50] DBN has become increasingly widespread during the decade following. DBN can be performed by time steps and repeating the network structure, or by nodes containing a number of different time states. In Section 7.7, we shall present and discuss a number of applications of BNs to risk problems with free or semi-free available software among which a DBN is an example. In Section 7.8 we shall discuss Petri net. This models resources, transition event triggers, and flow rates; it allows randomness and cyclic processes.

7.6 UNCERTAINTY, FUZZY SETS

7.6.1 UNCERTAINTY

In Chapter 3, we saw how different analyst teams come to very different risk figures. This is caused by many uncertainties taking various forms. One distinguishes *parameter, model*, and *completeness uncertainty*. Given a distribution, parameter uncertainty is represented by the standard deviation, or rather the *coefficient of variation*, which is the standard deviation normalized by the distribution mean. Parameter uncertainty affects equipment failure probability values and frequencies of operational upsets. In consequence analysis, limitations of the physical models cause uncertainty in effects, while given a physical effect uncertainty in damage parameters is often represented by probit distributions. Scenarios usually are incomplete and defective of organizational influences and human acts, which increase uncertainty further. Given a scenario, one can calculate the *extent of damage* that will be sustained. But, it is intrinsic to probability, that due to the many parameters of which the effect is not or not fully accounted for there is a fair chance damage is less than predicted or worse. For the *expected occurrence* of the scenario it is even harder. The probability of the event occurrence within a time interval, also expressed as event frequency or event rate, provides little certainty if asked, can it happen tomorrow? Given the event rate time-to-event would follow a Poisson distribution, but that gives no certainty for an answer to that question. Yet, we must work with the concept and make decisions based on it. In a relative sense, comparing events and their frequencies is certainly useful. In Chapters 9 and 12 we shall touch upon the issue of how to cope with uncertainty in decision making.

In a scientific sense, the uncertainties inherent to risk analysis results are distinguished in *aleatory* and *epistemic* ones. Aleatory uncertainty, or rather variability, is due to random fluctuations of properties and conditions leading to different outcomes and to ranges of outcomes. So, if a parameter is measured and the outcome values differ from one measuring instance to the other, it is aleatory. If the parameter is part of a model describing a process, the model is called aleatory, or probabilistic or stochastic, as opposed to deterministic. The variability in outcomes can be reduced only through more knowledge about the underlying phenomena, for example, mechanical or chemical mechanisms, or by using covariate models that relate one or more model parameters to one or more measured, correlated quantities.

Epistemic uncertainty is lack of knowledge on underlying fundamentals or total ignorance of, for example, a possible alternative scenario. One can be ignorant of the existence of an influencing parameter, or a known parameter appears to have two different values and one does not know why. It can also be caused by incompleteness or quality of data, available knowledge, or a simplification in modeling, by a mistake, inconsistence, or by confusion. But it also happens often that one does not want to see a possible event as realistic and by psychological displacement process even rules out the possibility!

Elizabeth Paté-Cornell[52] explained the difference between aleatory and epistemic uncertainty as shown in Figure 7.5. The figure represents an estimate of epistemic uncertainty about potential future losses resulting from an event by a spectrum of risk curves based on alternative hypotheses on possible threats. Using all available information and expert opinion, probability values are assigned to each fractile curve. Bayesian updates are expected in the future. These cumulative frequency distribution functions are thus conditional on the hypotheses. Aleatory

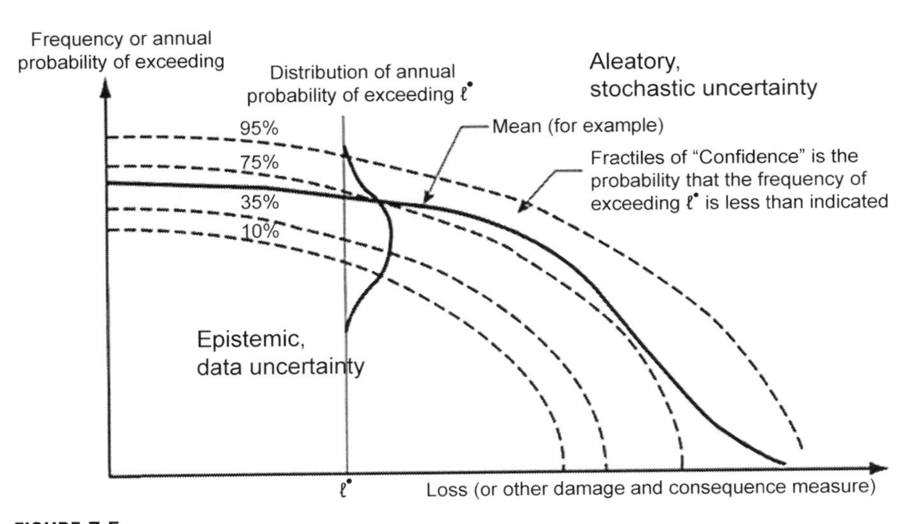

FIGURE 7.5

Way of coping with epistemic and aleatory uncertainties in a risk analysis, according to Paté-Cornell,[52] as explained in more detail in the text. The example concerns a risk analysis of a radionuclide waste isolation pilot plant in New Mexico.

uncertainties are kept separate. If all aleatory and epistemic uncertainties are aggregated to a "mean," this mean curve does not need to correspond in shape to one of the fractile curves. As a result of skewness of the distribution functions, the mean curve tends to be situated above the median of the fractiles family.

Methods exist to investigate uncertainty, which add some information but of course never enough to draw a sharp line or obtain a sharp cutoff. In case of epistemic uncertainty one can devise alternative scenarios, models, or parameter values; one can put alternative possibilities in a logic tree and assign probabilities, or rather weights, or even probability distributions to the possibilities and derive an overall probability distribution of outcomes performing a type of sensitivity analysis. (Strictly speaking, sensitivity analysis has as the objective to find out the effects of parameters on the outcomes, which parameter is important, what simplifications can be made, where errors are, hence what is the *robustness* of the model.)

Four formal methods to treat uncertainty are listed and referenced by Terje Aven and Enrico Zio,[53] discussing uncertainty in risk prediction of rare, extreme industrial disasters (which is the very subject of this book). These can be characterized as: (1) probability-bound analysis—combining probability analysis and interval analysis, (2) imprecise probability and robust Bayes statistics area, (3) the Dempster–Shafer random sets approach, and (4) possibility theory, which goes back to Lofti Zadeh's Fuzzy sets to be treated briefly in the next section. Interval analysis is a method of putting bounds on measurement errors. Bayesian statistics is a basis for the second and third method. The Dempster–Shafer method goes beyond Bayesian subjective probability of belief by introducing the concept of plausibility, which is the upper value the belief can have, because there is no evidence that goes against it. We shall come back to uncertainty related to rare events in Section 7.7.3, and when discussing decision making under deep uncertainty in Chapter 9, Section 9.3.

Because of lack of certainty in explicit knowledge in risk analysis, expert judgment is an important input and therefore expert elicitation a rather common mechanism to exploit human intuition or in popular terms "gut feeling." Human intuition for major decisions is best supported by a predictive decision model. The procedure to elicit expert knowledge can, however, procedurally and in treatment of results be rationalized. Experts can be "calibrated" against results known by the project initiator. Roger Cooke,[54] Delft University of Technology, has laid a basis for it. Subjective probability, belief, is very well supported by Bayesian statistics, and Cooke makes much use of that in processing expert elicitation results.

7.6.2 FUZZY SETS

As in many risk analyses hard data are missing, expert knowledge must be brought to bear. The first time this became possible in a tractable way was through application of Lofti Zadeh's[55] (University of California, Berkeley) fuzzy set theory published in 1965. Against "fuzzy" stands "crisp." Fuzzy sets allow vague concepts to be defined in a mathematical sense. This is particularly advantageous, when so-called linguistic statements must be processed, like X is larger than 10, but smaller than 20 and most likely to be 17. A fuzzy set can therefore be viewed as a *possibility distribution*. This provides a measure of uncertainty that needs not to sum to 1 and is therefore flexible but not coherent. In formal mathematical terms a fuzzy set A is defined as the elements of some universe X that satisfy the membership requirements of A. However, the elements have varying degree of membership (the truth value), $\mu_A(x) = 0$ for nonmembership, and 1 for total membership, resulting in a set of ordered pairs $\{x_1,\mu_A(x)\}$. Two classes of membership grade functions are commonly used—the triangular and the trapezoid—but others such as Gaussian shapes are possible. The triangle set is defined by a lower bound or minimum and an upper bound or maximum value of x at $\mu = 0$ and in between as top a modal value of totally possible at $\mu = 1$. The trapezoid set has a flat part at $\mu = 1$. Representation of expert estimates expressed in a linguistic variable on, for example, a five-point scale such as max, high, medium, low, min, can easily be realized by five triangles each overlapping its neighbor from the middle of the base.

After fuzzification of input variables inference can be accomplished by logic implication operations performed on the membership functions of the input variables represented by AND, OR, and NOT statements. The result of intersection of sets in an AND can be defined as taking the minimum or as the product of intersecting membership values, while OR follows from a union of sets and can be defined as the maximum or as the sum minus the product. NOT yields the complement, but other sense making relations can be defined. The logic statements are embedded in the antecedent of the IF-antecedent-THEN-consequent rules. The consequent shapes the output set contour, upon which an arithmetic defuzzifying operation, for example, by the centroid method, produces a crisp output value. Fuzzy set theory contains much more than the brief description given here and references are easily found. MATLAB® provides an extensive fuzzy set module. There are many papers on fuzzy set applications in risk analysis, for example, Adam Markowski et al.[56] applied the theory to bow-tie, LOPA, and full risk analysis. However, as the output is a point value, useful information on variability is lost. Uncertainty in the data given by the distribution of output values represented by only one point as a crisp outcome is ignored. As we shall see in the next part, BNs for which software was developed only over the last decades can easily absorb triangular or other more realistic probability distributions. Convolution will retain all uncertainty information. This makes working with fuzzy set somewhat devious, while BNs also fit better in a system approach.

The integration of technical and organizational/human aspects in risk analysis methods has been a long-standing wish. In the many cases in which available space for a production or storage site is tight and opposing interests exist, much wrangling would be avoided if safety distance lines could be drawn with a high degree of certainty. Because many aspects require specialism to really know "ins and outs," an integrated approach, or at least the building of an interoperable set of models, would require a modeling infrastructure allowing merging of different types of information while keeping data integrity. BN offers such a possibility. In the following sections a number of examples and developments will be presented.

7.7 SOME APPLICATIONS OF BN

7.7.1 BN LOPA

We shall start with a LOPA as discussed in particular in Chapter 3, Sections 3.6.6.2 and 3.6.6.3. A LOPA event tree can easily be modeled by a BN. As we have seen in the previous section, each node of the net structure represents a random variable. The initiating event (IE) is a parent node. All other nodes are conditional on nodes higher in the net to which they are linked. GeNIe offers also the possibility to turn a discrete BN into an influence diagram by including a node that provides options to select from, here a base case and two possible improvements, 1 and 2, and further a sensitivity node enabling certain confidence limits for an important variable, here high and low estimates of the IE frequency, and cost nodes, summing cost amounts of various options. The system data of the first example with all variables as discrete probability values are given in Table 7.6, and the corresponding BN is shown in Figure 7.6. The setting up of the problem with the DSL GeNIe software,[41] as explained in more detail by Pasman and Rogers,[48] is in this case still a rather easy task, and modification of the data and an additional calculation for updates are almost effortless. If desired, the network can be expanded to describe the cause—effect relations of CFs in a fault tree fashion leading to the top or initiating event. In that case, not the initiating event (leaf node) but the basic faults, become the parents (root nodes).

Although the sensitivity node provides a possibility to obtain an impression of the effect of spread in the value of the frequency of the initiating event, it is still preferable for a system approach to take full account of uncertainty information and introduce entire continuous distribution functions into the calculation. To solve a

Table 7.6 Input Data of BN Variables Used in the LOPA Described (CoO = Cost of Ownership)

	Probability	Damage	CoO Base Case	Improvements 1 and 2	
LOPA	**Base Case**	**k$**	**k$/yr**	**Prob. 1; 2**	**CoO 1; 2, k$**
Initiating event (IE)	0.03–0.1–0.2/yr	0, if IE = 0			
Layer 1 succeeds	0.80	35	10	0.94; 0.99	120; 900
Layer 2 succeeds	0.94	200	150		
Layer 3 succeeds	0.67	5000	45	0.67; 0.90	45; 90
Layer 3 fails	0.33	150,000			
Spurious trips	1.0/yr	30			

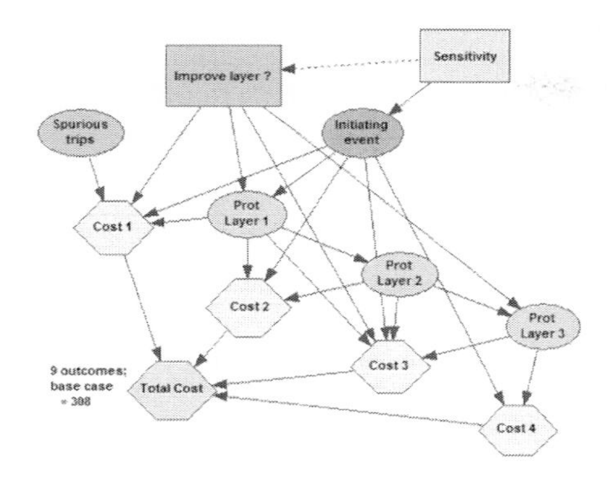

Sensitivity High, Initiating Event 0.03 /yr						
Spurious trip	No					
Initiating event	Yes					
Prot. Layer 1	Succeeds			Fails		
	Base Case	Improve 1	Improve 2	Base Case	Improve 1	Improve 2
Costs Layer 1, k$/yr	45	155	935	11	121	901
Initiating event	No					
Prot. Layer 1	Succeeds			Fails		
	Base Case	Improve 1	Improve 2	Base Case	Improve 1	Improve 2
Costs Layer 1, k$/yr	11	121	901	11	121	901
Spurious trip	Yes					
Initiating event	Yes					
Prot. Layer 1	Succeeds			Fails		
	Base Case	Improve 1	Improve 2	Base Case	Improve 1	Improve 2
Costs Layer 1, k$/yr	45	155	935	40	150	930
Initiating event	No					
Prot. Layer 1	Succeeds			Fails		
	Base Case	Improve 1	Improve 2	Base Case	Improve 1	Improve 2
Costs Layer 1, k$/yr	40	150	930	40	150	930

Final results									
Sensitivity	High IE 0.03 /yr			Medium IE 0.1 /yr			Low IE 0.2 /yr		
	Base Case	Improve 1	Improve 2	Base Case	Improve 1	Improve 2	Base Case	Improve 1	Improve 2
Failure freq./yr	1.2E-04	3.6E-05	1.8E-06	4.0E-04	1.2E-04	6.0E-06	7.9E-04	2.4E-04	1.2E-05
Total costs k$/yr	257	355	1196	308	378	1256	382	410	1342

FIGURE 7.6

Discrete probability value BN in GeNIe software (DSL)[41] of a three-layer protection system of a base case (BC) and two possible improvements. Initiating event (IE) frequencies are 0.03, 0.1, and 0.2/yr for sensitivities high-medium-low. Applying data of Table 7.6 produces nine final results with respect to overall failure frequency and total annual costs; see table below the figure. The effect of a spurious trip of 1/yr in layer 1 is included. The protection layer improvements appear not to be cost effective. Medium total annual BC cost is 308 k$/yr with differences of 51 k$/yr between high-medium sensitivity and 74 k$/yr medium-low, while the final failure probability is 4×10^{-4}/yr with differences of resp. 3×10^{-4} and 4×10^{-4}/yr. As not all detail probability and cost data can be shown, just for illustrational purposes, the table to the right of the figure shows results for node Cost 1 at high sensitivity with and without spurious trip.

continuous BN, one must resort to another calculation method. As we mentioned in the previous section there are different approaches available. As mentioned the continuous version of GeNIe applies Monte Carlo sampling to solve the convolution of distribution equations. It needs repeating runs to derive a mean. So far, though, GeNIe turned out to be the most flexible for engineering-type risk calculations. Data on the probability of failure on demand or frequency may be available in various distribution types.

In many cases with no existing data, experts may be consulted to give estimates of the minimum, top, and maximum value of a triangular distribution (or a mean and variance of a truncated normal, also called TNormal distribution available in AgenaRisk). Many reliability data of equipment are given as exponential, log-normal, or Weibull distribution, which can all be accommodated. Further, a Bernoulli, binomial, beta, uniform, and normal distribution are available, only a gamma distribution, the generalization of the exponential, is missing. Data assumed for this continuous case are collected in Table 7.7. In Figure 7.7, a GeNIe continuous mode

Table 7.7 Probability Distribution Data for a LOPA with Continuous BNs

Distributions		a	M	b	Mean	Std. Dev.
Initiating event/yr	Triangular	0.03	0.075	0.20	0.10	0.036
IPL1	Triangular	0.60	0.85	0.95	0.80	0.074
IPL2	Log-normal		−0.06	0.05	0.94	0.865
IPL3	Normal		0.67	0.10	0.67	0.100
Spurious trips/yr	Normal		1	0.20	1.00	0.200

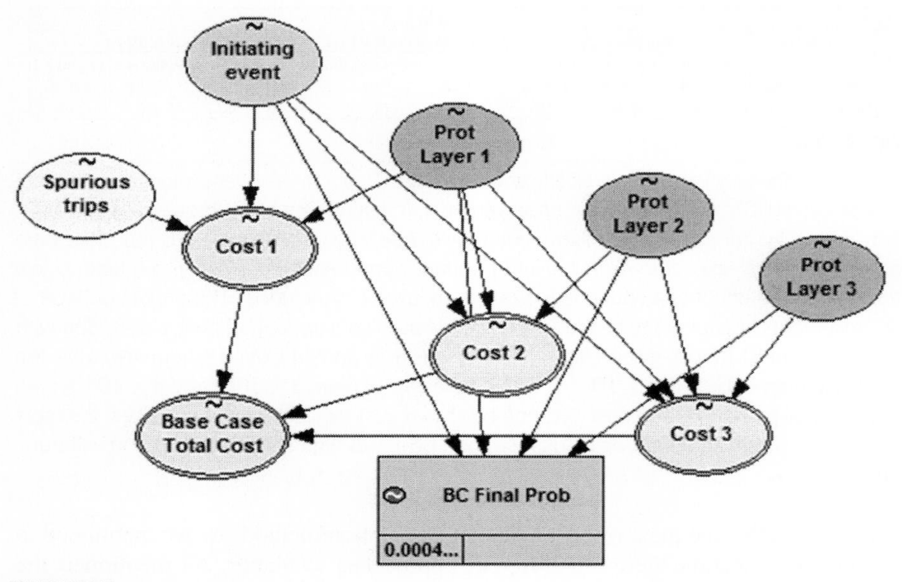

FIGURE 7.7

Continuous probability distribution function BN in GeNIe (DSL, 2010)[41] according to data of Table 7.7.

BN is shown. The mean total cost figure of the run (307 k$/yr) is the same as in the discrete case, but now also a standard deviation value (72 k$/yr) is produced. At each run the values vary slightly. GeNIe can deliver graphs of all distributions; see Figure 7.8 for an example.

One can easily model in BNs OR and AND gates, but also the more sophisticated exclusive OR: XOR, or the NAND (not AND), and further a probabilistic dependency gate, warm, and cold spares with or without repair, and others, all depending on the conditional relation between cause and effect. Continuous GeNIe provides a number of logical/conditional functions also in a direct fashion as an

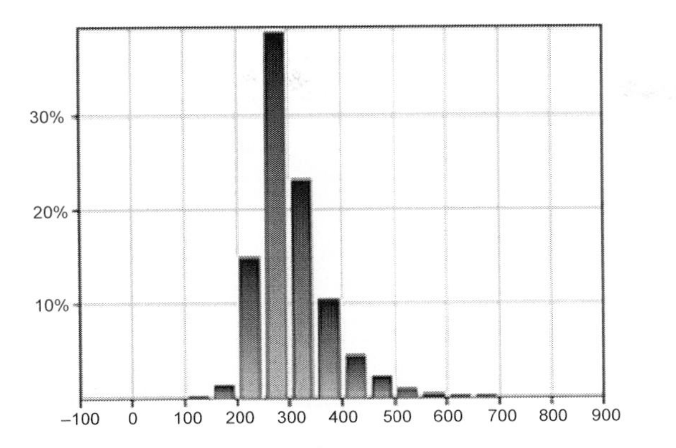

FIGURE 7.8

GeNIe BN continuous mode graphical result of total annual cost, applying data of Table 7.7.

IF-THEN-ELSE statement, OR, AND and XOR, Switch, MIN, and MAX. In addition, a variety of arithmetic, trigonometric, and hyperbolic functions can be used.

GeNIe does not allow "soft" relations between nodes. This is enabled in nonparametric BNs applying the rank correlation solution developed by Cooke and his group at the University of Delft, explained in the previous section, resulting and embodied in the UniNet software.[57] In UniNet, a distinction is made between probability nodes and functional nodes. The former can be connected by arcs on which the variable rank correlation between nodes can be varied. Here, the correlation is set to zero. Arithmetic expressions are evaluated in the functional nodes. The result of the LOPA previously modeled in continuous GeNIe is shown in UniNet in Figure 7.9. UniNet allows a transparent setup and presentation of results. The outcome of total cost (307 ± 44 k\$/yr) and overall system failure probability ($4.1 \times 10^{-4} \pm 2.7 \times 10^{-4}$) are as expected. The runs are reproducible, and the standard deviation calculated is smaller than the ones found by GeNIe. Common cause failure (CCF) can also easily be modeled. The effect of a 5% CCF contribution is relatively strong as presented in Figure 7.10. The effect increases the total cost to 397 ± 98 k\$/yr due to more than 10 times increase of the overall system failure probability ($5.5 \times 10^{-3} \pm 2.9 \times 10^{-3}$). UniNet computes rapidly and is very stable.

The third software package used is AgenaRisk developed by Fenton and Neil.[42] This software can be applied to a large variety of diagnosis, risk management, and decision reasoning problems in medical application, finance, economy, and law in addition to engineering. This is also true because of the ease to include new evidence and the elaborated graphics. A strong property of the software is that it allows the use of truncated normal distributions. Procurement of the referenced book allows downloading a free, light version with, for example, the limitation of saving a model of maximum 10 nodes. However, for accomplishing a LOPA calculation, the triangular distribution of the first layer used previously (Table 7.7), available in the software, had to be approximated by a truncated normal distribution and the log-normal one of

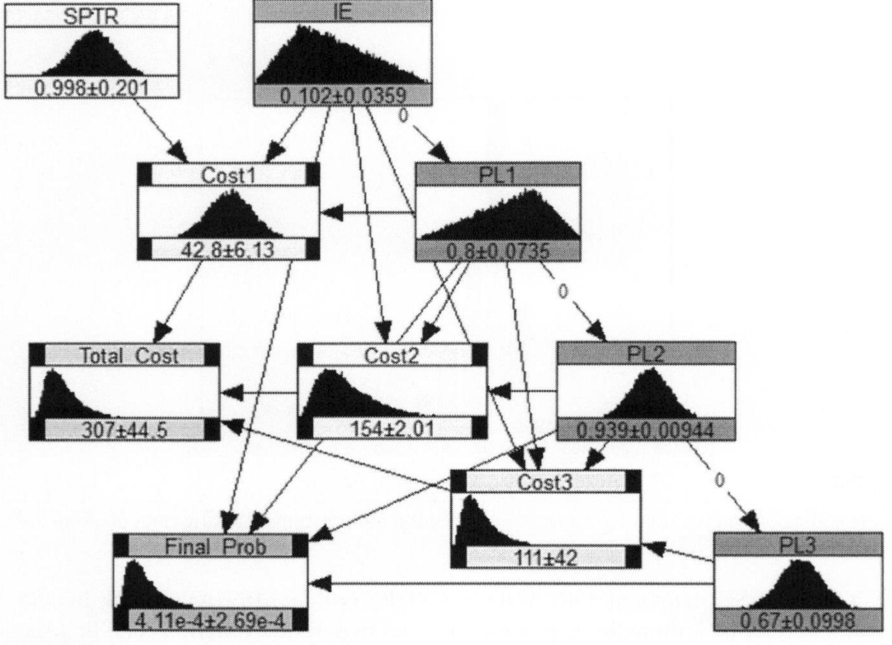

FIGURE 7.9

The same LOPA as in Figure 7.7, but now analyzed by means of UniNet software[57] and shown as distributions. SPTR, spurious trip; IE, initiating event; PL, protection layer.

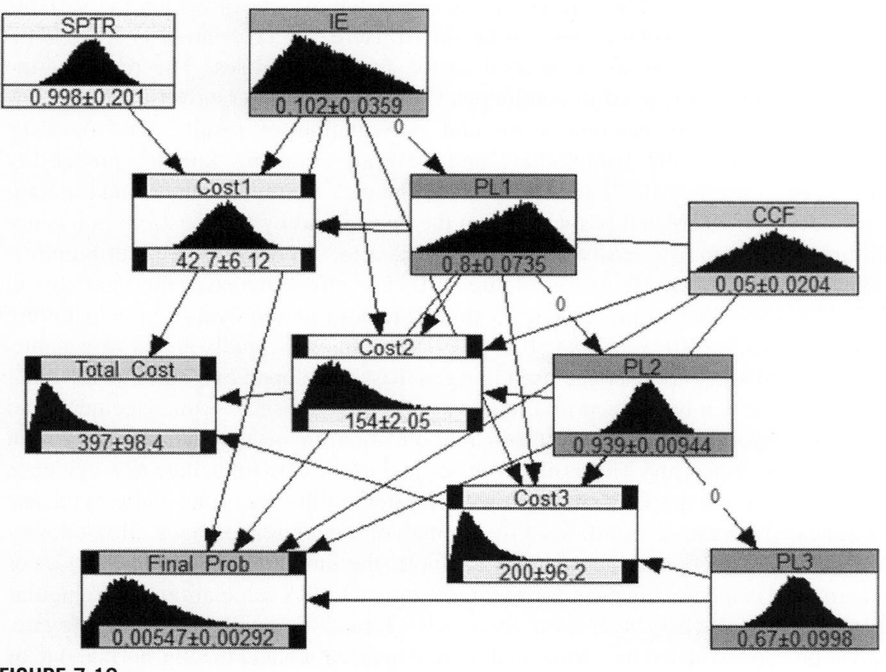

FIGURE 7.10

LOPA as shown in Figure 7.9, but with the effect of an average 5% common cause failure (triangular distribution, minimum 0%, maximum 10%).

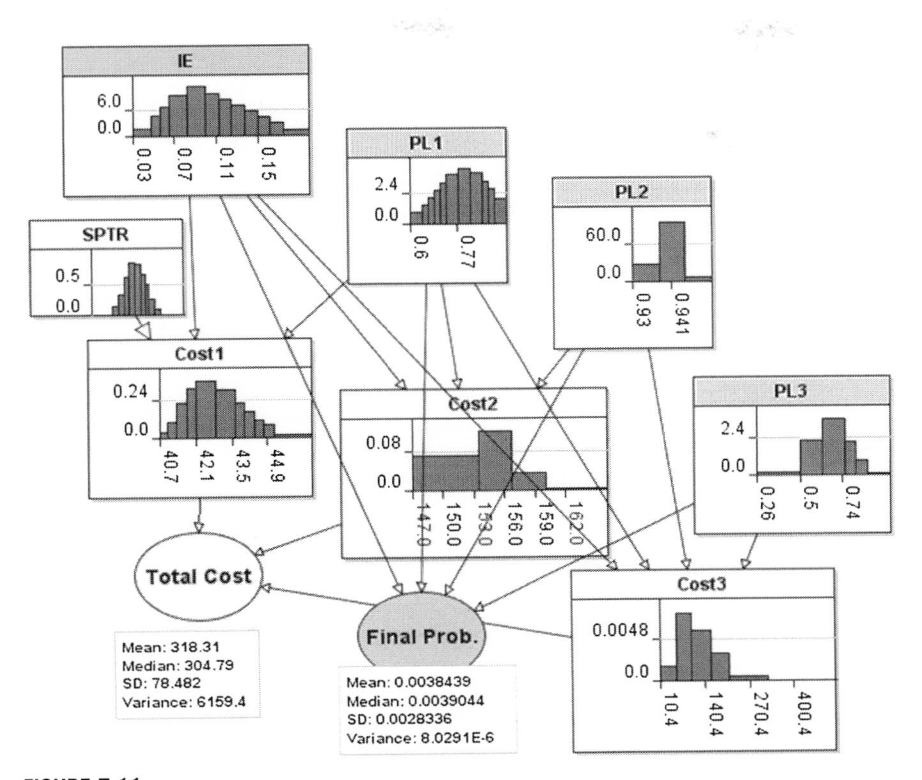

FIGURE 7.11

LOPA as in Figures 7.7 and 7.9 now performed by AgenaRisk software (light version 6.0, meanwhile 6.1).[42] Due to approximation of the triangular distribution of PL1 by a truncated normal and of the log-normal of PL2 by a normal, results deviate slightly from previous ones. AgenaRisk is well suited for solving many types of problems but seems less attractive for more complex engineering ones.

the second layer by a normal distribution. The results are not much affected as can be seen in Figure 7.11. AgenaRisk is on the other hand well suited to derive a probability density distribution function from data based on observed system successes and failures as shown in Figure 7.13.

Slow degradation of the effectiveness of protective layer systems, and hence increase of risk, may be another point to investigate. BNs can be made dynamic (DBN) by time-stepping to look at this in more detail, but again this will need data on degradation rates from system monitoring to work with. In Figure 7.12 an example on assumed data is given of how these can be performed. Other applications would be the modeling of typical dynamic fault tree problems such as upon failure switching to component spares, priority AND gate with preassigned order of failures, or functional dependency gate where a trigger event blocks the

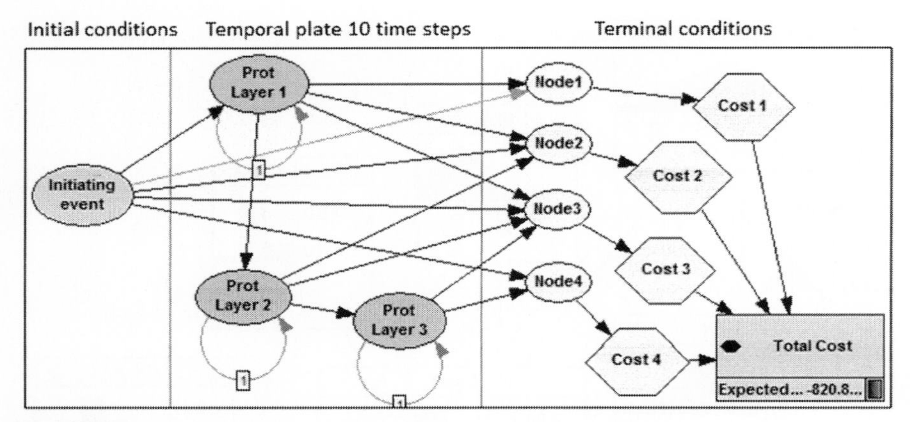

FIGURE 7.12

Dynamic Bayesian network of a time stepped LOPA applying discrete mode GeNIe (DSL),[41] which allows a temporal plate. In the example based on the data of Table 7.6 and an Initiating Event frequency of 0.1/yr, at each time step representing, for example, one year, reliability of each layer decreases 5%, a value arbitrarily chosen. (Each layer can be given a different degradation factor if desired.) Also the number of time steps can be easily varied. With 10 steps the total annual cost increases from 307 to 820 k$/yr.

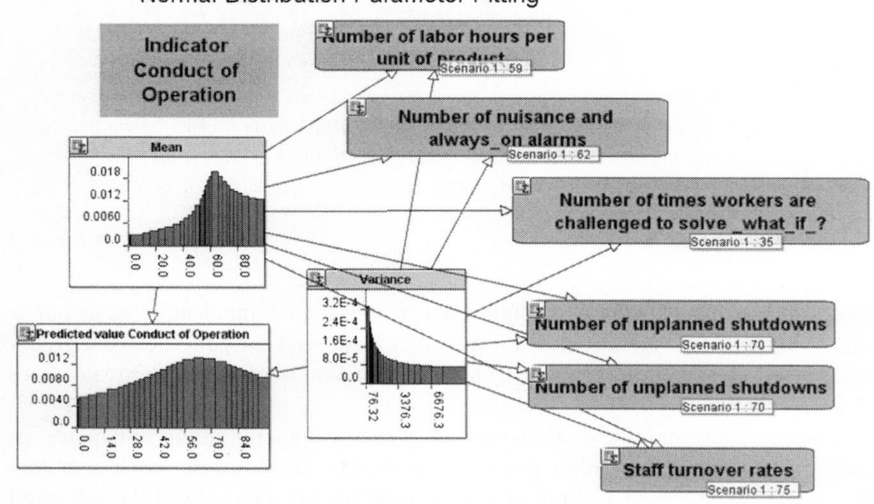

FIGURE 7.13

Aggregating process safety performance leading indicators to the level of one of the principal indicators, for example, Conduct of Operation. AgenaRisk[42] BN composes a predicted value as a (posterior) truncated normal distribution with a mean and variance calculated from the distributions of the individual indicators and the new (Scenario) evidence, based on uniform distribution priors.

functions of dependent gate components. Examples can be found in Khakzad et al.[58] As mentioned in Section 7.3.3 combination of HAZOP and DBN as shown by Hu, Zhang, and Liang[28] and even more clear by Hu et al.[59] opens perspectives not only for partly automating HAZOP but also for predictive modeling of degradation and predictive maintenance. Pérez Ramírez and Bouwer Utne[60] described a well-elaborated example of selecting alternative solutions for life extension of an aging system and finding out what can be done to mitigate degradation during continued operation after a life-extension repair. This was accomplished all by applying DBN and making use of operational data before and after repair. Reliability (PFD) was optimized versus cost, while even CCFs and maintenance parameters were taken into account.

Pasman and Rogers[61,62] and others published more intricate examples of, for example, mapping a BN on to a bow-tie of a gas-oil separation pressure vessel on board of an offshore platform or of a complete risk analysis of hydrogen refilling tank stations. To keep overview it is recommended to draw the bow-tie first and only then develop the BN. In the following, an application will be presented showing how PSPIs can be made more productive.

7.7.2 BN APPLICATION TO PSPI METRICS

First, aggregation of collected indicator values will be shown. This can be done by BNs quite easily. For this application, AgenaRisk proved to be most convenient. One of the 21 principal leading indicators proposed by CCPS is depicted in Figure 7.13; see Chapter 3, Figure 3.3, "Conduct of Operations." CCPS suggested under the heading of this principal indicator 38 leading indicators categorized in six groups. Here, for the example, a selection is made of one indicator out of five for each of the six groups, although aggregation could have been done as easily with all 38 leading indicators. Selected are: Number of labor hours per unit of product (from the group Control Operations Activities); Number of nuisance and always-on alarms (Control the Status of Systems and Equipment); Number of times workers are challenged to solve "what-if" scenarios (Develop Required Skills/Behaviors); Number of unplanned shutdowns (Monitor Organizational Performance). The latter indicator has been doubled in weight because of its seriousness. The fifth indicator is Staff turnover rates (Group: Maintain a dependable practice).

The collected data in number of times a count occurs per period of, for example, 3 months, must be normalized, that is, brought on a scale, for instance, from 0 to 100. This is accomplished by normalizing on a mean of 50. Suppose the count of the last period is 3 and the mean over previous periods is 4.0, then the normalized value is $3/4 \times 50 = 37.5$. Subsequently, the complement is taken from 100, because a better result means a higher number. The figures are then introduced as new evidence in respective nodes (e.g., as Scenario 1:59 for the top one) containing all previous information in truncated normal distributions. Next, the aggregated principal indicator distribution mean and variance are derived applying the AgenaRisk BN, as shown in Figure 7.13. Note that because the aggregated information consists of distributions, the data mean, data variance, and predicted Conduct of Operations are distributions.

A further step is to exploit a relation between indicator metrics and their influence as risk factor via safety attitude, human error proneness, and management system effectiveness, on the overall risk level. Because there will be weight factors and nonlinearity in the relations, validation based on historical data will be a must. However, as we shall see, with the exception of the nuclear power community and the Norwegian offshore, such data are not generally available.

Following a paper by Pasman and Knegtering,[63] we shall show first the result of a simplistic indicator effect model on the performance of an operator. The indicators are aggregated to three, namely a personnel, an operations, and a maintenance integrity indicator. When a reactor upset occurs due to a certain temperature increase, the operator should initiate emergency cooling, which can save the operation or at least result in a safe shutdown. However, the alarm can fail, the operator may not notice it, but may still be alerted by the temperature increase at an indicator (on his screen display). When cooling starts too late, fire will break out, and when the cooling is not activated an explosion occurs. Aggregating the indicators of the plant to the three top indicators mentioned for assessing asset integrity follows a study on actual maintenance activities on oil and gas processing facilities by Jakiul Hassan and Faisal Khan.[64] The situation is shown in the continuous GeNIe BN of Figure 7.14. The results given integrity inputs are presented in Table 7.8 and Figure 7.15. A relatively small degradation in integrity seems to imply a drastic increase of probability of explosion.

The example BNs shown in this section consist of only a relatively small number of nodes. For more extended problems the number of nodes can go up into the thousands. For coping with such numbers the various software makers allow a *modular* structure, wherein modules can be tested individually.

7.7.3 RARE EVENT PROBABILITY ESTIMATION BY MEANS OF PRECURSOR FREQUENCIES

In quantitative risk analysis the real catastrophic, low likelihood event thought possible generates most of the discussion. The frequency expectation of that event follows from a multiplication of many CFs or other subevents, themselves uncertain and assumed independent, so that the outcome of occurrence frequency is always highly uncertain and arouses much discussion. A sufficient number of the same, or at least similar, events to extract a significant estimate is usually not possible. Such catastrophe may not have occurred anywhere else yet, so there is no statistic on it. Obtaining an expected frequency value of an identified rare event is a key problem, and an optimum solution is only accomplished through a system approach. In Chapter 9, Section 9.3 we shall discuss methods for decision making under so-called deep uncertainty. However, if frequency values are available of the more common near misses, also called near accident or accident precursor events, these can be used for a prediction of the rare catastrophic event failure frequency. This can be

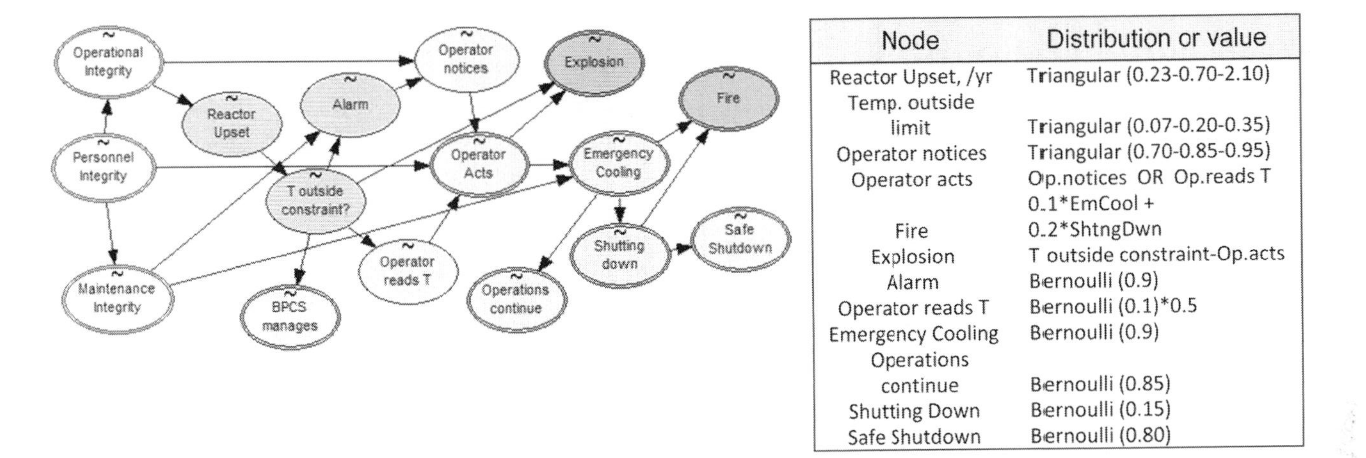

Node	Distribution or value
Reactor Upset, /yr	Triangular (0.23-0.70-2.10)
Temp. outside limit	Triangular (0.07-0.20-0.35)
Operator notices	Triangular (0.70-0.85-0.95)
Operator acts	Op.notices OR Op.reads T
Fire	0.1*EmCool + 0.2*ShtngDwn
Explosion	T outside constraint-Op.acts
Alarm	Bernoulli (0.9)
Operator reads T	Bernoulli (0.1)*0.5
Emergency Cooling	Bernoulli (0.9)
Operations continue	Bernoulli (0.85)
Shutting Down	Bernoulli (0.15)
Safe Shutdown	Bernoulli (0.80)

FIGURE 7.14

BN (GeNIe-DSL[41]) representing the variables of an operator having to act at an upset of reactor temperature, T. The net can cope with all variables as distributions. The data are given in the table at the right. The integrity nodes' default value is 1. Personnel integrity is dominating and influences the operator; operational integrity the reactor and operator alertness; and maintenance integrity the reliability of the alarm and the emergency cooling.

Table 7.8 Probabilities of Safe Continued Operations, Reactor Shutdown, Explosion or Fire

Integrity O, P or M	Reactor Upset/yr	T Outside Constraint/yr	Operations Continue[a]		Safe Shutdown		Explosion		Fire	
			Mean/yr	Std. Dev.	Mean/yr	Std. Dev.	Mean/yr	Std. Dev.	Mean/yr	Std. Dev.
1	1	0.21	0.13	0.08	0.02	0.01	0.04	0.07	0.02	0.01
0.9	1.13	0.23	0.09	0.07	0.01	0.01	0.10	0.09	0.01	0.01
0.75	1.34	0.28	0.06	0.05	0.01	0.01	0.18	0.11	0.01	0.01

[a] *Operations continue given temperature outside constraint.*

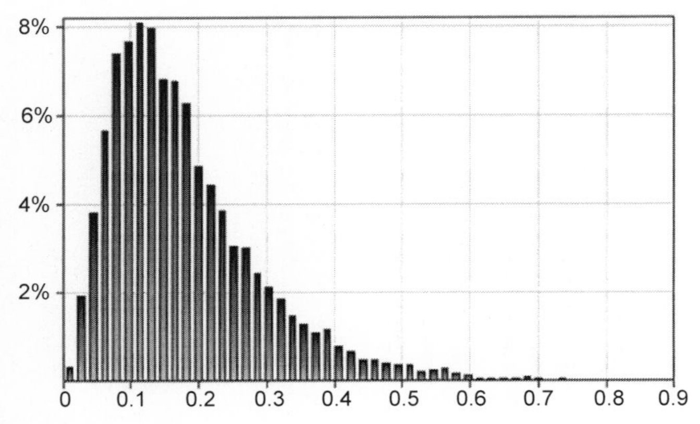

FIGURE 7.15

Distribution of explosion occurrence per year in case personnel integrity is 0.75. Mean is 0.18/yr. This is a factor 4.5 higher than at an integrity value of 1. However, there are multiple uncertainties associated with directly relating integrity to event nodes without weights. Hence, historic data and the following development over a period of time are needed to validate and obtain confidence.

achieved by applying Bayesian data analysis, even if that frequency is 100−10,000 times lower. This method called accident sequence precursor is a known program in nuclear safety and was described recently by Cooke et al.[65] for BOEMRE (now BSEE) in view of the Deepwater Horizon blowout, demise, and oil spill. Yang et al.[66] followed by Khakzad et al.[67] applied the method to offshore blowout accidents as at the Deepwater Horizon operation, but it has general application potential. The method is summarized below.

First of all, the near-miss frequency values are derived from counted events over a certain period as described by Bier and Mosleh[68] and later by Bier and Yi.[69] The precursors may be summed from a number of the same installations at a certain plant or site, but if figures are too low, data originating from the entire industry sector may also be used. Secondly, the near-miss events must fit into an event tree such as of Figure 7.16. A particular IE (or degraded condition) connects via branching nodes sequentially in time of various types of possibly occurring precursors (e.g., fault detections, barrier activations, or other intermediate phenomena) to a set of consequence end states of different severity including the catastrophic event. There may be more than one event tree involved for different IEs, but we shall focus on one IE. Subsequently, the estimate of the rare disastrous end state is derived by making use of the hierarchical Bayesian approach (HBA). HBA has been described among others by Dana Kelly and Curtis Smith.[70] HBA is using all available evidence by sequentially applying Bayes' theorem, typically in two stages. As we shall also follow Yang et al.[66] and Khakzad et al.[67] our nomenclature will be slightly different.

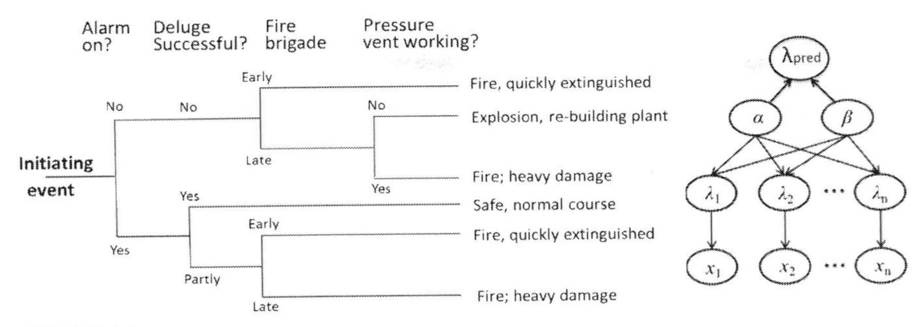

FIGURE 7.16

Left: Simple example event tree showing four types of end state events, which can be the result of a certain initiating event (IE). Explosion will be a rare, but most serious, end-state event. If numbers of other, more frequent intermediate and end-state events occurring over a period of say 10 years are counted, the hierarchical Bayesian approach (HBA) can produce an estimate of an explosion frequency. Right: A Bayesian network DAG representing a number of observed precursor events, x, occurring over the period, t, modeled by the exponential or Poisson distribution of various installations/sources 1, 2...n producing each an estimate of frequency, λ. The two-stage HBA with hyperparameters α and β determines the variability of the population and can predict a distribution for λ from the α and β values through aggregation of the observed time of failure data in the BN. The sources are assumed to be independent. The process results in a predicted frequency of the rare event, λ_{pred}.

Expert estimates and/or sector-established precursor occurrence data from different, but similar, sources over, for example, 10 or 30 years contain as a population uncertainty the source-to-source variability. For each source i out of k sources over a certain time, t, x_i events are observed. The probability to observe that number of events is following a homogeneous Poisson distribution $f(x_i|\lambda_i) = \frac{\{(\lambda_i t)^{x_i} e^{-\lambda_i t}\}}{x_i!}$, $x_i = 0, 1, 2...$ events, where λ_i is the rate parameter describing the mean number of events over recorded time, t. This rate is assumed to be randomly selected from the source-to-source population distribution. Prior belief is that the source-to-source variability in λ can be modeled with a gamma distribution (conjugate with the Poisson distribution). Shape and scale parameters α and β in this distribution are, however, unknown and assumed independent of each other before the data are observed. Hence, the population variability in λ, given the vector $(\alpha, \beta)^{\mathrm{T}}$, is represented by the gamma prior $\pi_1(\lambda|\alpha, \beta) = \beta^{\alpha}\lambda^{\alpha-1}e^{-\beta\lambda}/\Gamma(\alpha)$. The first stage prior for α and β is called a hyperprior $\pi_2(\alpha, \beta)$ and represents the uncertainty in the vector of the hyperparameters. For those gamma distributions can also be taken with suitable known parameters to make them diffusive. A solution will proceed by applying Bayes' theorem in a first step to obtain a posterior distribution of (α, β) given data E:

$\pi_3(\alpha, \beta|E) = \frac{L(E|\alpha.\ \beta)\pi_2(\alpha.\ \beta)}{\int_0^{\infty}\int_0^{\infty} L(E|\alpha.\ \beta)f(\alpha.\ \beta)\mathrm{d}\alpha\mathrm{d}\beta}$, and assuming the data, E_i, from the various sources i are independent, with the likelihood as: $L(E|\alpha, \beta) = \prod_{i=1}^{k} L(E_i|\alpha, \beta)$.

The evidence, E, consists of the set x, t. In a second step, the (posterior) predictive density for λ_i can be derived in analogy with Eqn (7.3) as: $\pi(\lambda_i|E) = \iint \pi_1(\lambda_i|E, \alpha, \beta)\, \pi_3(\alpha, \beta|E)\mathrm{d}\alpha\mathrm{d}\beta$.

This equation is a so-called continuous mixture of gamma functions and will hold for any particular source. Hence the equation can be generalized to produce the λ^*, the predictive distribution of precursor event rate or the population variability curve. If later new evidence becomes available, by applying again Bayes' theorem, further updates can be accomplished. For binomial distributed variables, the hyperparameters are assumed to be beta distributed (see also Khakzad et al.[67]). WinBUGS or OpenBUGS is used to solve the equations by means of MCMC. As mentioned, Fenton and Neil's[42] AgenaRisk, offering a so-called dynamic discretization algorithm, updates satisfactorily and quickly for simpler cases including cases with non-conjugate prior and likelihood distributions.

HBA in combination with DyPASI HAZID database searching may offer further perspectives, certainly when by updates it is made dynamic and would generate warning signals; see Paltrinieri et al.[71]

7.7.4 BN APPLICATION IN INTEGRATED RISK ANALYSIS WITH HUMAN ERROR PROBABILITY

Use of BNs for predictive risk models combining technical and human aspects started some years ago. An early project was Causal Model for Air Transport Safety (CATS). The aim of the project was to perform better and more rational cost–benefit trade-offs on safety measures. CATS made extensive use of BNs, historical accident data, and human performance expertise (pilots, traffic controllers). It also introduced management influences. Because of their versatility, BNs were preferred over fault and event trees. The final model was validated against available accident data. The conclusion was that results looked consistent but that more work still had to be done. The effects of maintenance had remained underdeveloped. Ali Mosleh, mentioned below, cooperated in the project with Roger Cooke, Patrick Hudson, Ben Ale (Delft University of Technology), and many others (NL National Aeronautics Lab) in work sponsored by the US and Dutch aviation safety agencies.[72]

There are three early on studies applying BNs to model human error probability more effectively: the first is the PhD study of Katrina Groth directed by Ali Mosleh, University of Maryland, USA; the second a project of Vinnem and coworkers of the University of Stavanger, Norway; and the third is application of Erik Hollnagel's ETTO principle by means of a BN to model breach of protective barriers as a preparatory project for the fourth-generation French nuclear power reactors.

Groth and Mosleh[73] advanced the field of human reliability analysis by developing a causal model based on a theory pivoting around performance-shaping factors, or rather named performance-influencing factors or PIFs, in operator decision making within a team context. As data sources, Groth had at her disposal the Human Events Repository Analysis (HERA) database of operator errors in controlling processes, developed over the years by the nuclear community, and the results of an application of the cognitive information-decision-action (IDAC) model. HERA works with 11 PIFs that are, however, further influenced by 250 PIF details to represent the possible states in which an operator must make a decision. In her dissertation, Groth[74] developed a collapsible hierarchical structure for the PIFs along the cross-section: organization-, team-, person-, situation/stressor-, and machine-based. This structure is further elucidated in Groth and Mosleh.[75] The PIFs have various states and are interdependent. The data are the metrics for PIF quantification. The framework helped to extract quantitative information concerning how the PIFs conditionally relate to each other from the discrete state data by applying tetrachoric correlation analysis (suited for binary variable data with an underlying normal distribution of which integration calculates the threshold the variable will change state), while factor analysis (minimum residuals) was used to identify patterns of PIFs linked to human error and capture synergistic interactions of PIFs. The pattern of PIFs showing this effect is called error contexts (ECs).

To limit the number of parent nodes in the Bayesian net as far as possible, the PIFs must be combined using the results of correlation and factor analysis. If desirable, expert intuition can be added. Application of the method developed by Groth and Mosleh[73] on available data resulted in nine representative parent nodes and four ECs, as shown in Figure 7.17. The four Error ECs, in the BN positioned as child nodes, are in decreasing order of importance. The ECs give much insight for the interrelations. Directed arcs were drawn between PIF nodes and ECs having a factor loading and correlation above |0.3|. Arc direction was determined by expert

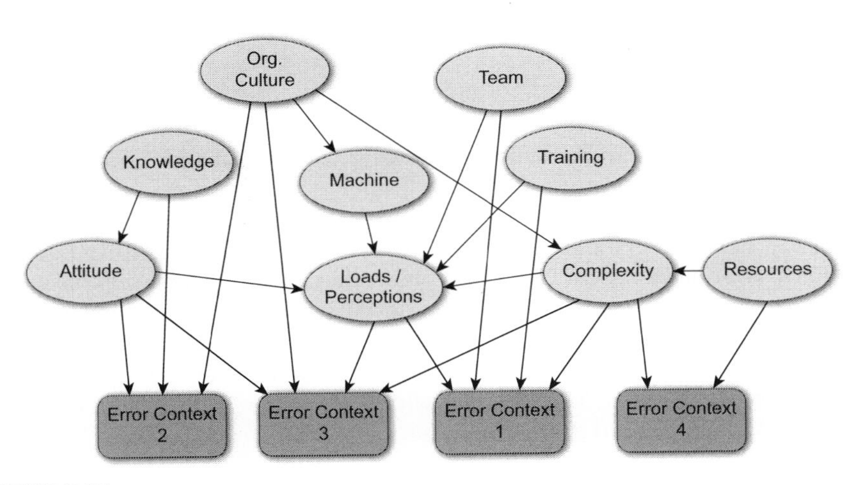

FIGURE 7.17

Causal model of performance influencing factor (PIF) nodes relating to error contexts based among others on PIFs correlation and factor analysis, according to Groth and Mosleh.[73] The net stores both qualitative and quantitative information on human reliability for scenarios of which data and other inputs are used.

judgment. In principle, given the BN for a particular scenario, human error probabilities can be calculated using the EC values derived by applying Bayes' theorem and HERA data on that scenario. This approach would provide a predictive human reliability model. However, data are missing, for example, on successful decisions given a developing accident. Future work should provide for this lack of information.

In an earlier paper, Groth et al.[76] proposed a so-called hybrid causal logic model connecting the human factor part via a fault tree to an event sequence diagram (similar to an event tree) and thus forming an integrated risk analysis model. The model was applied to an incorrect runway takeoff accident. As a follow-on, Azarkhil and Mosleh[77] developed for a control room team (equipment operator, decision-making shift supervisor, and technical consultant) based on the information, decision and action in crew (IDAC) context cognitive model, causal structures for role awareness, team cohesion, leadership, communication, and coordination. They merged it all together with a model of the process in CREWSIM (coded in MatLab Simulink environment), capturing the dynamics. For validation this can generate a dynamic event tree with branches representing distinct combinations of system and operator states associated with error. The abnormal situation of a pipe break in a nuclear power steam generator was simulated and resulted in "face validity." Also, results were compared with the Halden exercise (Section 6.3) of observing a team's performance of coping with the incident in a simulator. The simulation yielded the same trend in overall response time difference of the slowest and the fastest team.

The second study concerns the prediction of leaks at Norwegian offshore installations due to errors in maintenance performed by Vinnem et al.[78] and Gran et al.[79] (University of Stavanger, Norway). This theme comes closer to process risks. The goal of the work is again design of cost-effective risk-reducing measures. This group applies the rather straightforward, Reason classification of human error affecting failure rates leading to hydrocarbon release (error of execution: slip of mind/lapse, mistake, violation, and error of omission) and generic error probabilities drawn from several sources. The model uses scores of quality of work, comparable to performance indicator values, which act upon management risk influencing factors (RIFs) that in turn affect a number of worker level RIFs and through these adjust the human error probabilities. Figure 7.18 shows the structure. The scheme is quantified by years of collected scores of various installations, in addition weighting the influence of scores and RIFs on error rates by expert judgment. The model was validated against recorded leak numbers per platform-year applying the platform specific scores. It proved "capable of reflecting relative differences between alternative installations" to cite the authors.

In *Supervision and Safety of Complex Systems*[80] the French GIS 3SGS collection of projects is reported. The project is in anticipation of the fourth generation of nuclear reactors and is in a cooperative effort of a large number of French institutes. It is on new methods of reliability prediction, fault diagnosis, radioactive waste disposal, data fusion, robotics, signal processing, and various kinds of models including risk analysis. In various applications use is made of BNs. Chapter 15 of

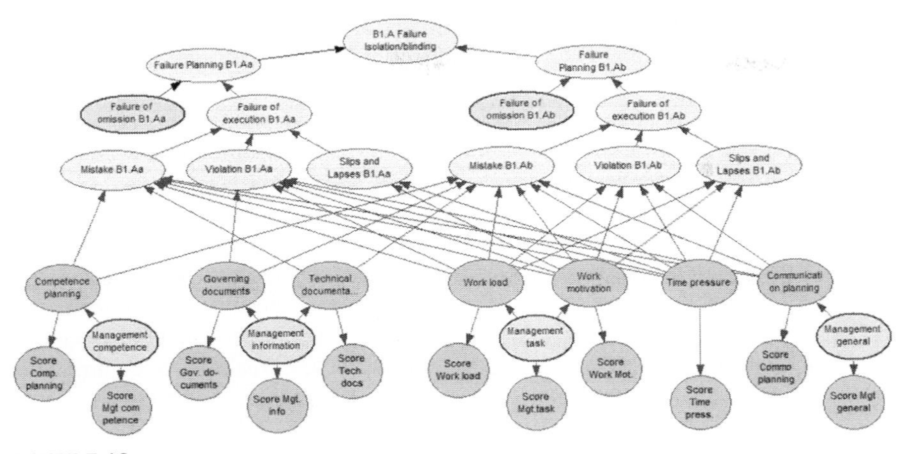

FIGURE 7.18

BN of the two-level (management and worker) risk influencing factor (RIF) structure by Vinnem et al.[77] for the example of a maintenance planning activity B1.A (failure is incorrect blinding/ isolation: Aa. planning; Ab. check on planning correctness). For other failures additional RIFs may appear, such as disposable work descriptions, human–machine interaction, design, supervision on worker level, and technical management. The bottom layer represents nodes of score observations obtained in audits and surveys, comparable to indicators.

the book is on integrating human and organizational factors in a risk analysis model by means of a discrete BN (influence diagram). The focus is on breaching barriers voluntarily. The model considers benefit–cost–deficit and is known as the BCD approach. For running a system human operators always will supervise it, but they are at the same time the largest source of errors. Infringement of barriers, including personal safety devices, is an important type of error.

A system must operate within its safety window delimited by different types of barriers. Beyond the barriers there is somewhere a borderline, passing of which is declared as unacceptable for functioning by the designer, but the limit a human operator may consider unacceptable is even deeper in the unsafe area. Barrier infringement can bring immediate benefit (convenience, less effort, time saving) with acceptable losses to the operator (costs: risk of injury, reprimand, or worse), but with the risk of a "deficit" or sustaining a system loss. The balance is then expressed in: $Utility = w_1B - w_2C - w_3D$, as the person taking the decision perceives it, in which w_1, w_2, and w_3 are weighting factors, while taking benefit, cost, and deficit values positive. The setting, in which the operator decides to infringe or respect the rule, is influenced by nine *human factors*, which in turn are influenced by seven *organizational factors* or management conditions.

Human factors that influence the individual's decision are (acronyms are abbreviations of French expressions): delegation of responsibility, aids, training (formation), experience, possibility of respecting guidelines, contextual factors (RWS), conducting dynamic management and group collective (CMGD), management and achievement of objectives (RTC), and feedback (FEx).

Organizational factors are: weakness in safety culture, failure in day-to-day safety management, weakness of monitoring organizations, poor treatment of organizational complexity, difficulty in implementing feedback, production pressures, and failure to reexamine design.

The example discrete nodes BN of Figure 7.19, structured as an influence diagram, represents two safety barriers for operating a printing press: personal protection equipment and an emergency stop when the machine needs to be cleaned. *B*, *C*, and *D* are composed of three factors—workload, safety, and time—each as benefit and deficit nodes and aggregated to a global value. Only three organizational factors are considered: weakness of organizational culture (TOC), weakness of control, and

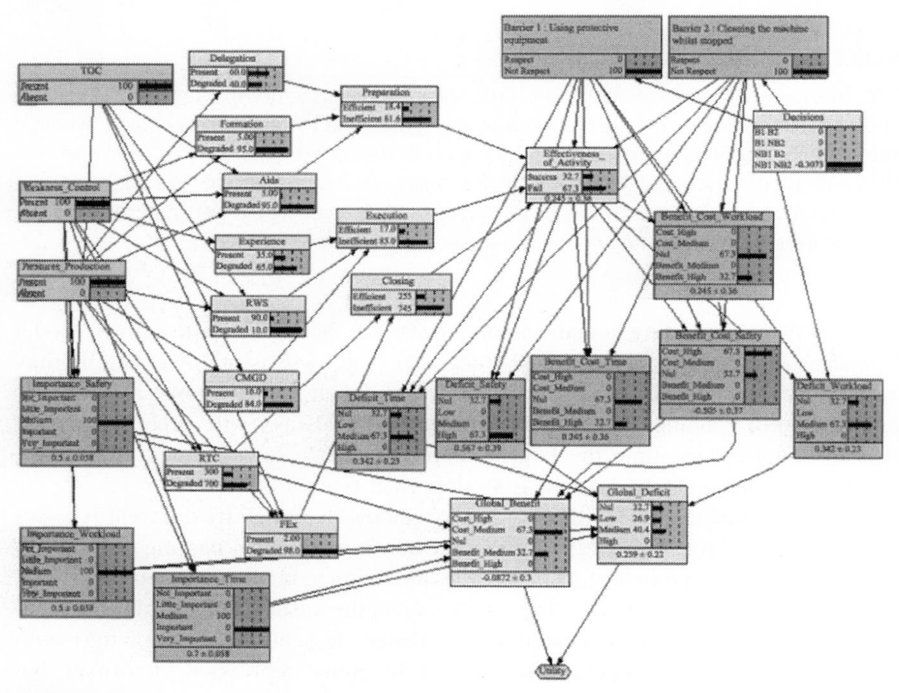

FIGURE 7.19

Impression of an example of a discrete BN (influence diagram) according to the benefit–cost–deficit model, developed in the framework of projects for the fourth-generation French nuclear power reactors,[80] accounting for (a selection of) organizational factors (far left), human factors (second column) on possible breaching of two barriers (wearing protective equipment and cleaning machine while stopped), resulting in an overall utility figure, which in the example is positive, meaning that there is on average a benefit for a worker to breach the barriers. (*The reproduction quality of the figure is regrettably low, but for the message it is just adequate.*)

production pressures. All human factors are represented, while the activity stages are split into three: preparation, execution, and closing. These three determine the effectiveness of activity node. To the far right is a decision node with four possible state combinations of respecting the safety barriers. Also in this case, probability values and weighting factors must be based on performance metrics and expert estimates. Figure 7.19 shows the situation when both barriers are ignored. The benefit probability at level high is 32.7% but cost of safety went up to high 67.3%, while deficit in time and workload medium is 67.3% and safety high 67.3%. Applying the weight factors it results in probability of Global Benefit medium 32.7, Cost high 67.3, while Deficit probability at medium level is 40.4% and at low 20.9%.

Erik Hollnagel was a member of the scientific board of GIS 3SGS. The BCD model appeals to a certain extent to his ETTO principle (see Chapter 6, Section 6.4), but it lacks still the variability that according to Hollnagel is essential and consistent with a system approach. This again will be a matter of collecting data.

Sponsored by Royal Dutch Shell plc, previously Ben Ale's group[81–83] at the Delft University of Technology made quite a few advances in applying BNs (in their jargon Bayesian belief networks, which as naming in view of subjective inputs is fully justified) to process plant predictive, operational risk assessments tying together technical, organizational, and human factors.

7.8 MERGING TECHNICAL AND HUMAN FACTOR: AGENT-BASED MODELING AND PETRI NETS

As is argued by several leading thinkers in the field—Perrow, Rasmussen, Leveson, and Dekker—in a sociotechnical system with many possible interactions among controlling agents, human or otherwise, risk is for a certain part emergent from the system and cannot be completely predicted by conventional sequential cause–consequence analysis methods on components, such as fault and event tree (bow-tie, etc.). In the beginning of this chapter, we have seen approaches, such as STPA and BLHAZID, that attempt to encompass all risk aspects, including the emergent ones, but which are not yet fully operational.

Sybert Stroeve et al.[84] reported a summary of interesting advanced modeling work by a research group in aviation safety at the National Aerospace Laboratory, NLR, and the Delft University of Technology in the Netherlands over a period of at least 10 years on risk prediction in a complex sociotechnical environment. The scenarios they studied have in common an aircraft ready for take-off, another one on a taxiway wanting to cross the runway, while runway controllers, ground controllers, and automatic alerting devices guard the situation. Many subscenarios can be derived. Crossing a runway is an event occurring many times at airports every day. A question is what is the collision risk, assuming all participants try to do their work well? The group had many interviews with pilots and others and the availability of historical data. They put against each other an analysis based on a conventional *event tree* (or event sequence diagram) and one with a multiagent dynamic risk model (MA-DRM).

Agent-based modeling (ABM) is the preferred choice for modeling system emergent phenomena as is well described by Eric Bonabeau.[85] Agents are autonomous decision-making entities. Citing Bonabeau: "ABM is more a mindset than a technology." It is technically simple but conceptually deep. This is because the modeling is in the agent interacting with others and not at an aggregate system level as a system dynamics model does, and it therefore is able to capture emergent phenomena. Individual behavior is complex and contains random features. Bonabeau notes that where agents may simulate humans, the latter can react irrationally, and subjective ABM model outcomes need validation by expert judgment.

In an ABM model the "macro equations" describing transition states must be adapted to those states for an individual element reacting on what he observes with his senses about other agents he is in contact with in an active way, or passively because they are in his field of view. Stroeve et al.[84] solved that by modeling individual agents by means of a Petri net.

Carl Adam Petri developed this network concept in a PhD project at the University of Bonn, Germany, in 1962. A Petri net distinguished two types of nodes: *Places* (P; as a circular or oval node) and *Transitions* (T; as a rectangle) connected by directed *Arcs* (arrows: to a transition is an input arc and to a place an output one). In Figure 7.20 (middle and lower figures right) an example representing a technical dynamic process is shown. Places each have a state, transitions represent a changing-state event, and an arc is the expression for how the state changes when an event occurs. The event can be triggered (also called firing or jump) by another event or just by the lapse of a time period, which can also be zero. A place represents resources of any kind, for example, data, information, or physical items, and a place can be marked with one or more tokens (as black dots) enabling its present state. At a transition, the token is moved in the arc direction, thus removed from an input place and deposited or produced in an output place. (Double arcs allow motion in two directions). The state of the system is given by a tuple with an entry for each place of which the value gives the tokens in that place. This is called the marking, and the model is defined by its initial marking. Reachability is the set of all possible markings a net may develop dependent on its initial marking.

A Petri net can be seen as a discrete event dynamic system. Although developed for solving capacity problems in (unreliable) communication networks, the concept appeared very fertile in solving many other problems, and other features were added. The firing time can be bound to rules, and the input can be deterministic (timed Petri net). If the firing rate is a constant and the delay time between enabling and firing is sampled from an (memory-less) exponential distribution, a Petri net will have the character of a Markov process and be called a stochastic Petri net (SPN). Deterministic and stochastic Petri nets (DSPN) have a constant firing rate and exponentially distributed timings. Generalized stochastic Petri nets (GSPN) include, besides the exponentially distributed transition times, immediate transitions, which have priority and reflect actions that are independent of timing specification. Kurt Jensen[86] at Aarhus University, Denmark, added color to the thus-far black token dots making it a (CPN). This addition upgraded the net to a high level

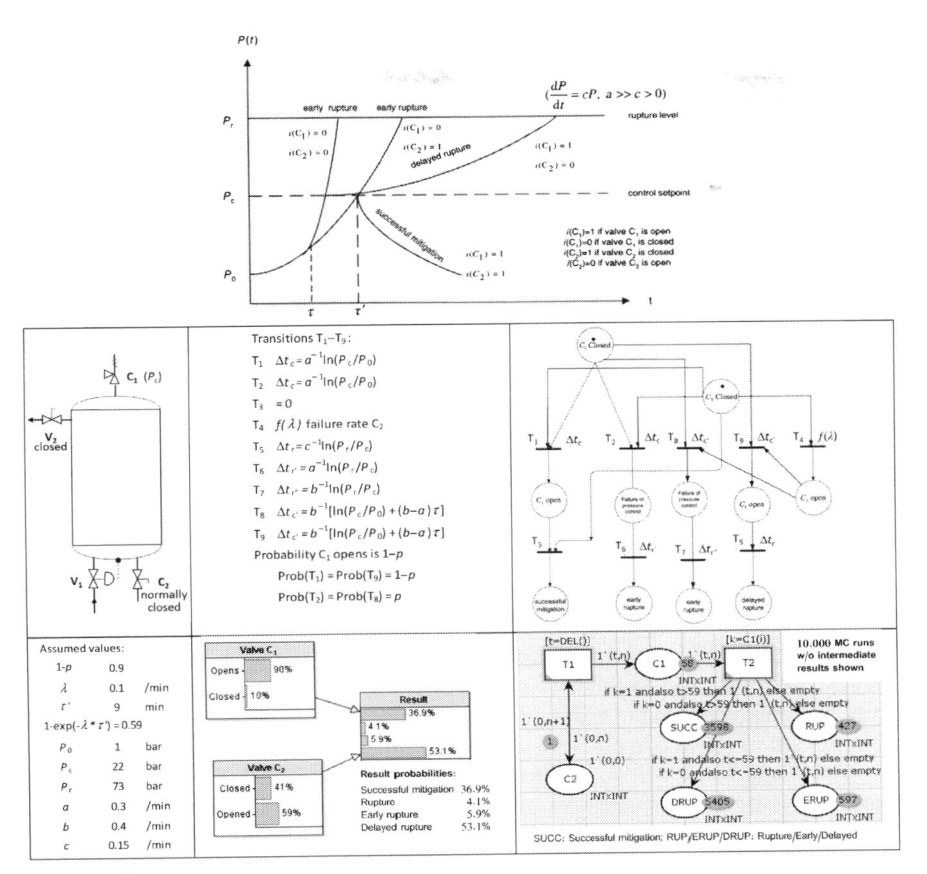

FIGURE 7.20

Middle row, left: A gas is pumped into a vessel until a preset final pressure, P_f, is reached and valve V_1 automatically closes. The example is inspired by Labeau et al.[50] Top: Possible pressure—time scenarios are shown in case valve V_1 is failing, so that at critical pressure, P_c, relief valve C_1 must open. If C_1 fails, the vessel ruptures at P_r. Also, during the operation C_2 can be opened by an operator ignorant of V_1 failing. C_1 cannot cope with the additional gas supply, so there will be a delayed rupture. If in that case C_1 also fails, rupture occurs early. At middle right is shown a timed Petri net and a legend according to Labeau simulating the scenarios. In the lower row for conditions assumed by this author is shown an AgenaRisk BN,[42] yielding outcome probabilities, and a modern CPNtools[87]-colored Petri net enabling Monte Carlo simulation of the process. The latter is far more elaborate to develop but obviously gives the same final result.

because the color distinguishes the tokens for a finite set of values or data they carry and determines which transition will be enabled. CPNTools[86] is a free available CPN, since 2010 as Web downloadable software managed and further developed by Wil van der Aalst (Eindhoven University of Technology, the Netherlands). Peter

Haas[88] of IBM extended the possibilities of random timing of the transitions depending on the color. Petri nets appeared to be very useful in developing computers and software. In contrast to Bayesian nets, Petri nets can be cyclic and are often used to model loop processes. So, they are also applied to model failure events in safety instrumented system (SIS) circuitry and overall SIS safety/dependability.[89,90]

As a further step, Everdij and Blom,[91] as part of the group of Stroeve et al.[84] at NLR, extended Petri nets first to dynamically colored Petri nets (DCPN). This was realized by two extensions: (1) the color is identified with the value of a Euclidian state space evolving from the solution of a differential equation, and influencing the firing rate; (2) new tokens obtain a color produced as random functions from the removed tokens and their color. As a final step, they added diffusion (also called Brownian motion) by tying the color of the dot to the outcome of a stochastic differential equation, which made it a stochastically and dynamically colored Petri net (SDCPN). Details on this development are given in the above-mentioned reference. Everdij and Blom[92] also showed that these enhanced hybrid-state Petri nets have the analysis power of generalized stochastic hybrid processes (GSHP). Hybrid means here the interaction between discrete and stochastic components, hence enabling simulation of, for example, networked control of distributed real-time, physical components with phased multimodal operation interacting dynamically with uncertainties and random elements. Components include human operators working according to procedures but afflicted with the human performance variability Hollnagel claimed (Sections 5.3 and 6.4). GHSP satisfies the strong Markov property (future states are conditioned only on the present and not on past states; strong if present state is defined by a random variable; as mentioned before an exponential distribution has a constant transition rate, so is "memory-less"). The SCDPN was adapted to application in ABM to obtain the MA-DRM.

Evaluating the event tree (ET) and determining the collision probability required rare event Monte Carlo (MC)-ing on the MA-DRM to speed up the calculations. This was achieved with two methods: (1) risk decomposition in a sequence of conditional MC simulations of which in the end results are combined and (2) the interacting particle system approach by introducing a sequence of intermediate aircraft encounter conditions preceding a collision and simulating the process reaching these conditions in several parallel copies. In both cases the final result follows from the conditional probabilities of the intermediates thanks to the strong Markov property.

Stroeve et al.[84] found in this way a value for the conditional accident probability of about 2×10^{-6} in case of application of the ET and 2×10^{-4} by applying the MA-DRM. In other words, despite the limited dynamics built in the ET events as early, medium, and late resolution, the MA-DRM reveals that the dynamic interaction effects emerging from the system, in particular among the humans involved, increases the calculated risk strongly. The 95% confidence interval covered in both cases more than two orders of magnitude (for ET even three). MA-DRM yields also different values for the contribution of risk reduction devices than from the event-sequence analysis. The new method opens in principle the possibility of analysis of system risk control and may therefore provide in the future a contribution to an "online," actual "risk barometer" measurement of an industrial processing activity.

7.9 RESILIENCE ENGINEERING

In Chapter 5, Section 5.3 we became acquainted with the concept of resilience engineering, we defined it, considered the various factors playing a role, and we distinguished an organizational and a technical side. The present chapter is about new methods to gain better control on risks, and therefore it is worthwhile to reconsider resilience engineering. Resilience is at a higher abstraction level and different from robustness. The latter can be measured by exerting given force and determining yield. Higher robustness will imply higher resilience, but resilience as an overall umbrella concept cannot be fully captured due to possible exposure to unknown, unforeseen, and unexpected stresses. These stresses may also include intentional attack.

Considering a running industrial facility, organizational resilience is of interest in the time after the plant has been built and has become operational, while technical resilience must be brought in for the greater part in the design and in fact must be borne in mind already at the specification stage. Hence, we are talking too about different groups of people. On the other hand, the two are tied together—operations people inherit hardware and control options from the designers. More than in the past, operations people should have models available in which the designer's thoughts are conserved. Running these models from time to time would help them to remain vigilant and be aware of weak points. Testing also will enable operations to check the state of resilience to various threats and to improve where possible and suitable the original design (but while applying management of change analysis!).

7.9.1 RESILIENCE IN THE TECHNOLOGY

So, while discussing new methods and models and resilience, we should reverse the sequence and start with the technical side because there is still much to gain, as we saw earlier in this chapter. Although not discussed explicitly here, there will be steady improvements in the consequence models described in Chapter 3. Computer power and insight will further develop so that damage predictions based on all of the relevant data will become more accurate. The system approach with STPA and BLHAZID tools will enable improvements in identifying possible scenarios. In turn, the system approach will make more accurate and precise predictions of future behavior using deterministic models, covariate models, and probabilistic models to support human decision making. It will further indicate ways to improve risk control.

BNs will enable more nuanced and refined event probability calculations, although this hinges for a fair extent on availability of historical data supplemented by expert judgment and updated based on events that occur as observed through system monitoring. Although more research will be needed, we have also seen how the influences of human and organizational factors via the BN can be introduced into the risk determination. This approach will enable operational use of the risk models by collecting PSPIs (or key performance indicators), aggregating and weighting them, apply them in the models, and examine trends in the risk level. In a relative sense safety performance indicator values will also reflect safety culture level. Modeling the risk of high hazard operations with highly interactive control instances, as we have seen in the previous section on runway safety, will take some time to further develop, but it is in principle possible and worthwhile to pursue risk contributions from emergent phenomena.

What we have not yet considered are process simulation models such as provided by ASPEN Tech, Aspen HYSYS, and COMSOL, which can be of great help to analyze the extent of resilience. Also, the extensive efforts in CAPE, as, for example, embodied in the European ESCAPE symposium series, and computer-aided design (CAD) software such as Plant Design Management System (PDMS, of Aveva), can heavily contribute. By the ever-increasing power of computers, process simulation has become a mature tool also able to cover nonidealities as encountered in practice. Simulation starts in the R&D/conceptual design stage and is employed for optimizing performance (see Figure 7.21), but simulation can be used in later stages as well to identify risks and to predict resilience to various kinds of disturbances both from internal and external origin. Further, as we have seen

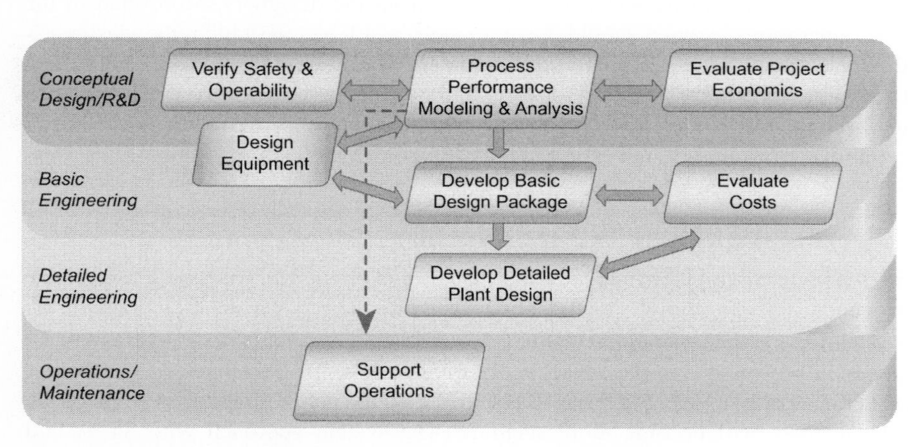

FIGURE 7.21

Process simulation and design packages such as Aspen can be used in all stages of plant development, while they are also useful for supporting operations. Such packages will also be helpful in analysis of resilience.

The figure is taken from an Aspen Plus brochure. (N.B. To play an effective role in risk analysis, Aspen must also cope with value ranges, probability distributions, and sensitivity analysis.)

when discussing HAZOP automation in Section 7.3.3, the higher abstraction multi-level flow modeling (MFM) process simulation developed by Morten Lind[93] may also have its merits, for example, for describing abnormal situations as proposed by Van Paassen and Wieringa.[94] Simulink simulating dynamic systems and control processes and Stateflow monitoring complex system state change events and making them visible, both in MATLAB environment provided by MathWorks®, together form another tool to explore.

7.9.2 RESILIENCE IN THE ORGANIZATION

In case the design is made sufficiently resilient, meaning in the operations that follow there are adequate reserves in redundancy and spares, escape possibilities and emergency measures, it comes down to the resilience and the culture of the organization. During the last 10 years, several books on resilient organizations have seen the light. One of them is that by Karl Weick and Kathleen Sutcliffe[95] on managing the unexpected in highly reliable organizations (HRO), already mentioned in Chapter 5, which in its second edition places emphasis on resilience. They put forward five principles for an HRO:

1. *Preoccupation with failure*, which is tracking and learning from failures (and near misses), in our case collecting lagging and leading indicators, analyzing them, and taking corrective action. For leading indicators, failure would mean failing to achieve earlier set targets.
2. *Reluctance to simplify*, also cast as resist oversimplification, which is an incitement to think and dig deeper. Nature and installations are complex, organizational mechanisms and "movements" too. Stay mindful.
3. *Sensitivity to operations*, that is, to stay informed about what is going on at the frontlines.
4. *Commitment to resilience*: Do not cut all redundancy and organizational slack, stay fault tolerant and nurture a capability for improvisation, containment, and recovery to survive upsets.
5. *Deference to expertise*, which means build and maintain diversity in expertise in the organization and foster knowledge, so that before decisions are finalized by the top managers, relevant experts can give their opinion without rank distinction.

The first three principles concern mindful anticipation: receive weak signals, interpret, and decide, which for a large part consists of creating the right climate in the organization for it. The last two principles have to do with successful containment of the effects of a disturbance. In fact, all five principles within the context of a system approach shall produce together the required overall mindfulness. One can remark that the principles to a certain extent counter a different adage, which is heard in times of cost cutting and reorganizations under pressure of competition in a crowded market place: "stay lean and mean" or perhaps better in wording of a system approach be "efficient and effective"!

Of course, to describe how a resilient organization should look is one thing and to determine how resilient an actual organization is, is another. Weick and Sutcliffe provide a checklist for auditing on the five principles. Meanwhile, several specialist organizations devised methods for benchmarking on resilience; one of them is a research organization in New Zealand, ResOrgs,[96] consisting of a cooperation of several universities, which presents their scorecard testing items as shown in Figure 7.22. An analysis of organizational resilience potential of small and medium businesses in the process industry applying fuzzy sets has been conducted by Alecsić et al.[97]

Introducing organizational resilience indicators in analogy with PSPIs has also been proposed. Øien et al.[98] developed a system of indicators to keep track of organizational resilience level in an operational plant by timely obtaining *weak warning signals*. The indicators are bundled groupwise to "general issues," which in turn are grouped to eight contributing success factors (CSFs): risk understanding, anticipation, attention (combined to risk awareness); response, robustness, resourcefulness/rapidity (combined to response capacity); and decision support and redundancy (combined to support). The approach resulted in the proactive resilience-based early warning indicator (REWI) method. Further work has been undertaken by Paltrieneri et al.[99] to investigate a relation of REWI with

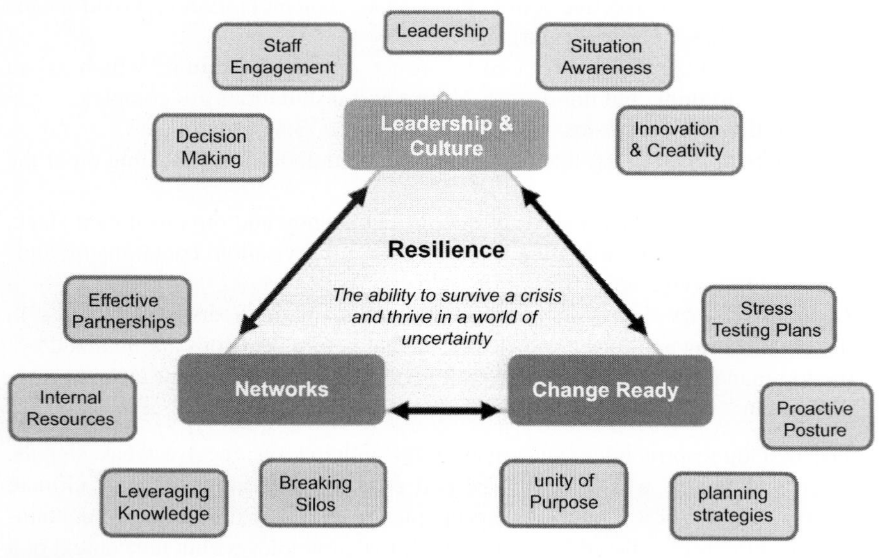

FIGURE 7.22

ResOrgs[96] model for organizational resilience. The three groups of outside boxes are indicator headings, on which an audit can be based. The three groups have strong interdependencies that add to the complex nature of the concept of resilience. It would be challenging but useful in a system approach to model this in Bayesian networks.

PSPIs as in HSE's dual assurance method (Chapter 6, Section 6.7). This was combined with the HAZID database search method of DyPASI[100] and the whole tried out on the "atypical" 2005 Buncefield tank overfill and explosion/fire case. The dominant role of knowledge in the CSF risk awareness was convincingly shown. (Knowledge, learning, and knowledge management will be discussed in Chapter 11). DyPASI proved to be very helpful to bring "unknown-knowns" to the surface.

Sidney Dekker,[101] summarizing gains made by discussions and new ideas during the initial symposium on resilience engineering in Söderköping, Sweden, in October 2004, on safety of sociotechnical systems, stressed the following points:

- "We have to get smarter in predicting the next accident." The new methods collected in this chapter can make significant contributions toward more accurate predictions, which is a primary objective of a system approach.
- "Detecting drift into failure that happens to seemingly safe systems *before* breakdowns occur is a major role for resilience engineering." As we have seen previously in Chapter 5 (Figure 5.9) and in Chapter 6 (Figure 6.8), organizational drift is lurking, but using indicator aggregation methods of a system approach, such trends can be more easily identified, tracked, and adjustments made. Rasmussen[102] recognized the various influences causing drift, as is clearly indicated in Figure 7.23.
- "Detecting drift is difficult and make many assumptions about our ability to distinguish longitudinal trends, establish the existence of safe boundaries, and warn people of their movement toward them." It is on the gap that can exist between the real situation and what is in the minds of the decision makers how they think it is. PSPIs, the suggested aggregation method, and applications in connecting technical and organizational/human factors can be developed into a tool for management at each level of the organization. Can we quantify Rasmussen's figure (Figure 7.23)?
- "Looking for additional markers of resilience, what was explored in the Söderköping symposium?" This is about a continuously updated indicator of resilience of an organization that keeps staff alert on risk in their internal communications, even though operations do not show any sign of lack of safety. With this same attitude, near misses, unusual behavior, and failures shall be examined and lessons applied.

Few attempts have been undertaken thus far to quantify resilience. The least difficult approach would be to define an index or a set of indexes depending on the characteristic chosen, but this has only a limited value. With respect to organizational resilience in a process plant, Shirali et al.[103] propose performing measurements by conducting surveys. Principal component analysis (PCA), a kind of factor analysis, and numerical taxonomy (NT), that is, a classification on a multiplicity of characteristics, are applied for analysis of results. The survey is focused on the following indicators: top management commitment, just culture, learning culture, awareness and opacity, preparedness, and flexibility. The data are related to various

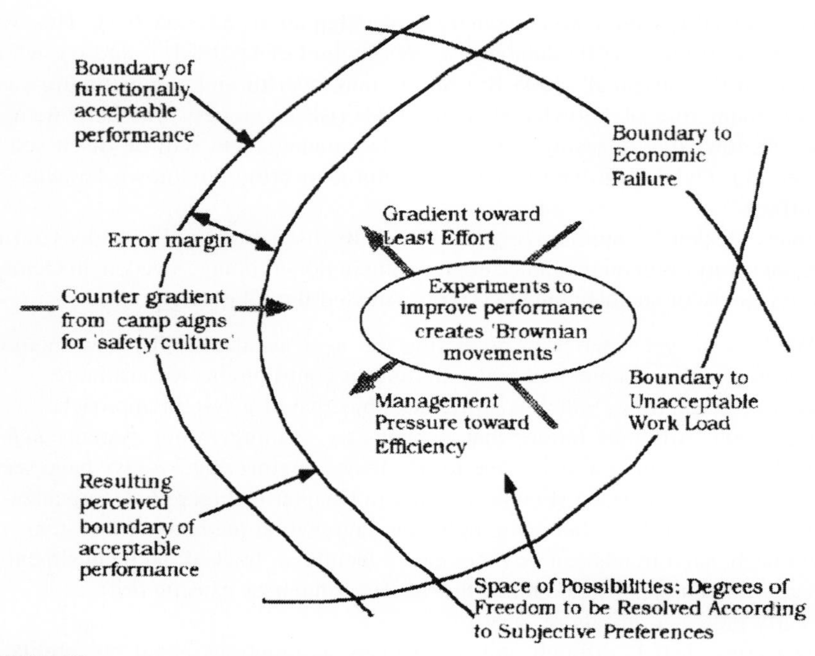

FIGURE 7.23

Representation of influences and pressures driving an organization toward the boundary of acceptable safety performance as Rasmussen[102] saw it in 1997. In addition, we can say that in a resilient organization the space for maneuvering within the boundaries should be sufficiently large. But one could immediately ask how large is sufficient? In practice, it depends much on the wisdom of leadership and the measures leadership takes. Rational quantification can only be performed probabilistically and will need continual monitoring and adjustment.

(11) process equipment units. The result is a ranking of the units, but the PCA and NT ranks differ considerably. The study is not very conclusive.

In the same 2004 Söderköping symposium, David Woods,[104] Ohio State University, contributed on the essential characteristics of organizational resilience (or as he mentions its absence, "brittleness"). He stresses that resilience of a system is its capability "to handle disruptions and variations that fall outside of the base mechanism/model for being adaptive as defined in that system." So, how well can it cope with uncertainties? For this, Woods states, the organization must monitor and assess risk distributions and how close it operates to safe boundaries. In that context, it must look at its buffering capacity, its flexibility, the margin, and the tolerance. With the latter, Woods means the system behavior distribution near a boundary—slowly degrading or quickly collapsing—and he further delves into the complexities of organizational processes influencing its resilience distribution. In his epilogue with Hollnagel, once again is emphasized that resilience requires knowledge and competence to anticipate a problem, to perceive it when it is there, and to respond adequately. It requires a "constant sense of unease that prevents complacency." Monitoring the gap between "work as imagined" and "work as practiced" is fundamental, but how this can be realized in practice is a problem of resilience engineering.

We discussed safety culture in the previous chapter. Perhaps, we should emphasize that sincerity is a first condition for any success. All systems can be misused and sabotaged. Fraud and deceit are unfortunately not uncommon. Also, less explicit, derailing can occur—hard facts can be kept concealed. Therefore, auditing shall be performed by fully independent parties. Favorable results can emerge from reports by parties that are biased by being financially dependent on the assigner. Unfortunately, with performance indicators cheating also is possible. If there is sincerity and transparency in an organization, and also by effective communication an atmosphere of trust, then motivation and loyalty will be created that in the end will be a major contributor to safety and will strengthen resilience. In all this, understanding of, and cooperation with shareholders and other stakeholders, and potential availability of external resources will increase organizational resilience.

7.9.3 RESILIENCE AND EMERGENCY RESPONSE

A matter that in a way is in between technical and organizational resilience is emergency response, which is affected by the resilience of nations. In 2005 the word *resilience* was used at a United Nations world conference on disaster reduction, held in Kobe, Hyogo, Japan. It called for "Building the Resilience of Nations and Communities to Disasters." It was a follow-on of the 1994 Yokohama Strategy for a Safer World guideline. It has of course the deeper goal for a society and economy to veer back after it has been hit by disaster, such as a storm and flooding as in New Jersey/New York by hurricane Sandy in October 2012, or in Tacloban, Philippines, by the typhoon Haiyan in November 2012. After such catastrophes, economies should be restored to normal as soon as possible to prevent worse outcomes. Governments startled by shocks and disaster have begun to promote a culture of resilience in organizations. Building resilience pays off because it assures as much as possible business continuity. According to Soczek,[105] recent studies have shown that for every US dollar invested in resilience and prevention, between four and seven dollars are saved. The insurance industry will therefore also have an interest.

In companies of significant size there will be plant emergency responders, partly professional, partly volunteers. The plant emergency response team must cooperate smoothly with the community emergency response. The latter has a proactive task in inspecting fire engineering aspects of facility design and layout, preparing for eventual action by scenario analysis (see Section 3.6.6.6) and contingency planning, for example, for evacuation, and repression in case of disaster. In action, emergency response comprises three elements that must cooperate intimately, which requires a field command and skillful leadership:

1. Containing and combatting the event, which is the traditional, repressive task of the firefighters.
2. Medical first aid staff collecting, examining, and transporting victims to hospitals by ambulance.
3. Police and investigators for controlling crowds, keeping access routes open, finding out what happened, and legal matters.

In many countries, emergency response has become organized on a regional basis as a permanent staff focused on longer-term interests, while the local squads shall be trained, practiced, and ready to deploy on short notice. In addition, there shall be an expert on hazardous materials available, who can judge an event with respect to the hazards the emergency responders are engaging and to impel evacuation of public around a site. In the case of a real threatening disaster possibly making many victims a crisis center must be set up in which the local mayor will "have the last word." Effective cooperation and communication between plant management and the firefighters' commander is of utmost importance. This will avoid wrong or ineffective actions from the side of the firefighters and later blame and reproach. The communication may concern the way and locations to attack in view of chemicals stored, potential of escalation, or estimating and upscaling of the alarm level. A few other technicalities have been described in Section 3.6.6.6.

7.9.4 RESILIENCE SUMMARY

Resilience is an umbrella concept. Good resilience will be achieved by bringing together such attributes as inherently safer design, robustness, error tolerance, adaptability, alertness, and necessary resources. Progress in safety will be by adopting a system approach including all aspects: human, organizational, and technical. Resilience is similar to the concept of safety in the sense that one never can reach the 100% assurance milestone because of the unknown scenarios that can emerge. Resilience analysis will have to make use of an ensemble of system approach tools. Helpful tools such as recognition aids of weak signals from different sources and data analysis, process simulation, human factors engineering, and risk assessment including scenario identification tools are available and certainly more will appear. We shall never be able to deterministically predict the future, but current probabilistic methods provide support for development of effective strategies to handle uncertainties in a quantitative statistics–based fashion. The resulting resilience distribution shall be presented as aggregated distributions of parameter values.

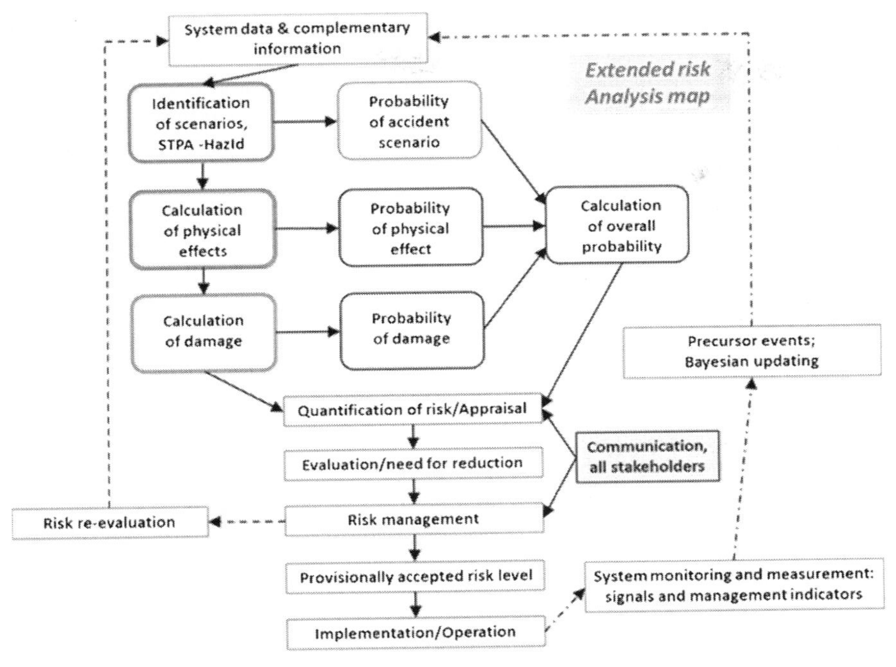

FIGURE 7.24

Extension of the risk analysis map shown in Figure 3.26 with updates, feedbacks, and new possibilities.

7.10 CONCLUSIONS

The new approaches and methods described in this chapter, being the fruit of many years of development and that in most cases are still further maturing, cover a few most vulnerable parts of risk assessment and control, namely hazard identification, system emergent risks, and human interaction such as STPA and BLHAZID. In addition, methods, techniques, and tools developed also for quite different purposes, such as Bayesian and Petri networks, agent-based modeling, and process simulation tools, will also play an important future role in risk and resilience assessment. Not to forget, rare event probability estimation with the aid of Bayesian statistics and Monte Carlo random number technique also contribute. Given HAZOP results and "Learning BNs" algorithm, optimal causal structures can be derived. Next, given data from observable process variables, dynamic BNs can track state changes and degradation and trace hidden causes of upsets. Together with the evolutionary development of consequence analysis, it will deepen our insight, improve event and behavior prediction, make available online as an actual risk indicator "dashboard," and by simulation strengthen resilience estimation. Figure 7.24 presents extensions to the risk analysis map as compared to Figure 3.26. Several mentioned references in this and previous chapters create possibilities to realize such a concept. In the next chapter on process control we shall meet some additional possibilities. However, trustworthiness of the methods will still depend on the availability of systematized historical data to enable validation, both with respect to technical failures as well as human and organizational ones. Indicators contribute but not enough by themselves. Industry must organize itself for these needs of data to enable further development.

Resilience considerations deserve much more attention, especially for high-hazard operations.

REFERENCES

1. Thomas J. *Extending and automating a systems-theoretic hazard analysis for requirements generation and analysis* [Ph.D. dissertation]. Massachusetts Institute of Technology; April 2013.
2. Seligmann BJ, Németh E, Hangos KM, Cameron IT. A blended hazard identification methodology to support process diagnosis. *J Loss Prev Process Ind* 2012;**25**:746–59.
3. Paltrinieri N, Tugnoli A, Buston J, Wardman M, Cozzani V. Dynamic procedure for atypical scenarios identification (DyPASI): a new systematic HAZID tool. *J Loss Prev Process Ind* 2013;**26**:683–95.
4. Seligmann BJ. *A functional systems framework and blended hazard identification methodology to support process diagnosis* [Ph.D. dissertation]. Australia: School of Chemical Engineering, University of Queensland; 2011.
5. Hollnagel E. Human reliability in context. *Nucl Eng Technol* 2005;**37**(2):159–66.
6. Embrey DE. SHERPA: a systematic human error reduction and prediction approach to modelling and assessing human reliability in complex tasks. In: Steenbergen, et al., editors. *ESREL 2013*. London: © 2014 Taylor & Francis Group; 2013, ISBN 978-1-138-00123-7. p. 311–6. Amsterdam, 29 September–2 October.
7. Embrey DE, Zaed S. A set of computer-based tools for identifying and preventing human error in plant operations. In: *Proceedings of the sixth global conference on process safety*. San Antonio (USA): American Institute of Chemical Engineers; 2010. On CD paper # 107e.
8. Németh E, Hockings K, O'Brien C, Cameron IT. Knowledge representation, extraction and generation for supporting a semi-automatic blended hazard identification method. In: *CHEMECA 2009: the 39th Australasian chemical engineering conference*; 2009. p. 10. On CD, paper # 227.
9. Németh E, Cameron IT. Cause-implication diagrams for process systems: their generation, utility and importance. *Chem Eng Trans* 2013;**31**:193–8. http://dx.doi.org/10.3303/CET1331033.
10. Kletz TA. Hazop–past and future. *Reliab Eng Syst Saf* 1997;**55**:263–6.
11. McCoy SA, Wakeman SJ, Larkin FD, Jefferson ML, Chung PWH, Rushton AG, et al. HAZID, a computer aid for hazard identification; 1. The STOPHAZ package and the HAZID code: an overview, the issues and the structure. *Trans IChemE (Process Saf Environ Prot)* 1999;**77B**:317–26.
12. Dunjó J, Fthenakis V, Vílchez JA, Arnaldosa J. Hazard and operability (HAZOP) analysis. A literature review. *J Hazard Mater* 2010;**173**:19–32.
13. Venkatasubramanian V, Zhao J, Viswanathan S. Intelligent systems for HAZOP analysis of complex process plants. *Comput Chem Eng* 2000;**24**:2291–302.

14. Vaidhyanathan R, Venkatasubramanian V. Digraph-based models for automated HAZOP analysis. *Reliab Eng Syst Saf* 1995;**50**:33–49.
15. Vaidhyanathan R, Venkatasubramanian V. A semi-quantitative reasoning methodology for filtering and ranking HAZOP results in HAZOPExpert. *Reliab Eng Syst Saf* 1996; **53**:185–203.
16. Vaidhyanathan R, Venkatasubramanian V. Experience with an expert system for automated HAZOP analysis. *Comput Chem Eng* 1996;**20**(Suppl.):S1589–94.
17. Srinivasan R, Venkatasubramanian V. Petri net-digraph models for automating HAZOP analysis of batch process plants. *Comput Chem Eng* 1996;**20**(Suppl.):S719–25.
18. McCoy SA, Wakeman SJ, Larkin ML, Chung PWH, Rushton AG. HAZID, a computer aid for hazard identification: 4. Learning set, main study system, output quality and validation trials. *Process Saf Environ Prot* 2000;**78**(2):91–119.
19. Khan FI, Abbasi SA. TOPHAZOP: a knowledge-based software tool for conducting HAZOP in a rapid, efficient yet inexpensive manner. *J Loss Prev Process Ind* 1997; **10**:333–43.
20. Khan FI, Abbasi SA. Towards automation of HAZOP with a new tool EXPERTOP. *Environ Model Software* 2000;**15**:67–77.
21. Zhao C, Bhushan M, Venkatasubramanian V. PHASUITE: an automated HAZOP analysis tool for chemical processes, part I: knowledge engineering framework. *Process Saf Environ Prot* 2005;**83**(B6):509–32. Part II: Implementation and Case Study 83(B6): 533–48.
22. Cui L, Zhao J, Qiu T, Chen B. Layered digraph model for HAZOP analysis of chemical processes. *Process Saf Prog* 2008;**27**(4):293–305.
23. Zhao J, Cui L, Zhao L, Qiu T, Chen B. Learning HAZOP expert system by case-based reasoning and ontology. *Comput Chem Eng* 2009;**33**(1):371–8.
24. Rahman S, Khan F, Veitch B, Amyotte P. ExpHAZOP+: knowledge-based expert system to conduct automated HAZOP analysis. *J Loss Prev Process Ind* 2009;**22**: 373–80.
25. Rossing NL, Lind M, Jensen N, Jørgensen SB. A functional HAZOP methodology. *Comput Chem Eng* 2010;**34**:244–53.
26. Wu J, Zhang L, Lind M, Hu J, Zhang X, Jensen N, et al. An integrated qualitative and quantitative modeling framework for computer assisted HAZOP studies. *AIChE J* 2014; **60**(12):4150–73.
27. Rodriguez M, De La Mata JL. Automating HAZOP studies using D-higraphs. *Comput Chem Eng* 2012;**45**:102–13.
28. Hu J, Zhang L, Liang W. Opportunistic predictive maintenance for complex multi-component systems based on DBN-HAZOP model. *Process Saf Environ Prot* 2012; **90**:376–88.
29. Hu J, Zhang L, Wang Y. A systematic modeling of fault interdependencies in petroleum process system for early warning, WCOGI2014, The fifth World Conference of Safety of Oil and Gas Industry 2014, paper OS8-4 1061598, Okayama, Japan.
30. Thunem H. The development of the MFM Editor and its applicability for supervision, diagnosis and prognosis. In: Steenbergen et al., editors. *ESREL 2013, safety, reliability and risk analysis: beyond the horizon*. London: © 2014 Taylor & Francis Group; 2013, ISBN 978-1-138-00123-7. p. 1807–14. Amsterdam, 29 September–2 October.
31. Zhao C, Bhushan M, Venkatasubramanian V. Roles of ontology in automated process safety analysis. *Comput Aided Chem Eng* 2003;**14**:341–6.

32. Morbach J, Yangl A, Marquardt W. OntoCAPE—A large-scale ontology for chemical process engineering. *Eng Appl Artif Intell* 2007;**20**:147–61.
33. Khakzad N, Khan F, Amyotte P. Dynamic risk analysis using bow-tie approach. *Reliab Eng Syst Saf* 2012;**104**:36–44.
34. Hu Y-Sh, Modarres M. Evaluating system behavior through dynamic master logic diagram (DMLD) modeling. *Reliab Eng Syst Saf* 1999;**64**:241–69.
35. Jaynes ET, Larry Brethorst G. *Probability theory: the logic of science*. Cambridge University Press; 2003, ISBN 0521 59271 2.
36. Christensen R, Johnson W, Branscum A, Hanson TE. *Bayesian ideas and data analysis: an introduction for scientists and statisticians*. Boca Raton (FL): CRC Press/Taylor & Francis Group; 2011, ISBN 978-1-4398-0354-7. 33487-2742.
37. Modarres M, Kaminskiy M, Krivtsov V. *Reliability engineering and risk analysis, a practical guide*. 2nd ed. CRC Press, Taylor & Francis Group; 2010, ISBN 978-0-8493-9247-4.
38. Project BUGS, Cambridge University, Medical Research Council, Biostatistics Unit. http://www.mrc-bsu.cam.ac.uk/bugs/.
39. Gregory PC. *Bayesian logical data analysis for the physical sciences, a comparative approach with* Mathematica® *support*. Cambridge University Press; 2005, ISBN 978-0-521-15012-5.
40. Pearl J. *Causality, models, reasoning and inference*. 2nd ed. New York (USA): Cambridge University Press; 2009, ISBN 978-0-521-89560-6.
41. DSL. *GeNIe (graphical network interface) and SMILE (structural modeling, inference, and learning engine)*. Version 2.0, software. Decision Systems Laboratory, University of Pittsburgh; 2010., http://genie.sis.pitt.edu/.
42. Fenton N, Neil M. *Risk assessment and decision analysis with Bayesian networks*. Boca Raton (FL): CRC Press, Taylor & Francis Group; 2013, ISBN 978-1-4398-0910-5. 33487-2742, USA.
43. Darwiche A. *Modeling and reasoning with Bayesian networks*. Cambridge: Cambridge University Press; 2009, ISBN 978-0-521-88438-9.
44. Neapolitan RA. *Learning Bayesian networks, Prentice Hall series in artificial intelligence*. 2003. ISBN-13:978-0130125347, ISBN-10: 0130125342.
45. Cooper GF, Herskowits E. A Bayesian Method for the Induction of Probabilistic Networks from Data. *Machine Learning* 1992;**9**:309–47.
46. Murphy K. (now, University of British Columbia), Software packages for graphical models, last updated 16 June 2014 12 February 2013. http://www.cs.ubc.ca/~murphyk/Software/bnsoft.html, [accessed December 26,27 January 2014].
47. Hanea AM, Kurowicka D. *Mixed non-parametric continuous and discrete bayesian belief nets, advances in mathematical modeling for reliability.* IOS Press; 2008, ISBN 978-1-58603-865-6.
48. Pasman HJ, Rogers WJ. Bayesian networks make LOPA more effective, QRA more transparent and flexible, and thus safety more definable! *J Loss Prev Process Ind* 2013;**26**:434–42.
49. Hanea DM. *Human Risk of Fire: building a decision support tool using Bayesian networks* [Ph.D. Dissertation]. The Netherlands, Delft: Delft University of Technology; 2009.
50. Labeau PE, Smidts C, Swaminathan S. Dynamic reliability: towards an integrated platform for probabilistic risk assessment. *Reliab Eng Syst Saf* 2000;**68**:219–54.

51. Innal F, Cacheux P-J, Collas S, Dutuit Y, Folleau C, Signoret J-P, et al. Probability and frequency calculations related to protection layers revisited. *J. Loss Prev Process Ind* 2014;**31**:56–69.
52. Paté-Cornell ME. Conditional uncertainty analysis and implications for decision making: the case of WIPP. *Risk Anal* 1999;**19**:995–1002.
53. Zio E, Aven T. Industrial disasters: extreme events, extremely rare. Some reflections on the treatment of uncer-tainties in the assessment of the associated risks. *Process Saf Environ Prot* 2013;**91**:31–45.
54. Cooke RM. *Experts in uncertainty, opinion and subjective probability in science*. Oxford University Press; 1991, ISBN 0-19-506465-8.
55. Zadeh LA. Fuzzy sets. *Inf Control* 1965;**8**:838–53.
56. Markowski AS, Mannan MS, Kotynia (Bigoszewska) A, Siuta D. Uncertainty aspects in process safety analysis. *J Loss Prev Process Ind* 2010;**23**:446–54; and earlier papers.
57. UniNet, developed by the Risk and Environmental Modeling Group at the Department of Mathematics of the Delft University of Technology. Installer software is downloadable at http://www.lighttwist.net/wp/uninet.
58. Khakzad N, Khan FI, Amyotte P. Risk-based design of process systems using discrete-time Bayesian networks. *Reliab Eng Syst Saf* 2013;**109**:5–17.
59. Hu J, Zhang L, Ma L, Liang W. An integrated safety prognosis model for complex system based on dynamic Bayesian network and ant colony algorithm. *Expert Syst Appl* 2011;**38**:1431–46.
60. Pérez Ramírez PA, Bouwer Utne I. Use of dynamic Bayesian networks for life extension assessment of ageing systems. *Reliab Eng Syst Saf* 2015;**133**:119–36.
61. Pasman HJ, Rogers WJ. How can we use the information provided by process safety performance indicators? possibilities and limitations. *J Loss Prev Process Ind* 2014; **30**:197–206.
62. Pasman HJ, Rogers WJ. Risk assessment by means of Bayesian networks: a comparative study of compressed and liquefied H_2 transportation and tank station risks. *Int J Hydrog Energy* 2012;**37**:17415–25. and Erratum 38 (2013) 1662.
63. Pasman HJ, Knegtering B. What process risks does your plant run today? the safety level monitor. *Chem Eng Trans* 2013;**31**:277–82. http://dx.doi.org/10.3303/CET1331047.
64. Hassan J, Khan F. Risk-based asset integrity indicators. *J Loss Prev Process Ind* 2012; **25**:544–54.
65. Cooke R, Ross HL, Stern A. *Precursor analysis for offshore oil and gas drilling-from prescriptive to risk-informed regulation*. 2011. http://www.rff.org/documents/RFF-DP-10-61.pdf.
66. Yang M, Khan FI, Lye L. Precursor-based hierarchical Bayesian approach for rare event frequency estimation: a case of oil spill accidents. *Process Saf Environ Prot* 2013;**91**: 333–42.
67. Khakzad N, Khan FI, Paltrinieri N. On the application of near accident data to risk analysis of major accidents. *Reliab Eng Syst Saf* 2014;**126**:116–25.
68. Bier VM, Mosleh A. The analysis of accident precursors and near misses: implications for risk assessment and risk management. *Reliab Eng Syst Saf* 1990;**27**:91–101.
69. Bier VM, Yi W. The performance of precursor-based estimators for rare event frequencies. *Reliab Eng Syst Saf* 1995;**50**:241–51.
70. Kelly DL, Smith CL. Bayesian inference in probabilistic risk assessment—The current state of the art. *Reliab Eng Syst Saf* 2009;**94**:628–43.

71. Paltrinieri N, Khan FI, Cozzani V. Coupling of advanced techniques for advanced risk management. *J Risk Res* 2014. http://dx.doi.org/10.1080/13669877.2014.919515.
72. Final Report *Causal model for air transport safety.* The Hague (The Netherlands): Ministry of Traffic and Water Management (Ministerie van Verkeer en Waterstaat); 2009. ISBN:10:90 369 1724-7; ISBN 13:978 90 369 1724-7, http://www.nlr-atsi.nl/fast/CATS/CATS%20final%20report.pdf.
73. Groth KM, Mosleh A. Deriving causal Bayesian networks from human reliability analysis data: a methodology and example model. *Proc IMechE Part O: J Risk Reliab* 2012; **0**(0):1–19. http://dx.doi.org/10.1177/1748006X11428107.
74. Groth KM. *A data-informed model of performing shaping factors for use in human reliability analysis* [Ph.D. dissertation]. Department of Mechanical Engineering, University of Maryland; 2009.
75. Groth KM, Mosleh A. A data-informed PIF hierarchy for model-based human reliability Analysis. *Reliab Eng Syst Saf* 2012;**108**:154–74.
76. Groth K, Wang Ch, Mosleh A. Hybrid causal methodology and software platform for probabilistic risk assessment and safety monitoring of socio-technical systems. *Reliab Eng Syst Saf* 2010;**95**:1276–85.
77. Azarkhil M, Mosleh A. Impact of team characteristics on crew performance: an object based modeling and simulation approach. In: Steenbergen, et al., editors. *ESREL 2013, safety, reliability and risk analysis: beyond the horizon*. London: © 2014 Taylor & Francis Group; 2013, ISBN 978-1-138-00123-7. p. 501–8. Amsterdam, 29 September–2 October.
78. Vinnem JE, Bye R, Gran BA, Kongsvik T, Nyheim OM, Okstad EH, et al. Risk modelling of maintenance work on major process equipment on offshore petroleum installations. *J Loss Prev Process Ind* 2012;**25**:274–92.
79. Gran BA, Byeb R, Nyheim OM, Okstad EH, Seljelid J, Sklet S, et al. Evaluation of the Risk OMT model for maintenance work on major offshore process equipment. *J Loss Prev Process Ind* 2012;**25**:582–93.
80. Matta N, Vandenboomgaerde Y, Arlat J, editors. *Supervision and safety of complex systems*. ISTE Ltd. – John Wiley; 2012, ISBN 978-1-84821-413-2. Chapter 15 and Figure 15.7.
81. Van Gulijk C, Hanea DH, Almeida KQ, Steenhoek M, Ale BJM, Ababei D. Left-hand side BBN model for process safety. In: Steenbergen, et al., editors. *ESREL 2013, safety, reliability and risk analysis: beyond the horizon*. London: ©2014 Taylor & Francis Group; 2013, ISBN 978-1-138-00123-7. p. 1867–73. Amsterdam, 29 September–2 October.
82. Ale BJM, van Gulijk C, Hanea D, Hudson P, Lin P-H, Sillem S, et al. Further development of a method to calculate frequencies of loss of control including their uncertainty. In: Steenbergen, et al., editors. *ESREL 2013, safety, reliability and risk analysis: beyond the horizon*. London: ©2014 Taylor & Francis Grp; 2013, ISBN 978-1-138-00123-7. p. 1839–46. Amsterdam, 29 September–2 October.
83. Lin P-H, Hanea D, Ale BJM. Modeling contractor and company employee behavior in high hazard operation. In: Steenbergen, et al., editors. *ESREL 2013, safety, reliability and risk analysis: beyond the horizon*. London: ©2014 Taylor & Francis Group; 2013, ISBN 978-1-138-00123-7. p. 335–40. Amsterdam, 29 September–2 October.
84. Stroeve SH, Blom HAP, Bakker GJ. Contrasting safety assessments of a runway incursion scenario: event sequence analysis versus multi-agent dynamic risk modelling. *Reliab Eng Syst Saf* 2013;**109**:133–49.

85. Bonabeau E. Agent-based modelling: methods and techniques for simulating human systems. *Proc Natl Acad Sci USA* 2002;**99**:7280–7.
86. Jensen K. *Coloured Petri nets. Basic concepts, analysis methods and practical use*. Volume 2, Analysis Methods, ISBN:3-540-58276-2; Volume 3, Practical Use, Monographs in Theoretical Computer Science. *Basic concepts*, vol. 1. Springer-Verlag; 1997, ISBN 3-540-60943-1. ISBN:3-540-62867-3.
87. CPNtools 4.0, www.CPNTools.org.
88. Haas PJ. *Stochastic Petri nets, modelling, stability, simulation*. New York (USA): Springer-Verlag; 2002, ISBN 0-387-95445-7.
89. Liu Y, Rausand M. Reliability effects of test strategies on safety-instrumented systems in different demand modes. *Reliab Eng Syst Saf* 2013;**119**:235–43.
90. Signoret J-P, Dutuit Y, Cacheux P-J, Folleau C, Collas S, Thomas P. Make your Petri nets understandable: reliability block diagrams driven Petri nets. *Reliab Eng Syst Saf* 2013;**113**:61–75.
91. Everdij MHC, Blom HAP. Hybrid Petri nets with diffusion that have into-mappings with generalized stochastic hybrid processes. In: Blom HAP, Lygeros J, editors. *Stochastic hybrid systems: theory and safety critical applications*. (Berlin, Germany): Springer; 2006, p. 31–63.
92. Everdij MHC, Blom HAP, Enhancing hybrid state petri nets with the analysis power of hybrid stochastic processes, National Aerospace Laboratory, Netherlands, Report NLR-TP-2008-402. http://reports.nlr.nl:8080/xmlui/.
93. Lind M. An introduction to multilevel flow modeling. *Nucl Saf Simul* 2011;**2**:22–32. Control functions in MFM: basic principles, Nuclear Safety and Simulation, 2(2011) p. 132–139; An overview of Multilevel Flow Modeling, Nuclear Safety and Simulation, 4 (2013) p. 186–191.
94. Van Paassen MM, Wieringa PA. Reasoning with multilevel flow models. *Reliab Eng Syst Saf* 1999;**64**:151–65.
95. Weick KE, Sutcliffe KM. *Managing the unexpected: resilient performance in an age of uncertainty*. 2nd ed. San Francisco: John Wiley; 2001, ISBN 978-0-7879-9649-9. Jossey-Bas.
96. Resilient organizations, a collaboration between research & Industry. http://www.resorgs.org.nz/.
97. Aleksić A, Stefanović M, Arsovski S, Tadić D. An assessment of organizational resilience potential in SMEs of the process industry, a fuzzy approach. *J Loss Prev Process Ind* 2013;**26**:1238–45.
98. Øien K, Massaiu S, Tinmannsvik RK, Størseth F. Development of early warning indicators based on resilience engineering, Paper presented at PSAM 10, June 7–11, 2010, Seattle, USA, p. 10.
99. Paltrinieri N, Øien K, Cozzani V. Assessment and comparison of two early warning indicator methods in the perspective of prevention of atypical accident scenarios. *Reliab Eng Syst Saf* 2012;**108**:21–31.
100. Paltrinieri N, Øien K, Tugnoli A, Cozzani V. Atypical accident scenarios: from identification to prevention of underlying causes. *Chem Eng Trans* 2013;**31**:541–6. ISBN 978-88-95608-22-8.
101. Dekker S. Resilience engineering: chronicling the emergence of confused consensus, chapter 7. In: Hollnagel E, Woods DD, Leveson N, editors. *Resilience engineering, concepts and precepts*. Aldershot (UK): Ashgate; 2006, ISBN 0-7546-4641-6.

102. Rasmussen J. Risk management in a dynamic society: a modelling problem. *Saf Sci* 1997;**27**:183–213.
103. Shirali GhA, Mohammadfam I, Ebrahimipour V. A new method for quantitative assessment of resilience engineering by PCA and NT approach: a case study in a process industry. *Reliab Eng Syst Saf* 2013;**119**:88–94.
104. Woods DD. Essential characteristics of resilience, chapter 2. In: Hollnagel E, Woods DD, Leveson N, editors. *Resilience engineering, concepts and precepts*. Aldershot (UK): Ashgate; 2006, ISBN 0-7546-4641-6.
105. Soczek ChA. Building resilience: a risk based approach to disaster response and business continuity planning. In: *16th annual international symposium Mary Kay O'Connor process safety Center, College Station*; October 22–24, 2013.

CHAPTER

Extended Process Control, Operator Situation Awareness, Alarm Management

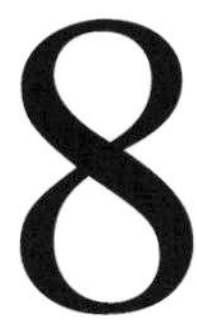

Simplicity is the extreme degree of sophistication.
Leonardo da Vinci, 1452–1519

SUMMARY

At the heart of keeping a process within its safe operating window is the control room operator and the technology at his or her disposal. The latter technology went through an intense evolutionary and sometimes even revolutionary development. In the 1970s, George Stephanopoulos (MIT, Cambridge, Massachusetts) and others laid the descriptive mathematical foundation for process control. In his book, Stephanopoulos[1] lists the following objectives of process control. Foremost is maintaining a safe operation, followed by keeping product specification, staying within environmental regulation, taking account of operational constraints given by the physics and chemistry of the processes, and finally satisfying market conditions by optimizing process economics, hence overall performance. Given a set of optimal conditions, the influence of external disturbances shall be suppressed and the stability of the process ensured. Over the years, the economic consideration received more emphasis, and as we shall see, economic optimization within safety and environmental constraints is now the foremost objective of control. This is in line with the shift toward an overall system approach enabling a prediction of optimum output in a noisy environment making use of probabilistic tools.

Achieving process control objectives is easier said than done because of various kinds of mutual interactive effects of physical and chemical processes on each other's rates. Heat transfer rates and concentration equalization rates of reactive chemicals in solution, as well as the velocity of reactions and therefore rates of heat production or consumption, all interact. This introduces dynamics in various forms with different time constants of change, possibly resulting in oscillatory response behavior and growing instability. The basics of control, given a disturbance, exist therefore in the first place in measuring process output variables such as pressure, temperature, and flow. Next, control exists in analyzing the values of these variables and, based on the analysis, in deciding on actions to manipulate

Risk Analysis and Control for Industrial Processes - Gas, Oil and Chemicals.

certain inputs to keep the process conditions at a set point level of steady state, or at least within desired limits. Hence, control is effectuated in a closed loop in which there is a *sensor* (measuring, e.g., temperature, pressure flow, or concentration), a (logic) *controller*, and an *actuator*, for example, a control valve.

The controller implements the control law. Today, these are programmable and are called programmable logic controllers. The two main approaches to serve the purpose are feedback and feed-forward control. The first just adjusts variables to their set point following observation of a change in output; the second is more of an anticipating nature, as it tries to measure a disturbance early on and taking into account time needed for changes (system time constants). It means that models are needed of the process, controlled by the loop, to make such prediction. This is also required in inferential control in case key variables cannot be measured directly. If in a system only one variable must be controlled, the control configuration is called single-input, single-output; if however, as is often the case in process industry, more than one output variable must be controlled simultaneously, the configuration is called multi-input, multi-output.

Response by a controller in feedback control to a disturbance, which can consist of a step function in change and felt only after a delay—the dead time—can be in several degrees of accuracy and speed. As control laws, one distinguishes only proportional response with gain proportional to the change (P-controller) with the risk of overshoot. Adding integration of the effect, smoothing it out, and bringing offset to zero improves this (PI-controller). Adding a derivative function proportional to the rate of change (PID-controller) is perfecting this further. If a process model can be described by a first-order differential equation, one calls the system first-order, while the more complex systems follow a second-order differential equation. The mathematics have been elaborated over the years, and sensor technology and instrumentation has become much more refined. The original analog controller hardware became completely digitized, and with all of that, the extent of automation rose tremendously.

Oscillatory behavior following a change in conditions or in input is analyzed by spectral methods, such as Fourier analysis determining eigenfrequency and power spectrum, and characterized by Bode plots of the ratio of output/input wave amplitude and also the wave phase lag, both as a function of frequency. Nyquist plots show the same in an alternative fashion. For linear, first-, and second-order control systems, analytical solutions of system descriptive equations are still possible and helpful in control design. A complication is that measurement of the variables is subject to uncertainty and noise. One therefore obtains a time series of measurement values shaped as wavy signals with superposed spikey "hair." Algorithms have been developed for processing and filtering the stochastic signals. One algorithm is the Kalman filter that assumes a linear dynamic system law and normally distributed (Gaussian) noise. The filter calculates in real time a best estimate of the measured variable using only covariances of current measurement and the previous calculated state. Covariances are derived by autoregression with the aid of least squares techniques. In case of high filter gain, more weight is placed on measurement of the current state with less weight on the previous state. Adoption of this in process industry began in the late 1980s.

The relatively reliable PID single-loop controllers became abundant in the ever more complex installations. But, because of the interactions among various plant sections and subsystems, overall control became a weak point. In case of a process upset at some location, interaction can lead to propagation of disturbances throughout the plant possibly presenting a hazard and certainly diminishing performance. Much effort has been spent to develop methods to mitigate this disadvantage with centralized model predictive control (MPC).

Development of MPC algorithms began in the late 1970s with application in refineries. Basically, MPC consists of a dynamic process model that is obtained by performing experiments with the plant measuring effects in outputs following changes in inputs. Initially, the models were linearized, but later nonlinear models were applied to achieve a better fit. Full digitization of signals and signal processing was introduced. Equations are solved by an iterative process making a prediction for a short-term time and a continually receding horizon when time progresses. Since the early 1990s, performance of an MPC algorithm was tested against the so-called Tennessee Eastman exothermic chemical model process.

Centralization inherits limitations. We shall make a jump in this chapter to the advanced distributive MPC with real-time optimization (RTO) that has become possible by the extensive use of computing in process control. Christofides et al.[2] recently reviewed and commented upon this approach and provided a future perspective. The newest network technology, wired or even wireless, will enable decentralized or distributed model predictive control (DMPC) for different plant sections, yet still together in optimizing the plant as a whole. Therefore, an operator is assigned the role of supervising the automated systems and managing abnormal situations.

Despite the sophistication of the control, there are problem areas potentially affecting safety. Besides failing or malfunctioning components, such as sticking pressure control valves, there are less explicit problems. In 2003, Venkatasubramanian et al.[3] made an extensive review of the problem field of operator process fault detection and diagnosis enabling intervention. If something is wrong, process pressure and or temperature quickly run up or start oscillating. Finding a fault and redressing a situation requires quick action, so that operators come easily under time pressure. This pressure will be exacerbated by alarm floods. Early warning and quickly understanding signals is important. Recent achievements in this respect will be briefly described.

A condensed description will be given of a most modern Supervisory Control and Data Acquisition (SCADA) architecture. Besides the above described control features and fault diagnosis, human factors also are taken into account. Apart from physiological ergonomic factors such as lighting, colors on screens, and information presentation on displays, operator situation awareness will be larger in case someone is actively involved in the control, but human performance will suffer if this workload becomes too high. Higher time pressure induces higher error rates.

Therefore, effective alarm management is an important topic. We shall review this type of problem and possible ways of improvement and optimization.

As already introduced in Chapter 5, at the end of Section 5.1, in 2011 Venkat Venkatasubramanian,[4] then still at Purdue University, wrote a vision article for the AIChE journal on systemic failures occurring and ending in disaster such as the Enron scandal, the financial crisis, but also the Deepwater Horizon demise and various process plant mishaps. These disasters can be ascribed to severe weaknesses in risk management, not able to produce convincing foresight of approaching collapse due to system complexity and lack of prognostic tools. He endorses Leveson's system approach and identifies the "need of real-time intelligent decision support systems that can effectively monitor various aspects of process operations, and detect, diagnose and advise operators and engineers about incipient abnormal events." He recommends pursuing multiperspective modeling, which means looking at a system from different perspectives: structure, behavior, and function. The combination with complexity science will then yield the support systems mentioned. In this chapter, we shall encounter a number of building blocks, which together with ones seen in earlier chapters, such as the blended Hazid method, will bring us closer to the objective he formulated.

8.1 PROBLEM ANALYSIS

As we have seen before in Chapter 3, from a safety point of view, distinction is made between a basic process control system (BPCS) and an instrumented protection safety system. The latter consists of safety instrumented systems letting off pressure or quenching the system and/or an emergency shut-down (ESD) system stopping flows and energy addition and bringing the installation to a full stop. The BPCS is physically separated from the safety systems and its components distributed over the plant (distributed control system or DCS). By the separation, the safety systems can function independently of a failing BPCS. Also, utilities for the safety system, such as instrumentation air and power, shall be independent of those for the BPCS. As a measure of last resort, ESD must be able to be activated manually.

Although the control architecture can provide a high level of reliability, there are still problems associated with the control as a whole. The first part of this has to do with unexpected and unforeseen events, which can be initiated externally disturbing the process or its instrumentation. Alternatively, the cause is internal by (partially) failing components, which are often not in the mainstream and have escaped attention for a while. A second part is due to system instabilities, which may be initiated by small disturbances or irregularities but grow and propagate through the plant and of which the cause is difficult to trace. Causes can be malfunctioning of components, such as valves or sensors, but they can also be changes in the chemical or physical processes. All this can be due to contaminated feedstock or deviating compositions, or even a plain error somewhere, for example, as a result of bad maintenance. Next, in Section 8.2 we shall explore in more depth the fundamentals and theory of model predictive process control, its strengths and weaknesses and future outlook. Section 8.3 will discuss present-day process control hardware and software.

A third problem category is associated with human factors. Operators usually work as a team, partly as control room or panel operators and partly as field or outside operators. Mostly on instigation of panel operators, the latter are closing or opening valves, checking particulars, and making close-in observations at equipment in the field in abnormal situations, or performing small, urgent repairs. Because of the progress in control technology in the newest systems, operators have only a supervisory role but may at the same time be responsible for more than one plant. After start-up, they have to come into action if some upset is developing or has already happened. In such case, recovery depends on their decisions, which will be prone to human error. In 1992, an abnormal situation management (ASM) team was formed by Honeywell with Amoco, Chevron, Exxon, and Shell oil companies as members. In 1994, the ASM team was converted to a consortium with a number of additional members. The ASM Consortium is still active and chaired by Honeywell. In Section 8.4, the human factor in control will be further explained and some of the achievements of the ASM Consortium will be summarized.

8.2 DEVELOPMENTS IN CONTROL THEORY

The drive to be best and most efficient fueled the wish to integrate processes with respect to energy (exergy concept), and exchange and subsequent conversion of materials of neighboring plants. In addition, product quality requirements and the sharpening of environmental and safety regulation placed tighter constraints. Changes in feedstock or desired product specification require increased flexibility. These developments demanded enhanced process control, which was already subject to difficulties. For example, exponential dependence of reaction rates on temperature and radiant heat transfer dependence on the fourth power of temperature causes nonlinearity of change. Hence, to correct an observed deviation from the right conditions requires a disproportional incremental alteration in input. Other limitations are imposed by the process model used for predictive control. Slight deviations of the fit from encountered process conditions under all circumstances will result in mismatches. Inherent delays in response, signal noise, and inaccuracy in quantitative responses of sensors, or design limitations of actuator reach and capacity (e.g., of control valves) add to the problem. Hence, besides nonlinearity, process control may also be plagued by time-varying uncertainty, all of which threaten stability.

The past 20 years have seen huge developments in mathematical signal processing and computing power to follow processes real time, also enabling automation. But there is still ample room for further improvements, as described by Christofides et al.[2] and by Christofides and El-Farra,[5] while there is not a single route to be explored. This all pivots around seeking stability via Lyapunov functions for the solution of differential equations describing dynamic, nonlinear systems. Around 1900 the Russian mathematician Lyapunov thought of modifications to a nonlinear system to obtain a linearly achievable stability near an equilibrium point. To understand broadly what it all means, and how it will lead to MPC serves the (over)simplified description below.

The continuous-time model describing the state-space of a *linear* process system (output roughly proportional to input) contains a state vector $\mathbf{x}(t)$ and its time derivative, an input vector $\mathbf{u}(t)$, an output vector $\mathbf{y}(t)$, all from sets of real numbers, and time variant matrices $A(t)$, $B(t)$, $C(t)$, and $D(t)$. Vector lengths are measured in space as Euclidian norm or metrics. Most generally, the state-space relations are formulated in the following so-called control dynamics and observer tracking equations:

$$\dot{\mathbf{x}}(t) = A(t)\mathbf{x}(t) + B(t)\mathbf{u}(t)$$

$$\mathbf{y}(t) = C(t)\mathbf{x}(t) + D(t)\mathbf{u}(t)$$

The system is usually considered only with time-invariant matrices. For a general *nonlinear* system, the equations can be written as just a function of time t, state x, and input u: $\dot{\mathbf{x}}(t) = \mathbf{f}(t, x(t), u(t))$ and $\mathbf{y}(t) = \mathbf{h}(t, x(t), u(t))$. If the functions are a linear combination of states and inputs, the equations can again be described in matrix form as above and also usually with time-invariant matrices.

As the system here will be steered at least on the short term to a desired, fixed output, for the control function the first equation is relevant. The values of x are affected by disturbances and uncertainty, which also can be described explicitly as a term separate from u. Therefore, following Christofides et al.,[2] for the various subsystems of the process coupled through the states (and not through the inputs), the control dynamics equation can be aggregated to

$$\dot{x}(t) = f(x) + \sum_{i=1}^{m} g_i(x)u_i(t) + k(x)w(t)$$

where f is a function aggregated over the various subsystems $i = 1 \ldots m$; g_i is a matrix relevant to subsystem i; w represents disturbances, assumed to be bounded; and k a matrix. It is assumed that these are locally all Lipschitz vector functions, which means that the ratio of the metrics or space distances between the vector elements of the functions of the state variable x, and the ones between the state variable elements, is equal or smaller than a constant, the Lipschitz constant. A further assumption is that the origin is a system equilibrium point unforced and undisturbed, so that $f(0) = 0$. By linearization and converting to the discrete time variant, approximated solutions of this equation for MPC have been proposed. However, Christofides et al.[2] propose a Lyapunov functions–based control solution, indicated as LMPC. This is by assuming a local Lipschitz control law making the origin of the closed-loop asymptotically stable and satisfying all state input constraints. Such control law covers the conditions of Lyapunov function based stability.

Overall control in a plant is hierarchical as it is realized in layers. The lowest layer is the PI and PID controllers for the local valves and other actuators. Above that is an MPC layer keeping main control variables at set-point values, while on top is an economic optimization layer for RTO. The latter is in high demand for coping with continual changes in market prices of energy and other resources. For the economics, optimization via MPC is required besides a model of the system, a performance index over a finite horizon, and a scheme that will have the horizon stepwise receding. The accuracy of the model to predict future trajectory without correction (open loop) determines the efficiency of the control, the performance index shall be minimized against the various constraints, and the receding scheme will provide the feedback to make up for model errors and disturbances. There is however the issue of stability of the scheme. Stability is achieved by combining here a nonlinear control law with the Lyapunov functions approach, although there are limitations to the length of the sampling time and the upper bound of the disturbances.

As mentioned, centralized MPC has the drawback of enlarged complexity, so that MPC decentralized to the subsystems, as shown in Figure 8.1, would enhance efficiency and flexibility. Depending on rates of change in the processes within a plant, separated MPCs can be on different timescales. If, however, the MPC units are not communicating, part of the advantage is lost. Making use of modern information technology (IT) networks (Ethernet, eventually at least partially wireless) enables communication between MPCs of different units, creating a distributed MPC or a DMPC. In each local MPC the cost function has to be optimized. This is realized by iteratively calculating the effect of its inputs on the plant as a whole assuming that the inputs of other MPCs are fixed at the optimum values of the previous iteration step. In a final step, a weighted sum of the individual MPC outcomes is determined and agreed upon.

Further, Christofides et al.[2] discuss asynchronous and delayed feedback effects in iterative DMPC. Future directions of research have the objective of covering other weaknesses such as optimal distribution of systems and optimal communication strategy between controllers. Another problem is coping with incompleteness of measurement data, while economic DMPC needs more attention. DMPC for controlling in part continuous time systems and for another part discrete event systems, the hybrids, is also a research topic deserving attention. Finally, DMPC provides the

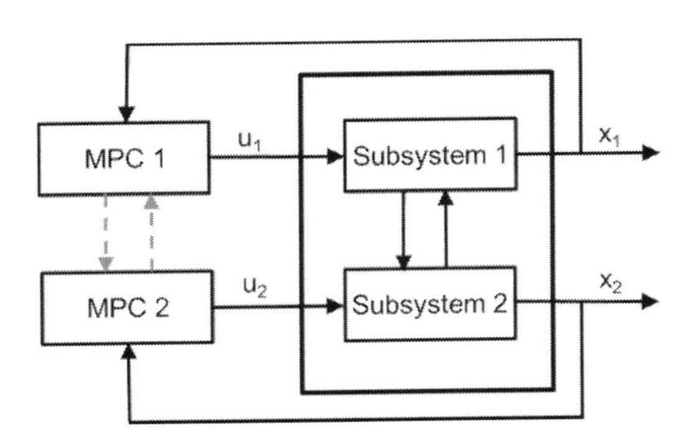

FIGURE 8.1

Decentralized or rather distributed but in parallel coordinating DMPC according to Christofides et al.[2] Only two units (LMPCs) are depicted.

possibility of monitoring and reconfiguring the system in case faults in the system are detected, so that the process can continue, while defect sensors or other components are being replaced. This kind of flexibility also offers the possibility to reduce data communication intensity in case of steady-state operation. For further reading on the background of MPC control see Christofides and El-Farra[5], and for integration of dynamic economic optimization (D-RTO) and MPC, see Ellis and Christofides.[6]

The latter authors describe a further development step. Where around 2000 RTOs were performed on an optimized steady-state model of the process, more recently a dynamic process model in the upper layer made it a dynamic RTO or D-RTO. Further, the simply weighted quadratic cost functions in the MPC were replaced by general economic ones, leading to feedback control economically optimized and integrated in one layer, the economic MPC or EMPC. Although this ambitious step proved possible, in the opinion of Ellis and Christofides[6] it cannot be absorbed easily due to necessary redesign of the control architecture and the requirement of fast response. Therefore, they propose a two-layered system of an optimizing EMPC on top, with Lyapunov stabilized MPC (LMPC) for controlling the process variables below.

8.3 FAULT DETECTION AND DIAGNOSIS AND FAULT-TOLERANT CONTROL

Fault detection and diagnosis in control and fault tolerance in control are important in general for running a process and maintaining the right product quality but also with respect to safety. The earlier detection and diagnosis, the earlier intervention will be possible. In 2003, Venkatasubramanian et al.[3] published an extensive review of methods of fault detection and diagnosis. As also becomes clear from the previous Section 8.2, computing led to much progress in control and enabled refined automation but added also to the complexity. In case of upset, which can arrive propagating from a remote part of the plant, the operator supervising the system must come in action to diagnose the fault and to take measures. Hence, abnormal event or ASM has not become an easy task. On the contrary, certainly younger operators with less experience in the mechanisms playing a role in the process may become overwhelmed by the number of causes that could be at the origin of the upset. Their mental image of the process may be incomplete or even wrong, leading to wrong decision making. Venkatasubramanian et al.[3] reviewed all methods that can help in such situations.

First, the authors specified the terminology. Fault is a deviation from the normal value of an observed variable or calculated parameter beyond the acceptable limit; underlying it is a (hidden) root cause or basic event, for example, a malfunctioning or failed component, but it can also be incompleteness in the model. A reliable diagnostic classifier with sufficient resolution but also speed would therefore be desirable. Such a tool should also be robust, not sensitive to noise, while known abnormalities

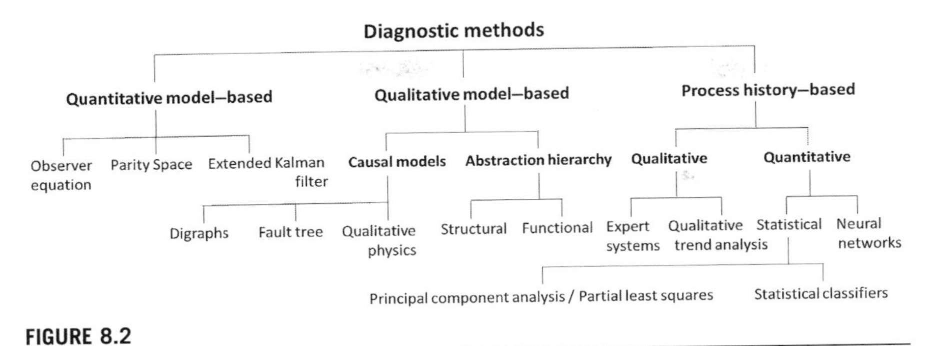

FIGURE 8.2

Diagnostic methods for fault classification.

Adapted from Venkatasubramanian et al.[3]

and faults shall be distinguished from unknown ones. Adaptability to changes in the process should be possible and an online decision support system should be able to explain the cause and the propagation with minimal modeling and minimal computational effort. Finally, the system should be able to identify multiple faults, which coincidentally appear, although this will be a difficult requirement.

Obviously, inputs to a diagnostic system come from measurements/observations. In a next stage, from the measurements features are extracted that can help in the diagnosis. Features are then fit into a decision space usually provided with an objective function to minimize error. This is performed by applying a statistical discriminant function analysis or by using simple threshold values. A discriminant function analysis differs from variance analysis in that it is tested whether a set of parameters determining a class fits the observation. Finally, the class of fault is derived. An overview of diagnostic methods is given in Figure 8.2.

8.3.1 QUANTITATIVE METHODS

Most of the quantitative model–based fault detection and isolation (FDI) occurs by the type of observer equations we have seen in the previous section. The methods are two-step ones: the first step generates residuals, that is, squared differences between expected and observed values, yielding possible inconsistency; the second step serves to decide what is at fault. In principle this can be achieved by hardware redundancy and comparison. However, because of cost reasons it is mostly realized by analytical, also called functional, inherent, or artificial redundancy. This is accomplished by comparing expected values of model-related process variables with observed ones. Direct redundancy is finding a faulty sensor by comparing sensor signals related via the model. Temporal redundancy analysis means relating outputs of different sensors and inputs of actuators to find out which component is failing. Venkatasubramanian et al.[3] provide equations for discrete and continuous, linear and nonlinear systems. As an example of the latter, Kazantzis et al.[7] developed a nonlinear observer typically suited for a chemical batch reactor or a continuously stirred tank reactor with multiple stable steady states.

Parity space relations consist of checks of the output of sensors against known process inputs. Residuals are never fully zero because of various disturbances: noise, model errors, bias, etc. Because disturbances have mostly a dominant random component, a third possibility is to apply a Kalman filter. Such a filter is an optimum predictor of the state parameters for a linear stochastic system in a noisy environment. (Note: today a Kalman filter can be substituted by a Dynamic Bayesian Network which will not limit application to linear Gaussian systems.)

In case of redundant hardware with voting as in high integrity systems, differences in signals may lead to fault isolation. Also, enhanced residual methods have been developed such as directional and structural residuals. Here, the essence is that a set of residuals is determined and that not only the existence of a fault is shown but also what fault, making the approach not only fault sensitive but also fault selective. "Directional" indicates the direction in which the fault should be found and "structural" links to a subclass of faults.

Although in principle there are possibilities by making use of the MPC gear already present, Venkatasubramanian et al.[3] report in their review that in practice several problems are encountered. Complexity of the system with associated nonlinearity, the high dimensionality of the vectors involved, and lack of good data complicate the development of a sufficiently accurate model for relying on residuals. There have been trials applying statistical treatment on the correlated diagnostic residuals by determining, for example, distributions, a moving average, and performing statistical tests. But the control on the basis of residuals, in particular in case of chemical processes which are by nature nonlinear, proved in general to be problematic, although some successful examples have been described.

A decade after the review mentioned above, Mhaskar et al.[8] assuming time-invariant, nonlinear systems proposed a fault-tolerant MPC stabilized by Lyapunov functions (LMPC), capable of FDI. After explaining in more detail the mathematics behind LMPC, first a single input nonlinear algorithm was derived capable of fault detection and fault tolerant control of input constraints due to failing actuators. Subsequently, a case was worked out of multi-input, multi-output subject to multi-failing actuators. The scheme introduces a fault detection filter with an identical state equation as the closed loop that is tuned first to a fault-free closed-loop state measuring a minimal residual. The filter runs parallel to the process and actual states are compared with the ideal ones. If the difference in residuals surpasses a certain threshold for a certain input variable, there is a fault in the actuator controlling this input. However, also below the threshold a fault can be present. By trimming the threshold as far as possible, the probability of missing a fault will be as low as possible. The magnitude of the difference of residual on- versus off-line is representative of the position in which the actuator failed.

Fault tolerance can be obtained by making use of a switching policy governed by a switching rule. Switching is executed by the plant supervisor, who can determine whether a "fall-back" control configuration can be established, which still can provide a stable closed-loop control around the nominal equilibrium point desired. However, in the case where this is no longer possible, it is proposed to obtain a "safe-park" point. This can be realized by establishing for the system a number of safe-park points off-line for the various positions of the potentially failing actuator, so that in the case it fails, the actuator can be repaired/replaced while the process need not be shut down. The fault information about which actuator is failing, and to what extent, is thus crucial for the selection of the safe-park point.

An even more challenging problem for nonlinear systems is locating a faulty sensor. To that end, a high-gain controller design is proposed and again residuals determined. Now, the difference between the state estimate of the high-gain observer and a sufficiently accurate predicted value is monitored. The information on which the latter is derived is again established initially when all sensors function. The residual will reveal the sensor configuration of which the faulty sensor is part. After isolation, redundancy will enable to switch off the faulty sensor. A second approach in case of sensor data losses or asynchronous measurements with loss of feedback is to have actuators that can store the last optimal input trajectory instead of going to zero or the last given value. For this, the Lyapunov-based control has been modified to deliver that trajectory information. The use of the methods was illustrated in example processes.

Because of the complexity of the method—the mathematics involved have not been detailed here—and the fact that one controls the process with the same system as one diagnoses, a second, independent method of diagnosis would be desirable, also to reduce the chance of false alarms. Failures can be other than failures of sensors and actuators, and it will not be simple to distinguish signals caused by true, maybe unknown instabilities from those caused by failure. Below are some candidate methods.

8.3.2 QUALITATIVE METHODS

The two structured methods for diagnosis, summarized by Venkatasubramanian et al.[3] in part 2 of their review, are signed digraph (SDG) and fault tree (FT) to link cause–effect chains. We could add here Bayesian network (BN), which became available more recently, adding also a quantitative element. We have described BNs in Chapter 7. In addition, there is the common sense reasoning of qualitative physics.

An SDG is a directed graph, which is a qualitative "cousin" of the BN. It consists of nodes connected by arcs directed toward the effect. The nodes represent variables or events; in an SDG the arcs have positive or negative signs attached to them depending on the direction of change they indicate. SDGs are efficient in showing graphically a cause–effect model, hence a causal graph. In contrast to Bayesian nets, SDGs can be in part cyclic representing self-amplifying or attenuating processes. In this respect, SDGs resemble system dynamic graphics. Venkatasubramanian et al.[3] present a simple example of a digraph of a tank being filled by flow F_1, while at the same time at the bottom flow F_2 exits. Rate of F_2 depends on the height Z of the liquid level in the tank. Hence, $F_1 - F_2 = \frac{dZ}{dt}$ and $F_2 = Z/R$, where R is flow resistance for F_2. The corresponding digraph is shown in Figure 8.3.

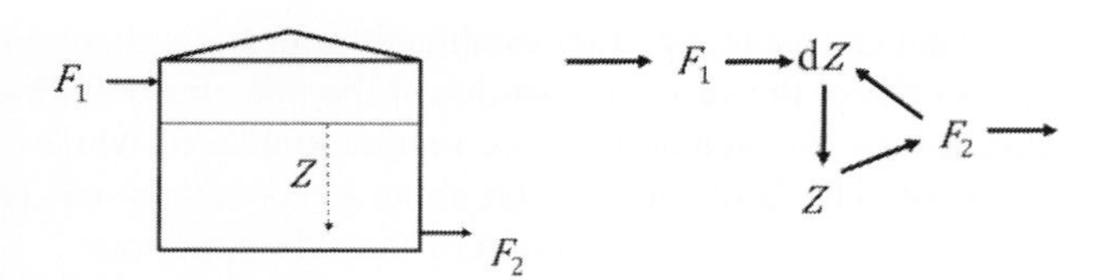

FIGURE 8.3

A simple digraph of a tank being filled, while at the bottom a flow exits.

A process can be modeled in an SDG with all material and information flows initiating or undergoing changes. Application in process control fault diagnostics started in the late 1970s and some quantification describing dynamics by means of ordinary and partial differential equations (as in system dynamics models) entered in the late 1980s. Combination with fuzzy logic and fuzzy reasoning (briefly described in Section 7.6.2) followed in the late 1990s.

FT analysis has been briefly described in Chapter 3 (Section 3.5.2.2). Given a failure event (top event), one asks continuously going further down in detail what underlying possible causes can yield the event, until one reaches basic failure events. So, in reverse one can see how a fault propagates to the top-event. The logic structure has an advantage over SDG as it can explicitly show that two or more underlying causes only together can cause a higher failure (AND-gate), or whether one or the other (OR) suffices, or one or the other but not simultaneously (exclusive or: XOR), or one but not in combination with the other (NAND). But like BN, also FT cannot represent self-amplification or attenuation as SDG can. FT is used quantitatively by attributing failure probability values to the nodes. Modeling temporal effects and partial failure in FT is, however, tiresome. On the contrary, SDG represents conditional processes easily. As we have seen in Chapter 7, Bayesian nets are more flexible than FT in, for example, modeling confounding causes and is able to handle probability distributions rather than only discrete values, and even to a certain degree able to model dynamics. But the rigidity of the FT structure often helps to obtain a quick overview.

SDG, FT, and BN are just methods to structure cause–consequence chains of which causes have been identified by different methods, while FT and BN can calculate top-event probability given quantified underlying causes. Methods to *identify* what can go wrong have also been treated in Chapter 3. The main ones are hazard and operability analysis (HAZOP) and failure mode and effect analysis (FMEA). HAZOP will be supported by common sense qualitative physics (CSQP) as mentioned by Venkatasubramanian et al.[3]

"Over-the-thumb" CSQP can be structured in two ways. The first is to derive qualitative, so-called "confluence equations" in terms of increase or decrease, and high or

[3]Note: As signed digraph in case of increasing flow F_1 all arrows will have a + sign attached, only the one between F_2 and dZ a - sign.

low, or order of magnitude change. The second is called "precedence ordering," which is ordering of variables in the direction of the information flow through them. CSQP can be elaborated into qualitative simulation or QSIM of the physics of a process.

The last qualitative methods according to the scheme in Figure 8.2 are the *structural* and *functional* abstraction hierarchy of process knowledge. These consider the hierarchical decomposition of the system, its subsystems and further, and how effects can be explained by their interaction. Structural decomposition looks top-down and represents connectivity (as we have seen in the system approach of Rasmussen and Leveson in Chapter 5), and functional decomposition concerns the means-to-ends relations bottom-up (as does HAZOP). In structural approach, pleas have been made to consider control loops for each subsystem (as Leveson with STPA advocates), while for functional analysis consideration of mass and energy flows at various functional levels has been suggested (multilevel flow models). This again is covered by HAZOP and even better by the recently developed Blended Hazid method, explained in Chapter 7 (Section 7.3), which is based on a functional systems framework.

Venkatasubramanian et al.[3] conclude discussion of qualitative methods by considering search strategies and mention the two types: *topographical* and *symptomatic*. The first is either by structural or functional search. Structural is by identifying the path information flows through a unit. In case of a fault found on this path, the fault is further localized in successive refinements. Functional search is by finding differences/mismatches between normal and actual functioning of all components connected. Once a subsystem is identified as suspect in a functional sense, the functional search is continued through further decomposition. In practice, structural and functional search are often combined. Symptomatic search consists of systematically applying input variations and observing outputs, while comparing with how these should be in the normal situation.

As we have seen here and before in Chapter 3, HAZOP and FMEA are main contributors for identifying faults and failures and their effects. The problem is that in general a process installation is large and complex and contains many components allowing many different ways in which control can be lost. Structuring identified mishap mechanisms and computerization is desirable for reliable storage and quick retrieval needed for diagnosis. Obviously, FMEA is much associated with FT for which software packages are commercially available. Digraph models have been applied by Vaidhyanathan and Venkatasubramanian[9] in an effort to automate HAZOP analysis (Chapter 7, Section 7.3.3).

However, in this connection the applicability of the new Blended Hazid method by Cameron (Chapter 7, Section 7.3) should be stressed. Computer storage of the results of the combined HAZOP and FMEA in Blended Hazid enables quick retrieval. The option of generating causal graphs by just entering in a computer running the software for an observed deviation will make this technique even more successful for fault diagnosis (in the sense of a functional search). As Bayesian network (BN) finds applications for medical diagnosis, also quantified Blended Hazid could fulfill that role by producing a probability ranking of potential causes. For example, in a BN portraying various modes of a disease of an organ with its causes and indications, based on historical data probabilities can be specified of possible different disease causes, conditional on patient history, gender, age, and many other preconditions. Over the years all these data may be known as means or distributions for a population as a whole. The quantified network then relates the hidden causes to indicators (e.g., blood constituents, temperature, and appearance). By performing an observation on an indicator for a particular patient, the most probable cause in that instance can be inferred. Similarly, this kind of symptomatic hidden cause search approach based on (deviating) observables could work for a process plant. In Section 8.5.5 we will return to this possibility.

8.3.3 PROCESS HISTORY–BASED METHODS

The third part of the fault diagnosis methods reviewed by Venkatasubramanian et al.[3] as shown in Figure 8.2 consists of five methods: The first two are qualitative: expert systems and qualitative trend analysis (QTA); the other three are quantitative: principal component analysis/partial least squares (PCA/PLS), statistical classifiers, and neural networks.

Expert systems have been around and are useful in a rather specialized field. The technique is rule based: if…then…else. Venkatasubramanian and coworkers have referenced a number of papers on applications of this method of feature extraction in the process industry.

Qualitative trend analysis can be considered as part of process monitoring and supervisory control; recognizing a trend will enable prediction. As process variable signals are blurred by noise, QTA uses filters, for example, autoregressive ones, to smooth the signal to detect a trend. Filtering introduces the risk that a true significant peak may not be observed, although patterns may be recognized. By using filters on different time-frequency domains, some relief of this drawback can be obtained. Most references concerning this topic stem from the 1980s to the early 1990s. In a follow-on article to the reviews, Dash, Rengaswamy, and Venkatasubramanian[10] discussed the application of fuzzy logic for a two-stage trend analysis. The first stage is only analysis in a qualitative sense but the second applies fuzzy set for a rough quantitative estimate that enabled identification of a faulty sensor, given knowledge of fault signatures.

In the quantitative, process history–based methods as PCA/PLS, the Gaussian time series of process signals (oscillatory with superposed white noise) are sampled online and analyzed much more rigorously. Multivariate analytical techniques (Pearson covariance matrix methods and correlation coefficient calculation) are applied to extract main features and separate correlated variables from uncorrelated ones. This technique differs principally from the quantitative fault-finding methods applying a process model (MPC). Numerous references on applications of PCA/PLS quoted are typically from the 1990s. Faults of, for example, sensors and valves should be detected as early as possible. Therefore, online sampling and statistical signal processing must be accompanied by threshold definitions for alarming. A design challenge is the development of a stopping rule to avoid false alarms as much as possible.

More recently, further analysis of wavy and noisy process variable signals has been undertaken. In these cases patterns resulting from nonlinearity can be recognized. As in many plants an operator has tens and even hundreds of control valves "under his wings," it is not only of interest to know whether one of those is malfunctioning but also which one. Shoukat Choudhury et al.[11] performed a thorough analysis in relation to valve stiction (sticking-friction). Nonlinear behavior can be induced by backlash, hysteresis, dead-band, and dead-zone of control valves caused, for example, by wear, corrosion, or design faults. Oscillations may show plant-wide propagation. The methods have been developed and tested in detection and diagnosis of these oscillations.

The signals may contain Gaussian noise but other distributions are possible. The signals may also possess an autocovariance property, which means a time-delayed autocorrelation of a stochastic system. (Autocovariance is not to be confused with autocorrelation. If the autocovariance is normalized by the product of the standard deviations at the two points of time—time and time delayed—the autocorrelation coefficient is obtained). Linear autocovariance is second-order moment related, but for nonlinear processes a third moment is needed for characterization, deriving the cumulant.

Analysis requires higher-order statistics. Fourier transform analysis is applied to reveal the dominant frequencies of a signal characterized by its power spectrum. This spectrum is the Fourier transform of the autocovariance function. The bispectrum is the frequency domain representation of the cumulant and derived as a double discrete Fourier transform. Bicoherence is the normalized bispectrum and can be plotted as an ordinate in a 3-D diagram versus two horizontal frequency domain axes, revealing nonlinearity (see Figure 8.4). Wiener filtering is applied to signal parts relevant in contributing to the nonlinearity. In addition, stiction is quantified by applying the fuzzy C-means clustering technique[13] on scattered groups of data points to derive an elliptic plot of just the controlled process variable output (*pv*) versus controller output (*op*). These plots have been made for various types of valve faults producing typical fault signatures, so that the method can be interpreted as a statistical classifier.

Although these methods are quite intricate, some striking results are shown. Valve stiction can easily be ascertained by singling out the valve and applying an invasive test. However, for an operating plant such a test is expensive. The afore-described fuzzy C-means clustering technique method was however successful in detecting stiction of a particular valve in a running plant and quantifying its extent by the developed algorithm independent of type of valve or loop, only requiring a set point, controlled process variable output, and controller output. The method was patented and is applied in industry. The methods described by Shoukat Choudhury et al.[11] were also effective in diagnosing plant-wide oscillations.

The last history-based method of the review of Venkatasubramanian et al.[3] for fault classification by feature extraction is neural network. Considerable work in this direction was done in the late 1980s and first half of the 1990s. One of the results is the fuzzy clustering algorithm mentioned above.

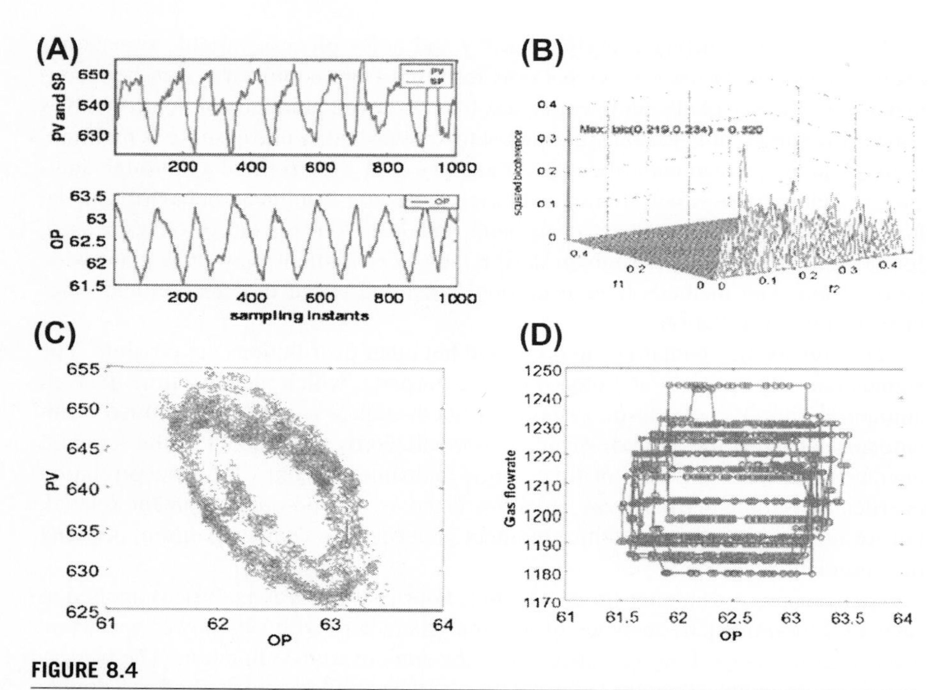

FIGURE 8.4

(A) Time series signal samples of the temperature control loop of a natural gas–fed combustion dryer system, (B) the bicoherence plot, (C) the derived *pv-op* (controlled process variable output-controller output) ellipse, and (D) the manipulated gas flow rate variable versus controller output after the 2004 work of Shoukat Choudhury et al.,[12] showing valve backlash and stiction.

8.3.4 COMPARISON OF THE VARIOUS METHODS

In comparing, we shall only highlight some salient methods. The recent quantitative method applying LMPC looks feasible but is too young to assess. According to Venkatasubramanian et al.,[3] industrial applications are mostly history-based statistical approaches with some emphasis on detecting faulty sensors. Fast online detection of abnormal situations with relatively easy-to-build statistical means is in demand. As we have seen in the previous section, recently, detection of valve stiction, applying higher-order statistics, has also become quite feasible. More refinement and depth of processing, however, requires more computational capacity. Older methods such as neural network and qualitative trend analysis have also been in use.

For the future, the instantaneous generation of causal graphs also seems promising to diagnose various kinds of defect and malfunction problems that can cause process upsets, as will be confirmed in Section 8.5.5.

8.4 TRENDS IN SCADA SYSTEM INFRASTRUCTURE

This section will be about SCADA, which integrates in its architecture the all-modular control equipment of a process distributed over the plant area (DCS). SCADA further centralizes all plant data and presents the information to the operator in the control room, at the so-called human–machine interface (HMI). Signal and logging data are given time stamps and stored for later use. Over the last decade security concerns with respect to the systems have been rising. This security is not only with respect to physical security of premises but in particular to cyber security.

The newest equipment, such as the Honeywell Experion® with universal channel technology, is enabling analog or digital, input or output on any channel and software configuration. As before, basic process control and emergency safety shut-down channels are fully physically separated and visibly distinguishable with blue and orange stickers on channels and a solver. Besides the in- and outputs, the safety modules also contain the safety logic solver, and they can be placed at the site of the equipment to be safeguarded. All devices are SIL3 TÜV certified, have a temperature range of −40 °C to +70 °C, and are hazardous area class Zone 2 (or Division 2) safe.

Field connections are by fiberoptic cable able to transport many different signals simultaneously, which simplifies the cable network considerably. The rest is within a fault-tolerant Ethernet. Wireless connection is in principle also possible (wide area network, WAN, and large WAN). Wireless device managers (WDMs) are for monitoring and are connected to servers. These WDMs are for actual functioning in a control loop not yet (summer 2013) sufficiently robust due to the battery capacity. Configuring field devices is realized securely with the control builder, while DCS system settings are protected by different types of interlocks, for example, for pre-shut-down conditioning related to safety systems. Control of the safety systems is via the safety manager.

Also, the instrumentation has been improved in stability, accuracy, and response time. Level measurement and chemical seal detection are more reliable; detection of pipeline vibration and plugging, and leak detection, will be possible in the near future. Signal communication after digitization in remote terminal units is to the field device manager (FDM), which supports up to HART 7 (Highway Addressable Remote Transducer) communication protocol for Foundation Field Bus and Process Field Bus (Profibus) data communication standards. The FDM enables the common tasks, such as loop tests, range updates, and following calibration procedures, but also less common ones, such as control valve stroke tests, drift analysis, and flow diagnostics. The control performance monitor (CPM) enables checking the performance of valves (stiction), sensors (drift, bias, failures), and controls (interactions, disturbances, service factors, and saturation/tuning issues). CPM produces management production key performance indicator reports and feeds the history database.

By the much enlarged computing capacity, the new system offers possibilities for virtualization up to DCS level enabling, for example, off-process developments and pre-acceptance, installing a (remote) backup control center, design and training, and many other activities. The system will help to make better use of the equipment and will save on the number of PC computers because on one computer other, virtual ones can be simulated. Large-screen technology (collaboration station) enables easy communication independent of distance. Another feature is a management of change application designed for preparing and keeping track of changes to the automation.

Much attention is given to cyber security measures, which continually will be updated. For a number of reasons, SCADA systems were believed to be relatively secure because of their specificity, but with the huge spread of IT knowledge this is not true anymore as the 2010 Stuxnet virus attacks have shown. Cyber security will be of increasing importance because of the evolution toward Internet use (TCP/IP-based equipment), cloud computing, and satellite communication. Drivers are cost savings and the enabling of remote unmanned operations. SCADA systems are crucial in keeping going processes that are the "blood, oxygen, and nutrients" for the economy, and for society cyber security is the contrary of a luxury. The International Society of Automation (ISA) is working on standards.

HMI including alarm management of SCADA will be discussed in the next section. The ASM Consortium, as mentioned earlier led by Honeywell, published several ASM Guidelines[14] on how to optimize HMI.

8.5 HUMAN FACTORS IN CONTROL, CONTROL ROOM DESIGN, ALARM MANAGEMENT

There is great pressure on companies to do more with fewer people. However, there is a limit to what people, given the best of intentions and motivation, can do. On the other hand, that limit shifts continually to higher level of capability as a result of improved and innovative technology. But even so, there are limits. This is also the case with the console operator taking process observations from his displays and making decisions about whether and how to make interventions. A crucial aspect in an operator's vigilance is their situation awareness, which we shall also discuss below. In acting, the operator must follow procedures, in some of which an error can evoke a large risk. Not only can it be tough to follow a procedure error free but writing a procedure that is interpreted unambiguously is not an easy task either.

8.5.1 PROCEDURES

The issue of automation of procedures was brought up at the 2013 MKOPSC Symposium by Thomas Williams,[15] chairman of the ASM consortium. For operating a plant there may be hundreds, even thousands of different procedures. Certainly when it concerns executing procedures in abnormal situations, apart from increasing the operator's work load, errors will hardly be unavoidable. Therefore, there is a trend to partly or fully automate procedures or at least to guide the operator executing those, although the investment in time to develop such automated procedures is not insignificant. For batch processes this has already being done for a while, and for which the ANSI/ISA-88 standard is available; for continuous processes, the ISA established committee ISA-106 to develop a standard. Benefits are particularly high when the procedure has to be used in an abnormal situation, when there may be time pressure and the operator does not have routine practice.

8.5.2 SITUATION AWARENESS AND EXTENT OF AUTOMATION

Mica Endsley[16] published a series of papers on the topic of operator situation awareness[a] culminating in two well-known papers in the mid-1990s. The key points from these have been summarized recently by Chris Wickens.[17] The construct of *situation awareness* (SA) can be defined pithily as "knowing, what's going on" and, more formally, as "the *perception* of the elements in the environment within a volume of time and space, the *comprehension* of their meaning and the *projection* of their status in the near future." In the latter, one can recognize the three main elements, also called levels of SA: perception, comprehension, and projection, distinction of which has consequences for system design and training. To a certain extent it resembles Rasmussen's operator knowledge-based behavior path shown in Figure 6.4.

SA is not the same as performance. At a high level of operation, performance may be good but the operator's SA may be low. Secondly, the time constant of SA is in seconds to at maximum some hours, while a *mental model* may be built in hours to years. Thirdly, the product of SA is not the same as the process of updating SA, but the distinction is fuzzy, certainly at the first level. Endsley enlightened SA by describing the effect of factors such as time, space, and team SA and also the links to decision making and performance. She further characterized the relation with other psychological concepts, for example, pre-attentive processing, attention, perception, working memory capacity, long-term memory, development of schemata and mental models, confidence level, automaticity, goals, plans, and scripts (outcomes).

Important to designers is the level of automation. In principle one can automate much, but it is known that the presence of a human supervisor is crucial in case of abnormal, not foreseen, or rare situations. If that occurs, he/she shall be alert and anticipate as much as possible. Decision time can be critical. However, with higher levels of automation, alertness decreases. So, to find an optimum is a dilemma as shown graphically in Figure 8.5. Quantitative information would be useful. Measurement attempts by various researchers have not produced clear answers. There are indications supporting case (b) but in some instances (c).

Another side of the same question of how far the ever more capable automation shall go has been analyzed by Flemisch et al.[18] This was accomplished in the light

---

[a]One finds perhaps even more often the expression *situational awareness*, but it concerns the awareness of the situation, so *situation awareness* is preferred here.

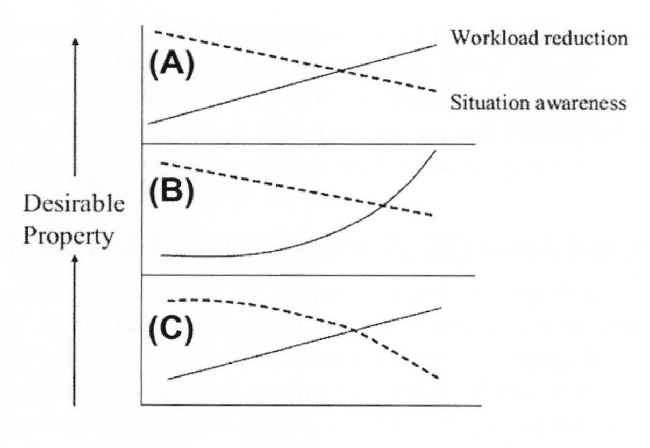

FIGURE 8.5

At higher levels of automation situation awareness (SA) tends to decrease, while at the same time workload reduces. The question for design is which of the three hypothetical curves is most real and where is the optimum? According to Wickens,[17] in case (a) it depends on the weighting of workload reduction versus SA; in (b) it will either be at the high or at the low end; and in case (c) it could be somewhere in the middle where the (absolute) values of the slopes are the same.

of maintaining consistency among four concepts: ability, authority, control, and responsibility, which are shared between the human and the machine in different ratios. Ability enables control, authority the right to decide, control the power to influence, and responsibility is being accountable for acting. The latter of course is relevant when something goes wrong and liability becomes an issue. Is the machine at fault or the operator? It turns out to be a delicate, even dynamic, balance which can be shifting in time, depending on the type of HMI and the work conditions. In full automation it will be all at the machine side, but in many situations the human keeps the authority and responsibility by having an essential part of the control. No clear, fixed rules can be given.

8.5.3 CONTROL ROOM ERGONOMICS

What does improve continually is the working environment from the point of view of ergonomics. Lighting, positions, console design, screens, the graphical user interface with (intuitive) display contents, colors, size are all optimized based on experience and experiment as indicated in the ASM Guidelines.[14] The way information is presented is also optimized with respect to decision making. The more is automated, the more complex are the questions remaining to be solved by the operator. According to Rasmussen's skill, rule, knowledge-rule (Chapter 6) this means a shift to higher cognition requiring more time for deciding to avoid error. On the other hand, the modern systems enable operators to call in data and communicate with each other via their screen regardless of distance, which facilitates decision making. But alarm and response of an operator is still considered to have on average a probability of failure on demand larger than 1 in 10. Hence, the combination cannot be classed as SIL 1. The above forms a compelling reason to prioritize alarms so that an operator can focus his mind on the problem requiring his attention most urgently.

8.5.4 TRAINING FOR OBTAINING THE MENTAL IMAGE OF THE PROCESS

All this increases the risk of operators losing physical contact with their plant. They don't smell it, and they don't hear it. This is certainly the case with plants remotely operated over large distance, underground, or out at sea. To build up a mental model of the process, necessary at the time of an abnormal situation, good knowledge, training, and physical experience are crucial. For training, simulation or "virtualization" with "human in the loop" shall be further developed but will not easily replace the real plant. A simulator for training shall be carefully developed to avoid being exposed to incomplete and simplified "reality" and learning a wrong procedure to solve a problem. In view of Leveson's system approach and operations management, not only a one-time training but also retraining shall be considered. The effect of training is fading out over a period of a year or so. Hence, training shall not be seen as static but as dynamic, in which personnel periodically combine, as with Bayesian updating. The goal of the retraining sessions is to attain progressively higher levels of effectiveness and reliability.

8.5.5 ALARM MANAGEMENT

As a result of the thousands of controlled variables in a plant, each equipped with Lo–Lo via Lo to Hi and Hi–Hi alerts or alarms, in case of a trip, breakdown of a component or other failure or abnormal situation or even achieving a designated target, alarms begin to annunciate on the screens or maybe alert loudly. So, an alarm's main task is to help keep the process within a safe and desired envelope and avoid ESD. Alarm annunciation is accompanied by data as time stamp, type of alarm, process variable tag, priority, and message. *False alarms* shall be avoided since they undermine safety culture. False-positive is just false; false-negative is when the alarm should have been activated. Because the many interacting and mutually influencing process variables (connectivity) propagate a disturbance, one alarm may be followed by many other alarms, up to dozens and higher. *Alarm flooding* increases operator's stress level, which may seriously decrease his ability to make the right decisions. This problem has been recognized since the 1994 accident at Milford Haven[19] in the UK, and there have been quite a few efforts to restrain the avalanches (some of these measures will be elucidated below).

Because of the seriousness of the problem, the Engineering Equipment and Materials Users' Association (EEMUA) published a guideline[20] and ANSI/ISA[21] published standard 18.2 in 2009. In addition, there is a host of literature available, amongst others a Guideline of the ASM Consortium[14] and by UK HSE.[22] The detailed EEMUA publication distinguishes types of alarms for a variety of purposes, how an alarm will be made effective, how it should be set, how alarms should be structured, how alarms can best appear on a display, priorities in handling alarms, et cetera. Highest priority shall be given to restraining alarm floods by rationalization/redesign. (Related) nuisance alarms and chattering (itself repeating) alarms should be suppressed. For prompt handling, it is recommended to reduce the alarm rate to *one per 10 min*. Much can be done to curb alarm flooding in the engineering stage by setting up an alarm database with settings and all other data. HAZOP-ing on P&ID and performing risk assessment will provide information for setting priorities, to identify groupings that can be automatically suppressed following the first alarm, which alarms should not be suppressed, et cetera. In an operational stage, the commonly too-large alarm frequency is addressed by "alarm management," that is, trying to identify clusters, taking account of connectivity, and analyzing the most frequent "bad actors."

In 2004, Brooks et al.[23] published an EPSC-awarded method applying a parallel coordinate transformation to display Hi—Hi and Lo—Lo alarm limits of all relevant process variables, together with an operating point for each of them and their operable ranges. By connecting the points in one display (reproduced in Figure 8.6) as zig-zag lines in outer red, center-blue, and a safe green envelope contour in between, respectively, an operator gets a more realistic overview. Shah and coworkers published several papers on graphical tools to trace related and redundant alarms in operating plant; a problem overview paper is by Izadi et al.[24] Kondaveeti et al.[25] developed based on the Jaccard similarity index applied to a high density alarm plot of, for example, one week of operation on an alarm similarity color map. Another one by Yang et al.[26] correlates points of time of specific tagged alarms by first superposing at each point a suitable Gaussian kernel function so that a wavy time series arises. Subsequently, the maximum cross-correlation coefficient is determined with the series shifted by a time lag. Other possible similarity analysis

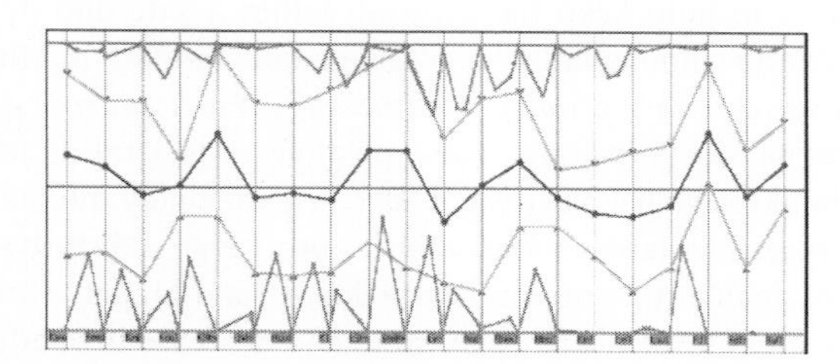

FIGURE 8.6

Display of process variables by Brooks et al.[23] for a distillation column example to keep a controlling wall temperature at a desired level: center plotted line; in an operating window: gray middle lines; within alarm dead-band limits: outer plots (on a screen the plots would have the colors blue, green, and red respectively).

methods are also described in the literature. A more direct alternative method to remove chattering alarms after detection is by introducing a suitable delay time. This is proposed by Wang and Chen.[27] The delay time is calculated by making use of previously observed patterns.

However, there remains the abnormal situation that needs to be detected and dealt with. In a technical sense if individual alarm settings can be made to avoid false and chattering alarms, the root of the remaining problem looks to be the connectivity. If in the cause—effect chain the right basic cause can be found quickly, in principle just one alarm can be activated and the connected cluster does not need to join. Signed digraph has been suggested to model connectivity. This could make a difference when combined with a quantifiable model digraph network in which time and probability are represented. At first sight, Petri net with its capability of simulation of timing seems better suited for this purpose than Bayesian net. Petri net was also the choice of Chao and Liu,[28] who proposed in 2004 a timed Petri net (named by them a Time Constraint Petri Net or TCPN). At closer look no PN would be needed, if a hidden cause of a safety critical upset can be established based on observables fast. Cause-effect with respect to alarm connectivity is predetermined by design and process models and can be called on when the cause is identified.

As we saw in Chapter 7, after a Blended Hazid once an upset shows up a causal graph can be called up by an operator without delay. However, for prioritizing repair actions historical data of probability values of different causes for a particular upset would be desirable. Meanwhile, Hu et al.[29] based on multi-level flow modeling supported HAZOP (see Chapter 7, Section 7.3.3), historical process upset data, a dynamic BN structure learned from the data and a two-step inference achieved the desired objective. Thus, the relatively easy-to-construct BNs will be helpful for both improved alarm management and situation awareness. The latter is shown by Naderpour et al.[30] for two examples of processes in which hazard can develop. The authors constructed dynamic BNs based on the process bow-ties and called it a situational network. Root nodes are fed by observable process variable values from the SCADA system producing risk indicators, while risks of abnormal conditions as operators would assess them are calculated by applying fuzzy logic. An alarm will sound when a risk threshold is passed that is still lower than critical, so that the operator can anticipate the way the abnormal situation may develop, to identify the cause from a number of different causes possible and prepare his decisions for recovery. This will relax the time pressure the operator will sustain otherwise. An alternative idea is to run a simulation of the process in parallel with the process. In the case of an incipient deviation of a known or presumed cause, the effects can be accelerated in the simulation. Here, though, one has to be sure about the cause—consequence scenario, otherwise confusion may result.

Adhitya et al.[31] performed a human factors study to investigate operator accuracy in establishing a diagnosis given an upcoming abnormal situation detected by a system called Early Warning. This system predicts a critical alarm shortly before it actually annunciates. The research applied third- to fifth-year chemical engineering students instead of (experienced) operators. To compensate, the process systems chosen were kept relatively simple. The students were given brief training while they were also given a task survey to find out whether their knowledge level was assessed sufficient. Time to diagnosis and accuracy of diagnosis were measured with and without Early Warning decision support. Although diagnosis with the support occurred faster, accuracy was not better and varied, depending on the scenario, roughly between 50% and 90% correct. Hence, not only early warning but also guidance to the right cause of upset is needed.

8.6 START-UP, SHUT-DOWN, AND TURN-AROUND

The "abnormal" situations of in particular start-up and turn-around, or unscheduled shut-down, are subject to a very substantial part of the incidents occurring in a plant. At start-up this is due to the rapidly changing conditions and possible lack of routine, while at turn-around there are many parallel maintenance activities, which are not all routine, and many exposed workers are present. Little systematic analysis has been done with respect to these process episodes. Only Ostrowski and Keim[32] addressed the issue in a very useful series of articles. The Transient Operation HAZOP that they developed focuses on operational tasks and procedures. They make a number of very practical recommendations. Guide words are *Who*, *What*, *When*, and *How long*. The exercise shall be preceded by a tour of the facility.

As already mentioned, errors in executing procedures in abnormal situations may lead to serious accidents. As mentioned in Section 8.5, Williams[15] reported on an effort to improve this situation by computer assistance to the operator or partly or fully automating procedures. However, a systemic error in an automated process shut-down may also be detrimental. Van Paassen and Wieringa[33] quote Lind's Multi-level Flow Modeling, which we briefly described as applied in HAZOP automation in Chapter 7, Section 7.3.3, as an excellent tool for reasoning when developing procedures for start-up and shut-down, whether or not automated.

8.7 CONCLUSIONS

Process control development is toward Lyapunov function–based distributed model DMPC, which besides improved control effectiveness and economic optimization also to some degree enables fault diagnosis of valves and sensors. The latter topic has been the subject of an extensive review by Venkatasubramanian et al.[3] Over the years, many methods have been proposed. Lately, higher-order statistical treatment of time series determining bicoherence in nonlinear systems offers much perspective.

A brief description of the newest control equipment showed many improvements, also with respect to security. In addition, system and control room ergonomics have improved the operator's work environment, which also stimulates situation awareness. Alarm management and the avoidance of alarm floods is still a "hot" topic. In recent years, a great variety of approaches have been proposed to relieve the problem. More insight in the causal structure of an abnormal situation, obtained by applying methods such as Blended Hazid on the design of the plant, may help to unveil connectivity. Installing its result as computer software in the control room and reinforcing it in the future by a BN may enable penetrating via the causal structure into the core of the problem.

Together with the signals collected by the SCADA system, the above will provide a step toward the prognostic process control tool that Venkatasubramanian[4] foresees. This could be further supplemented with safety management system performance indicator value information. This would enable that not only the direct, short-term risk factor effects may be seen but also the risk factors with effects on culture over a longer term as suggested in the holistic approach by Pasman et al.[34] The prognostic tool would then become even more universal and could in a more popular way really be termed a "safety dashboard."

REFERENCES

1. Stephanopoulos G. *Chemical process control, an introduction to theory and practice.* Englewood Cliffs (N.J.): PRT Prentice Hall; 1984. ISBN 0-13-128629-3.
2. Christofides PD, Scattolini R, Muñoz de la Peñad D, Liu J. Distributed model predictive control: a tutorial review and future research directions. *Comput Chem Eng* 2013;**51**: 21–41.
3. Venkatasubramanian V, Rengaswamy R, Kavuri SN, Yin K. A review of process fault detection and diagnosis, part I: quantitative model based methods. *Comput Chem Eng* 2003;**27**:293–311. Part II: Qualitative models and search strategies. *Comput Chem Eng* 2003;**27**:312–26; Part III: Process history based methods. *Comput Chem Eng* 2003;**27**:327–46.
4. Venkatasubramanian V. Systemic failures: challenges and opportunities in risk management in complex systems. *AIChE J* 2011;**57**(1):2–9.
5. Christofides PD, El-Farra NH. *Control of nonlinear and hybrid process systems: designs for uncertainty, constraints and time delays*. Springer; 2005. Monograph.
6. Ellis M, Christofides PD. Integrating dynamic economic optimization and model predictive control for optimal operation of nonlinear process systems. *Control Eng Pract* 2014;**22**:242–51.
7. Kazantzis N, Kravaris C, Wright RA. Nonlinear observer design for process monitoring. *Ind Eng Chem Res* 2000;**39**:408–19.
8. Mhaskar P, Liu J, Christofides PD. *Fault-tolerant process control, methods and applications*. Springer; 2013. ISBN 978-1-4471-4807-4. http://dx.doi.org/10.1007/978-1-4471-4808-1. ISBN 978-1-4471-4808-1 (eBook).
9. Vaidhyanathan R, Venkatasubramanian V. Digraph-based models for automated HAZOP analysis. *Reliab Eng Syst Saf* 1995;**50**(1):33–49.
10. Dash S, Rengaswamy R, Venkatasubramanian V. Fuzzy-logic based trend classification for fault diagnosis of chemical processes. *Comput Chem Eng* 2003;**27**:347–62.
11. Shoukat Choudhury MAA, Shah SL, Thornhill N. Diagnosis of process nonlinearities and valve stiction, data driven approaches. In: *Advances of industrial control*. Springer; 2008. ISBN 978-3-540-79223-9.

12. Shoukat Choudhury MAA, Shah SL, Thornhill N. Diagnosis of poor control-loop performance using, higher-order statistics. *Automatica* 2004;**40**:1719–28.
13. Dulyakarn P, Rangsaneri Y. Fuzzy c-means clustering using spatial information with application to remote sensing. In: *Proceedings of the 22nd Asian conference on remote sensing, Singapore, 2001*; 2001.
14. ASM Guidelines from Abnormal Situation Management Consortium (www.asmconsortium.com) can be obtained from www.Amazon.com.
15. Williams Jr ThN. Procedural automation. In: *Proceedings 16th annual international symposium, Mary Kay O'Connor process safety center, October 22–24, 2013, College Station, Texas*; 2013. p. 126–33.
16. a. Endsley MR. Toward a theory of situation awareness in dynamic systems. *Hum Factors: J Hum Factors Ergon Soc* 1995;**37**:32–64;
b. Measurement of situation awareness in dynamic systems. *Ibidem* 1995;**37**:65–84.
17. Wickens ChD. Situation awareness: review of Mica Endsley's 1995 articles on situation awareness, theory and measurement. *Hum Factors: J Hum Factors Ergon Soc* 2008;**50**: 397–403.
18. Flemisch F, Heesen M, Hesse T, Kelsch J, Schieben A, Beller J. Towards a dynamic balance between humans and automation: authority, ability, responsibility and control in shared and cooperative control situations. *Cognit, Technol Work* 2012;**14**:3–18.
19. Health and Safety Executive. *The explosion and fires at the Texaco Refinery, Milford Haven, 24 July 1994: a report of the investigation by the health and safety executive into the explosion and fires on the Pembroke Cracking Company Plant at the Texaco Refinery, Milford Haven on 24 July 1994*. ISBN 0-7176-1413-1.
20. EEMUA. *Alarm systems, a guide to design, management and procurement*. 2nd ed. London: Publication No. 191; 2007. ISBN 0-85931-155-4.
21. ISA, *Management of alarm systems for the process industries*. 2nd ed. Technical Report ANSI/ISA-18. 2-2009 International Society of Automation ISA, Research Triangle Park, NC; ISA, Alarm Management: A Comprehensive Guide.
22. Health and Safety Executive. The management of alarm systems, prepared by Bransby automation Ltd and Tekton engineering, Contract Research Report, 166/1998.
23. Brooks R, Thorpe R, Wilson J. A new method for defining and managing process alarms and for correcting process operation when an alarm occurs,. *J Hazard Mater* 2004;**115**: 169–74.
24. Izadi I, Shah SL, Chen T. Effective resource utilization for alarm management. In: *49th IEEE conference on decision and control, December 15–17, 2010*. Atlanta (GA, USA): Hilton Atlanta Hotel; 2010.
25. Kondaveeti SR, Izadi I, Shah SL, Black T, Chen T. Graphical tools for routine assessment of industrial alarm systems. *Comput Chem Eng* 2012;**46**:39–47.
26. Yang F, Shah SL, Xiao D, Chen T. Improved correlation analysis and visualization of industrial alarm data. *ISA Trans* 2012;**51**:499–506.
27. Wang J, Chen T. An online method to remove chattering and repeating alarms based on alarm durations and intervals. *Comput Chem Eng* 2014;**67**:43–52.
28. Chao C-S, Liu A-C. An alarm management framework for automated network fault identification. *Comput Commun* 2004;**27**:1341–53.
29. Hu J, Zhang L, Cai Zh, Wang Y, Wang A. Fault propagation behavior study and root cause reasoning with dynamic Bayesian network based framework. *Process Saf Environ Prot*, in press, accepted manuscript, available on line 7 April 2015.

30. a. Naderpour M, Liu J, Zhang G. A situation risk awareness approach for process systems safety. *Safety Science* 2014;**64**:173–89;
b. Naderpour M, Liu J, Zhang G. Supporting operator's situation awareness in safety-critical systems: an abnormal situation modeling method. *Reliab Eng Syst Saf* 2015;**133**:33–47.
31. Adhitya A, Cheng SF, Lee Z, Srinivasan R. Quantifying the effectiveness of an alarm management system through human factors studies. *Comput Chem Eng* 2014;**67**:1–12.
32. Ostrowski SW, Keim KK. *Tame your transient operations, use a special method to identify and address potential hazards, chemical processing*; June 2010. http://www.chemicalprocessing.com/articles/2010/123/?start=0.
33. Van Paassen MM, Wieringa PA. Reasoning with multilevel flow models. *Reliab Eng Syst Saf* 1999;**64**:151–65.
34. Pasman HJ, Knegtering B, Rogers WJ. A holistic approach to control process safety risks: possible ways forward. *Reliab Eng Syst Saf* 2013;**117**:21–9.

CHAPTER

Costs of Accidents, Costs of Safety, Risk-Based Economic Decision Making: Risk Management

To Be or Not to Be……

Is it nobler to put up with all the nasty things that luck throws your way, or to fight against all those troubles by simply putting an end to them once and for all?
William Shakespeare, *Hamlet*, Act 3, Scene 1 (modern text)

SUMMARY

Numerous decisions are made daily on conducting economically relevant activities involving hazards, hence each of these activities implies a risk. Risk taking is something that we do as a person on an individual basis everyday. Crossing a road as a pedestrian with approaching traffic in sight is commonplace, as well as a driver overtaking a car in front of him on a provincial road with oncoming traffic in sight. In any of these decisions, there is a weighing up front: gains versus risk of losses. In these simple examples the cost of an accident is high, the action could take yours or another person's life, and the gains may not be very significant—just a few seconds in time saved.

With industrial accidents the situation is not different: given a certain process a shortcut on a procedure can be associated with a large safety risk—the company can go bankrupt if the risk event materializes. However, gains in effort or time may be rather small, although these may be significant in surviving the competition better. Of course, the scale makes a difference. But, there may be many lives at risk (and as a decision maker, probably not your own life) and damage may run up from millions to billions, and it may have long-term effects on employment, and degradation of quality of life of people and the environment. Costs of an industrial accident are not only those of the company in a direct sense but also the personal cost of people involved and not the least the social and economic cost of society as a whole. Complexity may easily obscure the largest possible consequences of a process or a project. And even when one realizes the possibility, intuitively one is inclined to

Risk Analysis and Control for Industrial Processes - Gas, Oil and Chemicals.

think "it will not happen to me." Usually, the likelihood is lowest, but nevertheless, if an occurrence is possible, one cannot predict when it will take place: it may be soon, it may be not in a lifetime.

Uncertainty and lack of information and knowledge may easily cause underestimating a risk, which is reprehensive and may be fatal. Likewise, overestimating a risk in an industrial context may be economically damaging and, depending on scale, may even affect society if it results in banning a technology. Also, if disaster hits, in the aftermath, the decision maker as a person, but in an organization the board or the office responsible, may have to justify the decision taken. The moment of time the decision was made to embark on the project may have been many years before. And once risks strikes, there is a tendency after an event to deny one already knew about the possible risk, and there have even been cases of written evidence been hidden. On the other hand, for learning and prevention it will be very useful if the motivation and the foundations of the decision to embark on an activity are recorded, including the associated risks.

All this makes it crucial to take certain decisions after having performed a thorough risk assessment and weighing of pros against cons consciously and recordable. In organizations with many players involved, that weighing may be in its final stage in the mind of the top person, but before that there will be discussions about the case in which arguments will be exchanged. Decision analysis may require facts and verifiable information. The level of analysis will depend on the financial risk involved, the complexity of the project, and important details. A decision may occur in stages over a period of time because initially it may appear that certain information is missing. The value of obtaining additional information against the risk of not having it should be part of the decision process. Meanwhile, there are software tools available to facilitate conscious and underpinned decision making based on consequences expressed in monetary value and expected frequency or probability of occurrence. Also, the theory of decision making has made progress.

As George Kirkland, vice chairman of the board of Chevron and executive vice president of Upstream, emphasized, every large project requires a thorough decision analysis in which business gains and risks are considered together with further critical factors. It is the basis of risk management.

In previous chapters we have looked at various building blocks of risk management such as identification of the risk, its consequences and likelihood, and possible risk-reduction measures, including the aspects of organization, human factors, and safety culture. In this chapter we shall focus on the capstone: decision making. We shall have a look at costs of accidents, costs of safety measures, and how decision theory and analysis contribute, such as expected utility theory, decision tree, and supporting software. We shall even touch upon approaches to tackle cases in which decisions must be taken under deep uncertainty. Finally, with decision making we shall complete the various steps of safety risk management with one exception. In the weighing process of decision making, risk perception of the workforce and the public and the effect of risk communication can enter as factors as well and not rarely even as dominating factors. These societal considerations will be the topic of Chapter 12.

Risk taking is fundamental in business and entrepreneurship, and enterprise risk management is quite a common instrument. Yet, with the financial crisis and the demise of several companies and banks, and the almost chronic budget runovers of large infrastructural and other projects, awareness has further grown that sound risk management is a responsibility of leadership toward all stakeholders. Therefore, we shall at the end of this chapter briefly place safety risk management in context with other risk management processes.

Another aspect is the issue of security. In previous chapters we devoted no special attention to it because with respect to consequences due to hazardous materials present and technical protection systems to contain effects, it largely falls within the same types of generic measures described for safety risks. Of course, additional and much different specific measures apply with regard to early warning, access control, and internal organization. Also, a fundamental difference with safety risk is probability of an event in relation to the intention of a terrorist or insurgent attacker. What does an attacker want to achieve: anxiety of people, or for a defender loss of strategically important material? Anyhow, in optimizing measures, preventive or protective, given a certain budget or a requirement of maintaining a certain protective level, methods may be of relevance that are treated in this chapter such as multi-attribute utility and game theory.

9.1 COSTS OF ACCIDENTS

Marsh Insurance is the source of property damage cost data of industrial accidents,[1] although it concerns only insured risk, so some notable incidents are missing. In its 1972–2011 issue on Energy Practice, the 100 largest losses over the period are collected with figures based on the Marsh Energy Loss Database. Thereby, five sectors are distinguished, each with an associated number of incidents: Refineries 30, Petrochemicals 24, Gas Processing 10, Terminals and Distribution 5, and Upstream 31. Most losses are due to vapor cloud explosion, fire/explosion, and fires. In Terminals and Distribution and Upstream, losses are due to mechanical causes and weather, whereas in Upstream well blowouts are an important cause. In addition, the issue contains a supplement on natural catastrophe accumulation such as earthquakes accompanied by tsunami, and hurricanes by which losses can run up to tens of billion USD.

In Figure 9.1 property damage losses of the first two sectors are depicted in 5-year intervals. As mentioned, these loss figures are only damage to assets and do not include uninsured costs, which are usually much larger than asset damage. Business interruption, penalties, paid-out liability and compensations, and reputation loss (share value loss) can easily exceed an order of magnitude of asset damage costs. For example, the April 2010 Macondo well blowout with the loss of the Deepwater Horizon offshore rig is mentioned in the survey as a loss of 590 million USD, while according to a March 2012 Reuters press release,[2] total costs may become as high as 65.5 billion USD. This is composed of individual liability $13.9 billion, operational

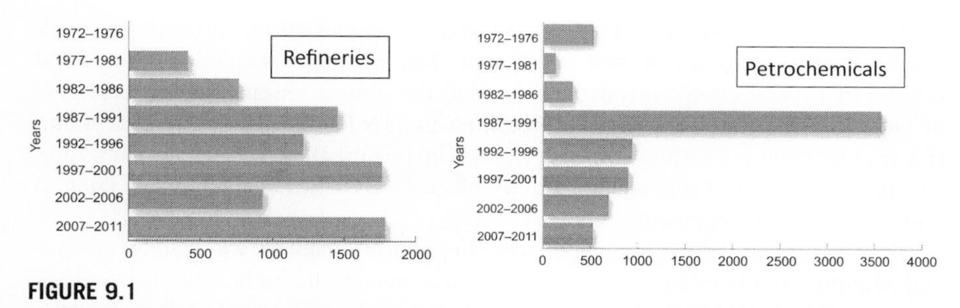

FIGURE 9.1

Two examples of sectors showing property damage losses in millions of 2011 USD over 5-year periods, according to Marsh,[1] The 100 Largest Losses 1972–2011.

response $14 billion, civil penalties $4.5–$17.6 billion, criminal penalties $5–$15 billion, and environmental damage $5 billion. (On September 4, 2014, BP was fined $18 billion for gross negligence.) Of course, due to the widespread pollution damage, this example will be rather extreme.

Despite periods of large peaks, there seems to be a certain trend of improvement in some sectors. Altogether, there is no reason to give process safety less attention, as also noted in the 1972–2011 Marsh publication preface by Judith Hackitt, chair of UK's HSE. On the contrary, where in personal safety there is a steady trend to lower accident frequencies and lower loss figures, with respect to the huge losses in major accidents there is no convincing downward trend. Because of the development toward higher performance in all sectors, partly competition driven, and partly because the easily exploitable resources have been harvested, there remains a need to strengthen process safety and risk management effort. As is mentioned in the introduction of the collection, long-term learning from incidents is rather difficult.

Kim et al.[3] reported about financial impacts of catastrophic safety events in the process industry. It turns out that the losses and damage to reputation greatly affect the share value of companies. Some effects are revealed only years later. The authors present a number of examples. One of the most striking examples is that of Union Carbide, the company that was blamed for the Bhopal disaster in India in 1984. As already summarized in Chapter 1, at a chemical site of Union Carbide India Limited, a producer of pesticides, a mishap due to water flowing into a tank with volatile methylisocyanate (MIC) caused the catastrophe. As a result of the heat due to reaction of water and MIC, a huge cloud of the very toxic MIC vapor spread in a low-wind condition during the early hours of night over a large area of makeshift housing caused the largest human loss by chemical accident so far. Thousands lost their life and hundreds of thousands were seriously injured, of which some never recovered. The mishap was the result of a complex design, human and organizational failures, while the aftercare by both the company and authorities did not deserve an award either. In fact, due to overproduction and economic malaise causing unemployment, the plant safety culture had declined. The financial losses that Union Carbide suffered due to the catastrophe were large but looked initially bearable. Yet, as

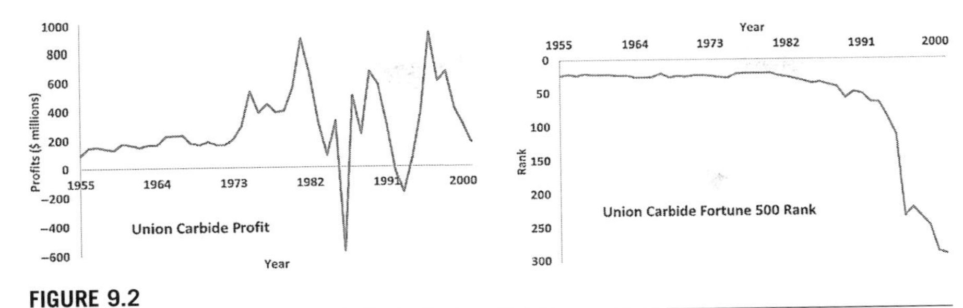

FIGURE 9.2

Left: Union Carbide profits from the 1950s when they became owner of the Bhopal plant until 2001 when it was acquired by Dow Chemical Company; and right: UC's Fortune 500 rank based on corporate revenue *(adopted from Kim et al.[3])*. The Bhopal disaster took place in 1984.

can be seen from Figure 9.2, during the longer term the company decreased precipitously in size and slipped down in rank order until it was taken over by Dow Chemical Company in 2001. Contributing to this downslide was also a toxic cloud (not MIC) that leaked from its plant in West Virginia in 1985, the year following Bhopal.

There are several examples showing a similar pattern. The stress that a company is exposed to and the resources to be spent on the aftermath cleanup, the loss of market share, and lawsuits force a company to shrink.

In the US, Hansen[4] emphasized the relation of the occurrence of major safety and environmental incidents and their impact on a company's financial soundness to the requirement for management arising from the 2002 Sarbanes–Oxley Act. This SOX Act requires publicly traded companies to audit and report their financial state periodically to the US Securities and Exchange Commission (SEC). Companies shall have effective management controls and accountabilities. The SOX Act also places requirements on auditors and auditing firms, and it will give protection to whistleblowers. The Act was to counter scandalous financial practices as in the Enron demise in 2001. As a consequence of the financial crisis in 2007 and 2008, the SEC applies the law more stringently. A failure that may significantly impact the organization's financial state shall be reported. According to the author, this may encompass serious failures of safety, chemical spills, and other failures. Installing safety and environmental management systems will be of great help to convince the public and regulators that controls are functioning.

All this does not consider the imponderable personal (casualties, pain, suffering, trauma, anxiety, stress) and societal costs of a major accident.

9.2 COSTS OF SAFETY

In the initial period of the Loss Prevention Symposia in the 1970s, it was not done, not ethical to talk about safety in any relation to money value. The title of a paper proposed by T. J. Webster "Safety as a Money Spinner" therefore had to be changed to "Safety is Good Business." Only after that change was it acceptable in the eyes of the Symposium Committee for the 1974 Loss Prevention in Europe, although the contents of the paper and hence the message remained the same. The message contained an analogy with Heinrich's accident pyramid. The top events are very costly but are few. Smaller-size accidents are numerous, their costs are quickly forgotten, but because of their sheer number, their total loss is high. Therefore, it pays to invest in safety. But does it make sense to eliminate any identified safety flaw by a measure? One can think of a cost–benefit trade-off, although cost may be disproportionately higher than the (monetary) benefits.

Before 2000 there was very little in the literature on costs of safety and the weighing against assessed risk, yet it was an important consideration in the minds of people. Therefore Pasman,[5] the author of this book, decided to develop and publish a paper in 2000 on investment for safety. The first lines of the abstract read: "During economic doldrums, decision making on investments for safety is even more difficult than it already is when funds are abundant." Although competition was already on the rise in 2000, there was no economic crisis in sight. So, funds for investment may be even scarcer now. The paper discusses risk assessment, and preventive and protective measures, as in this book in a more extensive sense in Chapter 3. It then carries on with elementary concepts of business and process economics, which can be found in Perry's *Chemical Engineers' Handbook*,[6] but which for a good understanding will be repeated in brief here in Table 9.1.

Pasman[5] mentioned an example of an installation life cycle of 20 years, a required capital interest of 10% per year, and a risk reduction measure costing 250,000 USD. For installing an economically attractive measure, the *EAL* savings should then be larger than 29,300 USD per year, where the annuity present-worth factor of 10% over 20 years is 0.117. Hence, suppose the original risk is a consequence damage of 30 million USD at an event frequency of once in 1000 years. After installing the measure, the remaining event frequency is estimated as once in 50,000 years, while the consequence remains equal. On average, the investment is adequate because the $\Delta EAL = 10^{-3} \times 30 \times 10^6 - 2 \times 10^{-5} \times 30 \times 10^6 = 29{,}400$ USD/year. Besides the uncertainty in the figures, there is quite a chance that the event will not occur in 20 years, but it can also happen the day after start-up. However, given a large number of the same installations the mean will materialize, so this kind of calculation is interesting for a very large industrial or insurance company.

After 2000, more attention was given to this topic. The American Society of Safety Engineers (ASSE) dedicated in 2010 a white paper[7] to the subject after a few articles in ASSE's journal *Professional Safety*. The article by Veltri and Ramsay[8] on making the business case for safety, health, and environment (SH&E) is quite comprehensive. It summarizes the literature on the economics of safety, which is not very extensive. The authors make some observations such as that SH&E costs do not play an explicit role in decision making. But, critical business decisions are incomplete if in the life cycle costing of products, technologies, and processes SH&E costs are not revealed. Accounting usually includes SH&E costs in the overhead and does not consider these as investment. SH&E investments, however,

Table 9.1 Financial Concepts in Connection with Investment in Risk Reduction

Concept	Meaning	Equation
Annual cash flow, here defined as:	Annual sales income, minus various types of annual expense, minus annual tax, and minus expenditures on investment capital.	A_{CF}
Investment profitability: (simple), e.g., Payback period	Project life cycle number of years, n required to accumulate a total cash flow equal to the amount of fixed capital cost, C_{FC}.	$n = \frac{C_{FC}}{\left(\sum_{i=1}^{n} A_{CF\,j}\right)}$
Net present value, *NPV*	Present worth of money, P is related to the value, F of that money, j years in the future through the discount factor, being the reciprocal of the annually compounded interest, i over j years.	$P = F \times f_{d,j};\ f_{d,j} = \frac{1}{(1+i)^j}$
Investment profitability, A over n years	NPV of the discounted annual cash flows, $A_{DCF\,j}$ from the year of investment ($j = 0$) until and including year n.	$A_{DCF,j} = A_{CF,j} \cdot f_{d,j}$; $NPV_{\sum CF} = \sum_{j=0}^{n} A_{DCF,j}$
Even more realistic is a discounted measure of profitability	Discounted cash flow rate of return (DCFRR) is the accumulated cash flow the project generates over n years after covering all expenses, interests and taxes, which repays the original investment capital, C_{FC}.	$NPV_{DCFRR} = C_{FC}$
Expected annual loss cost, EAL, and event risk reduction measure	EAL cost is risk expressed as the product of expected event frequency per year, p, and the damage consequences (impact) of the event in monetary units, D.	$EAL = p \cdot D$ A risk reduction measure results in: $\Delta EAL = p_0 \cdot D_0 - p_1 \cdot D_1 = \Delta(p \cdot D)$
NPV of *EAL* amount	In analogy with investment *NPV*, a discounted loss cost can be calculated.	$\Delta EAL_{D,j} = \Delta EAL_j \cdot f_{d,j} = \Delta EAL \cdot f_{d,j}$, as ΔEAL is constant over the years.
Payoff of risk reduction	Over the life cycle of the project of n years the discounted "savings" by lower risk shall be larger than the investment cost of the safety measures, $C_{FC,S}$ (although this does not need to be true in case the measure is due to regulation). The annuity present-worth factor, f_{AP} represents the interest expression.	$\sum_{j=0}^{n} \Delta EAL_{D,j} \geq C_{FC,S}$ As ΔEAL is constant, this simplifies to: $\frac{(1+i)^n - 1}{i \cdot (1+i)^n} \geq C_{FC,S}$ Or with the annuity present-worth factor: $\Delta EAL / f_{AP} \geq C_{FC,S}$

pay off at the longer term, whereas profitability of a project is often only judged on the short term. The authors contend that SH&E is usually seen as a regulatory constraint and not as an opportunity, as an enabler.

For someone who would like to go in the direction of a more explicit SH&E investment policy, Veltri and Ramsay[8] propose a procedure for setting up a cost–benefit analysis, the main line of which is quite similar to the one shown by Pasman.[5] The authors present an extensive example oriented toward occupational safety and health. For an analysis there will be various obstacles to clear such as obtaining realistic cost information, in particular, of damage due to exposure to hazards, which can break down in many different cost factors. Veltri and Ramsay mention hiring and replacement costs, turnover and absenteeism, lost time due to inefficiencies of new employees, production losses, cost of administrative handling of the claims including supervisor involvement in hiring and retraining, and the costs of lower morale and higher stress. Part of these costs is externalized to the society in social security, particularly in Europe.

Yingbin Feng,[9] University of Western Sydney, undertook around 2011 a thorough study of safety investments versus their benefits in the building construction industry in Singapore. Investment comprised staffing costs, safety equipment and facilities, compulsory training, so-called in-house safety training with daily safety orientation before work starts and with first-aid and emergency drills, safety inspections and meetings, safety incentives and promotion and safety innovation. The investment effects were assumed to depend on hazard level of the work or the safety culture, which was found to be true. Investment does not compensate a low-safety culture. It further turned out that investment in safety professionals, provision of personal protection equipment, and formal training courses is less effective by themselves than one's own voluntary accident investigations, safety inspections, safety committees, safety incentives, and in-house safety training and orientation.

A further issue in monetizing damage is how to account for a fatality and a serious injury that because of a residual permanent handicap lowers the quality of life. Kip Viscusi[10] (economist, then Harvard University) proposed for a statistical American life a range of 4–10 million USD with a mean value of 7 million USD (in 2005). As regards injury, much depends on the permanence and the degree of disability. A measure, also developed at Harvard University is the disability adjusted life years (DALY). It can be valued in court as loss of income and in addition compensation for suffering. On the same basis as for a fatality, Viscusi[7] estimates the implicit value of a worker's injury, which could have resulted in a fatality, as between 20,000 and 70,000 USD.

In a later study in the framework of offshore safety cases, Bolu[11] collected literature information on how various scholars came to derive a value of a statistical life (*VSL*). Bolu quoted Miller[12] as defining *VSL* as the amount a group of people will pay for a fatal risk reduction in the expectation of saving ones' life. Miller estimated for various industrialized countries the *VSL* value by a regression equation: $\ln(VSL) = a + b\ln(Y) + c\,Z$, in which Y denotes an income measure,

Table 9.2 Averaged Value of Statistical Life (*VSL*)

Country	Number of Values	Averaged *VSL* 10^3 US $ (1995)	Country	Number of Values	Averaged *VSL* 10^3 US $ (1995)
Australia	1	2126	South Korea	2	620
Austria	2	3253	Sweden	4	3106
Canada	5	3518	Switzerland	1	7525
Denmark	1	3764	Taiwan	2	956
France	1	3435	United Kingdom	7	2281
Japan	1	8280	United States	39	3472
New Zealand	3	1625			

Data derived by Miller[12].

Z a vector of explanatory variables, and a, b, and c regression variables. Values found by different researchers vary quite widely per country and also differ widely from one country to another. An impression of average values in thousands of 1995 USD is given in Table 9.2 based on Miller's data. The income influence is obvious, but cultural beliefs also play a role.

Anyway, in the case of a fatality in an industrial accident losses will be high, think of the associated physical damage, business interruption, reputation loss, personal imponderables and possible liabilities. Because of this and because discussion about *VSL* is a moot point, considering risk-reduction measures companies may have to decide about investing an amount to avert the possibility of a fatality (willingness to pay). Aven and Renn,[13] in their book *Risk Management and Governance*, mention the orders of magnitude of guideline values used by an oil company to avert a statistical life. If the measure can be taken for less than 1 million Euros it is effective, at 1 M€ it is considered effective if individual risk level is high, at 10 M€ it is considered "if individual risk levels are very high and/or if there are other highly appreciated benefits," and at 100 M€ it is "not socially effective—look at other options."

9.3 RISK-BASED DECISION MAKING

Decision making in view of risks and distributing scarce resources over risk-reduction options (budget allocation) are crucial tasks in risk management. We are used to making daily life decisions almost continually and intuitively, and for the most part not consciously. In Chapter 6, Section 6.4, we have seen Erik Hollnagel's descriptive psychology collection of experience-based techniques for decision making (heuristics). More about decisions and risk psychology can be found in a book review by Tony Cox.[14]

Important decisions, however, which can have long-term and dramatic consequences for a company or a society, are usually not made overnight. They will be made by an executive board or council, or by a leader or manager on his/her own. Before making the decision, studies may have been done proposing competing options, while arguments and counterarguments may have been heard in discussion meetings. Hence, a decision will be rationalized. In his book *Foundations of Risk Analysis*, Terje Aven[15] dedicates a chapter to it and Figure 9.3 presents the process as shown there, which looks rather straightforward. Aven addresses mainly the Norwegian oil and gas industry and notes that risk-informed decision making is now widely accepted in the case of potentially high consequences and large uncertainty.

Depending on leadership style, a leader's decision can have a character ranging from autocratic to fully democratic, where the optimum may lie in between, but not too far from democratic. Leadership shall have influence. Policy decisions can be made on criteria, although these usually are rather intuitive and vague. Project investment decisions should be clear and based on a quality–price ratio; there may be a specification of minimum quality requirements including safety, health, and environmental ones, a constraint of time to realize the project, and an upper bound to the funding. The latter is often overruling.

Whatever the project, there are always issues concerning uncertainty and risk. Optimizing against investment funds is seen as a task for the executive. Information and knowledge are important resources to minimize uncertainty. In economics, business management and finance methods are available to identify, to structure, and to quantify business and project risks. In earlier chapters of this book we have seen methods to determine safety risks with potential impact to life, assets/property, and environment. So, an optimal decision can be made given the overall project goal is well defined, all risks can be expressed in monetary terms, and their

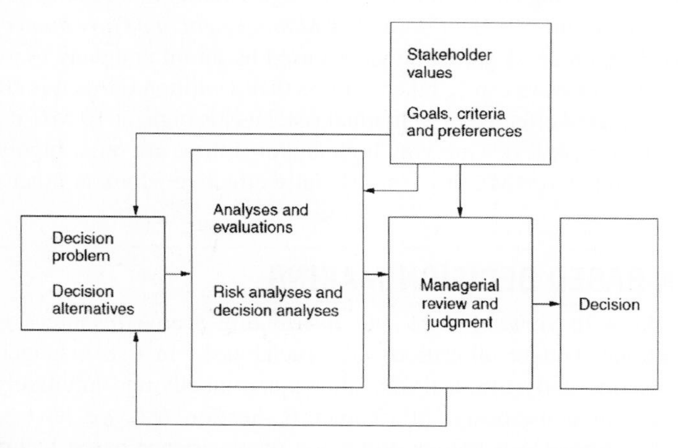

FIGURE 9.3

Process of rational decision making *(after Terje Aven[15])*. The development of alternatives is for an important part driven by constraints. These include among others stakeholder values, political views, and environmentalist objections. He further distinguishes decision making by the sharp end (close to the hazard) and the blunt end (remote from the hazard).

dependence on conditions and ways to realize a project can be determined. A great proponent of risk-informed decision making is George Kirkland, vice chairman of Chevron, already mentioned in the Summary to the chapter. He advocated decision analysis in a video.[16]

Theoretical decision methods and practical decision tools to determine the best direction for an organization based on financial and other inputs are plentiful. For the purpose of engineering decisions, the book by George Hazelrigg[17] (US National Science Foundation) provides a clear overview including relevant mathematics and economics. A few important tools will be briefly explained. The methods range from simple means such as Balanced Scorecard, developed by Robert Kaplan and David Norton in the early 1990s, to more structured ones as the Analytic Hierarchy Process (AHP), developed by Thomas Saaty in 1977. Another method in which various attributes as components of utility are made explicit is the MAUT. In case all factors/attributes can be expressed in monetary units and uncertainty is not made explicit, a simple cost—benefit optimization may support decision. If, however, investments are to be made with uncertain gains, economic theory provides an approach with the expected utility model. However, major decisions under uncertainty should be based on a logically coherent predictive model consistent with the axioms of probability. Tools are available that structure a decision process with all decision aspects or constituting components involved as branches of a "tree," branching out from a final decision to the more detailed components. Finally, methods for decisions under deep uncertainty are briefly mentioned. Below follow some more details.

Of course, economic considerations are only part of the decision-making process. Depending on magnitude of the risk, if lives of workers and residents become at stake (on- and off-site risk) there will be other decision criteria to be reckoned with. Societal considerations, for example, based on welfare economics, and regulatory risk acceptance criteria based on fatality risk and others such as As Low As Reasonably Practicable (ALARP) will be discussed in Chapter 12.

9.3.1 BALANCED SCORECARD

The original scorecards to plan, implement, and achieve business strategies are four: Customer satisfaction and requirements, financial requirements and performance, performance of internal processes, and learning, and growth. Safety will come under the last two cards. Each of the scorecards will encompass a number of performance indicators. There is ample literature on the subject; see, for example, KPIs and Balanced Scorecard,[18] and in relation to safety Tappura et al.[19]

9.3.2 ANALYTIC HIERARCHY PROCESS

An abundance of literature is available on AHP. In AHP a problem to be solved for achieving a certain objective is decomposed into a hierarchy of criteria or desired characteristics with underneath more or less fulfilling options or alternatives.

A popular example is the choice of a car type with its many characteristics and alternatives. First, a group of knowledgeable persons give their opinion (elicitation) on weights of defined criteria. Weights can be qualitative or quantitative. For that matter, *pairwise comparisons* can be made of one option over another one, which in a quantitative evaluation results in a matrix of which the eigenvector can be derived by squaring the matrix and normalizing, yielding a ranking order for the criteria. Next, similarly for each criterion by pairwise comparisons ranking orders of options are derived. By multiplying the option scores with the ranking order, a most preferred option is found. The process may be followed by a cost–benefit analysis, which will be explained later. Hazelrigg describes AHP in fair detail with an example but expresses a number of concerns, which leaves AHP of not much use in decision making about a design. His main concern is that the effects of a group process for decision are imponderable, but he mentions a few others. Indeed, critical decisions about a risky project should be resolved through support of probabilistic methods, cost calculation, and clear criteria, and should not be decided (for major or critical cases) based only on unaided human intuition and heuristics. Probabilistic methods are not only there for hypothesis testing on the basis of data.

9.3.3 MULTI-ATTRIBUTE UTILITY THEORY

As we have seen above in the AHP decision process, a decision can be dependent on a set of criteria or attributes that can be valued based independently on utility to a decision maker. The multi-attribute utility theory (MAUT) provides in principle a method to trade-off the attributes' utilities under uncertainty. In practice, independence appears to be a problem. Various value functions are possible. For example, assume in the simplest case that two attributes can be distinguished in a decision problem of which installation to buy, for example, quality of end product, Q and energy consumption, C. There are four possible candidate installations with different properties. The objective is to maximize Q and to minimize C. Q cannot be expressed in monetary value, so its utility must be estimated versus cost of energy, C, in a two-attribute utility function $u(Q, C)$. The next step is to determine the utility function, u. For this, combinations must be defined yielding: $u = 0$ and $u = 1$. Different ways to proceed are possible. Direct assessment is by experts estimating utility values of various combinations, plotting utility versus quality and versus energy cost in a three-dimensional coordinate system, and determining the optimum combination. More systematic is estimating the (marginal) utility of each component separately and the weights with respect to each other, here, for example, $u(Q, C) = w_Q u_Q(Q) + w_C u_C(C)$. Depending on conditions, not making it simpler, the additive form can also be multilinear or multiplicative.

9.3.4 STRAIGHTFORWARD COST–BENEFIT OPTIMIZATION

In Figures 9.4 and 9.5, two representations are depicted of, respectively, an overall operational life cycle safety cost optimization and one for a selection of alternative

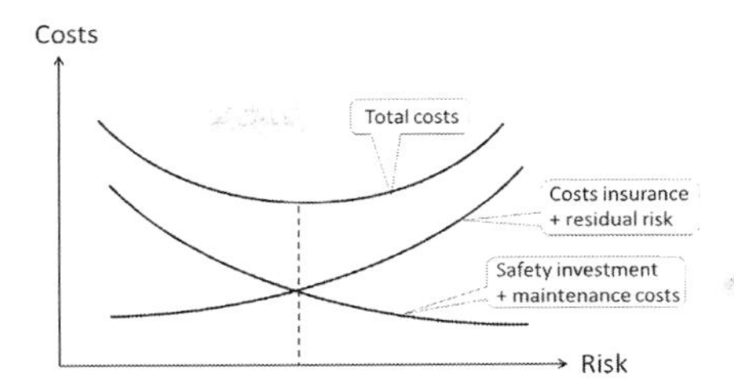

FIGURE 9.4

Overall operational life cycle safety cost optimization. Taking the notation of Table 9.1, total costs can be written as: $C_{tot} = C_{FC.S} + (C_M + C_{Ins} + EAL_{ResR})/f_{AP}$
Total cost = safety investment + maintenance + insurance + residual risk cost.
For enabling simple discounting annual cost figures are taken constant.

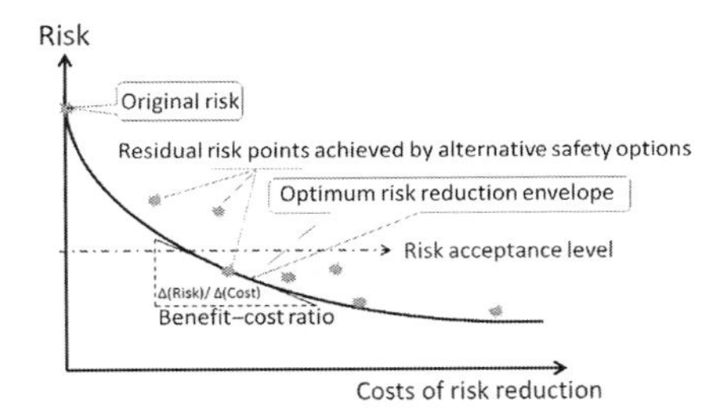

FIGURE 9.5

Besides selecting a risk-reduction option based on a minimum risk acceptance level, also a choice on the best cost–benefit ratio can be found. The optimum envelope line is drawn such that the tangent at the residual risk point on the line, with no other points below the line, represents the best ratio.

risk reduction options, both adapted from Pasman.[5] The graphical optimization is simple, but obtaining suitable data is less simple! In Figure 9.5, the optimum envelope clearly shows a diminishing return trend.

9.3.5 OPTIMAL BUDGET ALLOCATION AND GAME THEORY

The optimization problem represented by Figure 9.5 can become more aggravated in case a multiple of risks must be reduced and the budget is bounded. So, a trade-off will be necessary of overall safety benefit versus available investment. The budget constraint can be to such an extent that priorities of installing measures must be set with respect to the year of realization. Genserik Reniers and Kenneth Sørensen[20] have demonstrated that sorting the risks in a risk matrix of likelihood versus cost of impact (product is *EAL*), determining benefits and costs of risk measures, and applying the knapsack method works. The metaphor "knapsack" is used because the method is optimizing the benefit of items for a user, loaded in his knapsack with weight and volume constraints. The knapsack problem is solved with Mixed

Integer Linear Programming (MILP) software, which can even be obtained freely, such as LPSolve. The method can cope with complications such as interdependency between measures.

Reniers et al.[21] have also shown how managements of plants of different owners but clustered in an industrial area can optimize risk reduction options in view of domino effects, if they are willing to cooperate in safety matters and to communicate mutual risks posed by neighboring installations. Optimization solution is now sought by game theory (maximin concept) typically applied in situations of conflicting interests. Players are assumed rational and independent. This can also be applied to security problems, where conflicting parties are simply an attacker and a defender. Objective from a defender's point of view will be in that case the best strategic option, given physical/technical and/or budgetary constraints. Effectiveness of attack and defense measures expressed in loss values are formulated in multi-attribute utility equations calculating payoffs. A more extensive application of game theory is finding equilibrium situations in case both sides do not know about each other's decisions, and in the nastier case the defender sets its measures and the attacker responds. Optimization of both sides' payoffs is conducted by means of linear programming. This may now give guidance to find a theoretical best strategy for a defender in the spectrum of possible combinations of attacker and defender strategies. This game is described by Talarico et al.[22] applied to find a defender's best protection strategy for options of transportation of hazardous materials via different modes (rail, road, pipeline, or inland waterway). The Center for Chemical Process Safety[23] published a guideline on transportation risk management with a special chapter on security issues.

9.3.6 ECONOMIC UTILITY OF RISKY INVESTMENTS

In economics, decision makers use (normative) utility models to optimize investment strategy in view of risks. Utility functions are used to determine a decision maker's willingness to take risk (his/her risk appetite, with a similarity in the extreme of gaming or playing a lottery). For that matter, risks are monetized. The expected utility model developed by Von Neumann and Morgenstern[24] in the late 1940s became dominant. "Expected" is used here in the statistical sense of a weighted mean. As for example described by Chavas,[25] the question is how an individual making a risky decision behaves in view of an uncertain reward, represented by random variable, a. A factor in this decision making is the decision maker's initial wealth, w. One distinguishes the selling and bid price of a risky gain or income. In selling a risk, the price (R_s) is defined as the money a decision maker wants to receive as compensation for selling the risky prospect. Thus, the utility of obtaining a wealth $U(w + R_s)$ equals (or is rather indifferent to; symbol is $\sim$) the expected (mean) utility $EU(w + a)$. In buying a risky prospect, the price, R_b, a decision maker is prepared to pay follows from $U(w) = EU(w + a - R_b)$.

Next, is the concept of risk premium, or a person's willingness to insure for an amount, R. This is defined as the sure amount of money for which a decision maker has no preference for either the money or the risky prospect, or in formula: $EU(w + a) = U\{w + E(a) - R\}$. Hence, if $R = 0$ the person's risk aversion is neutral, but if R is positive the attitude is risk averse. The person is then willing to pay a certain amount to lower the risk by replacing random gain a by its mean. This can be further developed to a measure of risk aversion by taking a second order Taylor series expansion of $U(w + a)$ near $\{w + E(a)\}$, and subsequently a first-order one of $EU(w + a - R)$ with respect to R near $\{w + E(a) - R\}$. (By the definition of R at the point $\{w + E(a) - R\}$ itself, $R = 0$). This produces the first and second derivative of the utility function U with respect to wealth w, respectively, U' and U''. After some elaboration and substitutions, this results in

$$R \approx -0.5(U''/U')Var(a),$$

where Var denotes variance. Hence, near the point where the prospect produces on average a reward as expected, interestingly the risk premium turns out to be proportional to the risk variance, representing the uncertainty of what the prospect brings. The quotient, $r = -(U''/U')$ is known as the Arrow–Pratt coefficient of absolute risk aversion. It can be shown that under certain restrictive conditions, the relation holds not only near the point $R = 0$ but globally.

In the case the decision maker's utility function, given an initial wealth, increases linearly with the value of the prospect a as $U = w + a$, then U' is positive, but U'', and hence r and so R, are zero. This means the decision maker's attitude toward the risk is neutral, but when the utility increase is declining (U'' negative, Arrow–Pratt coefficient positive) as with most people, he/she is risk averse; only a few people are risk-seeking. This can be further elaborated. Suppose the Arrow–Pratt coefficient, r, is positive and constant. It turns out that the utility function in that case follows the equation: $U = -e^{-r(w+a)}$. This can be shown easily by (partially) differentiating twice with respect to wealth, w, keeping for that matter the prospect gain, a, at a constant value. (Author's note: this utility function is in the negative domain—utility is relative; it may be better configured as $U = 1 - e^{-r(w+a)}$. In any case, the function is convex corresponding to a risk-averse decision maker). So, at higher initial wealth, the utility of a given gain makes less impact.

If a person's initial wealth is larger, the demanded risk premium is usually smaller (rich people tend not to insure themselves). However, in such case of decreasing absolute risk aversion, r is not constant but decreasing, a utility function is less simple to derive. An additional condition becomes that the third derivative, $U''' \geq 0$. As a polynomial, a cubic equation would satisfy the conditions, but Chavas mentions two other functions describing utility in this case: $U(w + a) = (\alpha + w + a)^{\beta}$ with parameter $\alpha > 0$ and $0 < \beta < 1$, or $U(w) = \ln(\alpha + w + a)$. It would be interesting to conduct research on decision makers' preferences in process risk assessment following these types of reasoning.

In a frequently cited paper, Kahneman and Tversky[26] have argued that the expected utility theory has weaknesses, because human intuition makes bad judgments when it comes to larger numbers and uncertainty. The weighting of prospects with a probability creates the problem. They discuss a fair number of "problems" in which a person is asked to make a choice. For example, a number that seems certain is overweighted relative to one that is probable: in the case of a choice between A (2400—the authors do not mention a unit but it may be USD or Euros—with a chance of 0.33, plus 2500 with chance 0.66, and 0 with chance 0.01) or B (2400 for sure) 82% opts for B. They also identified other effects.

An aspect that also must be mentioned briefly here is "moral hazard." This is an old term referring to the phenomenon that when one is insured, one is inclined to take larger risks. In the past, it even had to do with unethical behavior, but today it just means that with respect to risk taking, behavior is less restrictive in case the event turns negative because somebody else pays.

9.3.7 DECISION ANALYSIS AND DECISION TREES

Decision making has a binary nature—we go for it or not. The primary objective of decision analysis is to identify the decision alternative that maximizes expected utility or expected monetary value with probability of occurrence as the outcome consequence weight factors. As mentioned in the introduction to this section, decision trees are a means to structure decision making taking account of the various aspects or components and motivate gathering the needed information. In computer science much use is made of binary decision trees. Binary refers to a Boolean basis: an aspect must be reckoned with or not, it is true or false, the value is one or zero (as in a truth table). In fact, it is embodying the "if-then-else" rule. Binary decision trees are very useful in development of digital systems. The tree that branches up to the final decision is a directed acyclic graph.

Here, we are more interested in decision trees that include uncertainty as required for a system approach. Ian Jordaan,[27] Memorial University of Newfoundland, described the field. Basic is the distinction of a (binary) choice node, which Jordaan calls a decision fork, followed at each branch by a probability node, or chance fork. In Figure 9.6, a simple example using point values is given of risk-based decision making use of the module PrecisionTree of Palisade@Risk MS Excel-based decision analysis software (easily found on the Internet). In a process under normal condition a light protective measure is adequate, but one needs a heavy protective measure if a coincidental process condition materializes. From the choice of protection four end states arise, called consequences or utilities because they represent values to the decision maker. Two of these end states can be classified as adequate protection, the other two as under- and over-protection. Hence, basically, the choice depends on how the decision maker perceives the chance or probability that the condition will occur. Here, an increase in occurrence probability of 0.08 to a value of 0.1 will change the preference from light to heavy protection.

An additional possibility is collecting more information about circumstances influencing the emergence of the coincidence. A value of information calculation can be cost-effective to lower uncertainty and reduce the risk of decision under uncertainty. Developing the knowledge by, for example, testing, requires funding, but the value of this information must be balanced against the gain in knowledge and reduction in uncertainty. The larger the uncertainty, the higher the value of

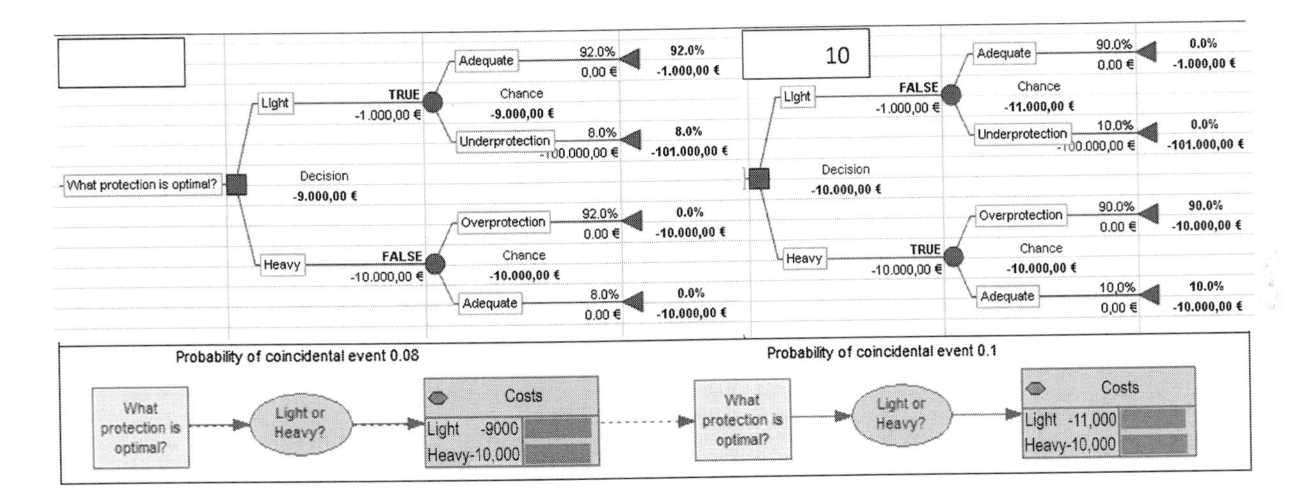

FIGURE 9.6

Top left and right: Shown is an example of a decision tree. As explained in the text, the decision is about choice of a protection system: *Light* costing €1000 but only adequate for normal process situation, or *heavy* €10,000. In case of underprotection, damage sustained by the installation is €100,000. Left: For a coincidental hazardous process condition estimated to occur 8% of times or occurrence probability of 0.08, light protection is the best choice based on minimum cost. Right: At occurrence probability of 0.1 or higher, heavy protection makes sense. The calculation to compare monetary value of the two decision alternatives was made by Palisade's Precision Tree. Bottom: The same calculation made with a Bayesian net by means of GeNIe v.2.0 of Decision Systems Laboratory of the University of Pittsburgh (see Section 7.5).

the information and at lower cost to obtain than when the uncertainty is at lower levels. This calculation will form a pre-decision node. In case testing would be very costly or not possible, improved information could be gained through estimation by employing so-called pre-posterior analysis. This is by simulating a posterior distribution by taking the prior and estimate the conditional probabilities of what you would observe in a test (this is a kind of contingency analysis). Another value of information is the value of control to reduce uncertainty of outcomes. Prior to finalizing a decision, both approaches can be simply added. The bad state probability may be strongly reduced. For this simple case, overview can be kept easily, but evaluating for example a complete bow-tie will be different. Apart from enabling to cope with complexity, the result of the calculation with the software offers clarity in team communication and later review.

By combining PrecisionTree with @Risk, or by calculation in Bayesian nets, uncertainties in the probabilities of occurrence and of consequences can be included. The Precision Tree software allows the tree to be converted to the appearance of an influence diagram. The GeNIe Bayesian Net can do this too, while it is at the same time more versatile and able to calculate a result using distributions of all relevant information.

9.3.8 DECISION MAKING UNDER DEEP UNCERTAINTY

In risk assessment the set of physically possible scenarios can be subdivided in four quarters of different combinations of "knowns" (K) and "unknowns" (U), namely: K−K, K−U, U−K, and U−U.[a] The K−K quarter concerns the risks where we know the possible scenarios, and we can estimate approximately the size of damage an event will cause and a probability of occurrence. Hence, there is uncertainty but it is bounded. In the U−K quarter we should know all this, and if we do our best to collect existing knowledge, we will know it, so that there is a shift from U−K to K−K. There are quite some means available to gain knowledge, for example, the DyPASI approach we have seen in Chapter 7. Below we are mainly treating the K−U cases in which uncertainty is large, while the U−U can only be guarded against with resilience measures, which are therefore critically needed as part of the system approach to overall system monitoring, measurement, and resilience modeling.

Decision making in case of a potential, rare catastrophic event that is believed by many as a nonexistent hazard, and apart from that, of which the estimated value of the probability is highly uncertain, always triggers much debate. In Chapter 7, Section 7.7.3 we have seen a method to use the more frequent precursor information to obtain an estimate of the rare event frequency, but often even that information is lacking, nevertheless it is important to look for it. Experienced risk analyst Tony

[a]As Wikipedia states, this distinction is well known from former Defense Secretary Donald Rumsfeld's famous words just before the start of the Iraq war in 2002 when he was commenting about the evidence of the Iraq regime letting weapons of mass destruction come into the hands of terrorist groups.

Cox,[28] University of Colorado, addressed the subject in view of decision making under deep uncertainty and presented an overview of methods developed over the last decade. It is important because a doubtful situation in an arena in which large interests are at stake can lead to highly emotional, political conflict, and large financial implications. Again, in Chapter 12 we shall discuss risk acceptance.

Cox first cited the criteria for the four levels of uncertainty: level 1 concerns a clear future, level 2 alternative futures with probabilities, levels 3 and 4, the deep uncertainty, with respectively a multiplicity of plausible futures and a fully unknown one. He further identified 10 methods to approach the problem. The first choice is to make use of the subjective expected utility (SEU) theory, in which here utility is rather disutility. In the most general case of an act, a giving rise to a set of possible scenarios, s, leading to potential consequences, c, with probabilities, Pr, is formulated as: $EU(a) = \Sigma_c u(c)[\Sigma_s \Pr(c|a,s)\Pr(s|a)]$, in which the probability values can be filled subjectively. A risk-averse, exponential utility function can be used of the type we have seen above: $u(x) = 1 - \mathrm{e}^{-rx}$, where x is a measure of wealth and r of risk aversion. The optimal act to choose is the one that minimizes the disutility. Cox lists four major obstacles in applying the theory, which essentially are all uncertainties of the elements of the risk management tuple $M = \{A, C, u(c), \Pr(c|a)\}$, where A and C are the sets of acts and consequences and the other symbols as above.

Cox[28] then goes into describing nine other methods. Superficially summarizing, these appear to have in common with the first choice, developing as many priors, plausible scenarios, or distribution, or other fitting models as possible and collecting also as many relevant data as possible. However, the way the methods develop models and arrive at a decision differ. Data may in our case be precursor frequencies as we have seen in Chapter 7, Section 7.7.3 and costs. Two main groups of decision methods may be distinguished: *robust* and *low regret* decisions. Robust decisions are based on model ensembles fitted to available data relevant for future consequences of the decision to be made now. The models may be generated by *resampling* techniques as boot-strapping (subdivide the data randomly in subsamples, fit a model to each, and average these models for a final one) and funnel down to conclusions by simply averaging forecasts. Another possibility is making use of more sophisticated learning algorithm methods as *adaptive boosting*, also called Adaboost. In contrast to resampling, in Adaboost data points, which are not well predicted by the assumed model, are given more weight (a bias). The model is then improved by fitting it to the points again and again. Experience shows that after many iterations a successful classifier can be obtained dividing points, for example, in good and bad risks. Further, *Bayesian model averaging* is rather successful. This technique uses statistical inference when one is uncertain about the right statistical model. Outputs (e.g., decision recommendations or predictions) of an ensemble of models $M_1 \ldots M_n$ are weighted according to their likelihood of consistency with observed data. The conditional probability a conclusion X is true follows from the law of total probability (see Chapter 7): $\Pr(X|Data) = \Pr(X|M_1)\Pr(M_1|Data) + \ldots + \Pr(X|M_n)\Pr(M_n|Data)$. For this expression, the models M_i must be mutually exclusive and collectively exhaustive.

In the case it is uncertain that available data are relevant, or even no data are available, the strategy must be different. Starting with plausible scenarios, one does not know how nearly correct these are. If the right model would be known, decisions could be evaluated on potential loss for the decision maker based on the value of the consequences. However, one has now to revert to comparing losses by decisions based on different models, while subsequently iterating to minimize regret. The latter is realized by weighting the models (or assigning to the models probabilities) according to yielding the decision recommendation with the best outcome, for example, minimized losses. Next, a new decision is made based on the weighted ensemble of models. The decision sequence is then minimized for regret. Further improvements are obtained by applying for the iteration artificial intelligence machine learning algorithms such as reinforcement learning (RL) of low-regret risk management policies for uncertain dynamic systems, (partially observable) Markov decision process or (PO)MDP, and the state-act-reward-state-act (SARSA) RL algorithms.

It is clear that this brief overview is only scratching the surface and that experience with these methods applied to industrial risk management decision making must be built up before any success can be claimed. There is, however, the Bayesian network method that can provide overview in support of decision making and make convenient computations based on estimates of "knowns" and more or less "unknowns." An example is the KUUUB factor network described by Fenton and Neil,[29] authors and developers of the AgenaRisk BN with whom we became acquainted with in Chapter 7. KUUUB is an acronym for K–U, U–U, and Bias. The following is an example adapted from theirs. In a company the financial losses due to undesired risk events are for the current year not precisely known yet but are estimated as a truncated normal distribution (TND) with mean 20 and variance 10. The question is to make an estimate for next year's budget in view of two new activities. Hence, three risk scenarios, called key risk indicators (KRI), are projected of probability of degrees of improvement and of degradation trend qualifications on a 7-point scale. KRI A is existing risk (weight 2.5); KRI B is an existing line with new product with different hazardous properties (weight 1.8), and KRI C is a new high hazard plant (weight 1.0), and its distribution is conditional on the precursor indicators of a previous test run. By multiplying the weighted mean distribution of trend scenarios estimate, E, with a Delta distribution, Δ, a KUUUB adjusted estimate is obtained: $K_B = E$. Delta parameter values expressing degree of uncertainty are conditional on each trend qualification according to Table 9.3.

In here is the crux: Delta is partitioned per trend qualification. Each partition is modeled as a TND with mean, variance, and upper and lower bound. The parameters are selected based on experienced expert judgment. If there is no change the Δ-value collapses to unity (or zero for variance). The compound Delta comes out as a distribution as shown in the figure. The improvements are designated K–U and thus relatively certain, while degradations are U–K and U–U, and spread is estimated to become quite large. The result can be seen in Figure 9.7. The Delta change adjusted projected loss distributions is less steep but has a long tail to higher losses, and hence a much larger value at risk (VaR). Distribution tails showing high uncertainty of outcomes to greater potential losses should be updated and followed to test measures designed to enhance organizational resilience to respond more effectively to unexpected upset

Table 9.3 Delta Truncated Normal Distribution (TND) Parameter Values Conditional on Trends

Trends/TND Parameters	Mean	Variance	Lower	Upper
Major improvement	0.1	0.2	0.1	1
Substantial improvement	0.5	0.1	0.1	1
Improvement	0.7	0.1	0.1	1
No change	1	0	1	1
Degradation	2	4	1	10
Substantial degradation	5	4	1	10

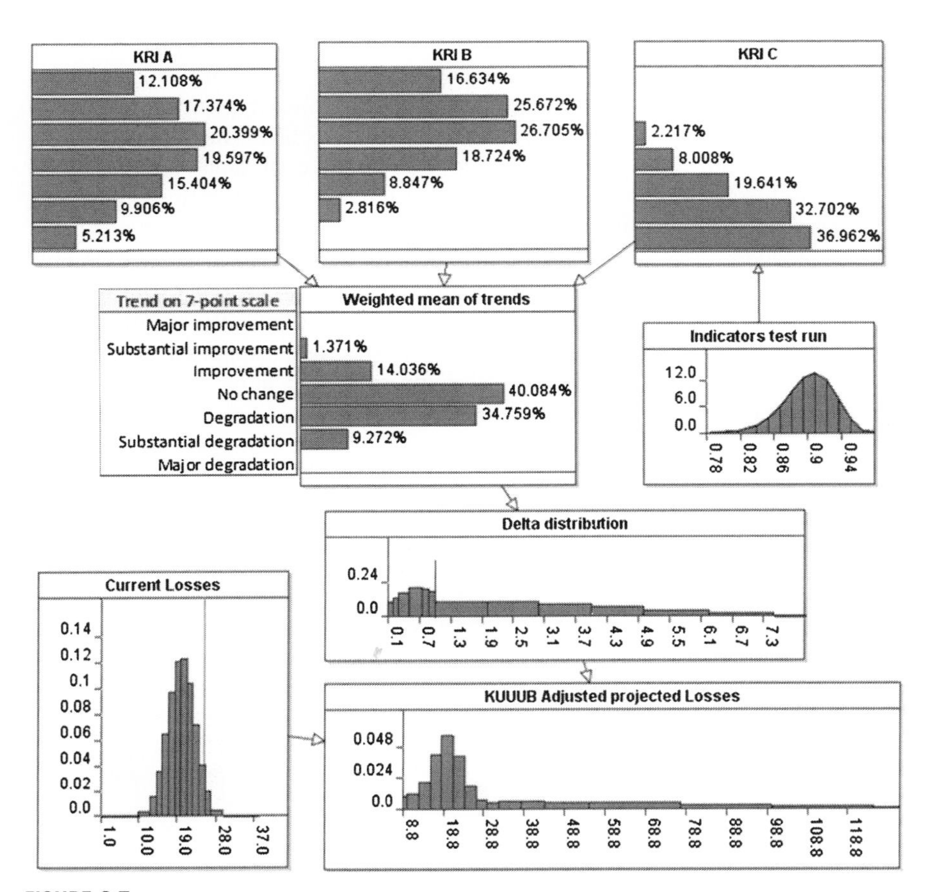

FIGURE 9.7

KUUUB factor example *(adapted from Fenton and Neil[29])* (Figure 11.15). Shown are key risk indicator (KRI) expected distribution scenarios for three plants A, B and C of improvement and degradation depending on stress and maintenance of equipment. KRI A concerns existing plant and product, KRI B an existing line with new product, and KRI C a new hi-hazard plant. The weighted mean trend estimates multiplied with Delta distributions depending on trend estimated by experts, produces the result as compared to current losses. Possible plant C risks cause a long loss tail.

events and losses. Given more information about potential precursor events and newly emerging scenarios, the well of uncertainty about unknown potential scenarios can be reduced.

9.4 SAFETY RISK MANAGEMENT IN CONTEXT

According to ISO Guide 73-2009 risk management is defined as "the coordinated activities to direct and control an organization with regard to risk." So, the concept is very broad, it can concern private and public organizations and a large variety of types of risk since following ISO 31000 risk is defined as "effect of uncertainty on objectives." In decision making on starting an activity the issue is always benefit versus cost and risk. One category of activity where this presents itself very explicitly is project management. This subject obtained much attention and there has been much written about project risk management. A general scheme of risk assessment for controlling project risk is shown in Figure 9.8. The similarity with the risk topics treated in this book is obvious. The assessment process as an incentive and as a means of communication is made clear, as well as the continuous monitoring of risk and the feedback. This is another stimulus to consider not only turnover and profit risks when discussing the balance sheet and reviewing business prospects with stakeholders such as investors and shareholders, but also to consider safety risks. Preferably, this should be in quantitative terms and in the context of internal factors such as level of competence in relation to process complexity, and of external factors such as environment and political, weather, and geographical stability. Because of the expertise involved, small companies shall have greater obstacles than large ones in doing this.

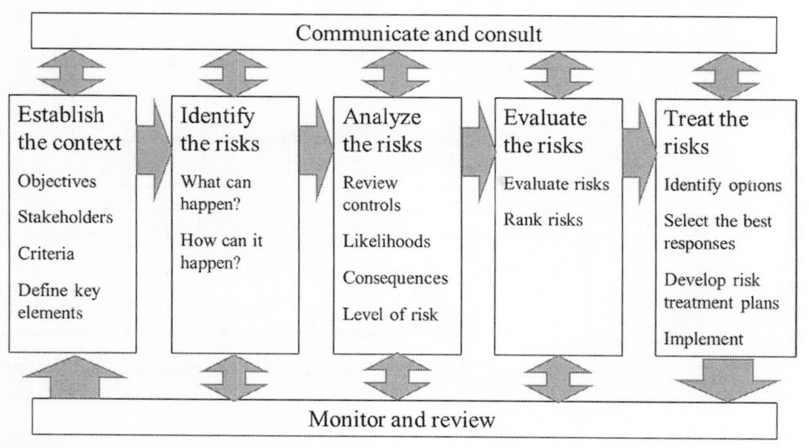

FIGURE 9.8

The general risk assessment process scheme as presented in a textbook on project management by Dale Cooper et al.[30] The similarity with plant and process risk assessment is striking.

9.5 CONCLUSIONS

Catastrophic industrial accidents, although being "rare events," may lead to the demise and disappearance of a once renowned company. Financial losses due to business interruption, liabilities, and reputation damage outstrip the direct physical damage. However, the sum of losses by smaller size incidents, which do not come out in the open, is also not negligible. It therefore pays to invest in safety measures.

An approach has been shown how returns on safety investments can be calculated and thus how a safety business case can be made. Also, methods for cost—benefit optimization of measures and for obtaining optimized limited budget allocation have been presented.

Finally, several methods of supporting decision making, qualitatively and quantitatively, including coping with uncertainty, have been explained. Straightforward cost—benefit optimization and decision analysis and decision trees offer the best perspective. For optimal decision making under deep uncertainty, which is critically important in many rare event risk assessment cases, directions were indicated how a fuzzy decision issue can be brought in better focus. At the end, the assessment process and management of safety risk was briefly compared with that of risk management in general and was found strikingly similar.

REFERENCES

1. Marsh Energy Practice, The 100 Largest Losses. *Large property damage losses in the hydrocarbon industry.* 22nd ed. Marsh & McLennan Companies; 1972—2011. https://usa.marsh.com/Portals/9/Documents/100_Largest_Losses2011.pdf.
2. Reuters, Factbox: What's BP's potential pricetag for Macondo?. http://www.reuters.com/article/2012/03/03/us-bp-costs-idUSTRE8220R320120303.
3. Kim BK, Krams J, Krug E, Leaseburge M, Lemley J, Alkhawaldeh A, et al. Case study analysis of the financial impact of catastrophic safety events. *J Loss Prev Process Ind* 2012;**25**:780—7.
4. Hansen MD. The Sarbanes-Oxley act & safety. *Proceedings of the 2008 ASSE professional development conference © 2008 by ASSE (American society of safety engineers).* www.asse.org/practicespecialties/bosc/docs/Mark%20Hansen%20Article.pdf.
5. Pasman HJ. Risk informed resource allocation policy: safety can save costs. *J Hazard Mater* 2000;**71**:375—94.
6. Maloney JH, Liley PE, Thomson GH, Friend DG, Daubert ThE, Buck E, et al. Perry's chemical engineers' handbook. In: Green DW, Maloney JO, editors. 7th ed. New York: McGraw Hill; 1997. ISBN:0-07-115448-5.
7. White Paper Addressing the Return on Investment for Safety, Health, and Environmental (SH&E) Management Programs. *ASSE council on practices and standards (CoPS) of the American society of safety engineers (ASSE).* 2010. http://www.asse.org/practicespecialties/bosc/bosc_article_6.php.
8. Veltri A, Ramsay J. Business of safety, economic analysis, making the business case for SH&E. *Prof Saf* September 2009;**9**(54):22—30. www.asse.org.
9. Feng Y. Effect of safety investments on safety performance of building projects. *Saf Sci* 2013;**59**:28—45.

10. Kip Viscusi W. *The value of life*. Discussion Paper No. 517 06/2005. Cambridge, MA 02138: Harvard Law School; 2005. http://www.law.harvard.edu/programs/olin_center/papers/pdf/Viscusi_517.pdf.
11. Bolu AG. *Economics of safety: an empirical study* (Master of Science thesis). University of Aberdeen; 2011.
12. Miller T. Variations between countries in values of statistical life. *J Transp Econ Policy* 2000;**34**(Part 2):169–88.
13. Aven T, Renn O. Risk management and governance: concepts, guidelines and applications. In: Mumpower JL, Renn O, editors. *Risk, governance and society*, vol. 16. Springer; 2010. ISBN:978-3-642-13925-3.
14. Cox LA. Decision and risk psychology: seven recent books. *Risk Anal* 2013;**33**(9):1749–57.
15. Aven T. *Foundations of risk analysis*. 2nd ed. Wiley; 2012. ISBN:978-1-119-96697-5; [chapter 5].
16. Kirkland G. video, How chevron makes decisions; 2010. http://www.youtube.com/watch?v=JRCxZA6ay3M.
17. Hazelrigg GA. *Fundamentals of decision making for engineering design and systems engineering*. © Copyright 2010 by George A. Hazelrigg. 2012. ISBN:978-0-984-99760-2, 0984997601.
18. Healthcare Business Media, Inc., KPIs and balanced scorecard, fall 2009 IDN summit, Peer-to-Peer Learning Exchange Research Reports. http://idnsummit.com/files/KPI.pdf.
19. Tappura S, Sievänen M, Heikkilä J, Jussila A, Nenonen N. A management accounting perspective on safety. *Saf Sci* 2015;**71**(Part B):151–9.
20. Reniers G, Sørensen K. An approach for optimal allocation of safety resources: using the knapsack problem to take aggregated cost efficient preventive measures. *Risk Anal* 2013;**33**:2056–67.
21. Reniers G, Cuypers S, Pavlova Y. A game-theory based multi-plant collaboration model (MCM) for cross-plant prevention in a chemical cluster. *J Hazard Mater* 2012:209–10. 164–176.
22. Talarico L, Reniers G, Sørensen K, Springael J. MISTRAL: A game-theoretical model to allocate security measures in a multi-modal chemical transportation network with adaptive adversaries. *Reliab Eng Syst Saf* 2015;**138**:105–14.
23. Center for chemical process safety *CCPS, guidelines for chemical transportation safety, security, and risk management*. N.Y.: John Wiley& Sons; © 2008. AIChE, ISBN: 978-0471-78242-1; [chapter 6].
24. Von Neumann J, Morgenstern O. *Theory of games and economic behavior*. Princeton, NJ: Princeton University Press; 1953.
25. Chavas J-P. *Risk analysis in theory and practice*. Elsevier Academic Press; 2004. ISBN: 0-12-170621-4.
26. Kahneman D, Tversky A. Prospect theory: an analysis of decision under risk. *Econometrica* 1979;**47**(2):263–92.
27. Jordaan I. *Decisions under uncertainty, probabilistic analysis for engineering decisions*. Cambridge U.K.: Cambridge University Press; 2005. ISBN:0-521-78277-5.
28. Cox LA. Confronting deep uncertainties in risk analysis. *Risk Anal* 2012;**32**:1607–29.
29. Fenton N, Neil M. *Risk assessment and decision analysis with Bayesian networks*. Boca Raton, FL 33487–2742, USA: CRC Press, Taylor & Francis Group; 2013. pp. 362–364, ISBN:978-1-4398-0910-5.
30. Cooper D, Grey S, Raymond G, Walker P. *Project risk management guidelines, managing risks in large projects and complex procurements*. John Wiley & Sons; 2005. ISBN:0-470-02281-7.

CHAPTER

Goal-oriented versus Prescriptive Regulation

10

Freedom and order are not incompatible... truth is strength... free discussion is the very life of truth

Thomas Henry Huxley, 1825–1895

SUMMARY

In case of industrial disaster the call of population to government is: "How could this happen?", "How is the safety situation in that branch?", and "Is regulation adequate?". As we have seen in Chapter 2, over the years these types of questions have been the driving forces to create, reinforce, or renew safety regulations. The name of the regulation reminds sometimes of the disaster that initiated it as, for example, in the European "Seveso" Directives. Indeed again, after the ammonium-nitrate detonation in West, Texas, on April 17, 2013, such questions arose. As indicated in Chapter 1, this explosion made the White House release an executive order on August 1, 2013, on Improving Chemical Facility Safety and Security, involving many regulating agencies. With ammonium nitrate, besides safety also security is not to be neglected. After noting that chemicals are essential to the economy but that risks adhere to these substances, the order announces the establishment of a working group. The tasks of the working group are collecting information, doing consultation, communicating, and coordinating with a multitude of government agencies and other stakeholders. It is to find out where improvements can be obtained in regulation (OSHA's process safety management (PSM) and EPA's risk management plan (RMP) rule), risk management, including emergency response. Results are forthcoming.

Prescriptive ruling tells an operator/owner exactly and in detail what is required. It specifies maximum temperatures, minimum wall thicknesses, material choices, et cetera. The basic problem of regulating prescriptively is that freedom for enterprises to find the most economical solution to comply may become impeded, while it is also felt to quantitatively increase the burden of inspecting and being inspected. In general, experience of the last decades is that the weak point is not so much the regulation but the inspection. In Chapter 2, Section 2.8.3, we have noted this already. A major cause of lack of compliance is absence, or too low frequency, of inspection, or at least of independent, external auditing. People involved in plant operations tend to lose sight of the hazards present, and nobody with

Risk Analysis and Control for Industrial Processes - Gas, Oil and Chemicals.

enforcing authority reminds them. This may be truest in small- and medium-sized companies. It also happens that when inspection occurs on different aspects at detail level, inspectors arrive at a site independently from each other, representing different regulatory bodies, which may result in contradictory conclusions. In addition, the multiplicity of detail rules and boundary conditions obscures the view on the essentials and makes life for those who want to comply rather complicated. Apart from plant personnel experiencing this as a nuisance, the system as a whole is not considered. Therefore, it is even possible that, for example, safety rules become conflicting with ones on environmental protection.

An obvious reaction to the classical *prescriptive* type of regulation is therefore to design regulation that is *goal oriented*, also called *performance based*. The rule sets the goal and the regulated party can freely decide how to achieve the objective. Whereas prescription drives production costs up, goal-based regulation may be more cost effective, and it will certainly be more innovative. Inspection shall in that case also be required but on a higher level. The inspector has to judge whether goals set are in agreement with the law and whether the complex of measures taken is adequate to pursue the objective. It requires inspectors to have a higher level of background knowledge and insight. However, this type of performance-based regulation also has its limitations and weaknesses as we shall see in this chapter.

Apart from prescriptive versus performance-based regulation, in the European Union (EU) around 2010 the concept of "smart regulation" was put forward (see, e.g., Marianne Klingbeil[1] and EU references). Similar to performance-based regulation, smart regulation aims for simpler laws. However, smart regulation's main aim is cost savings by "cutting red tape"; it is not limited to a specific area as safety but is general. On the other hand, regulation should remain "fit for purpose."

In information security compliance based protection is posed versus a risk based one. From the point of view of management managing its risks there is a certain similarity with what will be discussed in this chapter but this is less from a regulator's position. We shall briefly come back to this near the end of the chapter.

This will be a short chapter but of interest for avoiding overoptimism that the one or the other solution will be a panacea, and for watching experiences in finding an optimal balance between the two antipodes. In all it is a rather complicated issue with many aspects and writing a hundred pages on the subject would not be too difficult, but the question is whether it would contribute that much more?

10.1 BACKGROUND AND LITERATURE SOURCES

In the world of regulators the discussion on possible advantages of goal-based, also called goal-setting, goal-oriented, or performance-based regulation over the classical prescriptive type is not new and has been an ongoing process since the 1970s. In fact, in the UK the 1972 Lord Robens report[2] on health and safety in the workplace, which among others led to the founding of the Health and Safety Executive (HSE), opened the way to goal-setting and what was called employers' self-regulation. The discussion is relevant to a spectrum of economic activities in which safety and risk form an issue, such as the construction industry. As we have seen in the sociotechnical system approach in Chapter 5, in society ultimately government has to create the right conditions. This is for public safety as well as for trust of workers and public in facilities or materials used to their advantage with respect to health and safety, and to integrity of the environment. But how can a government do that when economic activity needs sufficient freedom and maneuverability to flourish in a competitive world?

If government stimulates goal-oriented regulation, it shifts responsibility for a significant part to the operator. One can immediately see here a political factor arising: laissez-faire and market economy, or in contrast a rigidly state-controlled, so-called planned economy with stringent rules. Meanwhile, humankind has been able to find out by the suffering of many people that neither extreme is optimal. Of course, within their jurisdiction a government can try to provide a "level playing field" for competing companies, but what about outside that government's jurisdiction? Global organizations, such as the United Nations, have been instrumental in some respects, but the effect of "recommendations" is limited to, for example, borders passing traffic, transportation rules, and materials. Further complicating things is the advent of new materials and products, the rising complexity of technology, engineered constructions, forms of organization, and the increasing variety of possible disasters. This contrasts with the increasing emphasis in which the public requires health and safety and the conservation of the environment to be guaranteed.

Against this background in the first decade of this twenty-first century, several papers were published, which in a way welcomed performance-based regulation but also had an eye for its limitations. Cary Coglianese et al.[3] of the Center for Business and Government of the John F. Kennedy School of Government, Harvard University (Cambridge, MA), organized in 2002 a workshop on the theme of performance-based regulation in view of what has been mentioned above: risks to health, safety, and the environment. In the series Jerusalem Papers in Regulation and Governance, a working paper by Peter May,[4] University of Washington appeared in 2010 with similar messages.

In 2011, three reports were published related to the subject, all concerning regulation with respect to offshore drilling activity, clearly in the wake of the Macondo well disaster. These are a comparison by the Canadian Pembina Institute[5] on experiences with regulation of offshore drilling in five countries: Canada, the US, the UK, Greenland, and Norway. The report covers many aspects, but with regard to type of regulation the US is mostly prescriptive, Canada is hybrid, and UK, Greenland, and Norway are performance based. Further, Amy Jaffe and James Coan (US), and Alexander MacDonald and Nicholas Crosbie (Ca) of the Woodrow Wilson International Center for Scholars[6] shared views on deepwater offshore drilling. They conclude that compared to the US, Canada is better off with its regulatory system, so their recommendation to the US is to shift further to performance-based regulation. (It should be noted here that in this respect, with the launch of the Safety and Environmental Management System (SEMS, 30 CFR §250, subpart S) in November 2011, the US made an opening toward performance-based offshore regulation). The third report was published in December 2011 by a review panel on the oil and gas offshore regulatory regime of the UK, chaired by Geoffrey Maitland,[7] Imperial College. Although the goal-setting worked well as a whole with respect to the safety case system, there was some doubt whether the well operations plan based on the risk assessment was always fully implemented by the operator. The environmental regime was governed by EU Directives and therefore more prescriptive, hence less encouraging for the operator to seek innovation and be proactive. A third concern was that the offshore was under three authorities instead of one as is the case for onshore installations in the UK. A number of recommendations were made.

In 1999 the Australian Commonwealth Government commissioned the Australian Offshore Petroleum Safety Case Review,[8] which proposes future arrangements based on the findings of an independent review team. The team found that in view of the expansion of offshore activities, the existing legal situation in Australia was rather fragmented and inconsistent with differences from state to state. The states also lacked expertise. The team favored a central Commonwealth approach with one safety authority (which now incorporates environment management and is called NOPSEMA). The team was also in favor of goal-setting legislation, of a safety case with management system, and of monitoring indicators. Australia had already adopted legislation and a management system in 1992 after the Piper Alpha disaster. In 2008 an independent team reviewed the situation again and concluded that much progress had been made and gave recommendations to resolve some remaining problems.[8]

On the occasion of the investigation of the Chevron Richmond, California, refinery pipe rupture and fire in 2012, the US Chemical Safety and Hazard Investigation Board (CSB)[9] issued in 2014 a regulatory report that highlighted a negative aspect of goal-setting regulation. This report criticized the practice of the OSHA PSM Standard 29 CFR §1910.119 Process Safety Management of Highly Hazardous Chemicals and its implementation in California. The standard was intended to be performance based, but it became largely activity based. This means that for an operator (duty holder) it is sufficient to show that the activity is performed, for example, of process hazard analysis, but not how effective the activity has been. To some extent the same is true for the Environmental Protection Agency (EPA) Risk Management Program rule. The regulation remained unchanged for many years; the state lacked skilled inspectors (organization understaffed and not well paid). The CSB described the situation elsewhere in the world where goal setting is accompanied by a safety case (a risk assessment) with the requirement to risk reduction to As Low As Reasonably Practicable (ALARP, Chapter 12), a periodic update and improvement by application of best available technology, and adequate inspection. Although the practice is unsatisfactory, the criticism is not a disapproval of goal-setting per se.

The cited literature shall be further supplemented with the experienced views of Brian Meacham,[10] Worcester Polytechnic Institute, Massachusetts, in the building/construction industry and the critical review of Andrew Hopkins,[11] Australian National University, on risk management versus rule compliance. A rather clear overview of the situation in the United States at the end of last century regarding regulatory Environmental, Health and Safety (EHS) issues in the chemical industries, in particular the difference between large and small firms, has been given by Karen Chinander et al.[12] It is felt that the overview is still valid and that the situation in Europe is not much different. We shall therefore start the discussion with some observations from this paper.

10.2 DISCUSSION

For investigating the challenges to firms becoming subjected to performance-based regulation Chinander et al.[12] asked themselves the following questions: (1) How do firms determine their level of compliance? (2) What regulatory incentives encourage firms to move beyond the "bare minimum" required by the regulation? (3) How do firms create accountability and responsibility systems in their organizations to comply with regulations or standards? (4) How do the effectiveness and cost of performance-based regulations depend on the type of risk being regulated and the characteristics of the regulated firm or industry? Thereby, the focus is on the Risk Management Plan (RMP) rule in the context of the Clean Air Act Amendments (the requirements of PSM are not explicitly mentioned in the article but will fall also under the CAAA, see Chapters 2 and 3). The authors considered by the regulation affected stakeholders in the broadest sense: the firm, its shareholders, the regulator, and third parties, such as trade unions, insurance companies, external consultants, and further the local community and the general public. As regards response strategy of firms to the regulation, three stages are distinguished: reactive (compliance driven), in which will be most companies, particularly the small ones; proactive (in view of corporate growth); and proactive (because of long-term value creation). Most small firms get stuck in the reactive stage due to lack of available expertise or resources to obtain expertise. They do the minimum to comply with the regulation, so they have installed PSM tools but do not really use them to reduce risk; instead they may commit evasive action or even play tricks to reduce the regulatory burden. Large companies, members of the Chemical Manufacturers' Association, are used to being audited, have expert employees, may be committed to responsible care, and communicate to the work force and the public about their EHS activities. For these companies performance-based regulation will likely result in a win—win situation, reducing transaction costs, while small companies in general will be glad to have prescriptive regulation.

Coglianese et al.[3] (workshop Harvard), and May[4] (Jerusalem paper) collected many comments on both types of legislation. A parallel can be drawn with design standards prescribing exactly how a certain product shall be built or a performance standard only requiring a performance level to be achieved. Prescriptions for different objectives, for instance for safety and environment, may be mutually in conflict, while prescriptions always lag behind development. Therefore, inspectors can be brought into a dilemma because due to complexity not all prescriptions can be fulfilled at the same time. Although prescriptions are rather inflexible, still some, mostly smaller size operators, like it because how to comply is exactly described and accountability is more simply fulfilled.

In contrast, goal setting assumes a goal that can be well specified and also that the performance can be measured. If goals of performance-based regulation are set too narrow, these may "overshoot the target" of providing a certain margin to select the most cost-effective solution. Wider goals, however, cannot always be clear-cut and unambiguous. If the process shall be safe, one can ask how safe? And even when the goal is clear and, for example, a maximum risk of a fatality on a certain distance is specified, performance measurement may be difficult. For example, this will occur if predictive methods, such as quantitative risk assessment (QRA) needed to show the criteria for the goal to be fulfilled, are inaccurate and uncertain. This is particularly true when it concerns rare events, the frequency of occurrence of which cannot be measured and model validation is not feasible. Operating regulated parties may then satisfy themselves by taking the uncertainty to their advantage. In such cases, enforcement may become an issue, as some years ago it occurred with the ALARP criterion in the UK where additional guidance rules by HSE[13] appeared to be necessary. (In Chapter 12 we shall see more details on this.) Another solution with hazardous process operations is to combine performance-based (process shall be safe) with *risk-informed* regulatory approach (risk controls shall be adequate). Inspections can then be carried out on having operationally effective, ISO standard and regulation-based quality and safety management (metrics of which can be monitored), and safety integrity level (SIL) standard-based protective systems. Performance-based regulation is then supported by the rules of standards that essentially are prescriptive.

So, yes, prescription has to function as a "backstop" for goal-based regulation, and "hybrids" are to be a fact of life. On the other hand, also in prescriptive regulation there can be underlying uncertainty so that prescriptive safety may be built on quicksand. Another point to consider is that changing the character from prescriptive regulation to goal setting is an exercise that needs considerable effort and often will meet resistance.

A further difficulty is that with increasing complexity of technology and requirements the number of people knowledgably participating in regulatory decision making decreases, and both public and private decision makers will have to rely partly on outside experts and professional judgment. However, trust in the latter sources is not always justified. Meacham[10] mentions a number of cases, among others as a proven case in which a fraudulent expert falsified computational code results. After 20 years of experience with performance-based regulation in the construction industry, Meacham is not very optimistic. He states that both systems have their weaknesses but in performance-based regulation, the challenges may be greater. It allows innovation, but one cannot predict in what this will result. He has seen "misaligned performance and stakeholder expectations, an inadequate level of competency for the technologies, the inappropriate use of computational models, inadequate data, and incomplete oversight schemes have been shown to result in unfortunate and costly outcomes." He also mentions a few improvement measures. Licensing of practitioners will help in case it is a matter of sufficient professionalism. Another possibility is third-party quality review and auditing, which in the process industry is done by companies such as DNV-GL, Lloyd's Register, Bureau Veritas, and many consultants. Verifying and validating risk assessment software has been limited so far to gas dispersion computational codes according to the EU SMEDIS Model Evaluation Protocol (verify theory, test robustness, and determine accuracy; see also Chapter 3), but this is practiced much too infrequently. Most effort has been spent on fire models, but according to Meacham[10] there is room for extension of validated fire model scenarios. There is further a need for performance-oriented testing over a range of conditions, which in the opinion of the author of this book holds the same for process hazards.

Hopkins's[11] analysis on decision making by rule compliance versus decision making based on risk management has much relevance to predictive versus performance-based regulation. He points to the fact that a risk assessment in depth and in quantitative sense is possible only in a relatively very small number of cases. And even if it is realized, a criterion is needed to make a decision about acceptability, which embodies a rule (criteria are discussed in Chapter 12). In most cases so far, safety is weighed based only on qualitative arguments. Then, decision making is made strongly supported by rules, certainly when time pressures intrude. So, most risk management needs rules. In many decision situations, risk ranges from almost zero to a fair probability of a catastrophe. This leaves the question: Do I have to shut the process down or not? Fulfillment of one or more criteria will make the situation clear, though, not only to the decision maker but also further down in the community involved. It is also necessary to make a distinction between an operational decision by an operator and a nonoperational one by management. The former will be much more bound to rule than the latter, which does not mean that an operator does not need to really think and only must care about rule compliance. For example, unworkable rules should be brought to the attention of management. On the other hand, risk management decisions cannot be made without rules. Hopkins quotes Webb[14]: "risk-assessment is like torturing a spy. If you do it for long enough you get the answer you want!"

Hopkins[11] mentions two examples of successful rules: the Rule of Three discussed by Hudson (Chapter 6, Section 6.5) in which three instances of doubt means "orange," equivalent to "red," a no-go. Another example is the Trigger Action Response Plan developed in Australian mines. The latter means reacting to signs that a hazard has developed. He gives the example of the need to support the roof of a mine gallery in case loose pieces come falling down. The rule shall have the miners take action when the observation is made. He further cites Hale and Swuste[15] who distinguish three types of rules: (1) rules defining a goal to be achieved; (2) rules of how a decision about an action should be arrived at; and (3) rules that define an action or a required system state. Rule type 3 leaves the decision maker the least freedom. Hale and Swuste draw a parallel of rules 1–3 in reverse order with those of skill-, rule-, or knowledge-based decisions by an operator controlling his installation, as first proposed by Rasmussen, later adopted by Reason (see Chapter 6, Section 6.3). Good engineering practice (in the US, RAGAGEP or recognized and generally accepted good engineering practice) and standards are full of rules of types 3 and 2. As examples, Hopkins shows how ignorance or absence of rules of these types contributed to causing and exacerbating the consequences of the 2005 Texas City and Buncefield accidents. Hence, a regulator referring to following standards or good practice is relying on prescriptive rules, although the overall setting of the regulation may be performance based. Therefore, the point he wants to make is that although certainly for nonoperational decision making risk management and goal setting are beneficial, the importance of prescriptive rules in safety regulation has not diminished. To some extent management policy in goal setting will be similar to that in risk-based information security, in which compliance with the regulation is a basis but by risk management an organization will adapt its protection to threats in a changing world (in which non-compliance is also a risk!). However, it does not say anything about the character of the regulation to be complied with.

10.3 CONCLUSION

The conclusion can be short because the referenced, experienced scholars all come to the same result. Although performance-based regulation brings considerable merit, prescriptive ruling cannot be missed. An optimum balance of a higher-level goal setting with underlying prescription "to force the laggards in line" as Hopkins[11] expressed it, shall be sought. So, it is not goal-based versus prescriptive but a balanced hybrid of goal-based *and* prescriptive regulation together with a system-based QRA to guide the balance. Enabling effective inspection and the way the regulation is formulated remain important issues.

REFERENCES

1. Klingbeil M. Smart regulation. http://www.oecd.org/regreform/policyconference/46528683.pdf.
2. Browne AC. Safety and health at work: the Robens Report. *Br J Ind Med* 1973;**30**:87–94.
3. Coglianese C, Nash J, Olmstead T. *Performance-based regulation: prospects and limitations in health, safety, and environmental protection*. Regulatory Policy Program Report No. RPP-03. 2002. Administrative Law Review 2003;**55**(4):705–28.
4. May PJ. *Performance based regulation, Jerusalem papers in regulation & governance*. Working Paper No. 2. Seattle: Department of Political Science, University of Washington; April 2010. http://regulation.huji.ac.il/papers/jp2.pdf.
5. Dagg J, Holroyd P, Lemphers N, Lucas R, Thibault B, Assisted by several others. In: Franchuk R, editor. *Comparing the offshore, drilling regulatory regimes of the Canadian*

Arctic, the U.S., the U.K. Greenland and Norway: The Pembina Institute; June 2011. http://www.pembina.org/pub/2227.

6. Jaffe A, Coan J, MacDonald A, Crosbie N. *The risk and regulation of deepwater offshore drilling: American and Canadian perspectives*. One Issue Two Voices, Issue 14. October 2011. http://www.wilsoncenter.org/sites/default/files/One%20Issue_14_Offshore_FINAL.pdf.
7. Maitland G. *Offshore oil and gas in the UK – an independent review of the regulatory regime*. December 2011. https://www.gov.uk/government/uploads/system/uploads/attachment_data/file/48252/3875-offshore-oil-gas-uk-ind-rev.pdf.
8. Australian Government, Department of Industry, Sciences and Resources, Petroleum and Electricity Division. Australian Offshore Petroleum Safety Case Review, Future Arrangements For The Regulation Of Offshore Petroleum Safety (undated, according to the Catalogue of the National Library of Australia presumably 2001), http://www.nopsema.gov.au/assets/document/Future-Arrangements-for-regulating-Offshore-Petroleum-Safety.pdf; in 2008 followed by Department of Industry, Resources and Tourism, Review of the National Offshore Petroleum Safety Authority Operational Activities, February–March 2008, Ognedal M, Griffiths D, Lake B. Report of the Independent Review Team, http://www.industry.gov.au/resource/Documents/upstream-petroleum/safety/Report%20Independent%20Review%20Team_Final.pdf.
9. U.S. Chemical Safety and Hazard Investigation Board (CSB), Chevron Richmond, California refinery pipe rupture and fire, Richmond, California, August 6, 2012, Report No. 2012-03-I-CA, October 2014.
10. Meacham BJ. Accommodating innovation in building regulation: lessons and challenges. *Build Res Inf* 2010;**38**(6):686–98.
11. Hopkins A. Risk-management and rule-compliance: decision-making in hazardous industries. *Saf Sci* 2011;**49**:110–20.
12. Chinander KR, Kleindorfer PR, Kunreuther HC. Compliance strategies and regulatory effectiveness of performance-based regulation of chemical accident risks. *Risk Anal* 1998;**18**(2):135–43.
13. HSE, U.K. *ALARP suite of guidance*. Defence Health Safety & Environmental Protection; May 2013. Leaflet 6, Demonstrating that Risk is ALARP, JSP815 Leaflets, November 2013, http://www.hse.gov.uk/risk/theory/alarp.htm., https://www.gov.uk/government/uploads/system/uploads/attachment_data/file/255801/20131105-JSP815-Leaflet6.pdf.
14. Webb P. Prescription – a step on the road to dependence or a cure for process safety ills? In: *Hazards XXI conference, IChemE, Manchester, November 2009*, ISBN: 9780852955369.
15. Hale AR, Swuste P. Safety rules: procedural freedom or action constraint? *Saf Sci* 1998; **29**:163–77.

CHAPTER

The Important Role of Knowledge and Learning

11

A little knowledge is a dangerous thing
Alexander Pope in *An Essay on Criticism*, 1709

SUMMARY

It is evident that knowledge of hazards and risks present at a location is crucial to maintain its safety. Only with knowledge, insight of a situation, and a common sense notion can adequate measures be taken to solve safety problems or to stay outside the problems. Knowledge is needed to devise and set rules and to train for skills. Theoretical and practical knowledge is next to skills, behavior, and values an important ingredient of competence. Not only "know-how" but also "know-why" is required to cope with complexity and apply a system approach. In Figure 3.11 of Chapter 3, we have shown how *general* process safety knowledge can be structured: (1) material properties and test methods, (2) system safety and analysis methods, and (3) process technology and operation, the latter including human factor/organization/management. All of this comes under the umbrella of a system approach to risk assessment. Meanwhile, as a matter of progressive insight, risk assessment shall be encompassed by an analysis of resilience. To be rooted, safety knowledge needs a general theoretical basis of mathematics, statistics, physics, chemistry, engineering, and the humanities. So, we are distinguishing here three layers of knowledge: (1) general basis, (2) process safety including risk assessment/management, and (3) specifics with respect to the own process. Depending on the task level, parts or a whole of this knowledge shall be acquired by the people involved. This occurs in a learning process, which may take many years.

Although designers may never see and experience the result of their designs in operation, their decisions are important for how plant safety materializes. So, they need thorough background knowledge to ensure the highest possible extent of inherent safety, although cost constraints will often set a limit. On a daily basis, plant staff is involved in operating the installations safely, so their need of knowledge is obvious. Going up the hierarchical level, required knowledge will be broader and perhaps less detailed, but sufficiently present to ask the right critical questions when decisions must be made.

Indeed, besides general awareness and knowledge of methods and process know-how, *specific* knowledge on one's own installations and their behavior is required. This knowledge is used primarily to interpret data provided by signals from sensors and instrumentation in general, and other information (e.g., noises, smells, visible appearances) to be aware and control the instantaneous and technical state of a process and its equipment. In addition, for safety it is imperative to recognize within this information flow weak predictive signs, the *weak signals* that are warning for future trouble and possible strong damaging effects. For the longer term and the state of affairs of the organization, to these technical signals shall be added characteristic safety indicators continually monitoring the organization and probing the health of safety management. In addition, there will be a stream of information arriving from outside. The knowledge basis set out above will be needed for detection and recognition of the weak signals and understanding the significance of collected information. To discern some pattern as significant in association with an undesired event in hindsight is easy, but in foresight and with the convincing power of prediction to others, in particular to supervisors and management, one needs experience. Hence, to the knowledge complex above shall be added learning from experience with rare violent events, or at least low probability intense events, the signals they emitted and their technical, procedural, and organizational causation.

But how can we get the latter experience? In an ideal case of perfect safety we do not obtain any experience of this kind. And, if it has never been seen, it cannot be that bad, right? This is especially true for young people, who have never been involved in an accident situation and have never been physically injured when playing violent computer games and watching violent movies. The only ways to obtain this kind of knowledge are to analyze unusual behavior, near-misses, and learn from accidents elsewhere (case histories). Although it will help to have taken note of a case once, extracted knowledge may not endure. Even when the knowledge becomes ingrained in rules, there is no guarantee that it will not be ignored later when the cause has been forgotten. So, we need a system or model in which information that we extract from cases can be stored and easily retrieved, for example, for design, operational safety meetings, personnel training, but also for risk assessments.

In a variation on the words of Donald Rumsfeld (US Secretary of Defense in 2002) we work with known "knowns," and by further learning we bring unknown "knowns" within our knowledge realm. Moreover, in safety and certainly in risk assessment we have to deal also with known "unknowns," and even with unknown "unknowns," which can be catastrophic. So at least, we have to minimize the "unknowns" by lowering uncertainty through process monitoring, searching for potential precursor events, consulting databases about what has occurred elsewhere and updating through an overall system approach. And because of the remaining "unknowns," the resilience approach with also the principles of the high reliability organization (Chapter 7, Section 7.9.2) has now become topical. Accident databases are visited by people already having a clear query, and literature is consulted with a positive effect only when one has a notion of the problem and a curiosity for it to be solved. So the search process requires experience to go for a consultation. Although we have software that embeds much knowledge, a reliable system based on that knowledge that automatically warns for hazard when a certain situation occurs may only be available for local conditions but not yet for a system as a whole.

In the absence of accidents there is a tendency to underestimate the importance of safety management. This is inherent to human nature with a focus on progress, where maintaining safety is just paying effort for not falling back. Time spent on safety is considered easily as time lost. This is not only in industry or government but also in a young generation that is selecting what education to follow. If in addition, process safety is considered to be ranking second, a side issue, and expertise in that sector is not paid well, and not important for a career, there is no incentive to take courses in that direction. This was the development in the first decade of this century both in the US and Europe. There is a slight but noticeable reversal, certainly when it is brought under the less specific roof of risk management.

Pascale Sagnier and Maxime Le Floch[1] of AXA Investment Managers reviewed the trend of availability of experienced engineers for the oil and gas (O&G) industry, and they note the risk of a widening gap between demand and supply. This is especially true for middle management petrotechnical engineers, which around 2010 has become acute by the upcoming wave of retirements. After such dramatic events as the demise of the Deepwater Horizon offshore drilling rig in 2010, not only governments but also investors are more alert to risks induced by lack of insight and technical skill. The authors' analysis of the environmental, social, and governance (ESG) framework led them to conclude that in the second half of the 1990s on short-term financial grounds a "stop-and-go" policy in recruitment occurred. By this, young people were discouraged to take the education required for designing and operating O&G installations. Complex technological processes can house tremendous risks in case decision-making people do not have the background and the drive to learn from mistakes and mishaps.

Inference and forecasting in science and engineering require probabilistic methods, which should be emphasized with more examples throughout the engineering curriculum. Disciplines become cross-fertilized, and the body of generally applicable methods is growing. Depth of conceptual thinking, methodological sophistication, and use of mathematical statistics are on the increase. Although the importance may be seen, in current engineering courses still little can be found on probabilistic methods, good exceptions aside. There will always be a gap in available new knowledge as a result of research activity improving methods and the application of knowledge in the field. However, this gap should not grow as some feel it does now. In the first place, industry should keep sufficient time reserved for key employees to absorb new knowledge, but at the same time with the expanding body of knowledge, it becomes more essential to include in the curricula next to existing approaches also the fruits of new, useful methods to solve industry problems.

11.1 THE NEED FOR STRUCTURED KNOWLEDGE

In design identifying what can go wrong, also in a system approach by applying new tools, described in Chapter 7 as STPA (system-theoretic process analysis) and BLHAZID (Blended Hazid), and making use of automated HAZOP (hazard and operability study) techniques, in the end still depends on the quality of human knowledge. The same holds in an operational situation for recognizing events and signals as "precursors." In Chapter 7, notably in Section 7.7.3, we have seen that recognizing events as precursors in a system's event sequence leading to an array of possible consequences, made visible in an event tree, can assist greatly in identifying and quantifying potential disaster events and their probabilities. But also the "message" of less severe disturbances and deviations both technically and organizationally should be understood. To that end knowledge shall be structured, consolidated, and made transferable to apply in predictive models, methods, and simulations. As already set out a few times before, in process safety and system safety there are many facets to cover. Because the field is that broad, in the "specialism" of safety expertise inevitably subspecialisms will arise. This enables one to apply the expertise, to make progress in the field, and to transfer knowledge to other users and students. It is only by systematic structuring of knowledge that new experiences, data, and information can be fitted and packed into storable and transferable data, models, and methods, in principle usable by anybody and leading to wise decisions. This is fully in accordance with the so-called DIKW model, the Data-Information-Knowledge-Wisdom hierarchy or pyramid.

Obviously, the humanities and the technical sciences form rather different disciplines, so that knowledge on human factors, organization, and management facets is represented typically by a different community than the technical aspects represented mainly by a mixed engineers' community. The best "coatrack" for a knowledge structure of process *risk* in the technical sense may be the six steps in a risk assessment process as shown in the Ishikawa diagram of Figure 3.27. For the humanities, parts of some headings in Chapter 6 give guidance: human factors, human reliability, behavior-based safety, safety culture (climate and attitude), organizational structures, and management systems, all represented by psychologists, sociologists, and organization/management experts. A system approach, risk assessment, and resilience analysis will tie it all together. In the next section we shall look from a slightly different perspective, namely what causes advances in the field of process safety.

11.2 KNOWLEDGE SOURCES AND RESEARCH

Knowledge sources range from general to specific. In the general sources much progress has been obtained by advances in computing and the humanities, as illustrated in Figure 11.1. Applying the tools and methods more specifically for process safety results in advances, for example, in prediction of dispersion of gas releases, vapor cloud, gas, and dust explosion damage, progress of various types of fires, collapse of constructions, et cetera. Knowledge of all the chemical and physical

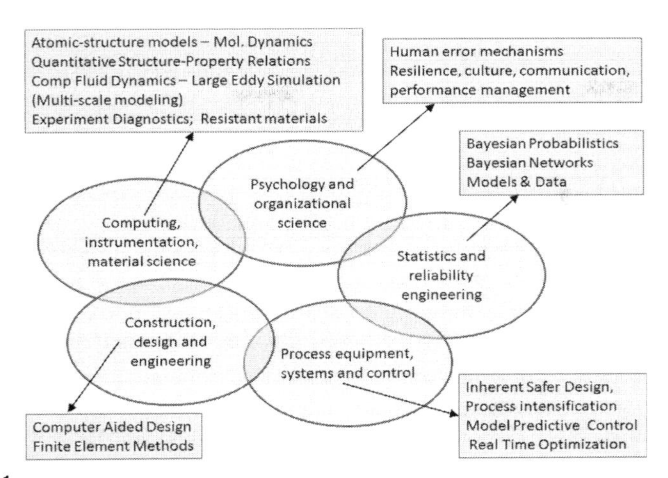

FIGURE 11.1

Benefits for process safety from advances in various areas of science and engineering of which computing and instrumentation are most important. After the Research Agenda[14] and adapted for the review of the past, present and future of the European loss prevention of safety promotion in the process industries by De Rademaeker et al.[20]

conversions enables improved countermeasures such as timely detection, cooling, diluting, closing, shutting down, protective deluge, and more.

Even more specific are the information and the data from accident case histories and failures. In the last few years accident reports of the Chemical Safety Board in the US and the Health and Safety Executive in the UK have become excellent sources. Books by Trevor Kletz[2] and Roy Sanders[3] contain many valuable stories of processes that went wrong and how they went wrong. Beside the books, there is the *Loss Prevention Bulletin*[4] published by the Institution of Chemical Engineers (IChemE) in the UK. Since 1974 with six issues per year, it contributes to process safety knowledge by publishing case histories and technical, analyzing articles. Equipment failure histories are usually extracted and collected in databases, such as the OREDA database (see references Chapter 3). These data fit into the predictive distribution models developed in reliability engineering on time-to-failure, availability, time-to-repair, et cetera.

Learning from case histories appears to be more difficult. Reading one history may leave an impression, but after the tenth the first will already be lost. The world knows several accident databases, in part also mentioned in Chapter 3. If not paid for by a government, however, their existence is fragile. The problem of learning from case histories has been extensively treated by Sam Mannan and Simon Waldram.[5] Most students in engineering, even in chemical engineering, only get a little education in process safety or even nothing at all. Later, when they need the knowledge, they do not know to find it or how to use the knowledge sources. And even if they do, there is so much variety—no two accidents are the same, nor are the entry systems of two databases the same—that without treatment of the data to a higher abstraction level in a system approach they may disregard the source as useless for their case. The authors make a plea for a worldwide effort to establish a unified database under effective leadership. This wish may to some extent be relaxed by the similarity-based search algorithm built into the DyPASI tool mentioned in Chapter 7, Section 7.3.1.

Even when an incident occurs in one's own plant, the adage that "company memory is short" will prove once again to be true. In France, *retour d'expériences* or learning from experience for high-hazard industries, also called experience feedback, obtained much attention. Bringaud[6] provides an overview of requirements to be successful in implementing a learning process in the organization. People first have to see the importance of it; further, a policy of "no sanctions" has to be applied, while a comprehensive approach is needed. The latter means that real deviations from a procedure in a given case shall be surfaced and that no fuzziness hides important information. Finally, it shall be made possible to discuss differing points of view openly and the approach shall be global and integrating. However, it is felt that for further improvement four fields must be enhanced: one related to motivation of the organization and management, one to storage and optimum use of data, deployment of the data, methods of a comprehensive approach, and how to support implementation.

To deepen *retour d'expériences* Eve Guillaume,[7] working with Andrew Hale in Delft, used accident cases to study how "weak signals" can be found and used in a better way. With weak signals, Guillaume meant signs of upcoming process disturbance. So, she tends to consider the more narrow subset of process-specific signals, while Luyk, mentioned in Chapter 5, Section 5.3.3, as part of company resilience management, looked at a broader signal spectrum. Guillaume's weak signals originate in the first instance from information collected in the operational daily routine state of affairs for keeping the process running. Relevant signals concern both technical operational anomalies as well as behavioral following procedures. Learning to recognize warnings that these signs can contain, helps operators to avoid abnormal situations and incidents. She based herself on the so-called "Delft model of risk control," that is, risk identification, building accident scenarios, defining countering barriers, which shall be monitored on effectiveness during their life cycle and shall be provided with sufficient resources to operate adequately, and finally continual learning and improving safety management. To detect patterns that you are not expecting in a flow of information is psychologically difficult. Also, in doubt, one is not inclined to admit suspicion of a developing high-consequence event. Besides this mental blocking or anchoring, there are other filters such as bureaucracy, gaps between hierarchical levels and fragmenting, and badly communicating "stove pipes" or "pillars" in the organization. There is further a tendency to structural secrecy of suppressing threat information, and there are of course work pressures that will block passing information on. In addition, there is the effect that if a signal does not fall into a recognized class it is ignored, while its evidence may become exaggerated (amplified) if it fits into a known category, with a risk of generating a false alarm. (Defense intelligence organizations have shown many instances of these mechanisms.) Information from recognized signals but failed to be analyzed for underlying causes of an anomaly can also be considered as lost, and even when it is sorted out but not adequately stored

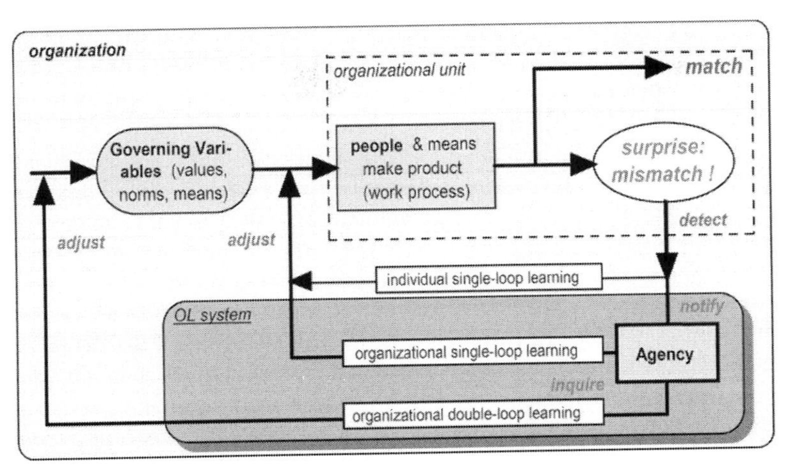

FIGURE 11.2

Scheme showing the difference between individual and organizational learning, and of the latter the difference of single versus double loop. Double-loop learning embeds the learned in an organization's values, norms, and means according to Koornneef,[8] inspired by the work of Argyris.[9]

into an organizational double loop learning it will still not maintain value. The double-loop learning is explained in Figure 11.2, which Guillaume borrowed from the dissertation of Floor Koornneef,[8] also of Hale's group, who in turn was inspired by Argyris's work; see, for example, Argyris and Schön[9] or Argyris's earlier publications.

By organizing focus group meetings in different companies, Linda Drupsteen and Peter Hasle[10] studied what the obstacles are to learn from own incidents. Common reasons are lack of priority, and of management commitment. Hence, no time is made available to acquiring information and conducting an investigation. Often the stage of reporting is not reached, so that planning intervention and acting do not even get close; in the daily workload no sense of urgency is generated, sometimes because the incident is thought to be an exclusive one-time event. And if it comes to implementation of measures, it is mainly technical and not improving culture. In a later paper Linda Drupsteen and Jean-Luc Wybo[11] defined indicators for the propensity of an organization to learn. Grosso modo these are identical to leading performance indicators.

11.3 KNOWLEDGE MANAGEMENT

Some years earlier Pasman[12] had also asked himself how effective case histories are, especially for extracting consolidated lessons, and whether we have made any progress in the right direction. The answers then were not very encouraging either. We not only need case histories to think of mechanisms and cause–consequence chains to help us identify scenarios under our own reach and to prevent those from occurring. But we will also need a modeling language that breaks the case down into elementary parts of damage potential, initiating action/trigger, preventive and protective barriers, and vulnerable receptors. In an ideal case, in design software programs as a "spelling checker" pop-ups can appear, drawing attention to hazards that will be introduced with inserting a certain component, material, or substance into an already defined environment. Development is certainly going in that direction with smart P&IDs, various process simulation models we have encountered in previous chapters and with HAZOP automation, but it is still unclear whether this goal will be achieved. Not going that far but still very useful would be to validate hazard identification tools against described cases. The difficulty there will be that many failures are latent and attributable to management, and that these do not appear explicitly in most case histories, although the US Chemical Safety Board and the UK Health and Safety Executive have developed in this respect much more complete reports on serious cases than on average. So to really learn, we shall also have to universally adopt an accident investigation, or rather root cause analysis method, to capture management failures such as Tripod Beta, although this does not classify management failures in standard terms. Milos Ferjencik[13] relates the cause—consequence chain in a scenario resulting in an incident to root cause deficiencies in the safety management system represented by the elements in CCPS′s Risk Based Process Safety guideline (see Chapter 3, Figure 3.3).

Knowledge management goes further than just organizing the primary sources. It is also more than preventing "owners" of knowledge incorporated in different groups/divisions/departments of an organization resisting or even refusing to share it for a common goal. Knowledge management has a strategic aspect. The flow diagram of Figure 11.3 shows how the knowledge investment "pump" keeps the knowledge reservoir filled. The level indicator is measuring its quality and quantity, so that one knows what knowledge it possesses, and when short of it, the flow can be increased. This second-order knowledge, however, must be steered by a research program that will find out as third-order knowledge that one should know.

In 2011 on the initiative of the Mary Kay O'Connor Process Safety Center, a global meeting was organized of process safety academia resulting in drafting a research agenda.[14] A priority list of topics was developed, and trends in the various topics identified and described. Further, a policy was conceived of how process safety could be promoted and brought under the attention of governing bodies and board members. Unfortunately so far, the response from those sides has been minimal, although due to the Deepwater Horizon tragedy in the US the foundation of the Ocean Energy Research Institute will certainly stimulate development of process safety methods and technology, while the demand for safety cases will boost risk assessment know-how. In Table 11.1, the topics priority list is reproduced with, left, topics of technical safety and, right, human and organizational ones. Topics with a strong relation have been brought together. Topics 3—7 cover common ground. Topic 17 is in view of the widening gap in theoretical knowledge between academia and industry process safety experts.

Meanwhile, German industry took the initiative of sponsoring a Center of Safety Excellence as part of the Karlsruhe Institute of Technology to conduct process safety

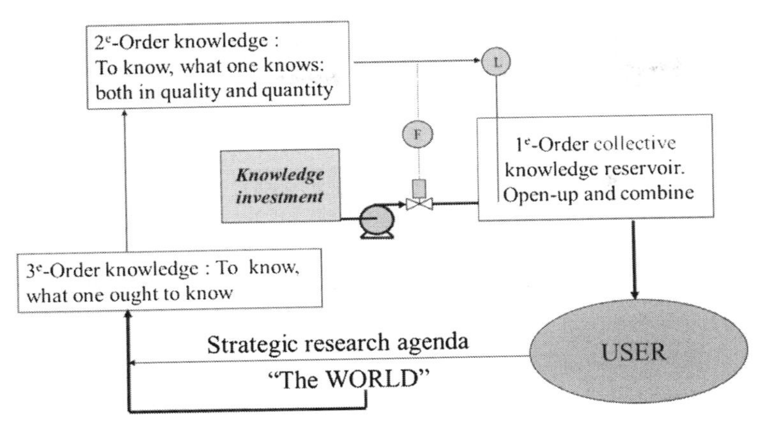

FIGURE 11.3

The knowledge management concept as seen from the top of an organization: various levels of knowledge to enable control of the process. *F* is flow control and *L* level control, according to Pasman.[12]

Table 11.1 Priority of Topics Research Agenda Twenty-first Century[14] and Theme Combinations. Topic numbers are according to priority and refer to the ones in the Research Agenda

Topic No.	Technical Safety Topics	Topic No.	Organizational Safety Topics
1, 11	Hazardous phenomena, properties of substances	8	Process and occupational safety
2	Inherently safer design	9, 10	Human factors, safety management, safety culture
12, 13, 18	Safety technologies, protection layers, drilling	14, 15, 17	Knowledge transfer, learning, standards, easy methods
3, 4, 5, 16, 19	Risk assessment, consequence analysis, NaTech (natural threats to plant)	3	Risk management, decision making
6, 7	Complex systems, resilience	6, 7	Complex systems, resilience

research starting in 2015. The Center is housed in the Fraunhofer Institute of Chemical Technology in Pfinztal, Karlsruhe. In 2013–2014 the Institution of Chemical Engineers founded the IChemE Safety Centre in Australia. The Centre for Risk Management and Safety Science at the Yokohama National University is a successful example of maintaining a safety knowledge network among Japanese universities, institutes, and industry. China, Columbia, and others are expected to develop such centers.

11.4 SAFETY EDUCATION AND TRAINING

Given the complexities of hazard phenomena and ways to prevent and to protect against those, it is essential that students in engineering, who will play a leading role in design, construction, and operation of industrial plants, will be educated in the basics of process safety, environmental protection, and sustainability. Experienced industrial process safety engineers retire. Meanwhile, industrial safety has become sophisticated and a science or discipline of its own with many facets, as we have seen in this book on several occasions. Many organizations offer specialized courses on selected safety topics, but without an academic level in which education and research form a synergy, knowledge advancements in depth and breadth will weaken. Lack of time to attend refresher courses on a sufficient level based on system thinking will further widen the knowledge gap between industry experts and academia noted above.

Where at least sustainability has a reasonably high profile, process safety does not. The Center for Chemical Process Safety of the American Institute of Chemical Engineers (AIChE-CCPS) saw the need early on and established a standing Safety and Chemical Engineering Education (SACHE) subcommittee. In 1999, Mannan et al.[15] made a plea for more emphasis on safety in chemical engineering curricula. In view of the relatively large number of serious accidents in the US in the first decade of the twenty-first century, the Accreditation Board for Engineering and Technology (ABET) in 2012 modified its criteria for chemical engineering. The program must now demonstrate that graduates are able to design and work with process equipment taking account of safety, health, and environmental requirements both from a technical as well as a regulatory point of view.

In various European countries, chemical engineering associations are active in education, and the European Federation of Chemical Engineering (EFCE) has a Working Party on Education. However, in view of retirements the demand for process safety engineers went up, while at the same time university resources dwindled, and in some countries chairs in safety disappeared. During the eighth European Congress of Chemical Engineering (ECCE 8) in Berlin in September 2011, a special two-day conference was held on the topic of process and plant safety competence of which proceedings have been published by DECHEMA via ProcessNet[16] and later this was followed by a model curriculum. IChemE has a program of course accreditation of chemical engineering degrees and accredited about 60 universities in 14 countries. IChemE published a guide[17] for university departments and assessors. This guide requires accredited universities to pay ample attention to sustainability, safety, health, and environmental issues and to teach a core program of process safety.

As safety teachings do not have the reputation of being thrilling, to say the least, it would be optimal if process safety engineering problems could also be mixed in with relevant applications of courses on thermodynamics, kinetics, reactor engineering, unit operations, equipment engineering, process technology, et cetera. This depends much on the lecturers involved and also on the extent to which it can be cooperatively and properly organized and thereby effective.

Generally, teaching *risk management* does appeal because it provides the feeling to the student of getting a grip on the problem and an overview of the problem field. Risk quantification requires a good insight, mastering of statistics and data analysis, as we have seen discussing the tools in Chapter 7. However, the investment in effort and time to develop and work with logical and predictive models using scarce data and delimiting uncertainty has a much wider application in life than just process safety engineering. Therefore, usually this topic raises interest. An example of risk management teaching is at the University of Queensland, Australia, in its fourth-year chemical engineering course (as personally communicated by Professor Ian Cameron).

Models, experiments, and videos demonstrating the dynamics of phenomena, such as gas dispersion, deflagration, and flame ball expansion in a closed vessel or an open pipe and blast propagation, help greatly to obtain a feel for physics and dynamics. Process simulation is another tool that develops quickly. It is already being applied and will become even more a great help in instruction. Shimon Eizenberg et al.,[18] Ben-Gurion University, Israel, show the advantages of conducting a HAZOP supported by a process model developed in Matlab in an example of a rather complex batch process of octanol-2 oxidation. The model makes the transient changes by invoking visible guide words.

Even more extensive is the use of simulation for educational purposes in the work reported by Tiina Komulainen et al.[19] These authors present a broad overview of the large variety of process simulation models used for education described in the literature. There is, for example, a distillation column simulated with Matlab, HYSYS, Aspen Plus, Visual Simulation Model, and Virtual laboratory. Their own work concerns a dynamic distillation simulation with D-SPICE® software. After defining learning goals for undergraduate and graduate courses, and students becoming familiar with the distillation, both FMEA (failure mode and effect analysis) and HAZOP analysis can be performed. A second example is a three-phase separator in use on oil offshore platforms, and a third a configuration of an upper and a lower tank, which by simulated controls should maintain a certain liquid level. Control algorithms and safety procedures can be checked.

As regards training and retraining (refreshers), there is an abundance of courses and workshops available organized by a variety of organizations for various topics: HAZOP, LOPA (layer of protection analysis), gas, dust, and vapor cloud explosions, QRA (quantitative risk assessment), accident investigation/root cause analysis, and safety management, to name a few. For training crew in operational or maintenance work, the technique of "augmented reality" by computer-generated pictorial or other input has proven to be very effective.

11.5 CONCLUSION

To maintain a high safety level in a complex process plant requires a broad spectrum of specialized expertise. In part this shall be with the staff in a company itself, for another part it will be called in when needed. However, knowledge is growing and the awareness of what one ought to know requires in fact one's own effort in research, as we have seen in Figure 11.3. This is one of the reasons why it is efficient that education and research go hand in hand.

Incident case histories are worthwhile to be stored and investigated, but the existing infrastructure is rather fragmented and fragile. Moreover, without extracting the information to a higher abstraction level and representing it in models, the usability of the information is rather limited. Time pressure and workload further impede learning from own incidents.

Industrial safety education within engineering curricula is crucial. The body of safety knowledge has grown over the past 50 years tremendously in depth and breadth. Also, the concepts need time of exposure to become ingrained in people and "safety made second nature" as the motto of the Mary Kay O'Connor Center reads. In teaching, there is quite an influential role for process simulation. Risk management teaching in a systems framework exhibits attractiveness.

REFERENCES

1. Sagnier P, Le Floch M. *Mind the gap: experienced engineers wanted, ESG insight AXA investment managers responsible investment*. March 2012. http://www.fundresearch.de/sites/default/files/partnercenter/axa-investment-managers/news_2011/20120608_axa_esginside_brochure_en_.pdf.
2. Kletz TA. *What went wrong? Case histories of process plant disasters and how they could have been avoided*. 5th ed. Elsevier Butterworth-Heinemann; 2009. ISBN 13: 978-1-85617-531-9; Still Going Wrong! Case Histories of Process Plant Disasters and How They Could Have Been Avoided, Elsevier Butterworth-Heinemann, 2003, ISBN: 0-7506-7709-0.
3. Sanders RE. *Chemical process safety, learning from case histories*. 3rd ed. Elsevier Butterworth-Heinemann; 2005. ISBN-13:978-0750677493, ISBN-10:0-7506-7749-X.
4. Loss Prevention Bulletin, https://www.icheme.org/lpb/about-loss-prevention-bulletin.aspx.
5. Mannan MS, Waldram SP. Learning lessons from incidents: a paradigm shift is overdue. *Process Saf Environ Prot* 2014;**92**:760–5.
6. Bringaud V. *Proposals for the establishment of an operating experience feedback organization*. In: Steenbergen RDJM, et al., editors. *ESREL 2014, safety, reliability and risk analysis: beyond the horizon*. London, Amsterdam: © 2014 Taylor & Francis Group; September 29–October 2, 2013, ISBN 978-1-138-00123-7. p. 509–16.
7. Guillaume EME. *Identifying and responding to weak signals to improve learning from experiences in high-risk industry*. Dissertation Delft University of Technology; 2011, ISBN 978-90-8891-264-1.
8. Koornneef F. *Organised learning from small-scale incidents*. Dissertation Delft University of Technology, Delft University Press; 2000, ISBN 90-407-2092-4.
9. Argyris C, Schön DA. *Organizational learning II, theory, method and practice*. Amsterdam: Addison-Wesley; 1996.
10. Drupstèen L, Hasle P. Why do organizations not learn from incidents? Bottlenecks, causes and conditions for a failure to effectively learn. *Accid Anal Prev* 2014;**72**:351–8.

11. Drupsteen L, Wybo J-L. Assessing propensity to learn from safety-related events. *Saf Sci* 2015;**71**:28–38.
12. Pasman HJ. Learning from the past and knowledge management: are we making progress? *J Loss Prev Process Ind* 2009;**22**:672–9.
13. Ferjencik M. IPICA_Lite - Improvements to root cause analysis. *Reliab Eng Syst Saf* 2014;**131**:1–13.
14. Process Safety Research Agenda for the 21st Century, A policy document developed by a representation of the global process safety academia, October 21–22, 2011, College Station, TX, USA, p. 80, ISBN:978-0-9851357-0-6, http://www.psc.tamu.edu/library/center-publications.
15. Mannan MS, Akgerman A, Anthony RG, Darby R, Eubank PT, Hall KR. Integrating process safety into ChE education and research. *Chem Eng Educ* 1999;**33**:198–209.
16. Process Safety Competence – European strength degrading to weakness? Booklet on the ECCE 8's special session, on process and plant safety, ISBN:978-3-89746-130-7, http://www.processnet.org/en/Documents-p-1000036.html.
17. IChemE (Institution of Chemical Engineers). Accreditation of chemical engineering degrees, A guide for university departments and assessors, based on learning outcomes, Master and Bachelor level degree programmes, http://www.icheme.org/accreditation.
18. Eizenberg Sh, Shacham M, Brauner N. Combining HAZOP with dynamic simulation—applications for, safety education. *J Loss Prev Process Ind* 2006;**19**:754–61.
19. Komulainen TM, Enemark-Rasmussen R, Sin G, Fletcher JP, Cameron D. Experiences on dynamic simulation software in chemical engineering education. *Educ Chem Eng* 2012;**7**:e153–62.
20. De Rademaeker E, Suter G, Pasman HJ, Fabiano B. A review of European loss prevention and safety promotion in the process industries. *Proc Safety Environ Prot* 2014;**92**:280–91.

CHAPTER

Risk, Risk Perception, Risk Communication, Risk Acceptance: Risk Governance

12

When it comes to effectively communicating the risks, a bond of partnership, a feeling of trust and an empathetic approach are every bit as important as the numbers.

John Paling, The Risk Communication Institute

SUMMARY

We arrive at the last chapter before drawing some conclusions. Now, all we have collected in knowledge in previous chapters shall be mixed together to support decision making in the public domain upon assessing the risk of an activity. In Chapter 9 we have placed emphasis on decision making from a company's business and economic perspective, but here we shall look at it predominantly in terms of acceptability from the public safety point of view. Where in Chapter 9 the process of weighing costs of installing preventive and protective measures against benefits of risk reduction falls under the heading "risk management," we shall speak in this chapter of the broader concept of "risk governance." This concept has been taking shape over the last decade. Risk governance fits in a systemic approach and can involve governmental and nongovernmental decision makers. It embodies besides management also risk assessment and risk communication. Here, emphasis will be on off-site risk.

In the case of a governmental decision maker, based on the assessment, a competent authority/regulator can decide to deny a license for a planned or existing activity, or to allow it but possibly imposing certain conditions. The decision can have important economic consequences for the region. In case of acceptance, it leaves the possibility of occurrence of a high consequence event although the expected probability of occurrence may be extremely low. Damage by the event can result in fatalities, it can cause many injuries and other massive damage, so accurate predictive power of risk assessment is crucially important.

The classical engineer's approach to presenting process industry risk is simply to multiply the magnitude of the consequence as the number of expected fatalities given the event with its probability to occur over a period of time, most often

Risk Analysis and Control for Industrial Processes - Gas, Oil and Chemicals.

1 year, and to compare risks on that basis. Consequences are then linearly weighed, as it were, by their probability of occurrence. It can make the risk of an event with light consequences but higher expected frequency of occurrence equivalent to a catastrophic one with a low estimated probability. One can ask whether this is justified from the point of view of risk acceptance.

A probability value of a possible scenario occurrence will usually be determined as a frequency value composed of component failures and critical event rates in which nominal human errors are included. It however remains just a theoretical measure, because in the huge majority of cases the event will never happen in a lifetime. Hence, it is not observable, and a predicted rare event frequency cannot be validated due to its low value and the uniqueness of the situation. Moreover, data, models, and identified scenarios contain relatively large uncertainties, but variance in risk assessment results is rarely considered. The spread is also due to differences in component failure mechanisms and the large variability in human error probability. It leads to the statement that rare events are unpredictable. It made many risk analysts cite Nassim Taleb's[1] book on "black swans," which were unknown in Europe before being spotted in Australia. Taleb was inspired by catastrophes in financial services taking many people by surprise, despite predictions of assets being safe based on (normal) probability distributions, because unknown outliers in the tails had not been accounted for. In addition, there is the possible rare confluence of known, but infrequent events pictured in the "perfect storm," after a shipwreck drama disaster written by Sebastian Junger.[11]

In contrast to rates of occurrence of rare events, consequences can be observed. At least they can have occurred in similar form somewhere else, although perhaps on a smaller scale, or can be seen entirely or partially in specially designed experiments. But even consequence size prediction still contains uncertainty, although generally to a lesser degree than the occurrence probability value. Anyhow, risk values contain relatively large aleatory and epistemic uncertainties, as described in Chapter 7, Section 7.6.1. It may explain in part that the public at large usually is far more concerned about the high consequence than welcoming the low probability.

There is another factor why probability is not viewed the same way as consequence. It is in the way the human mind functions, the way we are involved in the activity voluntarily or are reaping benefits from it. This involvement will be briefly reviewed in this chapter together with other factors that influence different persons to perceive a risk value completely differently. This often results in heated debate and irreconcilable disagreement on what is acceptable. It may create conditions in which political influences, even manipulation and conflict, will be lurking. In risk perception and hence decision making, various factors play a role, and this has been a topic analyzed by several well-known psychologists and risk analysts. We shall briefly summarize some of this information.

To avoid emotional disagreement and to arrive at even better decisions, good risk communication and stakeholder participation in decision making are of essential value. Good risk communication is an art. It has been shown that broad participation of nonexpert discussion partners will result in better decisions. In particular, the International Risk Governance Council (IRGC) has published guidance of how to reach a decision in the best harmony. Scientific assessors and governing decision makers shall be different persons. Before being presented to decision makers, risk assessment results by scientists shall be elaborated in a risk characterization step. In all, the whole problem field has evoked a rich harvest of articles and books.

12.1 INTRODUCTION, RISK AS CONCEPT, AND RARE EVENTS

A wealth of literature is available about the construct of *risk* in the context of societal acceptance, or rebuff, of a new technology (e.g., nanotechnology), a new infrastructural project (dam, bridge) or an industrial activity (nuclear power plant, chemical plant). Terje Aven[2] presented an extensive overview of the risk concept, also in a historical sense. Psychologist Erik Hollnagel[3] stressed the changes in the nature of risk. In this introduction we shall revisit briefly some definitions of risk, and in subsequent sections see how risk plays a part in public debate.

The oldest thoughts on a risk definition within the context of technology originate from the nuclear safety community. Rather early on though, Stan Kaplan and John Garrick[4] presented a definition from an engineering point of view. This was done in quantitative terms after first some basic qualitative statements, which are sometimes forgotten but have been recently stressed again by Aven.[5] The notion of risk consists of uncertainty and damage, or in the case of a potential for damage, that is, a hazard, how this is reduced by safeguards. There is a subjective element in risk, which depends on a person's knowledge of what can occur and how likely it is to occur. So, a risk perceived by one person may not to be perceived as a risk under the same conditions by another. Californians Kaplan and Garrick give an example of a rattlesnake hidden by someone in a mailbox. On the question whether it is a risk to put your hand in the mailbox, only a person with knowledge by the observation the snake was put there will be sure there is a risk. To this can be added that others not having the knowledge will have only a certain degree of belief there is a risk, based on trust they have in the warning by an observer. This is exactly the problem that we shall see when it comes to public decision making. Risk is not something absolute. The foregoing also brings together, as put forward by Aven (see Chapter 5, Section 5.3.2), how risk is a function of a hazard-initiating event, barrier-dependent consequences, uncertainty, probability, and knowledge. The larger the relevant knowledge, the lower will be the uncertainty. Charles Vlek[6] summarized risk as two-sided (objective-subjective), dynamic (it changes with the situation), and multidimensional (parameters of consequence, barrier, vulnerability, uncertainty, probability).

Kaplan and Garrick[4] defined the triplet: What can happen or go wrong (*scenario*), how likely is it (*chance, probability*), and how large is the (expected) damage (*consequences*). As we have seen in Chapter 3, a risk assessment consists of analyzing a (large) number of such triplets of which for a particular hazardous object (installation/plant) results can be summarized in individual and societal/group risk figures. However, something more must be explained about probability, which as a concept must be used, because one lacks precise information and because at branch points in the cause–consequence tree possible outcomes differ due to influencing factors that cannot be controlled.

In the 1970s–1980s, the time of Kaplan and Garrick, the debate was still active between "frequentists" or "objectivists," and the "subjectivists." The former consider probability as an absolute measurable quantity (throwing a dice n times to observe how often the "six" comes up which determines the frequency of hitting the "six"). Subjectivists see probability as an expression, a measure, of a state of knowledge or confidence. They are the "Bayesians," who consider probability determined on (prior) existing older knowledge or a belief updatable (to posterior) as new (likelihood) knowledge becomes available, as described in fine detail by Edwin Thompson Jaynes[7] and mentioned here briefly in Chapter 7. The belief can also be expressed as an occurrence per unit of time and hence as a frequency. So, one has to be careful in interpreting the word *frequency* in risk assessment context; is there a measured basis or is it a belief? In principle, a certain scenario, for example, failure of a pipe under process conditions and subsequent release, occurring over a certain time period a number of times, is measurable and produces a frequency of occurrence per unit of time or a rate[a]. Of course, the measurement is awkward and takes a long time, and depending on the extent of the experiment there remains uncertainty about its exact value. The experiment can be repeated a number of times to result in a distribution of rate values with a mean and a variance (see Chapter 3, Section 3.6.4). Even more exact, the rate values found may be fitted to a suitable probability density function of which there are several to choose from, for example, a log-normal or a Weibull distribution function. (The distribution exhibits aleatory uncertainty, although there also may be differences in mode of pipe fracture with a significant effect on the release: longitudinal or guillotine-like. In case the causing mechanism is unknown and there is no control over it, this yields epistemic uncertainty. Failure probabilities of both modes have to be measured independently.) So, the mean rate value of the event one uses in the analysis in a predictive sense has a certain probability of occurring over the specified period of time, or in other words the value used is within a certain interval in which one can have, for example, 90% confidence that the mean is indeed within these bounds. The triplet for a given scenario is therefore associated with a probability distribution of frequency values of basic and initiating events, and along the same lines there will be other distributions for the size and nature of physical

[a]Students are inclined to complain that they can find little or no data on failure rates but that is exactly what the concept of probability is for: we know the possibility of failure cannot be excluded, but there is lack of knowledge on how often it occurs. Yet, if they find or assume data, they are also inclined to present results obtained with an applied QRA software package to several decimal places. This apparent accuracy is in almost all cases totally unjustified.

effects and of damage. The damage can take many forms. Hence, the predictive power of the analysis is limited to showing possibilities of consequences with a mutual ranking order of probability, usually expressed as a metric or a measure of, for example, an event occurrence per year, hence frequency. Inger Lise Johansen and Marvin Rausand[8] provide an overview of various risk metrics.

The rare event with the highest consequence will usually be the least probable and thus lowest at the ranking order. As effect intensity decreases with distance, consequence figures, such as probability of a fatality, given a person would be present at that location at the time of the event, drop with distance from the risk source. Also, when the analysis is done well, a figure for the confidence that one can have in the prediction will be specified. Unfortunately, even some professional risk analysts draw risk contours or plot risks of a scenario in a risk matrix without any indication of confidence they have in the result given the available data. This degrades and worsens decision making! Imagining the myriad of possibilities determining the course of scenarios, limitations to the accuracy of predictions by the many assumptions to be made and the availability of failure and other data will be clear. The power of the analysis is realizing what can go wrong, up to what distances effects can be anticipated, and what can be done to prevent and protect against the occurrence and/or its consequences. Finally, it is necessary to determine how low a chance of the specified damage should be acceptable to the public at large. For that, we shall first consider psychologists' views on risk perception and risk communication.

In 1984 from a psychologist's perspective, Baruch Fischhoff et al.[9] mention that the first step in defining risk is determining the consequences, and they also point out that risk definition is inherently controversial. There are aspects of objectivity and subjectivity in risk estimation that in addition are not necessarily aligned between experts and the lay public. There are sometimes considerable differences in judgment, with that of scientists not necessarily being better than judgment of the public. The dimensionality of risk may vary. In Figure 12.1 the possible dimensions of consequence are reproduced as shown by Fischhoff et al.[9] Consequence "utilities" may be benefits/costs, environmental effects, deaths, morbidity, and concern. The

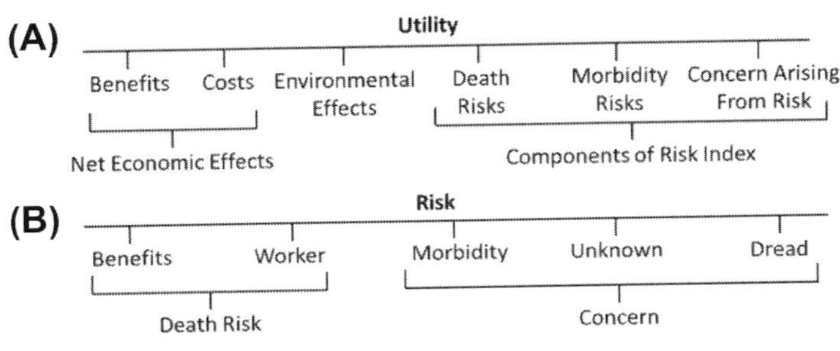

FIGURE 12.1

Possible dimensions of consequence: (A) for decision making and (B) for risk index according to Fischhoff et al.[9]

latter three form elements of a risk index. Given an accident, the death of workers executing tasks at the site usually receives a lower weight than death of members of the public outside. Concern varies from "unknown risk" to "dread," and may cause stress that results in deterioration of health. A concern score may vary between 0 and 100 with attributes as shown in Table 12.1. A known pair of components of "dread" is voluntary and involuntary risk. Voluntary means that an individual can decide to expose him/herself to it, for example, in playing a risky sport or driving a car, but involuntary is the risk imposed on a citizen by a plant in his neighborhood. Mountain climbing is known as voluntary at a score of 0 in dread risk, and exposure to a nuclear reactor accident is involuntary scoring very high (90) on the attribute dread. We shall continue discussing all this in the next section when we consider risk perception.

Public debate in connection with risk-based decision making is almost always on potentially catastrophic, rare events. As already alluded to in the Summary, Elizabeth Paté-Cornell[10] distinguishes two kinds of such events: the black swans and the perfect storms (the unknown "unknowns" and known "unknowns." respectively, see also Chapter 9, in particular Section 9.3.8). The former are the subject of Taleb's book,[1] who as a trader, generally refers to financial events. It concerns (epistemic) "outliers," which figure in the unknown tails of the distributions and are therefore considered unpredictable. The latter designation is inspired by the title of the book by Sebastian Junger.[11] It regards a coincidence of two or more similar low-frequency conditions making the rare storm consequences extreme. Hence, the uncertainty in this case is aleatory. In the reasoning for a risk analysis and the identification of scenarios including taking account of external conditions, the perfect storm should be part of it. In Chapters 7 and 9, we discussed some methods to estimate probabilities of rare events, although with significant uncertainty. In the experience of the author of this book, examples of perfect storms in industrial disasters are also plentiful further complicating prediction.

Nassim Taleb means with his "black swans" the totally unexpected, "never seen before" events (as mentioned before, it was the "discoverers" of Australia who were amazed seeing these birds), and examples are much more difficult to find than perfect storms. As Paté-Cornell mentions, human behavior/error should be part of risk assessment and the resulting accident can only be considered as a black swan when organizational interactions involving various levels would lead to a fully unexpected result. An example of a black swan may be the totally unexpected violence of the vapor cloud explosion after the overfilling of the storage tank at the Buncefield fuel depot[12] near London, UK, in 2005. The explosion of a vapor cloud after turnover in a slops tank of layers formed of different temperatures and composition causing evaporation of hydrocarbon waste, which occurred at the Shell Pernis refinery[13] in Rotterdam, the Netherlands, in 1968, was at the time certainly a black swan. An unconfined vapor cloud explosion was unknown and not recognized yet as a phenomenon, although later investigation revealed that a few explosions of vapor clouds in the open had occurred in Europe and the US before. For instance, twice there had been explosions of vapor clouds at BASF in Ludwigshafen in the 1940s, where they were not recognized as a new type of event. For many years no detailed explanation existed of the explosion mechanism, but as explained in Chapter 3 in the 1980s and 1990s modeling became possible. However, as mentioned the blast violence in Buncefield was again a surprise. And as explained in Chapter 3, the gasoline cloud explosion a few years later in Jaipur, India, has clearly been identified as a detonation instead of a deflagration despite a low degree of congestion. No one would have dared to predict such a possibility before. Of course, in hindsight also a black swan can usually be clearly explained, even if new theories have to be developed to do that. In such cases, if there is a reoccurrence then it can no longer be branded as "unexpected." Knowledge has increased, Bayesian update is possible, and it should now be part of the analysis of every possible QRA/PRA (quantitative risk assessment/probabilistic risk assessment) scenario.

12.2 RISK PERCEPTION AND RISK COMMUNICATION

Psychometric research based on questionnaires asking for preferences of both lay people and experts provided input for risk perception studies. The result of Slovic's[14] psychometric research based on the risk score components of *dread* (factor 1) and *unknown* (factor 2) mentioned in Table 12.1 is presented in Figure 12.2. It is called the multidimensionality of risk. The desire to regulate risk increases with the extent of dread and of risk being unknown.

Earlier in 1978–1979, Charles Vlek (University of Groningen, NL) and Pieter-Jan Stallen (TNO)[15] performed a psychometric study in the Rotterdam industrial area in the Netherlands and found a positive correlation between personal benefit of an activity and acceptance, more so than with riskiness. Probability as a concept is in general badly understood. Men see a relation between "riskiness" and "low degree of organized safety," while women are more sensitive to the overall size of a potential accident. Managers and technical people see large-scale production activity and benefit as positive for acceptance, but all other professions consider these as negative. A surprising finding was that people living nearby the industrial area show a higher degree of acceptance than people living at a considerable distance (50 km). Maybe this is because they, or their friends, are more likely to be employed at the facility or in associated industries and therefore they are better informed than the public at large?

In 2004, when discussing risk perception, Slovic et al.[16] distinguished three ways how people react to risk: (1) *risk as feelings*, (2) *risk as analysis*, and (3) *risk as politics*. There is no need here to elaborate the second way; in this book that has been done repeatedly. The first is the way humankind from primitive life onwards reacts to danger, fast, instinctive, and intuitive. It comes down to the *affect heuristic* (experience-based mental rule). This means when exposed to a hazard getting immediately a good or feeling of like, as opposed to a bad or feeling of dislike, associated respectively with benefit or risk. It concerns the so-called *experiential* or intuitive mode of thinking. This type of thinking also has an influence on wrongly interpreting probability. For instance, if people are offered a reward when they choose a red bean, while being allowed to draw once from a pot filled

Table 12.1 Fischhoff et al.[9] Risk Score Characteristic Components of the Attributes "Unknown" and "Dread Risk"

Attribute	Unknown Risk		Dread Risk	
	Score 0	**Score 100**	**Score 0**	**Score 100**
	Observable	Not observable	Controllable	Uncontrollable
	Known to exposed	Unknown to exposed	No dread	Dread
	Effect immediate	Effect delayed	Not global catastrophic	Global catastrophic
	Old	New	Consequences not fatal	Consequences fatal
	Known to science	Unknown to science	Equitable	Not equitable
			Individual	Catastrophic
			Low future risk	High future risk
			Easily reduced	Not easily reduced
			Decreasing	Increasing
			Voluntary	Involuntary
			Doesn't affect me	Affects me

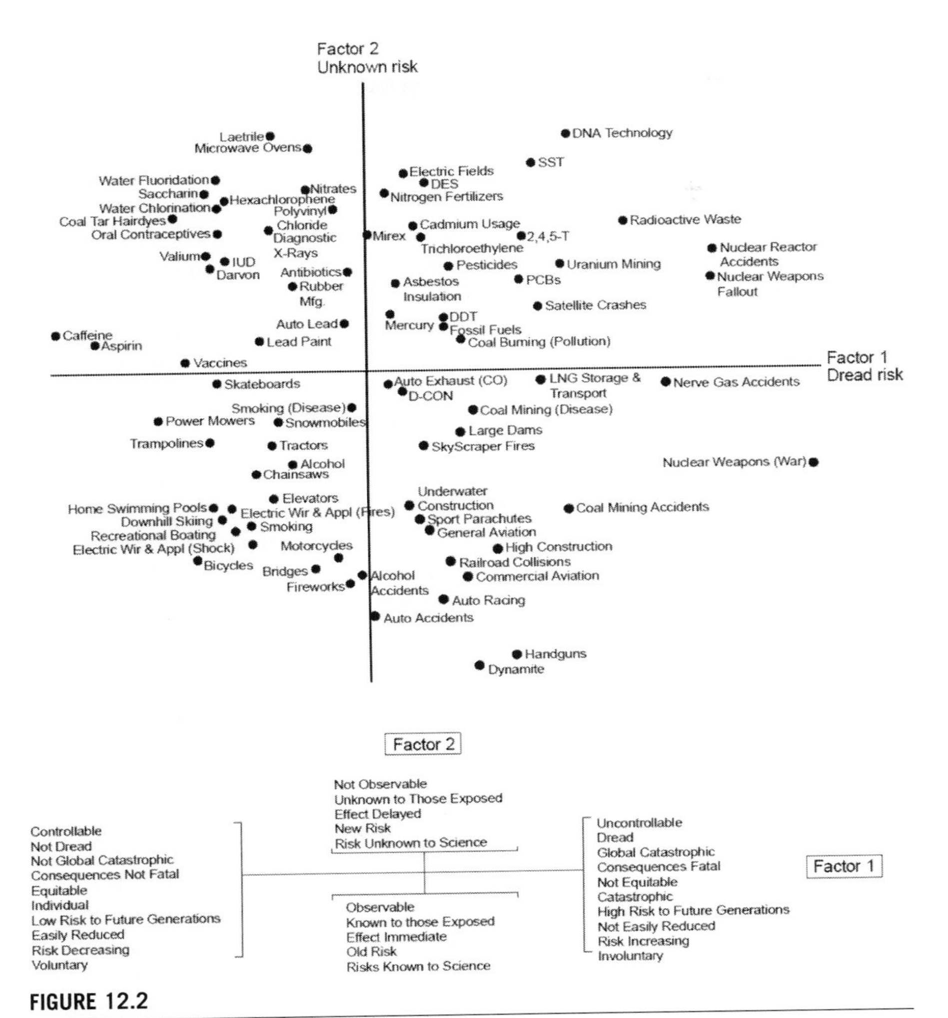

FIGURE 12.2

Results of psychometric research based on preferences locating 81 hazards based on characteristic properties of dread risk (factor 1) and unknown risk (factor 2), according to Slovic.[14] Some characteristics are highly correlated such as "voluntary" and "controllable." "not observable" and "uncontrollable" as for radioactive material or odorless, toxic gas are feared. Small accidents involving these agents are highly publicized and feed the anxiety of the public.

with an additional 100 white beans, or alternatively given the choice to do it from a larger pot with 1000 white beans and 10 red ones, they tend to opt for the latter, because they think they have then a better chance of the reward. Many other examples can be given, and the affect heuristic experience means also that a frightening scenario or a quip impresses more than a sober notice. We may readily go for a gamble that promises a high reward, but at the same time we will be inclined to take insurance for a possible damage with the same low probability. It is the possibility in those cases that dominates. Summarizing, the experiential thinking mode can put us on an illogical and wrong footing. It mixes in with rationality. So, it may overrule risk as analysis by risk as a feeling. For scientists and engineers performing risk assessments to the best of their abilities, his conclusions cannot be described as encouraging.

Slovic et al.[16] continue by stating that "the perception and integration of affective feelings, within the experiential system, appears to be the kind of high level maximization process postulated by economic theories since the days of Jeremy Bentham. These feelings form the neural and psychological substrate of utility." Bentham (1748–1832) was a thinker on human relations in society: individual legal rights, freedom of expression, and individual and economic freedom. This became the foundation of utilitarianism, a theory in normative ethics and a precursor of the present welfare economics. The underlying moral principle is that of "greatest happiness," so meant is maximization of utility. In welfare economics, the Kaldor–Hicks efficiency criterion plays a role, which in case of an investment requires balancing the total costs of "losers" with more than the total gains of "winners." Hence, a region may profit from a new industrial project, but the direct neighbors of the site may sustain a certain loss, for example, by a drop in their house value. Kaldor–Hicks is used to find out whether a project improves the Pareto efficiency. The Pareto optimum is a state when everybody gains and nobody is worse off. Various effects may be of influence such as information asymmetry between "winners" and "losers."

Governments have the task to provide conditions for a safe society. But the safer the life, the higher become the safety demands of citizens (and it seems at the same time as if individuals take on larger risk voluntarily). This trend was seen early on in industrialized countries but can be observed again in 2014 in China. However, our societies cannot function without essential facilities and plants, which also with the best of safety effort impose on- and off-site risks on their environment. As the risk may be life threatening, ethical principles come into play but also welfare economics. In the case of planning application of an industrial activity with hearings it may evoke such exclamations as NIMBY (Not In My Backyard) or even BANANA (Build Absolutely Nothing Anywhere Near Anything). Risk assessment and a democratic acceptance process are needed in such cases in order to arrive at a just decision. Utility will provide a foundation for such decision making.

In an earlier (1999) paper, Slovic[17] elaborated the third point: risk as politics or rather political worldview, where he contrasted egalitarian and individualism worldviews besides fatalism, hierarchy, and technological enthusiasm. When advantage is apparent, individualism is inclined to accept larger risk than egalitarian worldview. These worldviews, along with ideologies and values, influence risk acceptance too. For example, Slovic[17] mentions the remarkable finding that American white males are significantly less sensitive to risk than other people in the US, males or females. In general, risk sensitivity drops with higher income and education. The effect is really brought about by a group of 30% of white males; the remaining are the same as all the rest. Risk sensitivity somehow relates to people's worldviews; the 30% white males were found to trust institutions and authorities more (perhaps because they are part of them) and are inclined to antiegalitarian attitude. In fact, Vlek and Stallen's[15] findings mentioned above are in line with this. Trust plays an important role. Once destroyed by perhaps a silly mistake, it is hard to restore.

As shown in previous chapters, a further complicating trait of risk assessments is that they are subject to considerable uncertainty, and that they contain judgment that most often is not presented explicitly. Team A obtains results that differ from ones concluded by team B. Views and premises contain subjective elements and therefore differ. The result is that the trust of the public is undermined by the statements made. Also, the way and the units in which the results are presented influence a decision. This is grist to the mill of politics in either supporting or denying the project. Politicians use their power to decide and tend to take arguments based on physics as secondary. In addition, special interest groups succeed in influencing the outcome. Once distrust is present, more scientific studies only intensify public concern, as shown for example by the risk studies on electromagnetic radiation from, e.g., mobile phones and overhead power lines.

Kasperson et al.[18] discuss why the public is sometimes much concerned about a risk that scientists view as minor; they call this social risk amplification. Underestimation, hence risk attenuation, occurs too (e.g., not wearing car safety belts). The main reason of risk amplification is that people are not experiencing the risk directly but learn about it from others or from the (social) media. Information flow acts as an amplifier. Debates among experts reinforce the effect. Public fears are dramatized. Also, signal values of events are important, such as a news report that an awaited risk assessment report is still not finalized, which may be "interpreted" as the authorities are attempting to hide risks.

As regards *risk communication* between experts and the public, it is good to be aware of the seven rules the US Environmental Protection Agency (EPA)[19] conceived:

1. Accept and involve the public as a legitimate partner.
2. Plan carefully and evaluate your efforts.
3. Listen to the public's specific concerns.
4. Be honest, frank, and open.
5. Coordinate and collaborate with other credible sources.
6. Meet the needs of the media.
7. Speak clearly and with compassion.

The American National Research Council (NRC) defines risk communication as "an interactive process of exchange of information and opinion among individuals, groups, and institutions." Baruch Fischhoff[20] characterized the developmental stages of the risk communication process with the public by the adage of Ernst Haeckel: "Ontogeny recapitulates phylogeny," or in other words an embryo develops to an adult by going through the successive evolutionary stages of its remote ancestors:

1. All we have to do is get the numbers right
2. All we have to do is tell them the numbers

3. All we have to do is explain what we mean by the numbers
4. All we have to do is show them that they've accepted similar risks
5. All we have to do is show them that it's a good deal for them
6. All we have to do is treat them nice
7. All we have to do is make them partners
8. All of the above

According to the this author's experience, the course of a communication process between experts and lay people is well depicted with the above wording. Although Fischhoff paints it in many tones, in brief, meanings may be summarized as follows: (1) It will still take a long time before risk numbers are accurate as we have mentioned in this book in several instances. (2) Handing over the computational results to the public just generates more questions. Fatality risk numbers are interpreted differently by experts and by the public. (3) To get a feel for the risk the public may like to have entirely different information than risk numbers, for example, about how the process works. Uncertainty can be explained wrongly as evasiveness. (4) Risk comparisons do not convince and can even worsen the situation. (5) A benefit may make the risk acceptable. It depends also on the way it is presented. People like to hear what makes sense to them. (6) People want to be treated with respect and want their concerns to be taken seriously. How the message is presented can make or break the issue. In communication there are many pitfalls. (7) The public may be prepared to be more actively involved and have more influence on the process. Participation may have a positive influence on their willingness to accept risk. (8) There is no escape from going through the whole process.

Although the communication bill may be significant, it could head off the larger cost of project cancellation. Controversy and mistrust are easily born yet difficult to remove. Communication shall transmit and eventually build credibility and competence. The Organization for Economic Co-operation and Development (OECD)[21] issued a useful guidance document on risk communication for chemical risk management. Further practical hints for the communication between scientists and the public can be found in articles by Peter Sandman,[22] for example, those from 2004 prepared for the *Synergist*, the journal of the American Industrial Hygiene Association.

12.3 PUBLIC DECISION MAKING, STAKEHOLDER PARTICIPATION

As already briefly mentioned at the end of Chapter 3, the IRGC, located in Geneva, Switzerland, has made quite an effort to make recommendations how to carry out a risk assessment and appraisal, and how to make sometimes difficult decisions in consultation with a large range of participating stakeholder groups. The framework for risk governance as published in their 2005 white paper[23] is reproduced in Figure 12.3. It distinguishes sharply the world of scientists/risk analysts from that of management in which decisions are taken and actions agreed. Terminology and

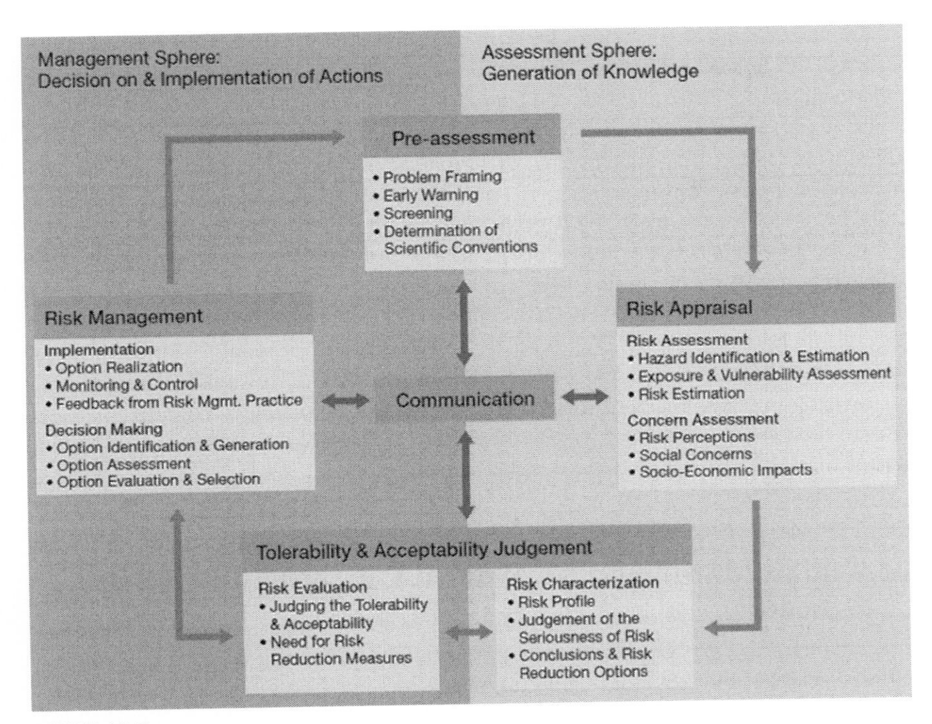

FIGURE 12.3

IRGC's framework of risk governance[23] showing various stages and activities. Risk appraisal encompasses a scientific assessment of the risk: first in a physical sense (natural and technical) and then in a second stage socially and economically.

understanding of the concepts is further explained in *Risk Management and Governance* by Terje Aven and Ortwin Renn.[24]

The most controversial and critical stage is the tolerability and acceptability judgment prepared by risk characterization and finished by risk evaluation, while resulting in a proposal of options and possible risk-reduction measures for decision making. *Risk characterization* is a kind of translation and preparation of the technical analysis for a decision maker, so that (s)he will understand the issues and will know what to do. The IRGC paper distinguishes four categories: (1) simple routine risk assessments, (2) complexity-induced problems to be tackled by risk-informed decision making and robustness focused solution; further, (3) uncertainty-induced problems that need a precaution-based approach by considering hazards, applying containment, ALARP (see next section) and a resilience focus, and finally, (4) ambiguity-induced risk problems to be resolved in participative discourse. The latter consists of methods to achieve resolution and to reach consensus with the stakeholders. Much more on the analytic-deliberative process of risk characterization can be found in a publication of the American National Research Council.[25]

The effectiveness of stakeholder participation on the quality of decision making has been scrutinized by Thomas Beierle[26] for the US Department of Energy because some analysts feared "that stakeholder processes may sacrifice the quality of decisions in pursuit of political expediency." After a detailed analysis of 239 cases, Beierle concluded that stakeholder participation in general improved the quality of decisions. Risk communication is considered of utmost importance and is centrally presented in Figure 12.3. It shall be a continuous activity with stakeholders. It will be the basis of reconciliation of conflicting interests and viewpoints. Risk management will be the topic of the next section.

A few years later, in 2010, IRGC[27] published a policy paper on risk governance deficits and how to improve. Ten types of deficits have been identified that all come down to lack of knowledge about the system: the hazard, early signals, factors influencing the system, knowledge about stakeholders, model weakness, cognitive barriers of imagination, etc. Boosting system resilience to absorb surprises is an important part of the solution.

12.4 RISK MANAGEMENT, RISK ACCEPTANCE CRITERIA, ALARP

Risk management has found many applications, for example, in financial organizations, project management in general, insurance, and of course, in process industry. In 2009, an international standard ISO 31000:2009[28] was published describing risk management definitions and principles. For the case of process risks, Terje Aven,[29] University of Stavanger, Norway, commented extensively on the standard's definition of risk as "effect of uncertainty on objectives," relative to the one on management as "coordinated activities to direct and control an organization with regard to risk." In industrial process activities the objective of safe operation can be disturbed by uncertain events with severe adverse consequences (see also Chapters 3 and 5). Risk management serves to reduce the likelihood of such events and/or to reduce their impacts. As a systematic process, risk management has all the features of the Deming cycle described in Chapter 3, Figure 3.7. The ISO standard provides a list of principles for good risk management and a framework in which it will operate. A key attribute is continual improvement. In management culture, organization and human factors are decisive.

As noted before, an objective of risk management will be to bring risks within acceptable boundaries. In 2001, Health and Safety Executive (HSE)[30] in a UK document drafted by Jean Le Guen issued a rather fundamental policy with respect to acceptability of risk and risk management as it launched the terms *broadly acceptable*, *tolerable*, and *unacceptable* for a risk. Tolerable refers "to a willingness to live with a risk so as to secure certain benefits and in the confidence that the risk is one that is worth taking and that it is being properly controlled." Acceptable means that risk has become reduced to such an extent that it has become negligible. In IRGC's white paper[23] this is represented by a so-called "traffic light" graph (the name is

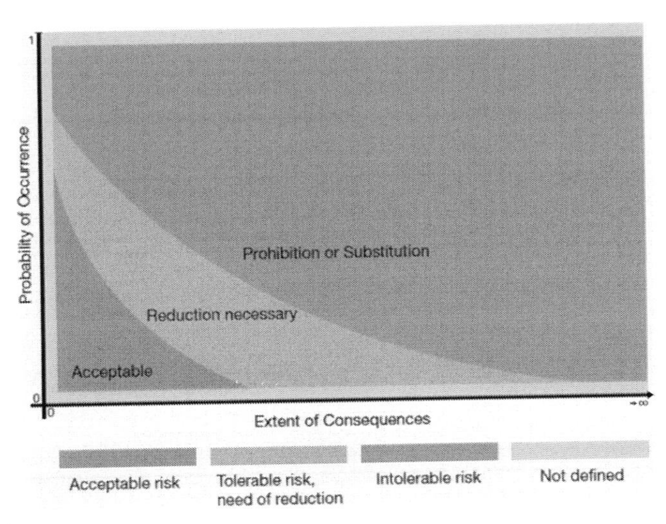

FIGURE 12.4

Qualitatively, areas in which a risk is considered acceptable (left, originally colored green), tolerable (middle, yellow/amber), and unacceptable (right, red). In the tolerable part (middle) reduction is needed.

After IRGC's white paper.[23]

because of the colors; the graph has coordinates as in a risk diagram. Incidentally, earlier in Chapter 6 we saw the concept of the traffic light stopping rule but that only shares part of the name). Figure 12.4 reproduces this graph in gray tones. It shows clearly the widening of the (yellow) middle area of risk "Reduction necessary" going toward low "Probability of Occurrence," high "Extent of Consequences" events. Hence for low probability—high consequence risks more is necessary to become acceptable than for high frequency—low impact ones. As consequence tends to zero—and one could speak of near misses or precursors—there are still probability values above which risk needs to be reduced or is even unacceptable.

Uncertainty must be considered at the difficult and sensitive transition stage of risk characterization to risk evaluation at the tolerability/acceptability step of Figure 12.3. This is most true in the IRGC's category (4) ambiguity induced risk problems mentioned in Section 12.3. Depicting uncertainty on potential loss as in Figure 12.5 can assist in resolving differences in opinion. It is showing effect of uncertainty on probability of exceeding loss L versus loss L itself. It has been proposed by Grossi and Kunreuther[31] for earthquake damage and generalized for other risks in IRCG's white paper.[23]

Indeed, this way of thinking about uncertainty is perfectly in line with the ideas promoted throughout this book. Uncertainty and subjectivity in risk assessment results must be made explicit, so that after adding social and economic value deliberations into the evaluation step and judgment of needed risk-reduction options, decision makers have a clearer overview and stakeholders will have increased

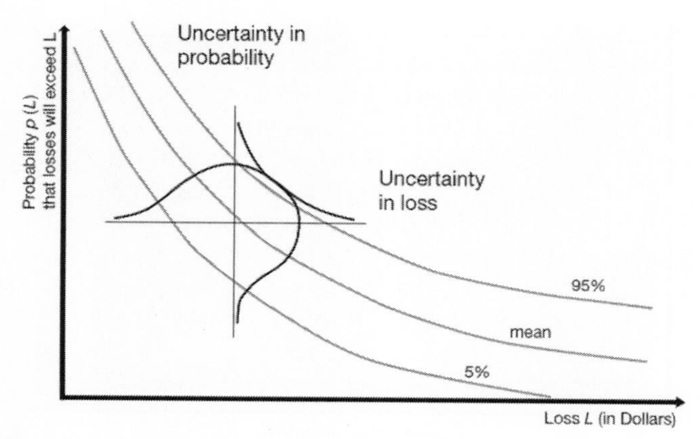

FIGURE 12.5

Probability of loss exceedance, P(L) versus loss, L with indication of uncertainty, showing distributions of probability values (vertically) and of loss values (horizontally) fitted in. Losses are here expressed in currency units but can be in any unit, for example, fatalities.

After IRGC's white paper.[23]

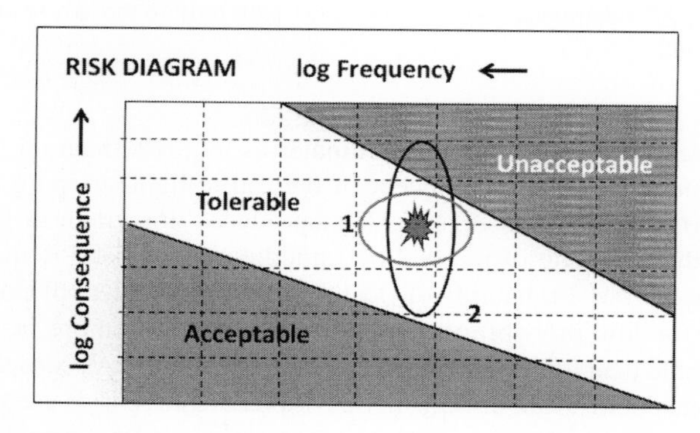

FIGURE 12.6

Risk diagram schematic showing a point representing two scenarios of equal risk but different uncertainty ranges. Risk of scenario 1 will look more preferable to a decision maker than risk of scenario 2, although in both cases ALARP (see text below) may require the risk to be further reduced. In most cases preference may not be that obvious and will depend also on other conditions. (Note that the errors are portrayed as, e.g., ellipses with axes of three times the standard deviation of normal distributions of logarithmic values of consequence and probability, while the errors are normally obtained relative to the values itself making the area in a log–log plot very irregular in shape).

confidence in the veracity of the decision makers. Figure 12.6 illustrates this, although only a fraction of cases will be so evident. Conditions such as magnitude of severity of consequence, emergency response capacity, economic importance, and the prospect of future improvement of risk reduction may influence the decision.

To quote Terje Aven and Enrico Zio[32] (respectively, 2014 vice chairman and chairman European Safety and Reliability Association): "disguised subjectivity of risk assessments is potentially dangerous and open to abuse if it is not recognized," while also should be prevented that "precise numbers are used as a façade to cover up what are often political decisions," which in the end appear to have a money focus. Also in the experience of this author, unrevealed uncertainty seriously undermines the credibility and value of risk assessments. Another point is the oversimplification by comparing a point value risk calculation result with a fixed quantitative criterion as described below and deciding for a go or no-go depending only on whether the risk number is higher or lower than the criterion. In addition, assuming that a risk based on the product of a small consequence and high probability event is equivalent to a risk with the same number but composed of a high consequence and low probability is another oversimplification. The latter is often unjustified as can be seen in Figure 12.4. The acceptability situation for such cases can be entirely different.

Much has been written about risk acceptance criteria. It is not the purpose of this book to present a complete literature overview. Rather fundamental contributions, also relating individual and societal/group risk criteria, have been given by Carter and Hirst,[33] Hirst and Carter,[34] Vrijling et al.[35] and by Jonkman et al.[36]

A general rule that has gained global popularity is the British ALARP or reducing a risk to "As Low As Reasonably Practicable" (or sometimes SFAIRP, "So Far As Is Reasonably Practicable," but the latter is used more in an occupational safety context and has then a slightly different meaning). ALARP is different from the older ALARA, stemming from the nuclear industry, which means "As Low As Reasonably Achievable." ALARP sounds rather simple, but going into detail quite a few questions arise pivoting around "What is reasonable?", "What is practicable?", and "In what stages of a project or points of time the principle has to be made true?" HSE has issued an ALARP Suite of Guidance[37] consisting of six guidance documents to assess compliance. These are meant for the guidance of industrial duty holders, HSE inspectors, and government enforcers. ALARP must be applied as a holistic approach during all project phases of design, construction, commissioning, and operation. To this the use of best available and safe technology (BAST) can be added. The foundation is the application of HSE-recognized good practice, standards, and codes. Where possible it further requires the use of inherently safer solutions. It also requires an effective operational safety management system. By applying measures, assessed risks shall first of all be reduced to at least below the level of being tolerable in terms of criteria of individual and societal risks; subsequently ALARP will be applied. For ALARP, risks shall be further reduced until the costs of continued risk reduction are disproportionately (grossly) higher than the benefits, or until a level is reached below broadly acceptable. Also, the higher the residual risk the higher the disproportion that will be required. To prove the disproportion, a cost–benefit analysis must be made. To that end a number of monetary values are prescribed as, for example, the value of a prevented fatality (on a 2003 Q3 basis) is a value £1,336,800 (times 2 for cancer), a permanently incapacitating injury £207,200, and a serious injury £20,500, et cetera. These figures will each be multiplied with a probability of an event to occur within the lifetime of the plant. Inflation and net present value rates of investments are given. Disproportion shall be at least a factor 3 but less than 10. Periodical review must be carried out to make sure ALARP remains true, while also site inspections will take place.

Australia has accepted ALARP in its Model Work Health and Safety Bill,[38] in which Part 2 Health and Safety Duties, Subdivision 2, is formulated as follows:

"In this Act, reasonably practicable, in relation to a duty to ensure health and safety, means that which is, or was at a particular time, reasonably able to be done in relation to ensuring health and safety, taking into account and weighing up all relevant matters including:

(a) the likelihood of the hazard or the risk concerned occurring; and
(b) the degree of harm that might result from the hazard or the risk; and
(c) what the person concerned knows, or ought reasonably to know, about:
  (i) the hazard or the risk; and
  (ii) ways of eliminating or minimising the risk; and
(d) the availability and suitability of ways to eliminate or minimise the risk; and
(e) after assessing the extent of the risk and the available ways of eliminating or minimising the risk, the cost associated with available ways of eliminating or minimising the risk, including whether the cost is grossly disproportionate to the risk."

Acceptable risk levels are described in Chapters 2 and 3. Best known and oldest are the British and Dutch regulatory levels of individual risk (IR), in the Netherlands instead of "individual risk" called "location bound risk," or in Chapter 3 also indicated as localized individual risk per annum (LIRA) and societal, also called group risk (GR). In Chapter 3 definitions are given as well as the general method of application. Ben Ale,[39] emeritus faculty of Delft University of Technology, NL, wrote a comparative essay on the history of the British and Dutch regulations. In 2007, HSE[40] issued a consultative document meant to initiate a review. For individual risk it restates a value from work activities at the *boundary between tolerable and unacceptable* as 10^{-3} per annum for voluntary exposure of risk workers at a site and 10^{-4} p.a. for involuntary risk for members of the public. For comparison, the 2010 death rate by accident in general (external cause) in developed countries such as the US[41] and Germany[42] is about 4×10^{-4} p.a. Lower down, at 10^{-6} p.a., hence at less than 1% of the death rate by external cause, industrial risk level becomes *negligible* for both workers and public (*the boundary between tolerable and acceptable*). This risk level is now assumed acceptable by many countries around the globe, although the way it shall be applied is still developing. This is even more the case with group risk. Questions come up such as: is such a criterion needed, what should be its value, and how shall it be used? In an earlier document, HSE[30] had mentioned for group risk a certain, fixed point: a disaster with 50 fatalities should at most have a frequency of 2×10^{-4} p.a. There had been proposals too for a straight line in a log Frequency p.a. exceeding N fatalities versus log N (F, N

curve) or societal risk plot with a slope of −1.4 or even −2, the latter equal to the heavy risk-averse Dutch approach. After extensive debate it was decided in the UK in 2010 not to prescribe a slope value but to let it depend on conditions.[43]

In 1993 in the Netherlands, the societal or group risk criterion (10 fatalities, maximum 10^{-3} p.a.; 1000 fatalities, 10^{-7} p.a., hence slope −2, and 5 times more stringent than a proposed British criterion at 50 fatalities) became an "orientation value" instead of a mandatory one. The objective was to have a decision maker judge whether a crisis after an event was still controllable, but it did not work that well because the concept of group risk was often misunderstood. Policy to modernize the approach is currently being developed.

Although in the Netherlands an individual risk criterion of 10^{-6} p.a. (or 10^{-5} p.a. at existing facilities, temporarily) at contours around a static risk source is well established, a QRA-based decision may be overruled. This occurred for instance with the Shell Barendrecht carbon dioxide (CO_2) sequestration project.[44] (The compressed gas would be permanently stored in an empty natural gas cavity beneath the town of Barendrecht. The heavier-than-air gaseous CO_2 is invisible and lethal for 50% of the population at a concentration of above 10% in air for a 10 min exposure.) The possible consequences imposed dread-fueled concerns that local property prices might drop. Aven[45] reports a similar occurrence in Norway with a planned liquefied natural gas plant. Dutch criticism on the norm criterion can be summarized as an inflexible application of probability multiplied by consequence in terms of fatalities. It lacks consideration of inaccuracy/uncertainty in risk assessment (uncertainty is not made visible by, for example, regions with fading colors on a map), and it results in less drive to reduce risk as is the case when ALARP must be applied. There is also insufficient public participation in the process, and a certain mistrust about possible "maneuverability" to reach a desired outcome despite standardization of applied model and data. In addition, licensing and occupational safety are too often treated separately as "external" and "internal" safety, which does not help either. Changes may be forthcoming. Also, procedures and criteria for hazardous materials transportation routes are still further developing.

It should be noted that following the Toulouse ammonium nitrate detonation disaster in 2001, in France a system was introduced with some interesting features partly meeting the shortcomings felt elsewhere. The main characteristics of the French criteria including physical effect thresholds are described in Chapter 2. Consequence and event probability are judged separately, taking account of the total number of people exposed, and the dynamics of the event (explosion, fire, or toxic dispersion) with the possibility of taking evasive action such as evacuation. A possible drawback of this system is the effort required to conduct an analysis.

Risk tolerability and acceptability judgment and taking account of divergent opinion in decision making will also differ from one country to another. Geert Hofstede (mechanical engineer and social psychologist, IBM and University of Limburg, Maastricht, NL) became known for his psychometric survey based on studies of national cultural differences as they appear in management, organizations, and societies. In the sample taken from his 1983 study[46] shown in Figure 12.7, differences can be noted among 50 countries with respect to the

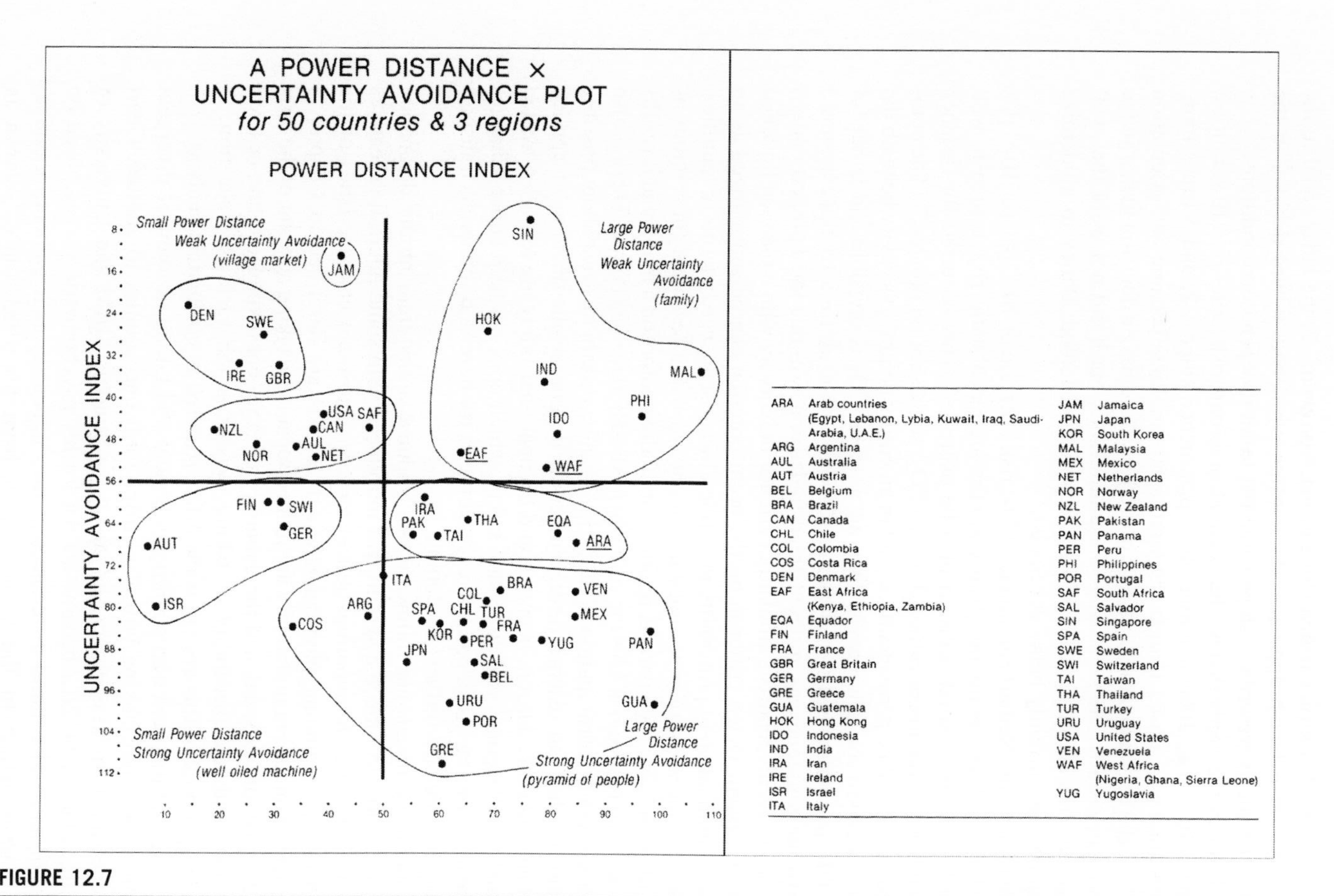

FIGURE 12.7

Hofstede's[46] psychometric results. The two indices shown can be considered indicative of a desire to protect against risk (ordinate = uncertainty avoidance) and the way management operates: more democratic versus more autocratic/centralized (abscissa = power distance).

indices of uncertainty avoidance and distance to power. The first index, "Uncertainty avoidance," distinguishes people with less anxiety for an unknown future and hence less drive to protect against risk from people with higher nervousness, emotionality, and aggressiveness wanting to create security and avoiding risk. Making uncertainty tolerable can be realized by technology/constructions, by laws, or by religion. The second index, "Power distance," is a measure of the degree of inequality in a country as it demonstrates itself in the distribution of wealth and power. Larger power distance means for organizations more centralization and autocratic leadership. It tells us something about stakeholder participation and democratic decision making. This study was conducted in the early 1980s; meanwhile, some shifts have taken place in, for example, the degree of inequality in a number of countries increased. There are more influences on risk decisions than shown in these two indices, and an index is only a relative measure but at least it gives an idea.

In the case of when no data, or totally insufficient knowledge about a risk of an activity or a product, is available to make an assessment, then in particular in Europe the *precautionary principle* holds. This means that one should not embark on that activity or not use the product until more knowledge has been generated. Potential benefits may exert pressure to drive research for the purpose of acquiring better and more complete knowledge.

12.5 CONCLUSION

A brief, but comprehensive booklet about the nontechnical aspects of risks, induced by technology and industrial activity and the influences on decision making as in this chapter, has been written by Baruch Fischhoff and John Kadvany.[47] It treats the definition of risk ("threats to outcomes that we value"), its perception, communication, and risk in relation to culture and society. They conclude that "risk definitions reflect norms about how the world is and should be. Knowing that, risk analyses help people to handle dangers and live the good life as they see it." In that context they state: "the value of analysis depends on how well its limits are understood."

Despite the many shortcomings of risk analysis as a predictive tool for risk assessments of industrial process systems, we still very much depend on it. Fortunately, there are continuous attempts to improve. As mentioned near the end of Chapter 3, Rae et al.[48] in their 2014 article entitled "Fixing the cracks in the crystal ball," built a QRA maturity model. The proposed model identifies and discusses all possible weaknesses and error sources, so that in a given risk analysis by peer reviews all potential flaws can be addressed. In addition, enabled by the conceptual thinking behind full risk analyses, simpler and more limited risk assessments such as layer of protection analysis with the aim to investigate adequacy of protection have proven highly valuable in terms of the balance between costs and benefits.

Finally, ALARP as a risk criterion below a tolerable and above an acceptable limit is the best available at this time. Risk governance as a process has developed considerably over the past decades. When having to decide whether it is at all necessary to go through, it seems a rather elaborate and costly affair, but experience shows, it pays to follow the successive steps and to ensure sufficient stakeholder participation and good communication. Canceling a planned project may be much more costly. On the other hand, there is much space for further development of optimal and effective risk-informed decision-making procedures.

REFERENCES

1. Taleb NN. *The black swan: the impact of the highly improbable*. New York: Random House; 2007, ISBN 978-0-8129-7381-5. e-Book 978-0-6796-0418-1.
2. Aven T. The risk concept - historical and recent development trends. *Reliab Eng Syst Saf* 2012;**99**:33−44.
3. Hollnagel E. The changing nature of risks. *Ergon Aust J* 2008;**22**(1−2):33−46.
4. Kaplan S, Garrick BJ. On the quantitative definition of risk. *Risk Anal* 1980;**1**(1):11−27.
5. Aven T. A risk concept applicable for both probabilistic and non-probabilistic perspectives. *Saf Sci* 2011;**49**:1080−6.
6. Vlek Ch. Wetenschapswaardigheden, Nieuwsbrief NVRB, (2014-3) 4-5.
7. Jaynes ET (posthumous), Larry Brethorst G, editors. *Probability theory: the logic of science*. Cambridge University Press; 2003, ISBN 0521 59271 2.
8. Johansen IL, Rausand M. Foundations and choice of risk metrics. *Saf Sci* 2014;**62**: 386−99.
9. Fischhoff B, Watson SR, Hope Ch. Defining risk. *Policy Sci* 1984;**17**:123−39.
10. Paté-Cornell E. On "Black swans" and "Perfect storms": risk analysis and management when statistics are not enough. *Risk Anal* 2012;**32**:1823−33.
11. Junger S. *The perfect storm*. New York: W.W. Norton & Company; 2000, ISBN 978-0-393-33701-3.
12. Buncefield Major Incident Investigation Board. Final Report. *The buncefield incident*, vols. 1 and 2; December 11, 2005. © Crown copyright 2008, ISBN:978 0 7176 6270 8; ISBN:978-0-7176-6318-7.
13. Ministerie van Sociale Zaken en Volksgezondheid. *Rapport betreffende een onderzoek naar de oorzaak van een explosie op 20 januari 1968 op het terrein van Shell Nederland Raffinaderij N.V. te Pernis*. Staatsuitgeverij 's-Gravenhage; 1968. http://www.nbdc.nl/cms/show/id=668352.
14. Slovic P. Perception of risk. *Science* 1987;**236**:280−5.
15. Vlek Ch, Stallen PJ. Judging risks and benefits in the small and in the large. *Organ Behav Hum Perform* 1981;**28**:235−71.
16. Slovic P, Finucane ML, Peters E, MacGregor DG. Risk as analysis and risk as feelings: some thoughts about affect, reason, risk and rationality. *Risk Anal* 2004;**24**:311−22.
17. Slovic P. Trust, emotion, sex, politics, and science: surveying the risk-assessment battlefield. *Risk Anal* 1999;**19**:689−701.
18. Kasperson RE, Renn O, Slovic P, Brown HS, Emel J, Goble R, et al. The social amplification of risk a conceptual framework. *Risk Anal* 1988;**8**:177−87.
19. Covello VT, Allen FH. *Seven Cardinal rules of risk communication*. Pamphlet drafted by U.S. Washington (DC): Environmental Protection Agency; April 1988. OPA-87-020

20. Fischhoff B. Risk perception and communication unplugged: twenty years of process. *Risk Anal* 1995;**15**:137–45.
21. OECD, Organization for Economic Co-operation and Development. *Guidance document on risk communication for chemical risk management*, vol. 18. OECD Environment Directorate, ENV/JM/MONO; 2002. Paris 2002.
22. Lanard J, Sandman PM. Scientists and the public: barriers to cross-species risk communication. http://www.psandman.com/col/species.htm and http://www.psandman.com/col/columns.htm.
23. IRGC. *White paper on risk governance towards an Integrative approach*. Geneva: International Risk Governance Council; September 2005. www.irgc.org.
24. Aven T, Renn O. Risk management and governance - concepts, guidelines and applications. In: *Risk, governance and society*, vol. 16. Springer; 2010, ISBN 978-3-642-13925-3. e-ISBN 978-3-642-13926-0.
25. Committee on Risk Characterization. In: Stern PC, Fineberg HV, editors. *Commission on behavioral and social sciences and education, national research council: understanding risk - informing decisions in a democratic society*. Washington (DC): National Academy Press; 1996, ISBN 0-309-05396-X.
26. Beierle ThC. The quality of stakeholder-based decisions. *Risk Anal* 2002;**22**:739–49.
27. IRGC. *Policy brief, risk governance deficits analysis*. Geneva: International Risk Governance Council; 2010, ISBN 978-2-9700672-0-7.
28. ISO 31000. *Risk management – principles and guidelines*. 1st ed. Geneva, Switzerland: ISO; 2009.
29. Aven T. On the new ISO guide on risk management terminology. *Reliab Eng Syst Saf* 2011;**96**:719–26.
30. HSE U.K. *Reducing risks, protecting people, health and safety executive, U.K.*, ISBN 0 7176 2151 0.
31. Grossi P, Kunreuther H, editors. *Catastrophe modeling: a new approach to managing risk*. New York: Springer; 2005.
32. Aven T, Zio E. Industrial disasters: extreme events, extremely rare. Some reflections on the treatment of uncer-tainties in the assessment of the associated risks. *Process Saf Environ Prot* 2013;**91**:31–45.
33. Carter DA, Hirst IL. 'Worst case' methodology for the initial assessment of societal risk from proposed major accident installations. *J Hazard Mater* 2000;**71**:117–28.
34. Hirst IL, Carter DA. A "worst case" methodology for obtaining a rough but rapid indication of the societal risk from a major accident hazard installation. *J Hazard Mater* 2002;**A92**:223–37.
35. Vrijling JK, van Hengel W, Houben RJ. A framework for risk evaluation. *J Hazard Mater* 1995;**43**:245–61.
36. Jonkman SN, van Gelder PHAJM, Vrijling JK. An overview of quantitative risk measures for loss of life and economic damage. *J Hazard Mater* 2003;**A99**:1–30.
37. HSE U.K. ALARP suite of guidance, http://www.hse.gov.uk/risk/theory/alarp.htm.
38. Part 2 Health and safety duties, Subdivision 2 *Model work and health act, revised draft*. June 23, 2011. http://www.safeworkaustralia.gov.au/sites/SWA/about/Publications/Documents/598/Model_Work_Health_and_Safety_Bill_23_June_2011.pdf.
39. Ale BJM. Tolerable or acceptable: a comparison of risk regulation in the United Kingdom and in the Netherlands. *Risk Anal* 2005;**25**:231–41.

40. HSE U.K. *Proposals for revised policies to address societal risk around onshore non-nuclear major hazard installations, CD 212*. 2007. http://www.hse.gov.uk/consult/condocs/cd212.htm.
41. http://www.cdc.gov/nchs/data/nvsr/nvsr63/nvsr63_03.pdf.
42. Witt W. *Seveso III and the problem of industrial risk criteria in the European union*. Technische Sicherheit, Springer VDI Verlag, 6; 2013. Pages, see alternatively, https://www.destatis.de. 5.
43. HSE U.K. *Status summary of '23 issues*. February 2010. http://www.hse.gov.uk/societalrisk/technical-policy-issues.pdf.
44. Van Gelderen L. *Improving a risk assessment method for CCS: public acceptance of a risk analysis for the Barendrecht project*. Amsterdam: RIVM and Vrije Universiteit; September 2013.
45. Aven T. Practical implications of the new risk perspectives. *Reliab Eng Syst Saf* 2013; **115**:136–45.
46. Hofstede G. The cultural relativity between organizational practices and theories. *J Int Bus Stud* 1983;**Fall**:75–89.
47. Fischhoff B, Kadvany J. *Risk, a very short introduction*. Oxford University Press; 2011, ISBN 978-0-19-957620-3.
48. Rae A, Alexander R, McDermid J. Fixing the cracks in the crystal ball: a maturity model for quantitative risk assessment. *Reliab Eng Syst Saf* 2014;**125**:67–81.

CHAPTER

13

Conclusions: The Way Ahead

Sir, I have found you an argument. I am not obliged to find you an understanding.
James Boswell, The Life of Samuel Johnson

In many cultures the number "13" is regarded as unlucky and the harbinger of bad fortune, although even with this there is uncertainty—for instance, in Italy the converse is true, and thus when "13" occurs it is welcomed! Nonetheless, in the foregoing we have considered an abundance of approaches, methods, and tools to help to fend off "misfortune" with respect to process risks and to keep these risks under control. A few of the main observations follow below:

1. The main cause of major accidents in the past was often lack of knowledge (know-why and know-how); today it is rather more often cost considerations as a result of economic competition and time pressure impacting on decision making. This results in not taking time to think what can go wrong, to bring in expertise where necessary, or loss of overview due to complexity. It all leads to wrong decisions, to decisions taken too late, or no decisions at all. This decision making can be at any hierarchical level. It also encompasses lack of competence, but the knowledge how to do it well need not be far away; there are many knowledge sources within Internet reach. On the other hand, it would be arrogant to state that we know everything. Yes, much can still be learned, but the "vacuum" of technical and organizational process safety and risk assessment knowledge of the 1960s no longer exists. Often, today's accidents could have been prevented by timely knowledgeable intervention, and worse, experts such as late Trevor Kletz became upset about the number of times accident histories repeated themselves at different places in the world. A known example is overflowing or unattended leaks of gasoline storage tanks with ensuing vapor cloud explosion and fire as mentioned in Chapter 1 of the 2005 Buncefield disaster in the UK. Despite the wide publicity of the latter the accident repeated itself elsewhere in later years again a few times.
2. The main critical factor is the insight, attitude, and policy of leadership and the measures it takes to establish and maintain a healthy safety culture in the company. If leadership is not aware of what the hazards and risks are, is underestimating or ignoring the risks and is not prepared to invest in protection against those, safety chiefs in the company will be "fighting a losing battle." If it is the case that such investments are registered only as costs, the company is

Risk Analysis and Control for Industrial Processes - Gas, Oil and Chemicals.

already on the wrong track. Leadership determines and defines the company's culture. It should assure itself by active and systematic inspecting and monitoring that will convince them that the business is safe and sound. Just to have the operations audited is insufficient. The Organisation for Economic Co-operation and Development (OECD)[1] issued a guidance document for leaders of high-hazard industries. For assurance, it strongly recommends monitoring corrective actions following an audit or incident.

3. If there is preparedness at the top to minimize the chances of accidents but at the same time to survive in the economic climate, a systems approach is the way to go. According to the business dictionary[2] a systems approach means: "A line of thought in the management field which stresses the interactive nature and interdependence of external and internal factors in an organization. A systems approach is commonly used to evaluate market elements which affect the profitability of a business." But apart from the "market elements," it also serves to control risks and maintain safety in an often complex and hierarchical system because it considers the whole and not just the constituent parts. Lack of safety also "affects the profitability of a business." A systems approach considers internal and external threats as well as component defects and human failure. It will also anticipate faulty interactions of otherwise healthy components and outcomes that according to the procedures might erroneously be classified as right decisions. The first to comprehensively describe accident causation in a sociotechnical system and what we should do to manage its risks was Jens Rasmussen.[3] It all starts with knowing where the safety boundary is and for that risks have to be reliably predicted.
4. Next, Nancy Leveson,[4] while stressing that safety is an emergent system property, showed a way forward for risk prediction to progress. She developed a query method to identify and analyze possible failures of the system's risk control loops that can breach the safety constraints. There are however millions of details to cope with if one descends to loops at the lowest level of technology and organization at which risks have to be controlled. Therefore, we shall need an automated, or at least semiautomated, scenario information generation, storage, and retrieval system of the kind like Blended Hazid that Ian Cameron[5] proposes, as we have seen in Chapter 7. This still requires further development effort.
5. We have seen an array of relatively newly applied tools such as Bayesian networks (BNs), Petri nets, and various kinds of process modeling and simulation tools (such as system dynamics, signed digraph, multi-level flow model, D-higraph), while also mention was made of AspenTech and HYSYS, for example, for the purpose of determining resilience. In particular a series of applications of BNs have been shown to assist in solving problems of cause—effect chains, extrapolation of precursor frequencies to rare event ones, and process safety performance indicator data analysis. These tools will facilitate risk prediction and hence improve risk-informed decision making. A dynamic BN of stored causal paths can help to interpret weak signals, enabling action to be implemented before propagation to a more serious upset occurs.

6. Accident scenarios have often not been predicted by risk analysis as has been shown many times in the past. In part this can be explained by not sufficiently analyzing what can go wrong in operation and maintenance, by the fact that in a system an accident can evolve from an unforeseen interaction between components (as mentioned in conclusion #4 above) or because of unforeseen domino effects, or external threats and coincidences. We have seen how Elizabeth Paté-Cornell[6] distinguishes between two kinds of such rare events: Taleb's black swans and the perfect storms (the unknown "unknowns" and a kind of known "unknowns," respectively; the perfect storms are coincidental "synergistic" events). Apart from trying to perform better in risk analysis by means of the new methods mentioned, it is also reason to pursue resilience, prepare for the unforeseen, and reckon with uncertainty by not removing all reserves or safety factors.
7. Resilience is the ability of a system to cope with unexpected shocks and disturbances. It may be damaged due to the load, but it should restore from abnormal back to normal operation as soon as possible. It is a property of both organization and technology. It counters cutting away reserves and it has a positive notion in the sense one should do things "right" instead of "not wrong" as in safety. Hollnagel et al.[7] emphasized in the "resilience engineering" paradigm the organizational and human sides of resilience. They opened new avenues to approach organizational issues with respect to safety. Looking further and considering the whole system, resilience requires an adequate emergency response and a sufficient flexibility and controllable system. This also presupposes a well-functioning safety management system and a maximum extent of built-in inherent safety. But above all, it ought to receive, interpret, and if necessary act upon weak signals of a wide spectrum of increasing risk and approaching threat. Such alertness is key in maintaining minimal damage.
8. Weak signals shall be collected from various sources. Precursor signals of acute upcoming abnormal situations in the technology can be produced by the control system, and component failures by sensors directly or via the control system and diagnostic tools. Precursors or antecedents and near misses should be recognized and recorded. Short-term risk fluctuations may occur due to things such as weather conditions, feedstock impurities, maintenance operations, fluctuations in numbers of exposed people, and incidents creating risky conditions. Long-term degradations both in a technical as well as organizational sense shall be signified by process safety performance indicators reflecting the effectiveness of the safety management system. So far, the information these indicators contain is used by management only to correct organizational issues. However, establishing the relationship among indicators and risks will enable the extraction of additional information. Feeding the various kinds of signals to a plant system model, preferably dynamic, shall enable continuous risk monitoring. Application of a BN will accommodate dependencies as mentioned in conclusion #5. This approach will also reveal risk emerging from the system, which in turn may lead to the ideal of a *risk*

dashboard. Such a device shall also recommend the type of action to counter the identified threat.

9. The changes we see in powering the world and in raw materials for industry cause a gradual shift from oil-based liquid energy carriers (hydrocarbons) to gaseous ones such as natural gas and hydrogen. In view of the associated hazards, large-scale storage and distribution of these gaseous energy carriers requires even more careful planning, design, construction, and maintenance of installations than was the case before with liquids such as gasoline/petrol and diesel fuel. It will also require specific competencies, hence acquiring knowledge, learning, and training.
10. Knowledge is continually expanding both from experiences in the industry and by scientific progress in various fields. The latter is the fruit of specific process safety and risk assessment methodology research, but for an important fraction it is also enabled by developments, for example, in computing technology, or by making use of results in fields such as organizational science and systems control. To use the new knowledge effectively, first it has to be learned; the learning shall take place on various levels, and for specific topics it may take the form of training and retraining. Progress entails sophistication, and if insufficient effort is paid to training it will cause the knowledge gap between industry and academia to grow. The importance of learning and education is emphasized here because due to a certain loss in the 1990s of the attractiveness of engineering and safety education in particular, there has been a decline in safety teaching capacity. At the same time there have been warnings that due to retirements a lack of competence is likely to occur, which even may deter investors' interest in large-scale high-hazard projects.
11. There has been a tendency to shift from prescriptive regulation to goal-oriented, also called performance-based, regulation. The latter bestows more responsibilities on industry as well as allowing it extra freedom to choose the most cost-effective solutions with respect to safety measures. However, it appeared that although performance-based rules certainly have advantages, they also come with limitations. In particular, small- and medium-sized enterprises may lack the knowledge to make wise choices themselves and therefore may prefer to rely for compliance on prescription. Meanwhile, regulatory authorities, due to lack of personnel numbers with adequate competence, may also have a problem with inspections associated with good enforcement of goal-oriented legislation. In addition, for guidance goal-oriented regulation tends to refer to standards, codes, and good practice. These are prescriptive by their very nature though they may not necessarily be legally enforceable unless explicitly recognized by the competent authority.
12. Risk management is a matter of optimizing distribution of resources over needed safety measures and the decision about what level of residual risk to accept. Distinction has been made between risk-informed decision making within companies and the processes of risk governance, where the safety of workers and public are assessed and the options are weighed and decided upon

by a competent authority. A variety of tools for both types has been described. In the case of in-house company decisions, cost–benefit considerations will play an important role. To a certain extent economic utility theory provides guidance in decision making. Although there shall be intensive communication between all stakeholders, in the case of risk governance experts conducting the risk assessment shall not also be the decision makers. There exist quite a few risk criteria in various parts of the world, but the one developed in the UK is gaining the most popularity. According to this criterion, compliance-assessed risks shall be reduced to below the minimum level of unacceptability (tolerability borderline), while subsequently the principle of As Low As Reasonably Practicable (ALARP) will be applied. If by ALARP a level is reached below that of broad acceptability, no further reduction is required. For compliance a set of rules is given in which good practice and disproportional cost to obtain any further increase in benefit prevail. So, this gives an answer to the question, "How safe is safe enough?" Periodic review during the system's life cycle must be carried out to make sure ALARP remains true—advances in equipment, software, or understanding mean that ALARP is always a moving, and progressively more demanding, target. Risks shall be duly assessed. In view of the opinions expressed in this book, it certainly demands in cases of high-hazard potential the application of the best methods for scenario identification, severity of effect determination, and frequency of occurrence. But it also means demarcation of uncertainty by indication, for example, of coefficients of variation or of confidence intervals. Decision making has to occur keeping uncertainty in mind and trading off against circumstances, economics, emergency response capacity, and possibly other factors. For decision making in cases of so-called deep uncertainty, special methods have been briefly described.

13. Besides 13 chapters, there are 13 conclusions. Given all of the above, it should be possible to make the knowledge about risks transparent and to gain the trust of the public. Withholding important information or downplaying serious results will in the end always be counterproductive. Openness about process safety performance indicators and their trends can assist in generating and maintaining that atmosphere of trust. Honesty and integrity will always be invaluable components of risk management.

REFERENCES

1. OECD (Organization for Economic Cooperation and Development). Corporate governance for process safety: guidance for senior leaders in high hazard industries. In: *OECD environment, health and safety chemical accidents programme, Paris*; June 2012.
2. http://www.businessdictionary.com/definition/system-approach.html.
3. Rasmussen J. Risk management in a dynamic society: a modelling problem. *Saf Sci* 1997; **27**(2/3):183–213.

4. Leveson NG. *Engineering a safer world, systems thinking applied to safety*. The MIT Press; 2011. 608 pp., ISBN-10:0-262-01662-1, ISBN-13:978-0-262-01662-9.
5. Seligmann BJ, Németh E, Hangos KM, Cameron IT. A blended hazard identification methodology to support process diagnosis. *J Loss Prev Process Ind* 2012;**25**:746–59.
6. Paté-Cornell E. On "Black swans" and "Perfect storms": risk analysis and management when statistics are not enough. *Risk Anal* 2012;**32**:1823–33.
7. Hollnagel E, Woods DD, Leveson N, editors. *Resilience engineering, concepts and precepts*. Aldershot (UK): Ashgate Publ. Ltd.; 2006, ISBN 0-7546-4641-6.

Index

Note: Page numbers followed by "f" and "t" indicate figures and tables respectively.

C

D

E

F

K

L

M

N

Q

R

T

U

V

W

X

Y

Z

Printed in the United States
By Bookmasters